Marine Ecosystems and Global Change

EDITED BY

**Manuel BARANGE, John G. FIELD, Roger P. HARRIS,
Eileen E. HOFMANN, R. Ian PERRY, and Francisco E. WERNER**

Technical Editor: Dawn M. ASHBY

OXFORD
UNIVERSITY PRESS

OXFORD

UNIVERSITY PRESS

Great Clarendon Street, Oxford OX2 6DP

Oxford University Press is a department of the University of Oxford.
It furthers the University's objective of excellence in research, scholarship,
and education by publishing worldwide in

Oxford New York

Auckland Cape Town Dar es Salaam Hong Kong Karachi
Kuala Lumpur Madrid Melbourne Mexico City Nairobi
New Delhi Shanghai Taipei Toronto

With offices in

Argentina Austria Brazil Chile Czech Republic France Greece
Guatemala Hungary Italy Japan Poland Portugal Singapore
South Korea Switzerland Thailand Turkey Ukraine Vietnam

Oxford is a registered trade mark of Oxford University Press
in the UK and in certain other countries

Published in the United States
by Oxford University Press Inc., New York

British Library Cataloguing in Publication Data
Data available

Library of Congress Cataloging in Publication Data
Data available

Typeset by SPI Publisher Services, Pondicherry, India
Printed in Great Britain by
on acid-free paper by
CPI Antony Rowe, Chippenham, Wiltshire
1 0 ✕6614373
ISBN 978–0–19–955802–5 (hbk); 978–0–19–960089–2 (pbk.)

1 3 5 7 9 10 8 6 4 2

Preface

In the 1970s, oceanography was viewed from an intradisciplinary perspective. Physical oceanographers studied the physics of the ocean, chemical oceanographers studied the chemistry, and biological oceanographers studied the biology. Biological studies were largely descriptive and partitioned into specialized areas and quantitative analysis and modelling of ocean production was limited to a relatively few forward-looking studies. Research tended to focus on primary production, the formation of 'carbon' and the feeding characteristics and metabolism of zooplankton, particularly the larger zooplankton.

In addition to being intradisciplinary, studies of the ocean tended to be limited in space and time. Most work was local or regional and focused on single time periods or intra-annual variability. This limited understanding of sources of variability at the smaller space scales (e.g. small-scale turbulent flow and smaller eddies) and larger space scales (e.g. basin and global). It also limited understanding of higher-frequency time scales that are so important to plankton life history, particularly the smaller plankton on the one hand and on the other the decadal–century–climate scales that are important from both a scientific and practical point of view (e.g. ocean ecosystem forecasting, fisheries management, and climate change).

During the 1980s, oceanography continued to evolve driven both by intellectual curiosity and the needs of society. For example, the development of satellites provided important new insights. Satellite altimeters led to new understanding on the physical structure of the *global* ocean, and colour scanners led to an appreciation of the distribution of chlorophyll in the *global* ocean. All the while, high-speed supercomputers were becoming more and more accessible so that not only could the huge amounts of data acquired by satellites be stored, processed, and analysed, but models of higher and higher resolution (models which were not necessarily dependent on satellite data) could be implemented.

The availability of satellites and computers not only facilitated the study of global and basin scales and enabled the processing of huge amounts of data and higher-resolution models, they also contributed to a new way of conducting ocean science. The availability of global data led to an appreciation that the scientific problems associated with the ocean and atmosphere were so large and complex that a small group of scientists, operating more or less independently, in one or several institutes, could not address the issues in a coherent and coordinated way. This generated a need for enhanced scientific cooperation, which became implemented in programmes that attempted to have a global focus, the World Ocean Circulation Experiment (WOCE) and the Joint Global Ocean Flux Study (JGOFS).

Also at this time, societal awareness of anthropogenic forcing was accelerating. Important problems relevant to anthropogenic forcing continued to emerge. These problems involved, for example, various scenarios associated with increased production of photoactive gases and their impact on climate change. Concerns about the effects of fishing became paramount as extension of national jurisdiction to 200 nautical miles offshore stimulated the need to understand the relative effects of fishing and environmental variability in controlling the abundance of fish stocks. This variability was and still is very important to fishery managers in the sense that they need to know whether a decline or increase in abundance is a stochastic accident or whether it is a signal of long-term decrease or increase in productivity. At the same time, heavy fishing on many fish stocks constituted a great ecological experiment where population dynamics specialists could observe the effects of externally induced mortality (i.e. fishing) on natural populations and ecosystems.

All of these issues and questions relate to the causal relationship between forcing variables and the structure and functioning of ocean ecosystems. In other words, the impact of the forcing variables on ocean ecosystems became in itself an important scientific problem with societal relevance. The role of phytoplankton in affecting carbon production, the influence of pollutants on ecosystems, or the variability of fish stocks could not be addressed without significant improvements in understanding the dynamics of plankton, in general, as affected by physical forcing.

So to put it in its simplest terms, the state of biological knowledge in the mid-1970s and early 1980s was intradisciplinary and tended towards being descriptive in an environment where important new technologies were becoming available; the spirit of scientific cooperation and direction in global programmes was becoming evident, and society was asking new and important questions.

This enhanced intellectual environment and a simple idea contributed to the creation of the Global Ocean Ecosystem Dynamics project, GLOBEC. The underlying idea was that there was a large gap in knowledge concerning the variability of ocean ecosystems. This was true in part because the ecosystem *dynamics* of the ocean were hardly known. The acknowledged strong influence of multi-scale ocean physics and chemistry as sources of ecosystem variability needed to be defined. This definition would need to include an understanding of how ocean physics was coupled with the trophodynamics and vital rates (i.e. growth, mortality, and reproduction) particularly of the zooplankton, because of its key role in mediating the transfer of energy from primary producers to fish. The lack of understanding of this trophic level limited the ability to address the important questions that were being asked about ocean ecosystems and fisheries. In addition, observations and theory on how plankton vital rates drive the dynamics of populations and in turn how the dynamics of populations interact with the dynamics of ecosystems is critical. It is important to remember that at this time major components of the ocean ecosystem such as the 'microbial loop' were only beginning to be recognized.

To address these issues, the scientific community held a workshop in Wintergreen, Virginia in April 1988, which recommended addressing Global Ocean Ecosystem Dynamics in new and innovative ways. These involved: (1) concentration on first principles, (2) enhancing the biological–physical partnership, (3) investigating genetics and related biotechnology, and (4) utilizing and developing new sampling technology and techniques. The emphasis on first principles centred on understanding how various processes at the level of individual organisms control population abundance. This would require specific focus on the 'interactions among ocean physics, trophodynamics of plankton populations' (Rothschild *et al.* 1989) and that 'the interrelationships among the individual populations of organisms be understood at the most fundamental and mechanistic levels. Special emphasis needs to be given to assessing the little studied roles of ocean physics in feeding success, growth rates, reproductive output, mortality rates including losses to predators, and recruitment' (Rothschild *et al.* 1989). The physical–biological partnership (which was virtually non-existent at the time) was thought to be particularly important to the acquisition and modelling of physical and biological processes on matching time and space scales. The lack of matching data was thought to be a 'major impediment to oceanographers interested in understanding population regulation of ocean organisms' (Rothschild *et al.* 1989). Linked with this, the need for advanced sampling technology involving acoustic and optic sensors was seen as a way forward to replace chronic under-sampling and a lack of synopticity. Thus the stage was set to design an exciting and innovative new programme to move biological oceanography into a new modality: the study of ocean ecosystem dynamics.

The international GLOBEC programme was initially co-sponsored in 1991 by the Scientific Committee on Oceanic Research (SCOR) and the Intergovernmental Oceanographic Commission of UNESCO. Subsequently, it was adopted as a core project by the International Geosphere–Biosphere Programme (IGBP) and for the 10-year duration of GLOBEC the three co-sponsors provided valued support and guidance. Two Regional Programmes also benefited regional sponsorship from the International Council for the Exploration of the Sea (ICES) and the North Pacific Marine Science

Organization (PICES). Many national and international funding agencies provided support, but in particular the US National Science Foundation, the UK Natural Environment Research Council, and the Plymouth Marine Laboratory are acknowledged for their central role in providing funds for programme coordination.

This book sets out many of the important outcomes of the new way of looking at marine ecosystems that GLOBEC fostered. From the outset, GLOBEC research was organized around four foci: retrospective analyses and time series studies, process studies, predictive and modelling capabilities, and feedback from changes in marine ecosystem structure. The work was initially developed within four Regional Programmes: Southern Ocean GLOBEC (SO GLOBEC), Small Pelagic Fishes and Climate Change

(SPACC), ICES-GLOBEC Cod and Climate Change Programme (CCC) in the North Atlantic, and the PICES-GLOBEC Climate Change and Carrying Capacity Programme (CCCC) in the North Pacific. In the later stages of the programme two additional Regional Programmes were started: Climate Impacts on Oceanic Top Predators (CLIOTOP) and Ecosystem Studies of Subarctic Seas (ESSAS). At least the latter two, together with other continuing research elements of GLOBEC, will continue after the formal end of GLOBEC in 2009 as part of the IGBP/SCOR Integrated Marine Biogeochemistry and Ecosystem Research (IMBER) project. National funding supported active GLOBEC programmes in many countries (Ashby 2004), for example Canada, Chile, China, France, Germany, Japan, Korea, Mexico, Norway, Peru, Portugal, Spain, United Kingdom, and United

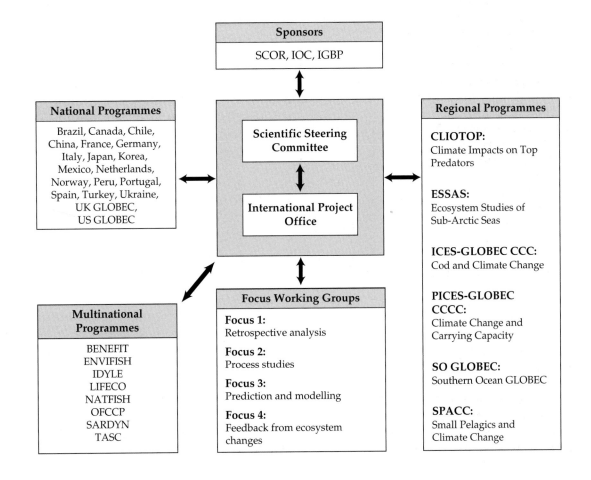

States. In total, scientists from 29 countries actively participated in national and regional GLOBEC activities.

The synthesis of a decade of research that this book represents could only have been possible through the active and enthusiastic involvement of a large number of GLOBEC scientists. Valuable contributions have come from all parts of the GLOBEC research community, the research foci and the Regional and National Programmes and all are gratefully acknowledged. The broad community wide basis of this synthesis is one of its strengths. It is by working together that we have advanced towards the GLOBEC goal: 'To advance our understanding of the structure and functioning of the global ocean ecosystem, its major subsystems, and its response to physical forcing so that a capability can be developed to forecast the responses of the marine ecosystem to global change.'

Brian J. Rothschild and Roger Harris

Acknowledgements

The editors would like to thank the International Geosphere–Biosphere Programme (IGBP) the Scientific Committee on Oceanic Research (SCOR), and the Intergovernmental Oceanographic Commission of UNESCO (IOC) for their support and sponsorship of GLOBEC over the duration of the programme. The valued regional sponsorship of the International Council for the Exploration of the Sea (ICES) and the North Pacific Marine Science Organization (PICES) is also acknowledged. Many national and international funding agencies provided support to GLOBEC, but in particular the US National Science Foundation, the UK Natural Environment Research Council, and the Plymouth Marine Laboratory are acknowledged for their central role in providing funds for programme coordination. The editors would like to particularly thank all the authors, for their efforts to integrate and synthesize the knowledge acquired during the life of GLOBEC. Almost 100 contributing authors from 17 countries have collaborated in this work and the outcome is a testimony to their hard work and enthusiasm. Completion of such a complex volume would not have been possible without the superb technical editing work of Ms. Dawn Ashby, of the GLOBEC International Project Office in Plymouth, United Kingdom.

Contents

List of Boxes

List of Abbreviations

ACC   Antarctic Circumpolar Current
ADCP   Acoustic Doppler Current Profilers
AMO   Atlantic Multidecadal Oscillation
AoA   Assessment of Assessments
AOGCM   Atmosphere-Ocean General
            Circulation Model
AOU   Apparent Oxygen Utilization
ARIES   Autosampling and Recording
            Instrumental Environmental
            Sampler
BASIN   Basin-scale Analysis Synthesis and
            Integration
BCM   Bergen Climate Model
BENEFIT   Benguela Environment and Fisheries
            Interactions and Training
            Programme
BIOMAPER   Bio-Optical Multifrequency
            Acoustical and Physical
            Environmental Recorder
BIONESS   Bedford Institute of
            Oceanography Net and
            Environmental Sampling
BM2   Bay Model 2
BMSY   Stock Biomass at Maximum
            Sustained Yield
BOFFFF   Big Old Fat Fecund Female Fish
CalCOFI   California Cooperative Oceanic
            Fisheries Investigations
CalVET   California Cooperative Oceanic
            Fisheries Investigations (CalCOFI)
            Vertical Egg Tow
CBD   Convention on Biological Diversity
CCAMLR   Commission for the Conservation of
            Antarctic Marine Living Resources
CCC   Cod and Climate Change
CCCC   Climate Change and Carrying
            Capacity
CCS   California Current System
CLIOTOP   CLimate Impacts On TOp Predators

COSEWIC   Committee on the Status of
            Endangered Wildlife in Canada
CPR   Continuous Plankton Recorder
CPUE   Catch Per Unit Effort
CTD   Conductivity, Temperature, and
            Depth
CUFES   Continuous Underway Fish Egg
            Sampler
CuSum   Cumulative Sum
DEPM   Daily Egg Production Method
DMS   Dimethylsulfide
DNA   Deoxyribonucleic acid
DNS   Direct Numerical Simulations
DVM   Diel Vertical Migration
EAF   Ecosystem Approach to Fisheries
EAM   Ecosystem Approach to
            Management
EC   European Commission
ECO-UP   Ecosystèmes d'Upwelling
EEZ   Exclusive Economic Zone
ENSO   *El Niño* Southern Oscillation
ENVIFISH   Environmental Conditions and
            Fluctuations in Recruitment and
            Distribution of Small Pelagic Fish
            Stocks
EOF   Empirical Orthogonal Function
EPICA   European Project for Ice Coring in
            Antarctica
EPO   Eastern Pacific Ocean
ERA   Ecological Risk Assessment
ESP   Eastern South Pacific
ESSAS   Ecosystem Studies of Sub-Arctic Seas
ESU   Evolutionary Sustainable Units
EwE   Ecopath with Ecosim
FAO   Food and Agriculture Organization of
            the United Nations
fhs   fish habitat system
FLEP   Fraction of Lifetime Egg Production
FlowCAM   Flow Cytometer And Microscope

GCOS	Global Climate Observing System	MOCNESS	Multiple Opening/Closing Net and Environmental Sampling System
GEOSS	Global Earth Observation System of Systems		
GLOBEC	Global Ocean Ecosystem Dynamics	MP	Management Procedure
GOOS	Global Ocean Observing System	MPA	Marine Protected Area
GPS	Global Positioning System	MPI	Max Planck Institut
GRT	Gross Registered Tonnage	MRS	Marine Resource System
HPLC	High Performance Liquid Chromatography	MSE	Management Strategy Evaluation
HTI	Hydroacoustic Technology Inc.	MSY	Maximum Sustained Yield
HTL	Higher Trophic Level	NAM	Northern Annular Mode
IBM	Individual Based Model	NAO	North Atlantic Oscillation
ICED	Integrating Climate and Ecosystem Dynamics	NAS	National Academy of Science (USA)
ICES	International Council for the Exploration of the Sea	NATFISH	Natural Variability of a Coastal Upwelling System and Small Pelagic Fish Stocks
ICNAF	International Commission for Northwest Atlantic Fisheries	NEMURO	North Pacific Ecosystem Model Understanding Regional Oceanography
ICOS	Investigation of *Calanus finmarchicus* migrations between Oceanic and Shelf seas		
		NEMURO.FISH	NEMURO For Including Saury and Herring
IDYLE	Interactions and Spatial Dynamics of Renewable Resources in Upwelling Ecosystems	NERC	Natural Environment Research Council
IGBEM	Integrated Generic Bay Ecosystem Model	NMFS	National Marine Fisheries Service
IMBER	Integrated Marine Biogeochemistry and Ecosystem Research	NOAA	National Oceanic and Atmospheric Administration
ING	Individual-based Neural Network Genetic algorithm	NORPAC	North Pacific Net
		NPGO	North Pacific Gyre Oscillation
IOC	Intergovernmental Oceanographic Commission	NPI	North Pacific Index
		NPZ	Nutrient–Phytoplankton–Zooplankton model
IPCC	Intergovernmental Panel on Climate Change	NPZD	Nutrient–Phytoplankton–Zooplankton–Detritus model
IPCC-AR4	Fourth Assessment Report of the Intergovernmental Panel on Climate Change	NSDW	Norwegian Sea Deep Water
		OGCM	Ocean General Circulation Model
IPO	International Project Office		
IWC	International Whaling Commission	OMP	Operational Management Procedure
JGOFS	Joint Global Ocean Flux Study		
LEP	Lifetime Egg Production	OPC	Optical Plankton Counter
LME	Large Marine Ecosystem	OSMOSE	Objected-oriented Simulator of Marine ecOSystems Exploitation
LOPC	Laser Optical Plankton Counter		
MDS	Multidimensional Scaling	OSSE	Observing System Simulation Experiments
MEI	Multivariate ENSO Index		
MEMS	Microelectromechanical systems	PCA	Principal Component Analysis
MOC	Meridional Overturning Circulation	PCR	Polymerase Chain Reaction
		PDO	Pacific Decadal Oscillation

PICES	North Pacific Marine Science Organization	SSC	Scientific Steering Committee
PICT	Plankton Interactive Classification Tool	SSH	Sea Surface Height
		SSMU	Small-Scale Management Unit
PISCES	Pelagic Interactions Scheme for Carbon and Ecosystem Studies	SST	Sea Surface Temperature
		STARS	Sequential T-test Analysis
PNA	Pacific-North America	SVM	Support Vector Machine
POM	Princeton Ocean Model	TAB	Total Allowable Bycatch
PSAT	Pop-up Satellite Archival Tag	TAC	Total Allowable Catch
qPCR	Quantitative PCR	TACIRIE	Transparent, Accountable, Comprehensive, Inclusive, Representative, Informed and Empowered
RAC	Regional Advisory Councils		
RAPID	Research on Automated Plankton IDentification		
RIS	Regime Indicator Series	TAPS	Tracor Acoustic Profiling System
RNA	Ribonucleic Acid	TASC	Trans-Atlantic Study of Calanus
ROMS	Regional Ocean Modelling System	THC	Thermohaline Circulation
ROV	Remotely Operated Vehicle	TOPP	Tagging Of Pacific Pelagics programme
RV	Reproductive Volume	TS	Target Strength
SAHFOS	Sir Alister Hardy Foundation for Ocean Science	UNCED	United Nations Conference on Environment and Development
SARDYN	Sardine Dynamics and stock structure	UNCLOS	United Nations Convention on Law of the Sea
SCOR	Scientific Committee on Oceanic Research	UNEP	United Nations Environment Programme
SEAPODYM	Spatial Environmental Population Dynamics Model	UNESCO	United Nations Educational, Scientific and Cultural Organization
SeaWiFS	Sea-viewing Wide Field-of-view Sensor	UNFCCC	United Nations Framework Convention on Climate Change
SIPPER	Shadowed Image Particle Profiling Evaluation Recorder	VLT	Vertical Life Table
		VPR	Video Plankton Recorder
SO	Southern Oscillation	WCPO	Western Central Pacific Ocean
SO GLOBEC	Southern Ocean GLOBEC	WG	Working Group
SOI	Southern Oscillation Index	WKEFA	WorKshop on the integration of Environmental information into Fisheries management strategies and Advice
SPACC	Small Pelagic Fish and Climate Change		
SRDL	Satellite Relay Data Logger	WMO	World Meteorological Organization
SRES	(IPCC) Special Report on Emissions Scenarios	WOCE	World Ocean Circulation Experiment
SSB	Spawning Stock Biomass	WSSD	World Summit on Sustainable Development

List of Contributors

Allison, Edward H. The WorldFish Center Malaysia Office, Jalan Batu Maung, 11 960 Bayan Lepas, Penang, Malaysia.

Badjeck, Marie-Caroline. The WorldFish Centre Malaysia Office, Jalan Batu Maung, 11 960 Bayan Lepas, Penang, Malaysia.

Barange, Manuel. GLOBEC IPO, Plymouth Marine Laboratory, Prospect Place, Plymouth PL1 3DH, United Kingdom.

Batchelder, Harold. College of Oceanic and Atmospheric Sciences, Oregon State University, Corvallis, OR 97331-5503, USA.

Beaugrand, Gregory. Universite des Sciences et Technologies de Lille 1- UMR CNRS 8187 (LOG), Lille, France.

Benfield, Mark. LSU Department of Oceanography and Coastal Sciences, 2179 Energy, Coast and Environment Building, Baton Rouge, LA 70 803, USA.

Botsford, Louis W. Department of Wildlife and Fish Biology, University of California, Davis, CA 95 616, USA.

Brander, Keith. DTU Aqua–National Institute of Aquatic Resources, Technical University of Denmark, Charlottenlund Castle, Jægersborg Allé 1, 2920 Charlottenlund, Denmark.

Buckley, Lawrence J. Graduate School of Oceanography, University of Rhode Island, Narragansett, RI 02 882-1197, USA.

Bucklin, Ann. University of Connecticut, Department of Marine Sciences, 1080 Shennecossett Road, Groton, CT 06 340, USA.

Campbell, Robert G. Graduate School of Oceanography, University of Rhode Island, Narragansett, RI 02 882-1197, USA.

Carlotti, François. Centre d'Oceanologie de Marseille, Universite de la Méditerranée, Rue de la Batterie des Lions F-13007, Marseille, France.

Checkley, David M. Scripps Institution of Oceanography, 2220 Sverdrup Hall, 8615 Discovery Way, La Jolla, CA 9203, USA.

Chiba, Sanae. Frontier Research Center for Global Change, JAMSTEC, 3173-25 Showa-machi, Kanazawaku, Yokohama, 236-0001 Japan.

Ciannelli, Lorenzo. Centre for Ecological and Evolutionary Synthesis, Department of Biology, University of Oslo, PO Box 1066, N-0316, Blindern, Norway.

Cochrane, Kevern L. Food and Agriculture Organization of the United Nations, Viale delle Terme di Caracalla, 00100 Roma, Italy.

Costa, Daniel D. Department of Ecology and Evolutionary Biology, University of California, Santa Cruz, Santa Cruz, California, USA.

Curchitser, Enrique. Institute of Marine and Coastal Sciences, Rutgers University, 71 Dudley Road, New Brunswick, NJ 08901, USA.

de Moor, Carryn. Department of Mathematics and Applied Mathematics, University of Cape Town, Private Bag, Rondebosch 7701, South Africa.

deYoung, Brad. Department of Physics and Physical Oceanography, Memorial University, St. John's, NF A1B 3X7 Canada.

Dickey, Tommy D. Ocean Physics Laboratory, Department of Geography, University of California, Santa Barbara, CA 93 106, USA.

Drinkwater, Ken. Institute of Marine Research, PO Box 1870, Nordnes, N-5817 Bergen, Norway.

Durbin, Edward G. Graduate School of Oceanography, University of Rhode Island, Narragansett, RI 02 882-1197, USA.

Field, David. Monterey Bay Aquarium Research Institute, 7700 Sandholdt Road, Moss Landing, CA 95 039, USA.

Field, John G. Marine Research (MA-RE) Institute, University of Cape Town, Post Bag X3, Rondebosch 7701, South Africa.

Fiksen, Øyvind. Department of Biology, University of Bergen, N-5020 Bergen, Norway.

Fogarty, Michael J. NOAA NMFS, Northeast Fisheries Science Center, 166 Water St., Woods Hole, MA 02 543, USA.

Frischer, Marc E. Skidaway Institute of Oceanography, 10 Ocean Science Circle, Savannah, GA 31 411, USA.

Gibbons, Mark J. University of the Western Cape, Bellville 7535, South Africa.

Gifford, Dian J. Graduate School of Oceanography, University of Rhode Island, Narragansett, RI 02882-1197, USA.

Hamilton, Lawrence. University of New Hampshire, Durham, NH 03824, USA.

Harris, Roger P. Plymouth Marine Laboratory, Prospect Place, Plymouth, PL1 3DH, United Kingdom.

Heath, Michael. Marine Laboratory, PO Box 101, Victoria Road, Aberdeen, AB11 9DB, United Kingdom.

Hofmann, Eileen E. Center for Coastal and Physical Oceanography, Old Dominion University, Norfolk, VA 23529, USA.

Hunt, George. School of Aquatic and Fishery Sciences, University of Washington, Box 35520, Seattle, WA 98195, USA.

Hurrell, James W. National Center for Atmospheric Research, PO Box 3000, Boulder, CO 80307, USA.

Irigoien, Xabier. AZTI - Tecnalia / Marine Research Division, Herrera kaia portualdea, z/g 20110 Pasaia, Gipuzkoa, Spain.

Ito, Shin-ichi. Tohoku National Fisheries Research Institute, Fisheries Research Agency, 3-27-5, Shinhama-cho, Shiogama, Miyagi 985-0001, Japan.

Jarre, Astrid. Marine Research Institute and Zoology Department, University of Cape Town, Private Bag X3, Rondebosch 7701, South Africa.

Kell, Laurence T. ICCAT, 8 Corazón de María, 28002, Madrid, Spain.

Kim, Suam. Department of Marine Biology, Pukyong National University, Nam-gu, Busan, 608-737 Korea.

Kimura, Shingo. Ocean Research Institute/ Graduate School of Frontier Sciences, The University of Tokyo, 1-15-1, Minamidai, Nakano, Tokyo 164-8639, Japan.

King, Jacquelynne R. Pacific Biological Station, Fisheries and Oceans Canada, 3190 Hammond Bay Road, Nanaimo, BC V9T 6N7, Canada.

Kiørboe, Thomas. National Institute for Aquatic Resources, Oceanography Section, Technical University of Denmark, Kavalergården 6, DK-2920 Charlottenlund, Denmark.

Kishi, Michio J. Faculty of Sciences, Hokkaido University, N18, W8 Kita-ku, Hokkaido 060 0813, Japan.

Köster, Fritz W. DTU Aqua-National Institute of Aquatic Resources, Technical University of Denmark, Charlottenlund Castle, Jægersborg Allè 1, 2920 Charlottenlund, Denmark.

Lehodey, Patrick. Marine Ecosystems Modelling and Monitoring by Satellites, CLS, Satellite Oceanography Division, 8–10 rue Hermes, 31520 Ramonville, France.

Lluch-Cota, Salvador. CIBNOR Mar Bermejo # 195 Col. Playa Palo de Santa Rita. La Paz, B.C.S., Mexico 23090.

Mackas, David L. Fisheries and Oceans Canada Institute of Ocean Sciences, 9860 West Saanich Road, Sidney, BC V8L 4B2, Canada.

Maury, Olivier. Institut de Recherche pour le Développement, Centre de Recherches Halieutiques Méditerranéennes et Tropicales, av. Jean Monnet, B.P. 171, 34 203 Sète cedex, France.

McKinnell, Stewart. M. North Pacific Marine Science Organization, c/o Institute of Ocean Sciences, P.O. Box 6000, 9860 West Saanich Road, Sidney, BC, V8L 4B2, Canada.

Miller, Arthur J. Scripps Institution of Oceanography, University of California, San Diego, 9500 Gilman Drive, La Jolla, CA 92093-0224, USA.

Möllmann, Christian. Institute for Hydrobiology and Fisheries Science, University of Hamburg, Grosse Elbstrasse 133, 22767 Hamburg, Germany.

Moloney, Coleen L. Zoology Department and Marine Research Institute, University of Cape

Town, Private Bag X3, Rondebosch 7701, South Africa.

Murphy, Eugene J. British Antarctic Survey, Natural Environment Research Council, High Cross, Madingley Road, Cambridge CB3 0ET, United Kingdom.

Neis, Barbara. Department of Sociology, Memorial University, St. John's, NF A1C 5S7, Canada.

Nejstgaard, Jens C. University of Bergen, Department of Biology, Box 7800, N-5020 Bergen, Norway.

O'Boyle, Robert. Beta Scientific Consulting, Inc. 1042 Shore Drive, Bedford, B4A 2E5, Canada.

Ohman, Mark D. Scripps Institution of Oceanography, University of California, San Diego, 9500 Gilman Dr., La Jolla, CA 92 093-0218, USA.

Ommer, Rosemary E. Institute of Coastal and Oceans Research, University of Victoria, Victoria, BC V8P 5C2, Canada.

Overland, James E. Pacific Marine Environmental Laboratory, NOAA, 7600 Sand Point Way NE, Seattle, WA 98115, USA.

Peterson, William T. NOAA Fisheries Service, Northwest Fisheries Science Center, 2030 South Marine Science Drive, Newport, OR 97365, USA.

Perry, R. Ian. Fisheries and Oceans Canada, Pacific Biological Station, Nanaimo, BC V9T 6N7, Canada.

Planque, Benjamin. Institute of Marine Research – Tromsø, Postboks 6404, N-9294 Tromsø, Norway.

Quiñones, Renato A. Center for Oceanographic Research in the South Eastern Pacific (COPAS Center) & Department of Oceanography, University of Concepción, Concepción, Chile.

Reid, Keith. CCAMLR, 181 Macquarie Street, Hobart TAS 7000, Australia.

Rose, Kenneth A. Department of Oceanography and Coastal Sciences, 2135 Energy Coast, and Environmental Building, Louisiana State University, Baton Rouge, LA 70803, USA.

Rothschild, Brian J. School for Marine Science and Technology, University of Massachusetts Dartmouth, MA 02744-1221, USA.

Runge, Jeffrey A. School of Marine Sciences, University of Maine, Gulf of Maine Research Institute, 350 Commercial Street, Portland, ME 04101, USA.

Saiz, Enric. Institut de Ciències del Mar, CSIC, Ps. Marítim de la Barceloneta 37-49, 08003 Barcelona, Catalunya, Spain.

Sakurai, Yasunori. Faculty of Fisheries, Hokkaido University, 3-1-1 Minato-cho, Hakodate, Hokkaido 041-8611, Japan.

Schwing, Frank. Environmental Research Division, Southwest Fisheries Science Center, NOAA Fisheries Service, 1352 Lighthouse Avenue, Pacific Grove, CA 93950-2020, USA.

Shannon, Lynne J. University of Cape Town and Marine and Coastal Management, P.O. Box X2, Roggebaai, Cape Town 8012, South Africa.

Sinclair, Mike. Bedford Institute of Oceanography, PO Box 1006, Dartmouth, NS B2Y 4A2, Canada.

Steffen, Will. ANU Climate Change Institute, The Australian National University, Canberra ACT 0200, Australia.

St. John, Michael A. Institute for Hydrobiology and Fisheries Science, University of Hamburg, Grosse Elbstrasse 133, 22767 Hamburg, Germany.

Sumaila, U. Rashid. Fisheries Centre, University of British Columbia, Vancouver, BC V6T 1Z4, Canada.

Sundby, Svein. Institute of Marine Research, PO Box 1870, Nordnes, N-5817 Bergen, Norway.

Tadokoro, Kazuaki. Tohoku National Fisheries Research Institute, 3-27-5 Shinhama-cho, Shiogama, Miyagi 985-0001, Japan.

Vadas, Flora. Center for Tropical Marine Ecology, University of Bremen, 6 Farenheit Street, 28359 Bremen, Germany.

van der Lingen, Carl. Marine and Coastal Management (MCM), P.O. Box X2, Roggebaai, Cape Town 8012, South Africa.

Werner, Francisco E. Institute of Marine and Coastal Sciences, Rutgers University, New Jersey, USA.

Wiebe, Peter H. Woods Hole Oceanographic Institution, Woods Hole, MA 02543, USA.

Wieland, Kai. DTU Aqua-National Institute of Aquatic Resources, Technical University of Denmark, Nordsøen Forskerpark, 9850 Hirtshals, Denmark.

Windle, Matt. Department of Biology, Memorial University, St. John's, NF A1C 5S7, Canada.

Wolff, Matthias. Center for Tropical Marine Ecology, University of Bremen, 6 Farenheit Street, 28359 Bremen, Germany.

Yamanaka, Yasuhiro. Faculty of Earth Environmental Science, Hokkaido University, N10W5, Sapporo 060-0810, Japan.

Yamazaki, Hidekatsu. Department of Ocean Sciences, Tokyo University of Marine Science and Technology 4-5-7, Konan, Minato-Ku Tokyo 108-8477, Japan.

Yatsu, Akihiko. National Research Institute of Fisheries Science, Fukuura 2-12-4, Kanazawa-ku, Yokohama 236-6848, Japan.

CHAPTER 1

Introduction: oceans in the earth system

Manuel Barange, John G. Field, and Will Steffen

1.1 The integrated blue planet

Our perception of the earth has changed dramatically over the last half century, and has been deeply influenced by observations of the earth from space. In 1966 the *Lunar Orbiter 1* spacecraft produced the first picture of the earth rising over the moon, and this, plus subsequent images from the NASA Apollo programme, electrified the world. A global perception started to emerge, recognizing that the earth was a single, interconnected unit, within which the biosphere is an active and essential component. All the elements of the earth, land, oceans, and atmosphere, appear not only connected but interdependent. Surely an impact on one of these components must reflect on the others and on the whole?

However, the extent to which the earth behaves as a single, interlinked, self-regulated system was not put into focus until scientists reconstructed the chemical history of the earth's atmosphere from 420,000-year-old Antarctica ice cores (the Vostok record; Petit *et al.* 1999), and more recently from the 650,000-year-old cores from European Project for Ice Coring in Antarctica (EPICA) Dome C site (Siegenthaler *et al.* 2005; Fig. 1.1). These data indicated that the CO_2 levels in the atmosphere had been constrained between 180 and 290 ppm by volume for over 500,000 years, with minima and maxima corresponding to the glacial and interglacial periods, respectively. Equivalent cycles from atmospheric temperature proxies demonstrated a strong coupling between climate and the carbon cycle as well as the stability of this coupling over the recorded time. These observations added weight to the initial sugges-

tion that the earth system must have some self-regulatory mechanisms to keep its composition within limits.

While most of these long-term climate records come from ice cores of Antarctica, we know that there is a linear dependency between climate events in Antarctica and Greenland, for example, linked by the dynamics of the deep Meridional Overturning Circulation (MOC) in the Atlantic Ocean (EPICA 2006). Put simply, a vigorous MOC is thought to deliver heat to the North Atlantic at the expense of the Southern Ocean. Increases and decreases in MOC strength thus result in a climate 'see-saw' between the Southern and northern Hemispheres (Stocker and Johnsen 2003). The role of the oceans on the earth system begins to take shape.

For some time scientists have suspected that the glacial/interglacial cycles that the earth has experienced every 100 kyr (Fig. 1.1) are caused by smooth changes in the eccentricity of the earth's orbit. However, these changes are too smooth to generate such sharp, non-linear responses in the chemical composition of the atmosphere, unless these are strongly modulated by biological, chemical, and physical feedbacks. The intimation is, therefore, that the earth system *responds* to the physical forces in complex ways, often non-linearly, but generally keeping the planet within limits. The evidence presented above led to the development of a new scientific discipline, earth system science, the study of the earth system, with an emphasis on observing, understanding, and predicting global environmental changes involving interactions between land, atmosphere, water, ice, biosphere, societies, technologies, and economies.

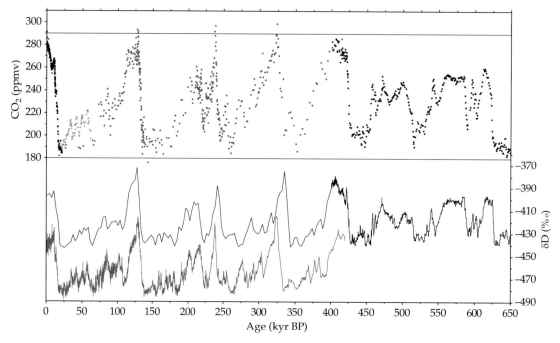

Figure 1.1 A composite CO_2 record of the earth's atmosphere over six-and-a-half Ice Age cycles, back to 650,000 years before present. The record results from the combination of CO_2 data from three Antarctic ice cores: Dome C (black), Vostok (blue), and Taylor Dome (light green). High resolution deuterium record (δD), a proxy for temperature, covaries with CO_2 concentrations, thus linking the climate and atmospheric chemical composition. (From Siegenthaler *et al.* 2005 reprinted with permission from AAAS.) (See plate 1).

1.2 The oceans in an earth system

The ocean is one of the major components of the earth system, providing 99% of the available living space on the planet. Water is essential to our existence, having secured life from the time of the primeval soup. It has been estimated that 80% of all life on earth depends on healthy oceans and coasts and more than a third of the world's population lives in coastal areas and small islands, even though they amount to less than 4% of the earth's land. Our planet, the only one where water exists in liquid form, should have been called planet water rather than planet earth.

Earlier in this chapter we described how the ocean has an important role in climate regulation. The ocean's heat capacity is about 1,000 times larger than that of the atmosphere, and the oceans net heat uptake since 1960 has been around 20 times greater than that of the atmosphere (Bindoff *et al.* 2007; Fig. 1.2). We have now direct evidence to conclude that the world oceans have warmed substantially

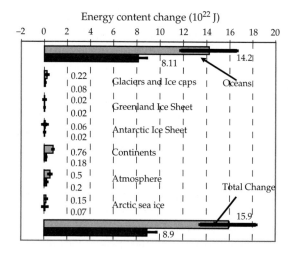

Figure 1.2 Energy content changes in different components of the earth system for two periods (1961–2003, light grey bars; and 1993–2003, dark grey bars). Positive energy content change means an increase in stored energy (i.e. heat content in oceans, latent heat from reduced ice or sea ice volumes, heat content in the continents excluding latent heat from permafrost changes, and latent and sensible heat and potential and kinetic energy in the atmosphere). All error estimates are 90% confidence intervals. (From Bindoff *et al.* 2007.)

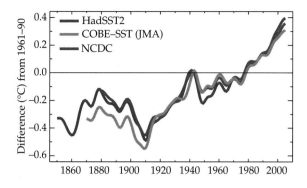

Atlantic Ocean

Indian Ocean

Pacific Ocean

−1 −0.8 −0.6 −0.4 −0.2 0 0.2 0.4 0.6 0.8 1 °C

Figure 1.3 (Left panel) Smoothed annual global SST anomalies, relative to 1961–90 (°C), from the UK Meteorological Office sea surface temperature data set (HadSST2) (Rayner *et al*. 2006), the USA National Climatic Data Centre (NCDC) (Smith *et al*. 2005; red line) and the Japan Meterological Agency COBE-SST dataset (Ishii *et al*. 2005; green line). (Right panel) Sea Surface Temperature (°C) annual anomalies across each ocean by latitude from 1900 to 2005, relative to the 1961–90 from HadSST2 (Rayner *et al*. 2006). (From Trenberth *et al*. 2007.) (See plate 2).

over the last half century, and that the warming accounts for over 80% of the changes in the energy content of the earth's climate system during this period (Bindoff *et al*. 2007; Domingues *et al*. 2008; Fig. 1.2). This global trend, however, has not been homogeneous or linear. Significant decadal variations have been observed in the sea surface temperature (SST) time series, and there are large regions where the oceans have been cooling (Bindoff *et al*. 2007; Harrison and Carson 2007; Fig. 1.3). Large-scale, coherent trends of salinity are also observed: global freshening in subpolar latitudes and a salinification of shallower parts of the tropical and subtropical oceans. These trends are consistent with observed changes in precipitation, implying that more water is transported in the atmosphere from low latitudes to high latitudes and from the Atlantic to the Pacific.

Although the oceans are interconnected, there are specific modes of response to forcing that indicate important differences as well (see Drinkwater *et al*., Chapter 2, this volume). The Pacific Ocean is dominated by a number of short-term *El Niño*-Southern Oscillation (ENSO) events and longer-term events modulated by the Pacific Decadal Oscillation (PDO), which provides ways of moving heat from the tropical ocean to higher latitudes and out of the ocean into the atmosphere (Trenberth *et al*. 2002). In the Atlantic, observations reveal the role of the Atlantic Multidecadal Oscillation (AMO), which is associated with the Meridional Overturning Circulation (MOC), which transports heat northwards, thereby moderating the tropics and warming high latitudes. Figure 1.3 presents latitude-time sections of SST across each ocean revealing, in the Pacific, a long-term warming trend punctuated by cooler tropical episodes. In the Atlantic, the warming from the 1920s to about 1940 in the northern Hemisphere was focused on higher latitudes, with the Southern Hemisphere remaining cool. This inter-hemispheric contrast is believed to be one signature of the MOC, as already mentioned. The recent strong warming appears to be related in part to changes in the AMO in addition to a global warming signal. The Indian Ocean also reveals a poorly observed warm interval in the early 1940s, and further shows the fairly steady warming in recent years. The multidecadal variability in the Atlantic has a much longer timescale than that in the Pacific.

The dynamic nature of the different oceans, exemplified by the varying heat exchange processes, is responsible for the diversity of marine environments and habitats. The properties of water generate different density layers and gradients. Tides, currents, and upwelling break this stratification and, by forcing the mixing of water layers, enhance primary production. This network of surface and deepwater currents tightly connects marine ecosystems and defines its dynamics.

1.3 Climate variability and change in ocean ecosystems

One of the pillars supporting the development of GLOBEC has been the recognition that marine ecosystems fluctuate at a multitude of time scales, due to a combination of climate variability and change, internal dynamics of the population, and predator-prey and competitive dynamics (Lehodey *et al.* 2006). This complexity of forcings and pathways makes it difficult to establish unequivocal connections between climate and ecological responses. Understanding the main patterns, mechanisms, and processes through which climate and physical forcing affect ecosystems from region to region has been a particular focus of GLOBEC.

A good framework for understanding the role of climate on marine ecosystems has been classifying them according to the directness of the impact (Ottersen *et al.* 2010). *Direct effects* are mechanisms that involve a direct ecological response to one of the environmental phenomena. *Indirect effects* are mechanisms that either involve several physical or biological intermediary steps between climate and the ecological trait and/or have no direct impact on the biology of the population. *Integrated effects* involve simple ecological responses that can occur during and after the year of an extreme climate event. *Translations* involve movements of organisms from one place to another. These alterations are based entirely on the physical changes produced by climate variability, resulting in both linear and non-linear responses at decadal (Schwartzlose *et al.* 1999; Chavez *et al.* 2003), multidecadal (Ravier and Fromentin 2004), and multicentennial (Baumgartner *et al.* 1992) time scales.

The other driving force behind GLOBEC's work has been the recognition that climate change has and will continue to have profound consequences for marine ecosystems (Harley *et al.* 2006; Barange and Perry 2009). It is expected that ocean warming will result in increasing vertical stratification, reduced vertical mixing and reduced nutrient supply, thereby decreasing overall productivity, but with large regional differences (Sarmiento *et al.* 2004). However, predictions for processes beyond phytoplankton depend not only on changes in net primary production, but also on its transfer to fish and mammals, about which there is low predictive confidence (Brander 2007a). However, changes in species composition (Bopp *et al.* 2005), process seasonality (Mackas *et al.* 1998; Edwards and Richardson 2004; Hashioka and Yamanaka 2007), and species distributions (Hawkins *et al.* 2003; Drinkwater 2005; Perry *et al.* 2005) are already observed.

1.4 Climate change and global change

In parallel to the realization that climate variability and change cause profound changes in marine ecosystems there has been increasing recognition that human activities have become so pervasive and profound in their consequences that they affect the earth at a global scale. This powerful human impact has been so dominant that a new geological era, the Anthropocene, has been proposed to describe the last few hundred years (Crutzen and Stoermer 2000; Steffen *et al.* 2007), and particularly the period since 1950. During these recent decades humans have changed the land surface of the planet more extensively than in any other comparable time in human history (Reid *et al*. 2005). The earth is entering its sixth major extinction event, with rates of species loss now at least 100 times background levels and expected to rise this century to about 1,000 times background (Pimm *et al*. 1995; Reid *et al*. 2005). Global change as a concept is useful because, until it was coined, the different problems it encapsulates were considered individually and treated with separate priorities and solutions. These problems include climate change, biodiversity loss, dwindling water resources, as well as the human activities that drive these environmental changes.

The oceans have not been immune to the relentless search for food and resources. World fish marine catches increased steadily from 16.7 Mt in 1950 to around 84 million Mt since 2000, with 76% of the marine fish stocks either fully exploited or overexploited (FAO 2007). Increasing human pressure has also been shown to profoundly impact marine ecosystems (Kaiser and Jennings 2002; Brander *et al*., Chapter 3, this volume). While historically there has been a tendency to pose ecosystem changes as a dichotomy, that is, whether (1) 'natural' climate variability or (2) exploitation patterns bear primary responsibility, both climate and other anthropogenic uses interact (see Perry *et al*., Chapter 8, this volume). Thus, climate may cause failure in a fishery management scheme and fishery exploitation may also disrupt the ability of a resource population to withstand, or adjust to, climate changes (Hsieh *et al*. 2006; Perry *et al*. 2010a; Planque *et al*. 2010). Understanding climate variability and change in the context of the broader impacts on marine ecosystems has been a priority in GLOBEC.

1.5 Marine ecosystem sustainability

Ecosystems are a collection of plant, animal, and microorganism communities, as well as their nonliving environment, interacting as functional units. Whilst humans are an integral part of ecosystems and not just an external force, we extract goods and services from ecosystems, and these can be economically valued. Although translating goods and services into a monetary value has many limitations, it has been a useful approach to understand their importance as a life support system. Costanza *et al*. (1997) estimated marine and coastal goods and services to value the equivalent of US$20 trillion ($10^{12}$), or two-thirds the value of the earth's biosphere as a whole. Individual sector valuations area also available: fisheries (captured and cultured) are worth approximately US$140 billion in 2002 (FAO 2007); offshore gas and oil produce US$132 billion in 1995; marine tourism, much of it along the coast, approximately US$161 billion in 1995; and trade and shipping US$155 billion (McGinn 1999). More than a billion people rely on fish as their main or sole source of animal protein, especially in developing countries, and the demand for food fish and various other products from the sea continues to grow, driven by population growth and human migrations to the coast. The message that these figures provide is that human societies need to both use and protect marine ecosystems with all their might, as their own sustainability is at risk.

The ultimate goal of GLOBEC's research is to underpin sustainable marine ecosystem management, recognizing that marine ecosystems are critical for global food security and for sustaining the well-being of many national economies, particularly in developing countries. At the 2002 Johannesburg World Summit for Sustainable Development signatories agreed to reverse the current trend in natural resource degradation as soon as possible, by implementing strategies which should include targets to protect ecosystems and to achieve integrated management of land, water, and living resources. These strategies require adequate understanding of the marine ecosystem and its response to climate and other anthropogenic forcings, as a policy requirement for a cross-sectoral, interdisciplinary, and science-based approach to marine management (see Barange *et al*., Chapter 9, this volume). Research must demonstrate the cumulative effects of global change and the additional difficulties these pose for sustainable development. It must clarify the local, regional, and global impacts of, for example, exploitation patterns, and quantify the scale differences in the human-environment relationship. Any attempts to achieve sustainability require learning to live with and adapt to global change, but, more fundamentally, to recognize and live within the biophysical limits of the earth system. This book aims at establishing the baseline scientific knowledge for a new generation of researchers and managers.

1.6 Objectives and structure of the book

The objective of this book is to explore what has been learned about the fundamental dynamics of marine ecosystems and their responses to anthropogenic changes (from climate change to overexploitation) as well as to natural variability. In doing this the focus is on work conducted under the Global Ocean Ecosystem Dynamics Programme (GLOBEC) and related activities, since the late 1980s. The book also draws upon the work of scientists not directly linked to GLOBEC but to ecosystem science and management agencies around the world. The book is divided into four sections that describe our current understanding of the marine ecosystem along four generic topics: how climate and humans impact and transform ecosystems (Part I), what is our current understanding of the structure and functioning of marine ecosystems as derived from both observations and models (Part II), the complex relationship between marine ecosystems and human societies (Part III), and predicting marine ecosystem futures and identifying future research needs (Part IV). A brief description of each part is provided below (Fig. 1.4):

1.6.1 Part I: the changing ocean ecosystems

Chapter 2 focuses on the role of climate variability and climate change in influencing marine ecosystems at large scales. It highlights the links between climate variability and ecosystem dynamics, and how they differ among regions. This chapter adopts a comparative approach across GLOBEC's regional programmes.

Chapter 3 reviews the evidence of human impacts on marine ecosystems from historic records and also the development of the concepts of stewardship and sustainability. Fishing is not the only human activity with an impact on marine ecosystems and most of these activities have increased over the past century to the point where global effects are apparent. Most capture fisheries have reached or exceeded the limits of sustainable exploitation and the ways in which this affects the productivity, diversity, and resilience of marine ecosystems are explored.

1.6.2 Part II: advances in understanding the structure and dynamics of marine ecosystems

Chapter 4 introduces the concept of target species, central to GLOBEC's retrospective and modelling studies on population dynamics. Target species are significant either for ecological, economic, or cultural reasons, and their study has provided GLOBEC with units to develop comparative studies across the globe.

Chapter 5 introduces how physical forcing affects the structure and dynamics of marine populations,

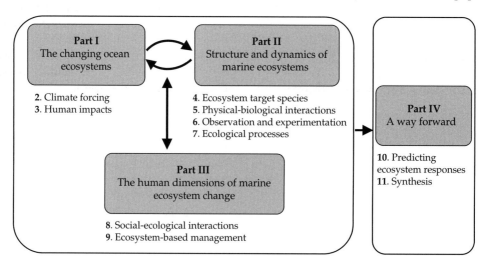

Figure 1.4 Structural outline of the book.

before focusing on the coupling of biology and physics in the marine environment and the development of physical-biological models, the cornerstone of GLOBEC's approach. It discusses the combination of model and field data and the challenges to modelling marine ecosystems and how modelling has improved our understanding.

Chapter 6 describes the observational (field and laboratory) approaches developed during the life of GLOBEC. It explores the unique challenges of sampling marine ecosystems and ecosystem target species and evaluates the successes and value of each of the applied methods. It particularly relies on the comparative approach among regions, ecosystems, and species and elaborates on the contributions to GLOBEC and its legacy.

Chapter 7 focuses specifically on the ecological processes that lead to ecosystem structure. It makes use of concepts such as target species and physical-biological interactions, introduced in previous chapters, and the observation and modelling approaches developed in GLOBEC, to address the response of the organisms and communities to physical forcings at different scales and across different GLOBEC regions. It concludes with the need to integrate ecological processes if we are to understand the consequences of global environmental changes.

1.6.3 Part III: the human dimensions of changes in marine ecosystems

In Chapter 8 we introduce the need to incorporate social sciences to understand the interactions between human societies and the marine ecosystem. These include literature, history, sociology, anthropology, political science, and economics, each with its own rich literature, traditions, and understanding. GLOBEC, being a programme rooted in the natural sciences, chose to focus on a narrower set of issues, but recognizing that one of the key elements is 'interactivity'; marine ecosystem changes impact on marine-dependent human communities, and vice versa. This chapter explores the reciprocal interactions between natural marine ecosystems and human communities within the context of global change.

Chapter 9 reviews how management needs have changed during the life of GLOBEC and assesses the projected needs of management over the foreseeable future. It specifically describes how ecosystem science and ecosystem management converged on the need for the application of an ecosystem approach to fisheries management. Examples are provided on how ecosystem science has led to changing management objectives, and on the application and requirements of the ecosystem approach.

1.6.4 Part IV: a way forward

Chapter 10 introduces the difficulties, approach, and state of the art in predicting ocean ecosystem responses to future global change scenarios. It addresses the limitations and uncertainties of these predictions and elaborates on future challenges.

Finally, Chapter 11 summarizes the contributions of the GLOBEC approach to marine ecosystem research. It uses the conclusions of previous chapters to encapsulate the responses of marine ecosystems to global change. The chapter identifies gaps in our knowledge and future research directions and discusses GLOBEC's contribution to securing sustainable marine ecosystems.

PART I

The changing ocean ecosystems

Ocean ecosystems are constantly changing on a range of scales. On the one hand, small space scales and short time scales, for example how physical turbulence affects microplankton and meso-zooplankton feeding on microplankton. On the other hand, global-scale ocean-atmospheric phenomena, such as the *El Niño* Southern Oscillation (ENSO), which affect fish stocks off Peru both directly and indirectly through their predators and prey. It also influences rainfall in Australia, California, and Africa, with large consequences for the freshwater and nutrient inputs to coastal zones. On an even larger temporal scale there are global ice ages and interglacial periods. These examples are caused by natural processes although their frequencies and magnitudes may change both naturally and through human interference in the Earth System. In all cases, physical changes and interactions reverberate through the food web and may affect many aspects of ecosystems upon which we depend for food,

transport and indeed most aspects of our lives. Chapter 2 deals with climate variability and change and how these interact with marine ecosystem dynamics. It also describes how these ecosystem responses and feedbacks are not uniform but differ from one Global Ocean Ecosystem Dynamics (GLOBEC) region to another.

In addition to climate change forcing ecosystem dynamics, there are also direct human effects on marine ecosystems, most notably through our harvesting activities. Chapter 3 reviews the historical evidence of human activity and shows how marine ecosystems have been impacted over time. Over the last century, fishing activities have intensified with the development of modern technology and the nature of fishing impacts has evolved from being purely local in nature to becoming global. This has affected not only target species but in many cases has cascaded down the food web affecting populations, spatial dynamics, biodiversity, and has damaged habitats with

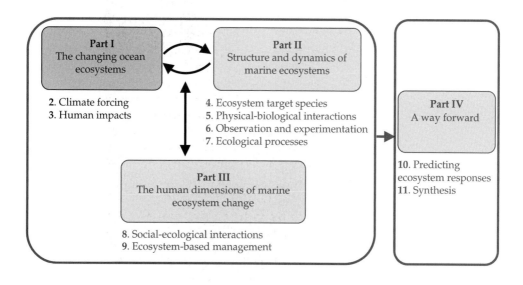

consequences for ecosystem productivity and carrying capacity. In recent years, we have also observed that the interaction between climate-induced changes and direct human exploitation patterns can result in unpredictable responses, such as dramatic changes in species dominance and composition, in what we now call regime shifts. Not only are there effects of climate change and variability on ecosystems, but there are also interactions between climate-induced ecosystem changes and direct impacts, which may lead to regime shifts and irreversible ecosystem changes. The twin concepts of stewardship and sustainability emerge as an historical consequence of human impacts on marine ecosystems.

CHAPTER 2

Climate forcing on marine ecosystems

Ken Drinkwater, George Hunt, Patrick Lehodey,
Salvador Lluch-Cota, Eugene J. Murphy, Yasunori Sakurai,
Frank Schwing, Gregory Beaugrand, and Svein Sundby

2.1 Introduction

One of the main aims of the Global Ocean Ecosystem Dynamics (GLOBEC) programme was to increase understanding of how the structure and functioning of the global ocean ecosystem and its major subsystems respond to physical forcing. Long before GLOBEC began in the 1990s, it was well understood that physical forcing strongly affects marine ecological processes. The American fisheries biologist, Spencer Fullerton Baird, recognized the importance of the environment in generating fluctuations in fish stocks as early as the 1870s (Lehodey *et al.* 2006). The expected links between the environment and fish stocks played a significant role in the strong emphasis on hydrographic data collection by the International Council for the Exploration of the Seas (ICES) from its inception at the beginning of the twentieth century. Around this same time a systematic search began for relations between physical conditions and different aspects of fish stocks (Helland-Hansen and Nansen 1909). A significant warming period in the North Atlantic during the 1920s and 1930s had demonstrated effects on the distribution and production of several important commercial fish species, as well as on other components of the marine ecosystem (Jensen 1939; ICES 1949; Cushing and Dickson 1976; Cushing 1982). By the mid-1900s, the mechanisms through which the physical environment affects primary production had been elegantly elaborated by Sverdrup (1953). New sampling methods began to reveal the extent of the ocean's variability in the 1960s and 1970s, and it became increasingly clear that physical variability and its influence on marine ecology occurred over a multitude of spatial and temporal scales. In spite of this progress, by the 1990s when GLOBEC was established, understanding of the role of physical forcing on marine ecology still remained incomplete and in particular the mechanisms linking the environment with ecological responses were often not well understood. Thus, GLOBEC began to tackle these issues in order to provide better advice on marine resource management issues (e.g. see Barange *et al.*, Chapter 9, this volume), and as a way to increase our capability to improve projections of what future marine ecosystems may look like, especially under human-induced global change (see Ito *et al.*, Chapter 10, this volume).

In the present chapter, we focus on the roles of climate variability and climate change on the ecological components of marine ecosystems, especially at large spatial scales, that is, regional and larger. We begin by defining climate, climate variability, and climate change. This is followed by descriptions of some of the large-scale climate indices and their connection to regional climate variability as well as a discussion of teleconnections across large spatial scales. After introducing the role of the oceans in the climate system, we describe the patterns and effects of climate forcing on marine ecosystems. We then provide several examples of the responses of marine ecosystems to climate variability and climate change from GLOBEC-generated studies. Further descriptions of the processes linking physical forcing to ecological responses are discussed by deYoung *et al.* (Chapter 5, this volume) and Moloney *et al.* (Chapter 7, this volume).

2.2 Climate forcing, climate variability, and climate change

Climate is defined by the American Meteorological Society (Glossary of meteorology, http://amsglossary.

allenpress.com/glossary/) as the slowly varying aspects of the atmosphere-hydrosphere-land system and is characterized statistically in terms of long-term averages and variability of climate elements such as temperature, precipitation, winds, etc. The period over which the 'average' is determined is typically 30 years, as recommended by the World Meteorological Organization (WMO). *Climate variability* is the temporal variation around this average state and is associated with time scales of months to millennia and beyond, that is, longer than those associated with synoptic weather events. Natural climate variability refers to climate variations due to such factors as changes in solar radiation, volcanic eruptions, or internal dynamics within the climate system, and pertains to any influence that is not attributable to, or influenced by, activities related to humans. Human effects on climate, such as those due to greenhouse gas emissions or land use, are termed anthropogenic influences. *Climate change*, on the other hand, is any systematic change in the long-term statistics of climate elements from one state to another and where the new state is sustained over several decades or longer (AMS Glossary of Meteorology). The new state might be due to changes in the mean level, the characteristic variability, or both. Climate change arises from both natural and anthropogenic causes. The United Nations Framework Convention on Climate Change (UNFCCC) has restricted its definition of climate change to causes arising directly or indirectly from human activity only, while it regards climate variability as changes attributable to natural causes. In recent years this definition has been often adopted by the media, in policy documents, and in some scientific literature. Climate change in this sense has also been used interchangeably with global warming. Finally, the term *regime shift* has also been used in the oceanographic literature in relation to changes in climate and marine ecology. While several different definitions have been proposed (see discussion by Overland *et al.* 2008), from the climate perspective it typically refers to a rapid shift from one climatic state, or 'regime', to another. Regime shifts, as defined, cover climate change as well as situations when the duration of the climatic state following the shift is too short (typically decadal to quasi-decadal) to be designated

as a climate change. Ecosystem regime shifts, also called abrupt ecosystem shifts (Beaugrand *et al.* 2008), usually refer to large time and space shifts in abundances of major components of marine biological communities (Bakun 2005; Moloney *et al.*, Chapter 7, this volume).

Examples of climate change include shifts between Ice Ages and the warmer interglacial periods. In historic times, it includes multicentenial changes in the northern Hemisphere from warm periods such as during the 'Medieval Warm Period' (890–1170 AD) to colder periods such as the 'Little Ice Age' (1580–1850 AD) (Osborn and Briffa 2006). On multidecadal scales, the shift in the North Atlantic from the generally warm period of the 1930s to the 1960s to the colder period in the 1970s and 1980s and a return to warm conditions in the 1990s has been labelled the Atlantic Multidecadal Oscillation or AMO (Kerr 2000). The AMO fits within the definition of climate change as it is multidecadal but perhaps fits better as an example of longer-term climate variability. Other examples of climate variability include the strong decadal fluctuations associated with changes in the atmospheric pressure systems such as the North Atlantic Oscillation (NAO; see Box 2.1) and the 2 to 3-year variability of *El Niño*-Southern Oscillation (ENSO) events in the equatorial Pacific. Examples of regime shifts include the changes in both climate and ecology in the North Pacific that occurred in the late 1970s (Francis *et al.* 1998; Hare and Mantua 2000) and in the north-east Atlantic, especially the North Sea, in the late 1980s (Reid *et al.* 2001a,b).

2.3 Large-scale climate variability patterns

By the early 1990s, at the onset of GLOBEC, researchers studying the responses of marine ecology to physical environmental forcing began to consider the large-scale climate variability (Forchhammer and Post 2004), typically expressed in terms of changes in climate indices related to atmospheric pressure patterns, such as the ENSO, NAO, or the North Pacific Index (NPI). The NPI is the area-weighted sea-level pressure over the region 30°N–65°N, 160°E–140°W. Other climate indices, such as the Pacific Decadal Oscillation (PDO) or the

AMO, were developed based on the intensity of sea temperature patterns. Further descriptions of several of these indices are provided in Box 2.1. While the observed climate-induced changes in marine ecosystems are indeed responses to local conditions, large-scale indices often account for significant portions of this local or regional variability. This is because the local climate changes are often a response to the large-scale processes, and also because large-scale indices are related to several physical elements (e.g. air and ocean temperatures, sea ice, winds, etc.) and as such they can be more representative of the climate forcing than any single local variable (Stenseth *et al.* 2003). Also, local indices can vary significantly while large-scale indices are smoother, with less noise. Thus, in some cases large-scale indices can account for as much, or at times more, of the variance of ecosystem elements than local indices (e.g. Drinkwater *et al.* 2003). Using large-scale climate indices potentially allows linking climate-induced ecological dynamics over a range of trophic levels, species, and geographic locations. Some examples of the ecological impacts of the variability in these indices are provided in Section 2.7.

Linkages between climate indices, weather events, or climate patterns at far distant locations are called teleconnections. Such teleconnections can be within ocean basins or between oceans. For example, Müller *et al.* (2008) found that at decadal to 20-year time scales, the NAO in the North Atlantic tends to be out of phase with ENSO in the equatorial Pacific and with the PDO in the North Pacific. Better known, through ENSO forcing, are the associations between heavy rains along the west coast of South America and droughts in Indonesia (McPhaden *et al.* 2006), or, in the Atlantic Ocean, between years of reduced ice in the Barents Sea and heavy ice off Labrador due to NAO variability (Deser *et al.* 2000). Because the ecological literature often puts emphasis on climate and ecosystem

---

Box 2.1 Climate indices

Several indices of large-scale atmospheric pressure patterns have been developed and used to compare with the variability of marine ecosystems. The most common have been the *El Niño-Southern Oscillation* (ENSO), North Atlantic Oscillation (NAO), Atlantic Multidecadal Oscillation (AMO), and the Pacific Decadal Oscillation (PDO).

ENSO

El Niño events are characterized by significant warming of equatorial surface waters over an area extending from the International Date Line to the west coast of South America. The atmospheric phenomenon related to *El Niño* is termed the Southern Oscillation (SO), which is a global-scale east-west see-saw in atmospheric mass involving exchanges of air between eastern and western hemispheres centred in tropical and subtropical latitudes. During an *El Niño* event, the sea level atmospheric pressure tends to be higher than usual over the western tropical Pacific and lower than usual over the eastern Pacific, and warmer-than-average SSTs cover the near equatorial Pacific. Warm ENSO events are those in which both a negative SO extreme and an *El Niño* occur together. The interval between ENSO events is typically 2 to 7 years. During the warm phase of ENSO, the increased upper layer sea temperatures in the central and eastern tropical Pacific shifts the location of the heaviest tropical rainfall eastward from its usual position centred over Indonesia and the far western Pacific. This shift in rainfall and the associated latent heat release alters the heating patterns that force large-scale waves in the atmosphere, producing an amplification of the Pacific-North America (PNA) pattern (Hoerling *et al.* 1997). Since the PNA pattern and PDO are linked, this has led to the PDO being described as either a long-lived *El Niño*-like pattern of Indo-Pacific climate variability or a low-frequency residual of ENSO variability on multidecadal time scales (Newman *et al.* 2003).

continues

Box 2.1 *continued*

The NAO

The long-term monthly mean surface atmospheric pressure pattern over the North Atlantic is dominated by a subpolar (Icelandic) Low and a subtropic (Azores) High (Barnston and Livezey 1987), which tend to intensify or weaken at the same time. This variability is known as the North Atlantic Oscillation or NAO (Hurrell 1995). The most commonly used NAO index is the average of the December to March surface atmospheric pressure difference between Iceland and Lisbon in Portugal (Hurrell 1995). In the positive NAO phase, the

Icelandic Low is lower than normal, the Azores High is higher than normal and the westerly winds across the Atlantic intensify with more storms of higher intensity (Box 2.1, Fig. 1). Stronger southerly flows produce warmer than normal air temperatures and higher precipitation over the eastern United States and northern Europe, whereas the northerly winds between eastern Canada and Greenland and in the vicinity of the Mediterranean cause cooler than normal air temperatures and drier conditions in these locations. The opposite pattern occurs during the negative phase of the NAO (Box 2.1, Fig. 1).

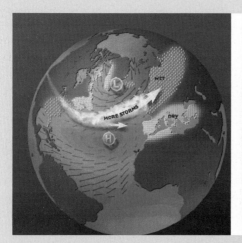

Box 2.1, Figure 1 The atmospheric conditions during the positive (*left*) and negative (*right*) phases of the NAO. (Taken from http://www. ldeo.columbia.edu/res/pi/NAO/). (See Plate 4).

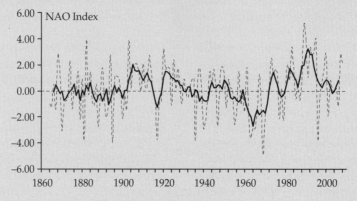

Box 2.1, Figure 2 The time series of the annual (dashed line) and 5-year running mean (solid line) of the NAO Index. (Based on Hurrell 1995).

continues

Box 2.1 *continued*

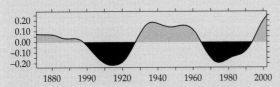

Box 2.1, Figure 3 The time series of the AMO (temperature anomalies in °C). The index is the deviation of the annual mean SSTs between 0–60°N and 75–7.5°W from its long-term mean, and low past filtered with a 37-point Hendersen filter and detrended. (Taken from Sutton and Hodson 2005 reprinted with permission from AAAS.)

There are periods when the NAO exhibits persistence from one winter to the next. For example, during the 1960s there was a long period of intense and principally negative NAO values followed by a period through to the mid-1990s of increasing values on top of strong decadal variability (Box 2.1, Fig. 2). The NAO has a strong influence on the atmosphere and hence the physical oceanography of the North Atlantic. Comprehensive reviews of ecological responses in the North Atlantic to NAO are provided by Drinkwater (2003) and Stenseth *et al.* (2004).

trend has been removed (Delworth and Mann 2000; Sutton and Hodson 2005). A positive AMO index is associated with higher-than-normal SSTs throughout most of the North Atlantic and a negative index with colder temperatures. The AMO index has varied between warm and cool phases with a period of about 60–80 years since the late 1880s (Box 2.1, Fig. 3). The temperature changes associated with the AMO can alternately obscure and exaggerate the increase in temperature due to anthropogenic-induced warming (Latif *et al.* 2004). The AMO index is correlated to air temperatures and rainfall over much of the northern Hemisphere, in particular, North America and Europe (Sutton and Hodson 2005). For example, the two most severe droughts of the twentieth century in the United States occurred during the positive AMO between the 1920s and the 1960s, that is the dust bowl of the 1930s and the 1950s drought. In contrast, extreme rainfall fell during this time in Florida and the Pacific Northwest. In the marine environment there are strong ecological signals associated with the AMO, examples of which are described by Sundby and Nakken (2008) as well as by Brander (2003a) and Drinkwater (2006) and references therein.

The AMO

The AMO (Kerr 2000) is defined by the annual mean SST averaged over the North Atlantic after the linear

The PDO

The PDO index is based on the observed monthly SST anomalies in the North Pacific north of 20°N

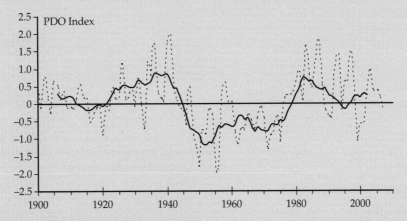

Box 2.1, Figure 4 The annual (dashed line) and 11-year running mean (solid line) of the PDO index. (from http:// www.beringclimate. noaa.gov/data/)

continues

Box 2.1 *continued*

after removal of the long-term warming trend. The monthly PDO indices are often averaged to create seasonal or annual indices (Box 2.1, Fig. 4). During the positive phase of the PDO, the Aleutian low pressure system deepens and shifts southward, winter storminess in the North Pacific intensifies, SSTs are anomalously warm along the coast of North America and cool in the central and western North Pacific, the subpolar gyre intensifies, the mixed-layer depth shoals within the Alaska gyre enhancing upwelling, the Oyashio pushes farther southward and there is an intensification of the Kuroshio Current (Box 2.1, Fig. 5). During the negative phase of the PDO, conditions are generally the opposite of those in the positive mode. Changes in the sign of the PDO have been associated with regime shifts (Moloney *et al.*, Chapter 7, this volume) in the North Pacific (Mantua *et al.* 1997).

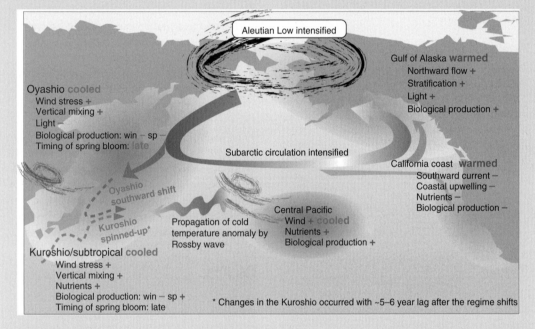

Box 2.1, Figure 5 The atmospheric and oceanic conditions as well as the phytoplankton response during the positive phase of the PDO. (Modified from Chavez *et al.* 2003 with perission from AAAS). (See Plate 5).

James E. Overland

teleconnections, we will discuss this in greater detail with a focus on northern Hemisphere atmospheric teleconnections.

2.3.1 Atmospheric teleconnections

The concept of atmospheric teleconnections and their influences on marine ecosystems has been explored by a number of authors inside the GLOBEC programme (e.g. Bakun 1996; Alheit and Bakun 2010; Overland *et al.* 2010; Schwing *et al.* 2010. Overland *et al.* 2010 noted that stationary atmospheric waves are formed in winter in the northern Hemisphere due to continent-ocean heating contrasts and the influences of large mountain ranges such as the Rockies and Himalayas. These waves generate regions of atmospheric low and high pressure which are linked, often with opposite conditions, over large spatial scales. The most prominent teleconnections over the northern Hemisphere are the NAO and the Pacific-North America (PNA) pattern (Barnston and Livezey 1987; also see Box 2.1). Analyses of northern Hemisphere pressure patterns indicate that the NAO, PNA, and similar large-scale climate patterns represent about one-half of the climate variability signal in the northern Hemisphere, the remainder being 'climate noise' (Overland *et al.* 2010). Although atmospheric teleconnections between ocean basins have been investigated (e.g. Honda *et al.* 2005), no consistent covariability between ocean basins is apparent when the full records for the twentieth century are considered (Overland *et al.* 2010). Evidence for atmospheric teleconnections within ocean basins is more convincing, in particular for marine populations across the North and South Pacific Ocean (e.g. Alheit and Bakun 2010; Schwing *et al.* 2010).

2.4 The role of ocean forcing on climate variability

The oceans play a key role in regulating the climate due in large part to their immense capacity to store heat compared to the atmosphere. Not only is the heat stored in the ocean, it is redistributed through transport by ocean currents and by heat exchanges with the atmosphere. The amount of heat in the upper ocean influences evaporation and precipitation patterns. How these are distributed geographically affects the distribution of atmospheric pressure systems, storm tracks and the development of severe weather such as hurricanes and cyclones, and can alter the timing and intensity of monsoons, droughts, and floods. The tropical oceans and their complex ocean-atmosphere interactions play a significant role in the global climate system. Variations in both the NAO (Liu and Alexander 2007) and the PDO (Lau 1997) have been linked to sea surface temperature (SST) variability in the tropical oceans. The oceans contribute to the formation, distribution, and melting of sea ice in polar and subpolar regions, and are subsequently affected by the climatic conditions they help to create.

The oceans are also an important storage medium for greenhouse gases such as CO_2. At the surface of the ocean these gases are in equilibrium with the atmosphere. Some of this CO_2 is transformed into dissolved carbon and transferred to the deep ocean through vertical mixing and circulation processes. Marine phytoplankton, which account for approximately half of the world's photosynthesis, also contribute to the removal of CO_2 because a proportion of the production is transferred to the deep ocean, whether by sinking detritus or faecal pellets, or via food webs. This is called the 'biological pump'. In this respect the oceans play a vital role in slowing down the build-up of CO_2 in the atmosphere, and hence the rate of anthropogenic-induced climate change (Fasham 2003).

2.5 Patterns of climate forcing on marine ecosystems

Individual organisms, populations, and communities in marine ecosystems fluctuate in response to a multitude of physical processes, variations in the dynamics internal to the population, and interactions with predator, prey, and competitive species (Harris *et al.*, Chapter 6; Moloney *et al.*, Chapter 7, both this volume). This multitude of forcing and pathways can often make it difficult to establish unequivocal connections between climate and ecological responses. It is thus important to understand the main patterns in how physics affects the biology, but in particular we need to know more about the actual mechanisms and processes involved.

A population may react immediately to a climate signal or it may have a delayed (lagged) response. The atmosphere tends to have a 'short memory' of external forcing, although because it responds quickly, the effects also dissipate quickly. The ocean tends to have a 'long memory' of external forcing, because it responds slowly but the effects persist. Ecosystem responses to atmospheric and ocean forcing tend to have features of both short and long memory: forcing effects are short at lower trophic levels which reproduce and turnover quickly, whereas they persist longer at higher trophic levels with slower turnover rates (Ottersen *et al.* 2010). The longer memory of higher trophic levels to climate forcing make them more predictable for management purposes when the climate signal can also be predicted (e.g. Perry *et al.* 2010).

Ottersen *et al.* (2010) identify four major classes of the effect of climate processes on marine ecosystems:

direct effects, *indirect effects*, *integrated effects*, and *translations*. *Direct effects* involve a direct ecological response to climate forcing with no or little time lag. The effect of climate on metabolic rates via temperature (e.g. Pörtner and Knust 2007) is one example. *Indirect effects* involve several physical or biological intermediate steps between climate forcing and the ecological trait. An example is the alternation in the abundance of two copepod species *Calanus finmarchicus* and *Calanus helgolandicus* in the North Sea due to changes in SST and phytoplankton production caused by variations of the NAO (Fromentin and Planque 1996). Indirect climate-induced effects can also occur through changes in bottom-up food web forcing from primary or secondary production or through top-down effects caused by changes in upper trophic levels and cascading food-web effects. Intensive fishing can also produce top-down effects (e.g. Frank *et al.* 2005, 2006).

Integrated effects of climate involve ecological responses that occur during and after an extreme climate event (Ottersen *et al.* 2010). Examples are provided by the gradual replacement of water masses in an area, and by the effects of climate warming causing spatial relocation of species distributions (e.g. Perry *et al.* 2005; Mueter and Litzow 2008). In contrast, *translations* involve movements of organisms from one place to another as for example, the advection of *C. finmarchicus* from the continental slope onto the continental shelf (e.g. Heath *et al.* 1999a).

Ecological responses to climate forcing may be linear or non-linear (Ottersen *et al.* 2004). The former occurs when the variability in the climate is reflected in the flora or fauna being investigated, either positively or negatively, and can be with or without time delays between the environment change and the response. Non-linear responses include dome-shaped relationships where the impact is at a maximum at some intermediate value. The dome-shaped relationship between pelagic fish recruitment and upwelling intensity (Cury and Roy 1989) or between Atlantic cod (*Gadus morhua*) growth and temperature (Björnsson *et al.* 2001) are two examples. Non-linear responses include regime shifts where climate events trigger major changes in ecological states (e.g. Francis *et al.* 1998; Reid *et al.* 2001a,b). Beaugrand *et al.* (2008) showed that marine ecosys-tems are not equally sensitive to climate change and revealed a critical thermal boundary where a small increase in temperature triggers abrupt ecosystem shifts seen across multiple trophic levels ranging from phytoplankton to zooplankton to fish. Another case of non-linear responses is when an ecological threshold is passed, for example, the temperature at which the symbiotic algae in corals will leave, resulting in coral 'bleaching' and ultimately coral death (Hoegh-Guldberg 1999). Coral destruction can lead to declines in reef community biodiversity as well as the abundance of a significant number of the individual species (Jones *et al.* 2004). Because physical forcing is often not the only factor acting on marine ecosystems, the strength of environment-ecological associations may vary over time, that is, they are non-stationary (e.g. Ottersen *et al.* 2006).

The response of a particular species to climate variability and change depends in part upon their life history strategies. For example, small pelagic fishes such as sardine (*Sardinops* spp.), anchovy (*Engraulis* spp.), and herring (*Clupea* spp.), among others, can respond dramatically and quickly to changes in ocean climate (Alheit and Niquen 2004). Most have short, plankton-based food chains, they tend to be short-lived (3–7 years), highly fecund, and some spawn all year round. These life history characteristics make them highly sensitive to environmental forcing and extremely variable in their abundance (Barange *et al.* 2009). Several orders of magnitude changes in abundance over a few decades are characteristic for small pelagics, especially in the upwelling regions, such as sardines in the California Current, off Japan and in the Benguela Current and anchovies in the Humboldt Current (see Section 2.6.2). Although not in upwelling regions, the abundance of herring in European waters also undergoes such variability (Alheit and Hagen 1997; Toresen and Østveldt 2000). In contrast, demersal (living close to the sea bed) species such as cod (*Gadus* spp.), haddock (*Melanogrammus* spp.) and red-fish (*Sebastes* spp.), etc. tend to be relatively long-lived (up to 15 to 20 years and beyond) and feed mainly on forage fish or other fish or invertebrate species. Before abundance changes can be perceived for demersal and other long-lived fish species, their populations usually need to be repeatedly affected by physi-cal forcing (biological inertia) or similar environmental conditions need to persist over many years (physical inertia) (Ottersen *et al.* 2004). In instances where fishing has reduced the mean size and age of a fish stock through removals of many of the older age groups, such stocks become more responsive to climate variability (Pauly 2003; Ottersen *et al.* 2006; Section 2.8).

The nature of the ecological response is also dependent on the frequency of physical forcing (Sundby and Nakken 2008). On very short time scales of hours to days (weather, not climate), the forcing tends to be on small geographic scales and acts on individuals or groups of individuals. Examples include changes in the feeding rates of fish due to variable turbulence levels (Rothschild and Osborn 1988; Lough and Mountain 1996) or through temperature-dependent swimming speeds (Pörtner 2002a; Peck *et al.* 2006). Under rare circumstances mortality can occur if the temperature or oxygen levels that fish are exposed to rapidly cross a potentially lethal threshold (Marsh *et al.* 1999; Hoag 2003; Lilly 2003). At interannual time scales, physical forcing operates on wider (regional) geographic scales and tends to be linked to processes in population ecology. These include phenomena such as individual growth rates (e.g. Brander 1994, 1995), recruitment (e.g. Baumann *et al.* 2006), transport of eggs and larvae (e.g. Lough *et al.* 1994; Hermann *et al.* 1996a; Vikebø *et al.* 2005), and spatial distributions of plankton and fish (e.g. Ottersen *et al.* 1998; Platt *et al.* 2003). On the decadal to multidecadal timescale, the physical forcing can be multi-regional and even basin- or global-scale and, as such, responses tend to be related to systems ecology. These responses include large changes in abundance (Lluch-Belda *et al.* 1997; Alheit and Hagen 2001), selection of spawning areas (Sundby and Nakken 2008), changes in spawning or feeding migrations (Vilhjálmsson 1997b), or large-scale distributional shifts (Drinkwater 2006). These can also lead to changes in the life history of fish stocks, species interactions, trophic transfer, and evolutionary ecosystem processes. Thus, shifts from individuals to populations to systems generally tend to occur with decreasing frequency of physical forcing.

2.6 Effects of climate on marine ecosystem processes

Drinkwater *et al.* (2010a) provide examples of ecosystem impacts in response to climate variability and the possible mechanisms by which the response is generated. It must be remembered, however, that each of these individual processes usually act in concert with several others to produce the observed response on marine ecosystems.

2.6.1 Sea temperature

Sea temperature is the variable that has received the most attention from researchers in terms of its effect upon marine ecosystems and its dependency on climate. While this may simply be because it is easily measured and often available, it can also be argued that temperature is the climate-influenced variable that is most dominant in terms of its influence on marine ecosystem processes. The following are some of the temperature-dependent responses.

2.6.1.1 Growth
Successful individual growth often occurs within a limited thermal range that can differ between developmental stages and even populations of the same species (Pörtner *et al.* 2001, 2005). One of the hypotheses explaining the out-of-phase stock oscillations at multidecadal time scales for anchovy and sardine species in the California, Humboldt, Kuroshio and Benguela Current systems (Lluch-Belda *et al.* 1989; Chavez *et al.* 2003) is based on differential thermal ranges for the two species (Moloney *et al.*, Chapter 7, this volume). The physiological background of these patterns requires answering why animals specialize on and perform in limited thermal ranges. Laboratory studies have focused on thermal adaptation and limitation at genetic, molecular, and stress levels, but have largely neglected organism performance. However, the concept of oxygen- and capacity-limited thermal tolerance as a unifying principle has been proposed as an integrative physiological concept linking climate to ecosystem characteristics and to ecosystem change under the influence of climate variability (Pörtner 2001, 2002b).

For widely distributed species, those inhabiting colder waters tend to exhibit slower individual growth than those in warmer waters. This conforms to the latitudinal compensation hypothesis that predicts local evolution should maximize metabolic efficiency and thus favour maximum growth under local thermal conditions (Levinton 1983). Faster growth results in reduced susceptibility to predation due to shorter durations during early development stages. Growth variations can lead to a greater than a 100-fold difference in survival probabilities during larval stages (Houde 1987). As such, even slight changes in temperature can induce growth variations that have the potential to result in dramatic fluctuations in recruitment success through cumulative effects of changes in stage durations and predation pressure.

2.6.1.2 Swimming speed and activity rates

Another impact of temperature is on performance through swimming speed and activity rate (Fuiman et al. 2005, 2006) and subject to modification through temperature-dependent evolution (Pörtner 2002a,b). Swimming speed affects both feeding success and anti-predator behaviour through changes in encounter rates with prey and predators, respectively. Temperature has two primary effects on swimming speed. First, it influences the viscosity of the water with increased viscosity at lower temperatures, which in turn increases the drag on the organism. Thus, higher energy expenditure is generally required to move through colder water with the overall effect of slowing the individual's swimming speed. This effect is principally on small organisms such as fish larvae and zooplankton (e.g. Müller et al. 2000, 2008). A second effect of temperature, primarily on adult fish, is related to the delivery of oxygen. Oxygen consumption rate and the organism's scope for activity tend to have temperature-dependent optima that are not only species specific but can vary between stocks of the same species (Lee et al. 2003).

2.6.1.3 Reproduction

Reproduction of marine organisms is also affected by temperature. Egg production rate has been found to be temperature-dependent for several zooplankton species in both the Pacific and Atlantic (Runge 1984, 1985a; Hirche et al. 1997) principally due to effects on the spawning interval of the zooplankton (Hirche et al. 1997). Temperature affects gonadal development of several fish species resulting in spawning times generally occurring earlier under warmer-than-normal conditions (Hutchings and

Myers 1994). The age-of-maturity in different stocks of Atlantic cod varies with temperature (Drinkwater 2000), and is believed to be caused by faster growth rates for those cod stocks inhabiting warmer waters. Egg size of Atlantic cod, Gadus morhua (Miller et al. 1995), and Atlantic mackerel, Scomber scombrus (Ware 1977), have been found to vary with temperature. It has been hypothesized that this is to match the size of the larvae to their prey at the time of hatching, the latter also being temperature-dependent (Ware 1977). Rapid changes in temperature also have been observed in the field to trigger spawning in some invertebrate species, such as the scallop, Placopecten magellanicus (Bonardelli et al. 1996).

2.6.1.4 Phenology

Temperature and light are the most important physical factors affecting phenology in the marine environment, with the responses being species-dependent. For example, recent observed temperature increases have resulted in earlier phytoplankton blooms in the Oyashio region (Chiba et al. 2008). In the North Sea, dinoflagellates have advanced their seasonal peak by nearly 1 month under increasing temperatures, while diatoms have shown no consistent pattern of change (Edwards and Richardson 2004). The latter is because their reproduction is triggered principally by light intensity that has not changed. Copepod responses in the North Sea have been more variable, but some species have had their seasonal maximum earlier in the year (Edwards and Richardson 2004). In the north-east Pacific, recent warming has caused the life cycle timing of the zooplankton species Neocalanus plumchrus to be earlier by several weeks (Mackas et al. 2007).

Earlier seasonal warming also leads to earlier migratory movements, for example, for Atlantic mackerel, Scomber scombrus (Sette 1950), and American shad, Alosa sapidissima (Leggett and Whitney 1972), off the east coast of North America; squid, Loligo forbesi (Sims et al. 2001), and flounder, Platichthys flesus (Sims et al. 2004), in the English Channel; and pink salmon, Oncorhynchus gorbuscha, in Alaska (Taylor 2008). In principle, such shifts in timing of migration may be understood as an earlier/later entry of ambient temperature into species-specific thermal ranges.

2.6.1.5 Distribution

The most convincing evidence of the effects of climate change on marine ecosystems comes from distributional shifts of marine organisms. They generally are most evident near the northern or southern boundaries of the species' geographic range with warming usually causing poleward movement and cooling an equatorward movement. Examples abound of northward distributional shifts in response to increasing temperatures, for example, *C. finmarchicus* and *C. helgolandicus* in the north-east Atlantic (Beaugrand *et al.* 2002c; Box 2.2); several zooplankton species in the north-east Pacific (Mackas *et al.* 2007); Atlantic cod off West Greenland,

---

Box 2.2 North-east Atlantic zooplankton

Continuous Plankton Recorder (CPR) surveys have collected plankton data since the 1930s (Reid *et al.* 2003a). The CPR is a towed body that funnels water past a moving filter of silk. Phytoplankton and zooplankton that are retained by the silk are later counted and identified to species. Species abundances can then be determined as a function of location knowing the ship's track and its speed. Through a concerted effort to maintain systematic sampling in space and time in over 70 years, CPR data have proven increasingly useful for describing seasonal changes, interannual changes, long-term trends, and species' distributional changes (Brander *et al.* 2003b). The longest and most complete spatial coverage of CPR data is in the north-east Atlantic.

The CPR data from the North Sea revealed significant changes in the plankton in response to climate variability in the late 1980s (Box 2.2, Fig. 1). A large influx of warm saline water entered the northern North Sea due to an increase in the northward flowing shelf edge current to the west of the British Isles (Reid *et al.* 2001a,b). In response, the North Sea phytoplankton biomass doubled (Reid *et al.* 2001a), the plankton community structure changed, in particular the calanoid-copepod diversity (Beaugrand and Ibanez 2002), and the phytoplankton blooms of several of the dominant species peaked 1 to 3 months earlier than usual (Edwards *et al.* 2002). Seabird abundances also increased and large numbers of horse mackerel (*Trachurus trachurus*) entered the northern North Sea subsequently resulting in high

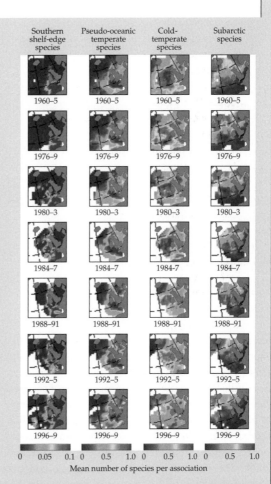

Box 2.2, Figure 1 The mean number of southern shelf edge, pseudo-oceanic temperate, cold temperature, and subarctic species in the north-east Atlantic for different time periods from the 1960s to the late 1990s. (From Beaugrand *et al.* 2002c.) (See Plate 6).

continues

Box 2.2 *continued*

catches by the fishing fleet (Reid *et al.* 2001a,b). These ecosystem changes led Reid *et al.* (2001a) to label this a regime shift.

The CPR data have also revealed longer-term trends in the plankton community linked to multidecadal climate change, in particular the AMO (Box 2.1). A decline in the zooplankton population of cold-water species such as *C. finmarchicus* and a rise in the neritic (coastal) species from the late 1950s to the end of the century were observed in the North Sea (Beare *et al.* 2002). West of the British Isles, the number of warm-temperate and temperate pseudo-oceanic species assemblages increased northwards by over 10° of latitude while the number of cold-mixed water and subarctic species assemblages decreased (Box 2.2, Fig. 1; Beaugrand *et al.* 2002c). All assemblages showed consistent long-term changes that appear to reflect the response of the marine ecosystems towards a warmer dynamic regime. Subsequent studies revealed that the biodiversity of calanoid copepods responded quickly to the SST rise by moving geographically northward at a rapid rate of up to about 23 km/year (Beaugrand *et al.* 2009). Increases in the occurrence of subtropical fish paralleled these changes and some studies

reported latitudinal movements of fish species distribution that have been mainly related to SST changes (Quéro *et al.* 1998; Brander *et al.* 2003a; Perry *et al.* 2005).

Consistent with this shift, along the continental shelf from the English Channel to Scotland and in large parts of the North Sea, *C. finmarchicus* were replaced by the smaller warm-water species *C. helgolandicus* (Beaugrand *et al.* 2002c). Helaouët and Beaugrand (2007) showed that among the various physical parameters they tested, the distribution of these two copepod species was primarily linked to temperature, although part of the temperature change was likely due to redistribution of water masses through changes in circulation. Beaugrand *et al.* (2003a) concluded that the shift from the spring spawning *C. finmarchicus* to the autumn spawning *C. helgolandicus* led to a mismatch of the spring-spawned Atlantic cod (*Gadus morhua*) larvae with their food. This mismatch, and an observed reduction in prey biomass and mean size of prey, are believed to have resulted in reduced survival of the young cod and contributed to the observed decrease in the population of cod in the North Sea.

Gregory Beaugrand

Iceland, and in the Barents Sea (Drinkwater 2006); capelin off the east coast of Canada (Frank *et al.* 1996), Iceland (Sæmundsson 1934), and the Barents Sea (Vilhjálmsson 1997b); and numerous fish species during recent warming in the North Sea (Perry *et al.* 2005), along the continental shelf from the Iberian Peninsula to west of Scotland (Brander *et al.* 2003a) and in the Bering Sea (Mueter and Litzow 2008). The spawning location of Atlantic cod off Norway has been found to vary in concert with the long-term temperature changes such that the cod tend to favour more northern spawning during warm conditions and more southern spawning under cold conditions (Sundby and Nakken 2008). The spawning areas of squid (*Todarodes pacificus*) off Japan expand in the Sea of Japan and East China

Sea during warm periods and retract to the East China Sea during cold years (Sakurai *et al.* 2000; Box 2.3). Temperature changes can have a strong, especially negative, effect on species whose distribution is tied to specific geographic features (Pershing *et al.* 2004) such as fish species that spawn on a few fixed submarine banks (Mann and Lazier 1996), seals whose breeding grounds are geographically limited, and the nesting grounds of marine birds (Graybill and Hodder 1985).

The reasons fish change their distribution is often unclear. Adult fish may actively move to seek their preferred thermal range, to follow their prey or to avoid their predators. Larval survival may improve in more northern areas with atmospherically induced temperature increases, but temperature

Box 2.3 Japanese common squid

The Japanese common squid, *Todarodes pacificus* (Cephalopoda: Ommastrephidae), is found in the Sea of Japan, the East China Sea, and in the Oyashio and Kuroshio current systems off eastern Japan. There are three subpopulations with different peak spawning seasons; summer, autumn and winter, with the later two the largest and most important (Murata 1989). Autumn spawning occurs mainly in the southern Sea of Japan, including in the Tsushima Strait between Japan and the Republic of Korea, while winter spawning occurs in the East China Sea off Kyushu Island in southern Japan. This species has a 1-year life cycle with the autumn-spawned squid migrating to the northern Sea of Japan and then back again to spawn before dying. Some of the winter-spawned squid migrate to feed in the northern Sea of Japan while others migrate in the waters off eastern Japan to the Oyashio region where they feed during the summer and autumn. These squid feed on large zooplankton and small fish, and can impact the recruitment of walleye pollack by feeding on their juveniles (Sakurai 2007). After

hatching, the squid paralarvae ascend to the surface layer above the continental shelf and slope and are advected into convergent frontal zones (Yamamoto *et al.* 2002, 2007). Laboratory studies reveal that hatchlings (<1 mm mantle length) ascend to the surface successfully at temperatures between 18 and 24°C, and especially between 19.5 and 23°C (Sakurai *et al.* 1996; Sakurai and Miyanaga, unpublished data).

Low-pass filtered SST data off south-western Japan show alternate warm and cold periods at decadal scales that correspond with the variability of the PDO. For example, there was a sudden decrease in SST in the late 1970s (shift to positive PDO) and an increase in SST (negative PDO) during the mid to late 1980s (Senjyu and Watanabe 2004). Note that the positive phase of the PDO corresponds to cool conditions on the western side of the Pacific and warm on the eastern side and vice versa in its negative phase. During the cool period around Japan in the late 1970s and early- to mid-1980s, the annual catch of Japanese common squid decreased and only began to

continues

changes may also be associated with modification in circulation patterns, which in turn may carry larvae into areas that they previously had not occupied.

One of the consequences of distributional shifts of individual species is the possibility of changes in the composition of ecosystem assemblages, as observed in the fish community around Britain (Attrill and Power 2002; Genner *et al.* 2004) and in the intertidal regions off California (Helmuth *et al.* 2006). These changes can occur due to differential rates of movement of various species in response to ocean climate conditions (Mueter and Litzow 2008), through invasion of new species (Stachowicz *et al.* 2002), or when some species disappear either due to the new temperatures exceeding the species thermal tolerance, a reduction of their prey, or an increase in their predators or competitors. These

changes in community structure can also result in changes in ecosystem function, that is, who is feeding on whom.

2.6.1.6 Recruitment

Recruitment levels of several species of fish vary with temperature during their first years of life (Drinkwater and Myers 1987; Ellertsen *et al.* 1989; Ottersen and Stenseth 2001; Pörtner *et al.* 2001; Sirabella *et al.* 2001). The response to temperature can vary between stocks of the same species. For example, Atlantic cod that inhabit the coldest water within this species' temperature range tend to show increasing recruitment with increasing temperatures, whereas those inhabiting the warmest waters show decreasing recruitment with increasing temperatures (Ottersen 1996; Planque and Frédou 1999; Pörtner *et al.* 2001; Sirabella *et al.* 2001). However,

Box 2.3 *continued*

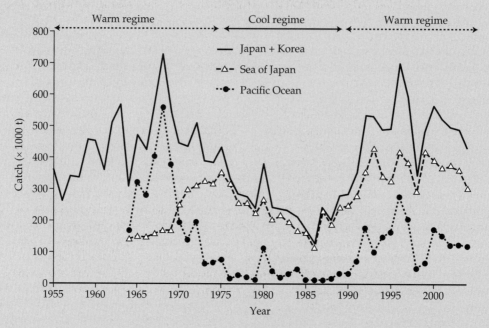

Box 2.3, Figure 1 Annual fluctuations in the catch by Korea and Japan of the Japanese common squid and periods of cold and warm regimes. (Adapted from Sakurai *et al.* Changes in inferred spawning areas of *Todarodes pacificus* (cephalopoda: ommastrephidae) due to changing enivronmental conditions. ICES Journal of Marine Science, **57**, 24–30.)

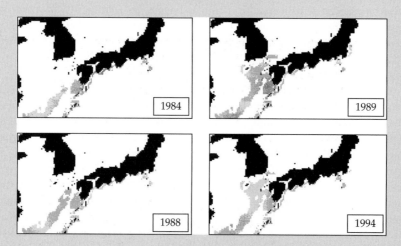

Box 2.3, Figure 2 The potential distribution of spawning squid during 2 cold years (1984 and 1988) and 2 warm years (1989 and 1994) based on temperature preference. The spawning locations are denoted by the grey areas. (Taken from Sakurai 2007.)

continues

Box 2.3 *continued*

increase after the shift to the warm period (Box 2.3, Fig. 1; Sakurai *et al.* 2000, 2002, 2003). The catch is considered a good proxy for abundance and hence reproductive success (Choi *et al.* 2008). One reason for the increased larval survival during warm periods might be related to an increase in the size of the spawning areas. In the warm periods, the winter spawning area extends northward into the Tsushima Strait and farther eastward to the continental slope (Box 2.3, Fig. 2). For the latter, the inner flow of the Kuroshio helps transport the hatchlings into the Oyashio region to feed. Thus, during warm periods more squid may be transported to the feeding grounds via this circulation. During cool periods there is less spawning along the

continental slope and the squid are deeper due to an increased mixed-layer depth caused by stronger winds. This may result in less squid being transported into the Oyashio region. Fluctuations in the strength of the northward flowing Tsushima Warm Current influences the area of the warm water region in the Sea of Japan and also matches the variability in squid catches from this region (Choi *et al.* 2008). The current's effect on squid is believed to be through its prey, as the changes of annual zooplankton biomass are also closely connected to the fluctuations in the Tsushima Warm Current. This too may contribute to the increased survival of the Japanese common squid in the Sea of Japan during warm years.

Yasunori Sakurai

temperature is often a necessary but not sufficient condition for good recruitment due to the influence of other factors.

2.6.1.7 *Mortality*

Temperature has an indirect effect on mortality, for example through its influence on growth rate and larval stage duration (Houde 1987; Cury and Pauly 2000; also see Section 2.1.1). However, on rare occasions, rapid temperature increases or decreases can cause direct mortality. Colton (1959) observed mortality of large numbers of cod larvae off Georges Bank due to thermal shock caused by transport of the larvae off the bank into much warmer offshore waters. Marsh *et al.* (1999) reported on the tilefish kill of 1882 along the continental slope off Georges Bank when millions of fish were found dead, floating on the surface, but showed no signs of long-term stress or disease. Their conclusion was that a rapid decrease in temperature led to the mass mortality of this bottom dwelling species. In Smith Sound in Newfoundland, Canada, adult Atlantic cod died in large numbers following the advection of extremely cold (−1.7°C) water into this overwintering region (Lilly 2003). Approximately 5% of the estimated cod at that time in Smith Sound died. Examination of the fish that survived showed they were supercooled

(i.e. their tissues were at a temperature below their freezing point), while those that died were frozen, likely due to contact with ice crystals (Lilly 2003).

Coral reefs provide habitat for a highly diverse ecosystem and short-term extreme water temperature can cause the symbiotic algae in corals to leave, resulting in coral 'bleaching' that may result in coral death (Hoegh-Guldberg 1999). For example, the elevation of temperature by 1–2°C above the climatological maximum for a period of weeks may trigger bleaching. When bleached corals do not recover, algae may grow over the corals, resulting in a shift to an algal-dominated ecosystem.

2.6.2 Vertical stratification and mixed-layer depth

The intensity of the vertical density stratification and the depth of the mixed layer have important implications for the marine ecosystem, especially primary production. Primary production depends on the amount of solar radiation and the availability of nutrients, both of which can vary with the intensity of upper layer stratification and the mixed-layer depth. Strong stratification tends to favour a pelagic-dominated system, where energy is recycled within the upper layers, while weaker stratification favours a more demersal or benthic-dominated system, especially on the continental shelves (Frank *et al.* 1990). Arguably, the most dramatic ecosystem effects of changes in stratification and mixed-layer depth occurs during *El Niño* events. These events result in a deepening of the thermocline and an intensification of the stratification between the surface and subsurface layers, as well as rapid warming of the waters off western South America. As a consequence of the increase in stratification, there is much less upwelling of nutrients. The *directly driven* biological responses include a decrease in the local primary production, collapses of a variety of small planktonic herbivore and low-trophic-level carnivore populations, and dramatic declines of the normally dominant Peruvian anchoveta population (Overland *et al.* 2010). Once an *El Niño* event has subsided, conditions begin to return to normal, although sequences of *transient responses* of varying duration can carry effects forward into ensuing

years (Overland *et al.* 2010). Such a process of transient responses to frequent *El Niño* events may represent a continuous 'resetting' of the Peru system by recurring *El Niño* episodes before internal *non-linear feedback responses* can arise, possibly switching the system into a different state (Overland *et al.* 2010). On the opposite side of the tropical Pacific during an *El Niño*, the thermocline shallows resulting in a vertical extension of the mid-depth temperature habitats of yellowfin and bigeye tuna, changes that are favourable for the tuna fisheries (Lehodey 2004; see Box 2.4). During *La Niña* (opposite phase to an *El Niño*), the reverse patterns prevail on both sides of the Pacific. These ENSO patterns also seriously impact the ecosystem in the northern portion of the North Pacific subtropical gyre. There, a *La Niña* tends to result in an increase in vertical stratification during winter leading to a drop in nutrients and subsequently plankton productivity (Polovina *et al.* 2008; Ottersen *et al.* 2010).

Box 2.4 Tuna in the equatorial Pacific

Fisheries for tunas and tuna-like species have existed for centuries and now extend throughout the tropical and temperate regions of the oceans and seas using multiple types of fishing gear. The tropical species of skipjack (*Katsuwonus pelamis*) and yellowfin (*Thunnus albacares*) tuna provide the bulk of the world tuna catch and are exploited by surface fishing gear. Both species are characterized by rapid growth, early age-at-first maturity (~9–15 months), and a short to medium lifespan (~4–6 and ~6–8 years for skipjack and yellowfin, respectively). Bigeye tuna (*Thunnus obesus*) have a wider habitat distribution (from tropical to subtropical and temperate latitudes), a longer lifespan (>10 years), and later age at maturity (2–3 years). Adult yellowfin and bigeye are exploited by subsurface longline fisheries for fresh fish and sashimi markets.

Climate variability affects the migration and abundance of tropical tuna as well as their catchability. For example, the spatial distribution of purse seine catches of tuna in the equatorial

Pacific clearly shifts during different phases of the ENSO (Box 2.4, Fig. 1; Lehodey 2000; Lehodey *et al.* 1997). Catches of skipjack tuna are highest within the western equatorial Pacific warm pool (surface temperatures > 29°C). During the *La Niña* phase of the ENSO, the easterly trade winds over the Pacific confine the warm pool to the western Pacific. At such times the purse seine fleet is mainly distributed west of 165°E where the catch per unit effort (CPUE) of skipjack and juvenile yellowfin tuna is high. With the development of an *El Niño*, the trade winds relax, and the warm pool spreads eastward towards the central Pacific (McPhaden and Picaut 1990). Tagging data show that the tuna follow the warm water eastward (Lehodey *et al.* 1997). The fishing fleet in turn follows the tuna and, although usually found west of the Date Line (180° meridian), significant numbers of tuna were caught east of the Date Line during the most intense *El Niño* (1997–8).

El Niño events have a strong effect on the vertical structure of the equatorial Pacific Ocean,

continues

Box 2.4 *continued*

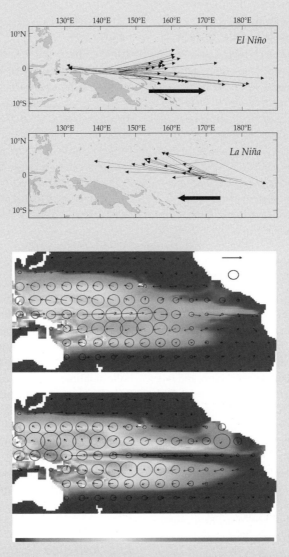

Box 2.4, Figure 1 ENSO and movements of skipjack tuna in the Pacific Ocean. *Left*: Observed movements of tagged skipjack during *El Niño* and *La Niña* phases. (Redrawn from Lehodey *et al.* 1997.) Tagging data were compiled from records of a large-scale tagging programme carried out by the Secretariat of the Pacific Community. *Right*: Predicted distribution of biomass of skipjack (age cohort 9 months) in November 1997 (*El Niño* phase) and November 1998 (*La Niña* phase) in the Pacific Ocean. Circles and arrows represent random (diffusion) and directed (advection) movements of population density correspondingly and averaged by 10 degree squares. (Redrawn from Lehodey *et al.* 2008 with kind permission from Elsevier). (See Plate 7).

in particular a deepening of the thermocline in the eastern Pacific during El Niño events and a rising in the west. As skipjack and yellowfin tuna tend to spread vertically over the entire upper mixed-layer depth, a deepening of the mixed layer reduces the concentrations of fish. This reduction is considered to be one of the factors responsible for the drop in catch rate of yellowfin by the eastern Pacific fleets during the 1982–3 *El Niño* and that led the US fleet to move into the western central Pacific,

continues

Box 2.4 *continued*

where the thermocline tends to rise during *El Niño* events and the mixed layer is shallow. Analyses of the CPUE time series in this region (Lehodey 2000) showed higher catch rates of yellowfin by surface gear (pole and line and purse seine) during the *El Niño* phases, which was attributed to increased concentrations of fish through the contraction of the vertical habitat of this species. The displacement of the United States purse seine fleet triggered by the powerful 1982/3 *El Niño* coincided with the development of the industrial purse seine fishery in the western central Pacific that is now the most productive tuna fishing ground of the World Ocean.

Thus, ENSO influences tuna species and their fisheries in the tropical Pacific but the impacts show different effects depending on the biology (e.g. lifespan) and ecology (e.g. habitat temperature) of the species, as well as the type of fishing gear used. A direct positive effect due to the vertical change in the thermal structure during *El Niño* decreases yellowfin catch rates of purse seiners in the eastern Pacific. For skipjack, large fluctuations in the catch and catch rates are driven by environmentally-induced changes in stock size with a short delay, and by horizontal spatial extension (*El Niño*) or contraction (*La Niña*) of the habitat.

Patrick Lehodey

2.6.3 Sea ice

In polar regions, changes in air and ocean temperatures produce variability in the seasonal abundance of sea ice, which in turn impacts their ecosystems. The timing of the ice retreat and its associated melt water affects the timing intensity, speciation, and fate of the spring bloom (Hunt *et al.* 2002; also see detailed discussion in Section 2.7.1). For example, sea-ice algae can produce intense blooms before zooplankton have developed, resulting in much of the production sinking out of the pelagic zone onto the ocean floor for use by the benthos (Carroll and Carroll 2003; Grebmeier *et al.* 2004). Along the Antarctic Peninsula, cold years tend to result in a krill-dominated ecosystem, benefiting from overwintering under the abundant sea ice. In contrast, warm years with limited sea ice result in fewer krill and a salp-dominated ecosystem (Atkinson *et al.* 2004). Sea ice also provides necessary habitat for many species of marine mammals, including several species of seals for breeding and protection of their young, walrus for resting areas, and polar bears for hunting and resting. Significant reductions in sea ice-coverage can pose hardships on these animals (Loeng *et al.* 2005).

2.6.4 Turbulence

Turbulence increases the contact rate between plankton predators and prey (Rothschild and

Osborn 1988), which in turn can increase their feeding rates (MacKenzie and Leggett 1991; Sundby *et al.* 1994; Saiz and Kiørboe 1995). At the same time, small-scale turbulence can change the behaviour of copepods between 'slow' and 'fast' swimming speeds (Costello *et al.* 1990; Margalef 1997). A parabolic relationship between feeding efficiency and turbulence exists with the shape and threshold limits dependent upon the species, their stage, and also the region (Cury and Roy 1989). Higher turbulence levels can also cause an increase in total metabolism (Alcaraz *et al.* 1994) and wind-induced turbulence and high temperatures during winter can reinforce the energetic imbalance of copepods and potentially decrease fecundity.

2.6.5 Advection

Dispersion of fish eggs and larvae from their spawning ground is considered a key aspect of recruitment success in several fish stocks as currents may carry them into or away from favourable nursery areas. Numerical models are well suited for the study of transport of fish larvae, for example, for gadoids on Georges Bank (Werner *et al.* 1993), in the North Sea (Gallego *et al.* 1999), in the Baltic (Voss *et al.* 1999; Hinrichsen *et al.* 2003), and in the Barents Sea (Vikebø *et al.* 2005) and for flatfish in the Bering Sea (Wilderbuer *et al.* 2002). These studies indicate that recruitment success has a strong dependency on wind-dependent drift. Advection through entrainment by Gulf Stream rings also has been shown statistically to lead to reduced recruitment levels of several groundfish stocks off the eastern United States and Canada (Myers and Drinkwater 1989). This is believed to be due to increased mortality of the larvae in the offshore waters through insufficient food quantity or quality or increased predation pressures (Drinkwater *et al.* 2000).

Zooplankton are also advected, leading to important ecological consequences. On the Faroe Plateau the transport of adequate numbers of *C. finmarchicus* onto the Faroese Shelf regions is required to ensure high recruitment of Atlantic cod (Hansen *et al.* 1994; Steingrund and Gaard 2005). The zooplankton overwinter in the deep waters of the Norwegian Sea Basin, rise in the spring and are then carried by the flows through the Faroes-Shetland Channel and the Faroe Bank Channel before reaching the vicinity of the Faroe Plateau (Gaard 1996). Similarly, advection of *C. finmarchicus* from the Norwegian Sea into the Barents Sea by the Atlantic water inflow impacts ecological production of phytoplankton and fish in that region (Skjoldal *et al.* 1987; Sakshaug 1997; Sundby 2000; Ottersen and Stenseth 2001). The strength of the Atlantic inflow is related to both local and the large-scale wind patterns.

Modification in circulation patterns can produce significant shifts in water mass distributions, which in turn can influence ecosystem dynamics. For example, the abundance of *C. finmarchicus* off northern Iceland has been linked to interchanges in the distribution of polar, Arctic, and Atlantic water masses. Higher *C. finmarchicus* abundance is generally associated with increased presence of higher nutrient Atlantic waters. The connection may be either through increased levels of associated primary production brought about from the increase in nutrient concentrations, direct transport of *C. finmarchicus* with the Atlantic water mass, or a combination of these (Ástthórsson and Gislason 1995).

Benthos also has been observed to undergo distributional changes in response to water mass shifts. Increased Atlantic water flow during the warm period of the 1920s to the 1960s resulted in the northward extension of its associated benthic fauna

and flora by 500 km along western Spitzbergen (Blacker 1957) and eastward into the Barents Sea (Nesis 1960; Cushing 1982).

2.7 Comparative studies of climate forcing on marine ecosystems

Previous sections have demonstrated the diverse and multiple impacts of climate on marine ecosystem processes. In this section we provide further examples, with an emphasis on insights gained using the comparative approach applied by GLOBEC.

2.7.1 Subarctic ecosystems

The subarctic ecosystems of the Bering (Alaska) and Barents (Norway) Seas are strongly impacted by climate, but there are significant differences in how they respond to climate forcing (Mueter *et al.* 2009), including different responses to sea-ice variability and possibly to differences in food web structures (e.g. Ciannelli *et al.* 2004).

Sea ice is a crucial aspect of the physical environment of the continental shelves in both the Bering and Barents Seas. The Bering Sea is typically ice-free from June through to the autumn when cold arctic winds cool the water and ice begins to form, first in the north and western shelves and later in the east (Pavlov and Pavlov 1996). Throughout winter, prevailing winds advect the ice southward into warmer water where it melts, cooling and freshening the seawater (Pease 1980; Niebauer *et al.* 1999). The date of the retreat of the sea ice, as well as the timing of the last major winter storms, determine the timing of spring primary production (Sambroto *et al.* 1986; Stabeno *et al.* 1998, 2001; Eslinger and Iverson 2001) and the ambient water temperatures in which grazers of the bloom must forage. In a typical light-ice year, the ice retreat occurs prior to mid-March, at which time there is insufficient light within the water column to support net primary production (Fig. 2.1; Hunt *et al.* 2002). The spring bloom is delayed until May or June, after winter winds have ceased and thermal stratification stabilizes the water column (Stabeno *et al.* 1998, 2001; Eslinger and Iverson 2001). In a heavy-ice year, melting is usually delayed until April or May and the ice-associated bloom occurs immediately after ice melt and hence earlier than in light-ice years. Although wind

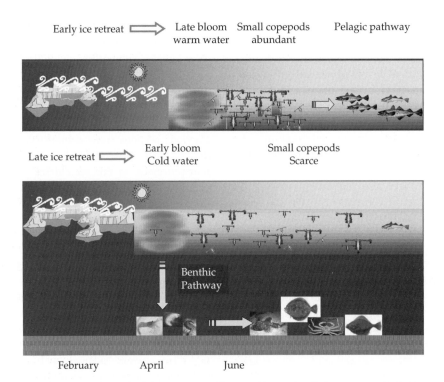

Early ice retreat ⟹ Late bloom Small copepods   Pelagic pathway
                     warm water   abundant

Late ice retreat ⟹ Early bloom   Small copepods
                    Cold water   Scarce

Benthic
Pathway

February   April   June

Figure 2.1 The relationship between the timing of the ice retreat and the spring bloom. When the ice retreat is early, the bloom occurs later, the water is warm, a large copepod biomass develops, and a pelagic food web is favoured (top panel). When the ice retreat is late, the bloom occurs earlier, the water is cold, much of the production goes into a benthic food web and the copepod biomass is small. (Modified from Hunt *et al.*, 2002 with kind permission from Elsevier; F. Mueter, University of Alaska, Fairbanks, personal communication). (See Plate 3).

mixing of the water plays a role in determining when the bloom will occur, the timing of the last winter storm is less important than the date of ice retreat. Peak primary productivity during warm years also consists of different species assemblages than those found in the ice-edge blooms (McRoy *et al.* 1986; Alexander and Niebauer 1989; Alexander *et al.* 1996). Thus, early blooms occur in heavy-ice years when the waters are colder and are related to ice-edge blooms, whereas late blooms occur in relatively light-ice years when the waters are warmer and the blooms are not related to the ice edge.

The late bloom under warm conditions leads to increased abundance of small neritic zooplankton (copepod) species while the early ice-associated bloom favours the larger shelf copepod species, *Calanus marshallae* (Baier and Napp 2003). The timing and ice-association of the bloom also influences the fate of carbon between the pelagic and benthic components (Walsh and McRoy 1986; Alexander

et al. 1996; Mueter *et al.* 2007). Under an early bloom, there is less zooplankton to crop the phytoplankton production and thus a significant portion of this production falls to the bottom of the ocean resulting in more food for the benthos (Fig. 2.1). On the other hand, if the bloom is late the primary production is mainly consumed by the zooplankton community and less makes it to the bottom. This hypothesis is supported by the inverse relationship between survival anomalies of walleye pollock (*Theragra chalcogramma*) whose young feed on zooplankton, and of yellowfin sole (*Limanda aspera*) that feed on the benthos (Mueter *et al.* 2007). In light-ice years with late blooms, more production remains in the upper layers, the populations of small copepod species increase and early stages of pollock thrive, while in heavy-ice years with earlier blooms, more of the production makes its way to the benthos and there are more yellowfin sole. However, in the very warm years with little ice, the large copepod, *C. marshallae*, has reduced recruitment, as is apparently also true for the shelf euphausiid, *Thysanoessa raschii* (Coyle *et al.* 2008; Hunt *et al.* 2008; Moss *et al.* 2009). Under these circumstances, in summer middle shelf populations of pollock of all ages lack zooplankton prey, and cannibalism of smaller pollock is increased. Thus an initial bottom-up limitation results in a subsequent top-down limitation of pollock recruitment. Survival of age-0 pollock is further limited in warm years because they invest more energy in growth than in storage, and therefore enter winter with very low energy supplies, leading to overwinter mortality (Moss *et al.* 2009).

In the Barents Sea, which lies approximately 15° of latitude farther north than the Bering Sea, approximately 40% is ice covered on an annual basis but with extensive seasonal variability (Loeng 1979; Vinje and Kvambekk 1991). Minimum ice coverage occurs in August/September but by the end of October new ice begins to form, eventually leading to peak coverage in March or April, although in some years not until early June. Approximately 60% of the sea is covered at the time of maximum extent, mainly in the northern and eastern regions. The interannual variability of ice coverage in the Barents Sea generally reflects local temperature conditions. In contrast to the Bering Sea, phytoplankton blooms in the northern Barents Sea tend to coincide with the disappearance of the ice, resulting in a northward progressing spring bloom, typically during June and July (Rey and Loeng 1985; Wassmann *et al.* 1999). The differences between ecosystems will likely depend on the relative timings of the increase in solar radiation, decline in wind strength (i.e. the number and intensity of the storms), and melting of sea ice. However, similar to the Bering Sea in cold years, a significant portion of the phytoplankton bloom in the seasonally ice-covered Arctic waters settles to the seabed (Sakshaug and Skjodal 1989; Sakshaug 1997). During extremely cold winters in the Barents Sea there is often a more southern distribution of ice than usual. If it extends far enough south to reach the warm Atlantic waters, it will melt and the resultant stratification will tend to initiate an earlier-than-usual phytoplankton bloom in these areas (Rey *et al.* 1987). As in the Bering, early blooms usually result in lower secondary production.

Another difference between the two seas is that in eastern shelf of the Bering Sea, most zooplankton are recruited from local populations, whereas in the Barents Sea advection plays an important role in the abundance of zooplankton. In the northern Barents Sea, the dominant copepods are Arctic species, which are advected in Arctic water masses, whereas in the southern Barents Sea, the advection of zooplankton, mainly *C. finmarchicus*, in Atlantic water plays an important role in their productivity, and ultimately the abundance of higher trophic levels (Skjoldal *et al.* 1992; Sundby 2000; Wassmann *et al.* 2006).

The Barents Sea ecosystem also differs from that of the eastern Bering Sea in that the upper trophic levels in the former appear to be strongly influenced by downward cascading trophic effects, that is, as defined by alternating patterns of abundance across more than one trophic level. Herring and cod prey upon capelin (Hjermann *et al.* 2004), and capelin heavily impact their zooplankton prey (Hassel *et al.* 1991). Since the mid-1960s, there have been alternations between periods when herring abundance was high and capelin abundance was low and vice versa. When capelin stocks were low, large zooplankton such as krill and amphipods increased to up to 10 times their former abundance owing to a release of predation pressure (Skjoldal and Rey 1989; Dalpadado and Skjoldal 1996; Mueter *et al.* 2009).

2.7.2 Upwelling regions

Small pelagic fishes (sardine and anchovy) in many of the large-scale upwelling regions of the world are one of the most remarkable examples of marine populations showing low frequency environmentally-driven multidecadal biomass variations. Kawasaki (1983) first noticed that catch time series of the world's largest populations of sardine, *Sardinops sagax*, from the Pacific (north-east, south-east, and north-west) Ocean showed large synchronous fluctuations. The Scientific Committee on Oceanic Research (SCOR) Working Group (WG) 98 investigating 'Worldwide Large-Scale Fluctuations of Sardine and Anchovy Populations' adopted the term 'Regime Problem' to describe an alternative to the hypotheses that the variability in fish stocks was caused either by the fisheries or by recruitment-induced variability. WG98 documented that some small pelagic fish populations grow for a period of 20 to 30 years and then collapse for similar periods until practically disappearing. Further, low sardine abundance regimes have often been marked by dramatic increases in anchovy populations (Fig. 2.2), and similar but opposite phase fluctuations were apparent in the Benguela system in the Atlantic Ocean (high anchovy abundance during sardine periods in the Pacific, and vice versa), suggesting a worldwide, rather than a basin-scale phenomena (Lluch-Belda *et al.* 1989, 1992).

A permanent concern in fisheries science is the use of catch as a stand-in for biomass, and no catch-independent biomass reconstructions exist for all upwelling systems over the same periods. Some independent evidence suggests that large multi-decadal fluctuations, and the alternation between

species, are natural phenomena. First, synchrony between fisheries occurring in geographically remote locations, and under different management strategies and fishing intensities, is likely only to occur as a consequence of similar global-scale forcing. Secondly, paleo-records of fish scales deposited in anaerobic sediments clearly show large deposition rate changes with periods of around 60 years, indicating large abundance fluctuations were happening well before any fishing pressure (e.g. off California by Soutar and Issacs (1974); Baumgartner *et al.* 1992). Also, Sandweiss *et al.* (2004) showed archaeological evidence of a shift from an anchovy to a sardine-dominated fishery at about AD 1500, after analyzing fish remains excavated at an Inca period (ca. AD 1480–1540) fishing site on the Peruvian coast. Further, they found evidence of fish assemblages from middle and late Holocene sites that also were suggestive of regime changes. Freon *et al.* (2002), however, noted that while statistical proof of teleconnections among marine ecosystems was available *within* regions, the contrary was not true for teleconnections *between* regions.

The underlying mechanisms of these alternating large-scale fluctuations remain a mystery, but several hypotheses have been proposed including ecological mechanisms, direct physical forcing, and recently some integrative frameworks. Ecological mechanisms include both top-down (predatory outbreaks) and bottom-up (primary productivity) trophic controls, sudden reductions in predation pressure resulting from environmental shifts that allow rapid increases of small pelagics (environmental 'loopholes'; Bakun and Broad 2003), competition between species (Matsuda *et al.* 1992), and the 'school-mix feedback' hypothesis to explain alternation between species and cycling though the schooling behaviour (Bakun 2001; see Moloney *et al.*, Chapter 7, this volume). While some of these mechanisms might be operating, they are extremely hard to test, and accounting for longer than interannual variability and synchrony between regions is problematic.

The first physical variable to be related to the observed fluctuations in small pelagics was temperature (Lluch-Belda *et al.* 1989), and detailed thermal habitat analyses were carried out for some of the systems and periods (e.g. Lluch-Belda *et al.*

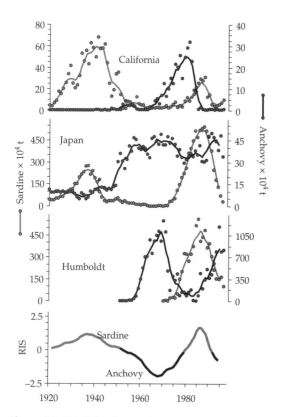

Figure 2.2 Historical catch series of sardine and anchovies from the main worldwide fishing systems and regime indicator series (RIS), a proxy of the synchronic variability.

1991; Hardman-Mountford *et al.* 2003). Takasuka *et al.* (2007) proposed that the dominant mechanism of fish regime shifts may be through direct temperature impacts on survival during early life stages. They documented that in the western North Pacific the temperature optimum range for sardine is colder and wider than that of anchovy, the opposite to what happens in the eastern North Pacific (Lluch-Belda *et al.* 1991).

Habitat expansion and contraction have been related to abundance fluctuations. For many years it was thought that when the Japanese sardine population increase, they expand offshore, whereas when they decrease, they are confined to a more coastal distribution (Kawasaki 1983; Watanabe *et al.* 1996). In contrast, in California and the other eastern boundary systems, sardine population growth coincides with latitudinal expansion poleward, and declines with area contraction equatorwards (Lluch-Belda *et al.* 1992). This notion was re-examined by Logerwell and Smith (2001) who postulated two principal spawning habitats in the California Current: inshore and offshore, the former resulting in continuing moderate sized cohorts, and the latter associated with oceanic eddies and fronts, allowing sporadic occurrence of large size cohorts. Recently, Barange *et al.* (2009) demonstrated that anchovy and sardine in the Benguela, California, Humbodt, and Kuroshio regions have a positive relationship between stock abundance and distributional area, but suggested that habitat availability may not be a prerequisite for sardine growth in some areas, while anchovy may require habitat to become available for its population to grow.

Ocean dynamics and large-scale circulation patterns have also been suggested as influencing the sardine populations. Lluch-Cota *et al.* (1997) proposed a hypothetical mechanism where shifts in large-scale ocean circulation and global average temperature conditions would drive low-frequency small pelagic regimes, with strong ENSO events as the possible trigger to shift from one regime to another. Support for this idea was provided by Parrish *et al.* (2000) who found relationships between the mid-latitude wind stresses and the major climate variations in the North Pacific and to the decadal variations in several fisheries, including sardines and anchovies.

Optimistically, we should see significant progress in the next few years in understanding the low-frequency fluctuations in sardine and anchovy populations and their synchronous behaviour based on available databases, modelling capabilities, and several interdisciplinary analysis and integration efforts presently underway.

2.7.3 Atlantic cod

The above examples have compared ecosystem responses to climate variability between similar types of ecosystems. However, important insights have also been gained within GLOBEC by comparing the response of a particular species within different oceanographic conditions. One of the primary species that has been studied is Atlantic cod (*Gadus morhua*), which inhabits the continental shelves of the North Atlantic, including those off the northeastern United States, Canada, West Greenland, Iceland, in the Barents Sea and south to the North Sea, the Irish Sea, and the Celtic Sea. One of the best studied of all of the world's fish species, evidence abounds that climate strongly influences cod abundance.

Fish recruitment is the number of young that survive through to the fishery and is typically measured for cod at age 1 to 3. The recruitment response as a function of temperature shows that some cod stocks exhibit increasing recruitment while others show decreasing recruitment or no relationship. By comparing the different cod stocks it was found that the temperature-recruitment relationship depended upon the mean bottom temperature the cod occupied. Increasing recruitment with increasing temperatures occurred for cold water stocks (mean annual bottom temperatures of approximately 0–5°C), decreasing recruitment for warm-water stocks (>8°C), and no relationship at intermediate temperatures (Ottersen 1996; Planque and Fredou 1999; Drinkwater 2005).

At multidecadal time scales, as expressed by the Atlantic Meridional Circulation (AMO; see Box 2.1), temperature has clear influences on the distribution, abundance and growth of Atlantic cod. For example, as temperatures warmed during the 1920s and 1930s, cod along West Greenland extended their distribution approximately 1,200 km northwards and

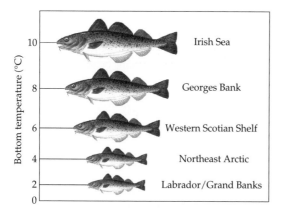

Figure 2.3 The relative size of an average 4-year-old Atlantic cod for different stocks as a function of their mean bottom temperature. (Drinkwater 2000 from data by Brander 1994.)

established new spawning sites on the offshore banks (Jensen 1949; Cushing 1982; Drinkwater 2006). A northward range extension also occurred around Iceland, in the Barents Sea and off western Spitzbergen (Drinkwater 2006), while more spawning occurred farther north off coastal Norway relative to earlier times (Sundby and Nakken 2008).

During the warm phase of the AMO (1930s–1960s) recruitment in the northern range of the cod's distribution (Barents Sea, Icelandic waters, and off West Greenland) was the highest on record (Drinkwater 2006). Also, since the growth of cod is temperature-dependent (Fig. 2.3), growth rates of individual cod increased throughout these northern regions during the warm period. For example, the weight of spawning cod off Norway increased, on average, by over 50% between the early 1900s and the 1950s (Drinkwater 2006). The successful recruitment and improved growth were linked not only to the warm temperatures but also to accompanying increases in phytoplankton and zooplankton production (Drinkwater 2006).

At decadal time scales, several cod stocks respond to the variability in the NAO. Ottersen and Stenseth (2001) demonstrated a positive association between the NAO and Barents Sea cod recruitment that accounted for 53% of the recruitment variability. The mechanistic link was thought to be through effects on regional sea temperatures and food availability. With a higher NAO index there is increased Atlantic inflow, which transports warmer waters

and more *C. finmarchicus* into the Barents Sea and hence more food for cod. On the other hand, the year-class strength of cod in the much warmer North and Irish Seas is negatively related to the NAO index, which is believed to result from a limitation in energy resources necessary to achieve higher metabolic rates during NAO-induced warm years (Planque and Fox 1998). Brander and Mohn (2004) found that the NAO has a significant positive effect on cod recruitment in the North, Baltic, and Irish Seas, and a negative effect at Iceland. Later, Brander (2005) showed that the effects on these four stocks plus the Baltic and west of Scotland stocks were significant only when the stock biomass is low. Growth rates also vary with the NAO. For example, the variability in the growth rate of northern cod off Labrador and Newfoundland is negatively associated with the NAO (Drinkwater 2002). The NAO accounted for 66% of the variance in the weight gain between ages 3 to 5 of the northern cod over the years 1980 to 1995 and is believed to be related through NAO influences on temperature and food, similar to the case in the Barents Sea.

2.7.4 Pacific salmon

Pacific salmon have also received attention from GLOBEC, with comparisons made between the dynamics of different salmon species and their response to climate forcing. Salmon spawn in freshwater, where they stay for up to 3 years, before moving into the marine environment. They remain in the ocean for 1 to 5 years depending on the species, during which time they mature sexually. They then return to their natal river to spawn and all Pacific species die immediately after spawning. Catches of salmon species in Alaskan waters (principally pink, *Onchorhynchus gorbuscha*, and sockeye, *O. nerka*) vary synchronously but inversely with stocks along the west coast of the United States, most notably chinook, *O. tshawytscha*, and coho, *O. kisutch* (Mantua *et al.* 1997). Salmon production is linked to variability in the PDO, with higher production in Alaska during the positive phase of the PDO (Mantua *et al.* 1997), when SSTs along the periphery of the north-east Pacific are relatively warm. Greater salmon production along the west coast of the United States coincides with the

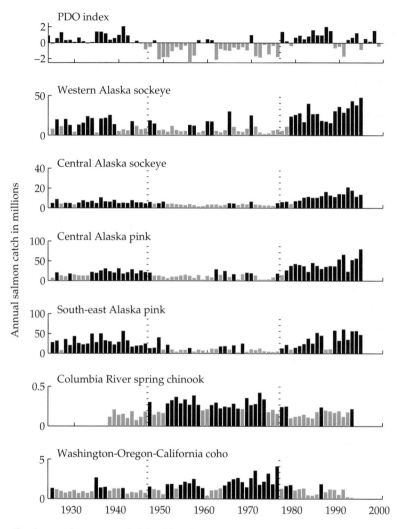

Figure 2.4 North Pacific salmon catch time series. Black (grey) bars denote catches greater (less) than the long-term medium for Alaska stocks. The shading convention is reversed for Washington-Oregon stocks (lower two panels). Vertical dotted lines denote regime shifts in the Pacific Decadal Oscillation (PDO; top panel) in 1925, 1947, and 1976. Alaska (Washington, Oregon) catches are higher during the positive (negative) phase of the PDO. (Figure from Mantua *et al.* 1997 © American Meteorological Society. Reprinted with permission.)

negative (cooler coastal) phase (Fig. 2.4). Thus, in the 2 decades prior to the mid-1940s, Alaska salmon production was relatively high, while west coast stocks were low (Fig. 2.4). From 1947 to 1976, the roles reversed as west coast production was high and the Alaska catches were low. Alaska salmon stocks rebounded rapidly in the mid-1970s. At the same time, west coast US salmon production dropped to levels requiring many stocks to be listed as threatened by US federal managers.

Gargett (1997) suggested the fishery responses in the late 1970s climate regime shift were due to bottom-up forcing by changes in nutrient availability that led to increased biological productivity at multiple trophic levels. Adjustments between the atmosphere and upper ocean due to heat exchange and wind-driven ocean currents contributed to the anomalously warm SSTs extending from the Bering Sea and Gulf of Alaska along the North American coast (Parrish *et al.* 2000). While basin-scale climate variations set the conditions for the decadal regime shifts, the ecosystem responded to local and regional perturbations in the environment. Adjustments in ocean mixing and basin circulation resulted in a 20–30% shallower mixed layer in the subarctic North Pacific after the 1970s, contributing to greater water column stability. This enhanced primary production and doubled the zooplankton biomass in the Gulf of Alaska (Brodeur and Ware 1992). The coincident deepening and strengthening of the thermocline and nutricline in the California Current (Palacios *et al.* 2004), reduced the availability of nutrients to the euphotic zone and led to a sharp decline in local zooplankton biomass (Roemmich and McGowan 1995). This contributed to the decline in salmon along the west coast of the United States. Peterson and Schwing (2003) suggested that this west coast salmon production shifted back to higher levels after 1997, coincident with a return to the negative PDO phase and higher ecosystem productivity in the California Current. This lasted up until 2002, after which conditions in the California Current have undergone year-to-year variability with no consistent periods of good or bad years for west coast salmon.

2.8 Influence of fishing on the responses of exploited ecosystems to climate forcing

Recently, investigators have recognized that we must understand and take account of the interactions between climate and fishing, rather than try to disentangle their effects and address each separately. Planque *et al.* (2010) and Perry *et al.* (2010a) reviewed how exploitation can modify the ability of marine populations, communities, and ecosystems to respond to climate forcing. They noted that fishing removes individuals with specific characteristics from the gene pool, and leads to a loss of older age classes, spatial contraction, loss of subunits, and changes in life history traits in populations. All of these make marine populations more sensitive to climate variability at interannual to interdecadal scales. Fishing also reduces the mean size of individuals and mean trophic level of fish communities, which decreases their turnover time leading them to track environmental variability more closely.

They noted that marine ecosystems under intense exploitation also evolve towards stronger bottom-up control and greater sensitivity to climate forcing. Overall, they concluded that a less-heavily fished marine system, and one which shifts the focus from individual species to functional groups and fish communities, would be likely to provide more stable catches with climate variability and change than would a heavily fished system.

2.9 Summary

The examples presented in this chapter demonstrate that marine ecology responds to climate variability and climate change over a myriad of time and space scales. It is important to note that these examples represent only a small fraction of the work done by GLOBEC on climate forcing on marine ecosystems and that further examples are provided in later chapters (see deYoung *et al.*, Chapter 5; Harris *et al.*, Chapter 6; Hofmann *et al.*, Chapter 11, this volume). GLOBEC science has also helped to show that, as the frequency of the physical forcing increases, there is a tendency for the nature of ecological changes to progress from local effects on individuals at synoptic weather scales, towards regional effects on population dynamics at monthly to decadal scales, and over basins and even across basins on systems ecology at multidecadal time scales and longer. The responses can be direct and immediate, or through predator-prey relationships and include delays between the physical forcing and the observed response. The nature of the impact is, in part, size- and age-dependent with generally greater and more rapid impacts on the smaller and younger individuals. The ecological effects can be on all trophic levels and include changes in production, recruitment, growth, abundance, distribution, catchability, and occasionally even mortality.

When GLOBEC began, the use of large-scale climate indices to link climate forcing with ecological responses was just beginning in earnest and GLOBEC researchers made major advances in documenting such relationships and improving our understanding of geographic extent of their impacts. In the North Pacific they identified the decadal scale climate variability and established the PDO index. They also championed the notion of regime shifts

Box 2.5 Penguins along the west Antarctic Peninsula

The Antarctic Peninsula is one of the most rapidly warming regions of the planet with an annual mean surface air temperature increase of nearly 3°C during the past 50 years (Vaughan *et al.* 2003). The seasonal changes have not been uniform, with those in winter being the most dramatic (increases of 5–6°C over the last 50 years). These increases have been associated with increased westerly winds. Over the period of satellite data from 1979, there has been a decrease in mean winter sea-ice extent in the area to the west of the Antarctic Peninsula of ~40% over 26 years (Ducklow *et al.* 2007). This reduction is the result of later formation of sea ice (by 3–6 days) across the region and earlier retreat (by 1–3 days). Over the last 50 years, the upper ocean to the west of the Peninsula has also warmed by approximately 1°C while salinity has increased by >0.25, probably a result of mixed-layer processes associated with the reduced sea-ice formation (Meredith and King 2005). Human-induced stratospheric ozone depletion and increases in greenhouse gases are required to explain the magnitude of the observed changes (Turner *et al.* 2007).

Determining the biological responses to these clear regional climate-driven physical changes is complicated by the multi-scale physical-biological process interactions that determine the dynamics of Southern Ocean ecosystems (Murphy *et al.* 2007). In the Antarctic Peninsula region, penguins are present in large numbers and are important higher trophic-level predators. Monitoring of their regional populations began in the 1970s (Fraser and Trivelpiece 1996). These long-lived predators are highly mobile so they integrate short-term and small-scale variability. Therefore, changes in their population dynamics or shifts in colony sizes reflect large-scale and long-term changes in habitat or prey availability (Fraser *et al.* 1992; Ducklow *et al.* 2007). The species of penguin present have different life histories and habitat preferences and it is the monitoring of the combination of species that gives such valuable

insight (Fraser *et al.* 1992). The Adélie penguin (*Pygoscelis adelieae*) is an ice-obligate species that since the 1970s has experienced more than a 50% population decline in the vicinity of Anvers Island (Fraser and Trivelpiece 1996; Ducklow *et al.* 2007). In contrast, populations of chinstrap (*P. antarctica*) and gentoo (*P. papua*) penguins, considered to be species that are more subpolar and less ice-obligate in habitat, have increased (Ducklow *et al.* 2007). Reduced sea ice and warm conditions along the Antarctic Peninsula may be pushing the habitat boundary between these species southwards. However, the exact mechanisms involved in generating the changes are unclear (Clarke *et al.* 2007; Ducklow *et al.* 2007). Reduction in winter sea-ice extent reflects the large-scale changes, but small-scale environmental processes also appear important in determining the observed biological changes. For example, the size of Adélie penguin colonies has declined faster where local rates of snow accumulation are enhanced compared to areas where snow accumulation rates are lower (Ducklow *et al.* 2007).

As well as direct impacts of changing physical conditions on life-histories of individual species, indirect impacts through changes in food webs also appear to be occurring, shifting the balance in favour of different species (Fraser and Hofmann 2003; Clarke *et al.* 2007; Ducklow *et al.* 2007). These indirect effects include changes in prey availability and community composition or in predator abundance. Krill abundance across the Atlantic sector of the Southern Ocean has declined by more than 50% over the last 30 years and is related to changes in sea-ice extent and duration (Atkinson *et al.* 2004). However, in the region around the west Antarctic Peninsula the effects of climate-driven changes in habitat on zooplankton species are unclear. Short-term reductions in krill associated with periods of low sea ice appear to favour salps (Loeb *et al.* 1997). A long-term warming trend across the region

continues

Box 2.5 *continued*

may, therefore, favour a shift from a krill- to a salp-dominated pelagic system generating major shifts in the whole food web (Clarke *et al.* 2007; Ducklow *et al.* 2007). However, data from the Antarctic Peninsula region do not show a simple decline in krill abundance over the last 20 years, but instead show fluctuating patterns of krill recruitment (Fraser and Hofmann 2003; Quetin *et al.* 2007). These fluctuations in recruitment are associated with Southern Hemisphere scale sub-decadal (2–6 years) fluctuations relating to the ENSO and possibly the Southern Annular Mode (Fraser and Hofmann 2003; Murphy *et al.* 2007; Quetin *et al.* 2007). The effect is marked interannual and sub-decadal fluctuations in krill population dynamics that make it difficult to assess trends given the relatively short time series (Fraser and Hofmann 2003; Murphy *et al.* 2007).

It is likely that increased warming differentially affects both pelagic and benthic species and further changes the competitive balance between species resulting in shifts in food web structure (Clarke *et al.* 2007).

The rapid regional warming around the Antarctic Peninsula is unusual as other regions of the Antarctic have not shown the same changes. Overall there is a small but significant increase in sea-ice extent around the circumpolar ocean during the last few decades and in some areas, such as the Ross Sea sector, there have been marked increases in sea-ice extent and duration (Ducklow *et al.* 2007; Smith *et al.* 2007b). Climate-related changes occurring in the Southern Ocean appear, therefore, to be highly heterogeneous.

Eugene J. Murphy

applied to both climate and ecology. This concept was useful in helping to understand the large changes in the abundance of small pelagics in upwelling systems as well as the ecological changes in the North Pacific, especially during the late 1970s (Moloney *et al.*, Chapter 7, this volume). It was later applied to physically induced ecological changes in the North Atlantic. GLOBEC science also improved our insights into the teleconnections between climate indices through studies of the low-frequency abundance changes of sardine and anchovy populations in upwelling regions located on both sides of the Pacific, and between those in the Pacific and those in the Atlantic off Africa. More details on the timing and nature of the shifts between anchovy and sardines in these same upwelling regions and their possible mechanisms have also been gathered within GLOBEC (see Checkley *et al.* 2009). The role of the NAO in forcing both physical and biological variability in the North Atlantic has been led to a large degree by GLOBEC scientists. NAO variability results in spatially dependent changes in the temperature, circulation, convection, and sea ice which in turn leads to changes in zooplankton production, as well as growth and recruitment of

several commercial important fish species. Studies on Georges Bank in the Gulf of Maine have shown the importance of large-scale processes operating through the NAO on the Bank's ecology. Hence, to understand the biological variability requires a broad geographic perspective. In more recent years, there has been increasing realization of the importance of the AMO in producing significant broadscale and long-reaching ecological consequences in the North Atlantic. On shorter time scales, GLOBEC researchers have demonstrated the importance of ENSO in the equatorial regions of the Pacific on tuna abundance, distribution, and catchability by the fishing fleets.

Other important research on climate effects has been carried out by GLOBEC researchers. In the Antarctic, they have documented ecological changes in response to recent warming, including those on krill and penguins (Box 2.5). They have also revealed the importance of the physical environment on the productivity and its variability in the Antarctic waters. In the subarctic regions, the role of sea ice and its variability on the timing and subsequent fate of plankton production in driving either pelagic or benthic food webs have been elucidated. Here too the ecological responses to recent warming have been well documented. The role of climate variability on cod dynamics has been one of the major foci within GLOBEC in the North Atlantic. In addition to the links with climate through the NAO mentioned above, this science has shown how the physical environment that cod stocks inhabit can help explain differences in growth rates, recruitment, and overall productivity.

The initial work linking the climate and ecological responses was largely done through retrospective analyses using correlation and regression methods, and while these still represent a substantial effort, new non-linear statistical approaches have been developed and used, and modelling has played an ever increasing role through the GLOBEC years (see deYoung *et al.*, Chapter 5, this volume). It is also important to note the importance of long-term datasets in establishing linkages between the physical forcing and the biology described in this chapter. To continue to document and understand the process linking the biology to the physical forcing, it is imperative that in the future such data series be maintained and, in several regions, expanded.

While this chapter has mainly focused on the climate influences on marine ecology, it is important to recognize that there are other forces acting along with climate. These include the internal dynamics of the biological components of the ecosystem due to internally generated trophic interactions and competition. Also, humans are a major contributor to changes in the marine ecosystems, primarily through fishing but also through their influence on pollution levels, the introduction of invasive species, effects on the acidification of the oceans, UV, etc. (e.g. Brander *et al.*, Chapter 3, this volume). It is often difficult to assign observed changes in any ecosystem to a particular cause because of the multiple forcing.

In the following chapters, further examples of the impacts of climate on the marine ecology of the world's oceans will be given, the multiple processes through which climate acts discussed, and the effects of and on humans explored.

Human impacts on marine ecosystems

Keith Brander, Louis W. Botsford, Lorenzo Ciannelli, Michael J. Fogarty, Michael Heath, Benjamin Planque, Lynne J. Shannon, and Kai Wieland

3.1 Introduction

Apart from the sea surface, the oceans are the most remote and least visited part of the planet by humans, yet marine ecosystems are under increasing threat from a range of human activities including pollution, mineral extraction, climate change, fishing, habitat modification, and shipping (which generates noise and transports introduced species). It is relatively easy to list these factors and to map their likely impacts (Halpern *et al*. 2008), but it is more difficult to assemble evidence of impacts 'in situ' (the terms 'on the ground' or 'in the field' are incongruous in an aquatic context) and to be confident that observed impacts are correctly attributed to specified causes. Among the reasons for this are our lack of familiarity with marine ecosystems, lack of data, high natural variability, and the fact that most of the life in the sea is microscopic and almost unknown.

The aim of the GLOBEC programme was to advance our understanding of the structure and functioning of marine ecosystems and their response to physical forcing. Although human impact was not in itself a topic of the programme, the effects of fishing in particular are so great, especially on higher trophic levels, that they must be taken into account when evaluating responses to physical forcing. Furthermore, some of the best information which we have on changes in marine ecosystems comes from monitoring fish catches. Other marine biota from phytoplankton to benthos to seabirds and whales have also been affected by anthropogenic factors, sometimes to the point of extinction, but these are not the main subject of the GLOBEC programme (Gifford *et al*., Chapter 4, this volume), whose principal taxonomic subject was zooplankton. Only some of the work described in this chapter was carried out as part of the GLOBEC programme.

Following the Introduction, Section 3.2 presents some of the ethical and utilitarian ideas which underlie our relationship with marine ecosystems; it gives a brief overview of historic evidence concerning past states of marine ecosystems and of the intensification of human impacts. Section 3.3 describes direct impacts of fishing on the demography and biomass of exploited species and on loss of vulnerable species and habitats. Section 3.4 looks at the interaction between fishing and environment in causing observed ecosystem changes; it deals with changes in trophic structure and consequences for resilience and productivity. In Section 3.5, a short summary is followed by emerging conclusions about human impacts, future human stewardship, and utilization of marine ecosystems. Five case studies are embedded within the chapter to illustrate some of the issues raised.

3.2 Human interaction with the natural world and evidence of impact on marine ecosystems

3.2.1 Stewardship and sustainability

Interaction with the natural world is fundamental to our existence as individuals, communities, and societies. We depend on nature for our atmosphere, food, fuel, shelter, recycling of waste, medicines, and recreation. Human attitudes to nature differ between communities and societies, depending on economic, cultural, and religious circumstances and attitudes change over time (Thomas 1983). What we judge to be acceptable or unacceptable behaviour has profound consequences for our use, exploitation, and often extermination of plants and animals. Such judgements form part of our spiritual and ethical values and inform our economic activities, diet, laws, regulations, and management.

The view that nature exists to serve man's interests has ancient roots in many cultures. Taming wild nature and domesticating plants and animals have been regarded as part of the advance of civilization although, as with most human actions, it is difficult to tell whether ethical justification precedes economic benefit or vice versa. Some cultures and religions take a more reverent view of nature than others, but it is evident that societies on all continents are capable of destroying their natural environment, with or without the aid of colonialism, industrialization, or religious conversion (Carrion et al. 2007). The doctrine of human stewardship and responsibility for nature has coexisted uneasily with the right to exploit 'inferior species' in most religions and cultures.

The terrestrial biosphere has now been transformed to a large extent for agriculture, urbanization, pastoralism, forestry, water management, and roads. Pests, weeds, predators, and competitors are exterminated and selected crops and animals are reared in enclosed, controlled conditions. The pressures of human population needs, food supply, and economic development dominate the landscape, although there is much concern over transformations which are taking place, such as loss of biodiversity, genetic engineering, and changes in land use (e.g. deforestation and loss of wetlands). Of the remaining natural world, some has been preserved by establishing areas which are protected to varying degrees from being transformed.

Coastal and shallow sea marine ecosystems have also been transformed by human pressures, and concerns about the depletion of fish stocks have been expressed for several centuries (Bolster 2008). However, it is only recently that historical awareness of such changes has acquired any prominence in marine science and fisheries management (Lotze

et al. 2006). This lack of information and historical awareness has given rise to what has been called the 'shifting baseline' syndrome, in which successive generations of scientists accept as the baseline the state of marine ecosystems at the start of their careers (Pauly 1995). In this way, the potential abundance, productivity, and diversity of marine species may be consistently underestimated (Rosenberg et al. 2005).

Our relationship with marine life is entirely different from our relationship with life on land. On land, we farm large herbivores (cows and sheep), but in the sea most herbivores are very small (zooplankton) and we capture only the larger herbivores (small pelagic fish) and carnivores (large pelagic and demersal fish). We do not live in the sea and have no direct experience of it. The vast majority of marine plants and animals can only be seen under strong magnification and have lifespans of a few hours, days, or weeks. Apart from coastal grasses and macroalgae, marine plants are motile and do not provide structure and shelter as terrestrial plants do. There is almost no cultivation of phytoplankton, although some investigation into their potential as biofuel is taking place (Haag 2007), as is the possibility of a role in mitigation of climate change by ocean iron fertilization (Browman and Boyd 2008). Over the past 3 decades the demand for protected areas in the sea has grown in order to preserve biodiversity and ecosystem functions, and to provide areas of refuge from overfishing and natural beauty for recreation.

There is a rapidly developing framework of national and international declarations, conventions, guidelines, and regulations on stewardship and sustainability of marine systems, which is discussed by Barange et al. (Chapter 9, this volume). The impulse for these developments comes from individuals, scientific and environmental organizations, and political leaders who express concern for the restoration of damaged ecosystems, preservation of biodiversity, and prevention of overfishing. The ecological pillar of the ecosystem approach to fisheries (EAF) aims to ensure sustainable yields of exploited species, while conserving stocks and maintaining the integrity of ecosystems and habitats on which they depend. There can be little disagreement that these are laudable aims and yet we have to ask whether they are compatible with other social and economic aims (the other two pillars of the EAF), in particular maximizing food production from the sea? Here the contrast with attitudes to terrestrial food production is stark. If agriculture were forced to reverse the trends of the past few hundred years, to restore ecosystems and preserve predators and pests, then food production would rapidly decline below the level needed to sustain the human population. The UN Food and Agriculture Organization (FAO) advocates an EAF, but not to agriculture (although there are many initiatives to make land use less ecologically damaging).

Many unresolved questions and dilemmas arise, which require an understanding of how marine ecosystems function. Could we stabilize and maximize food production from the sea, while preserving as much of the ecosystem as possible, by transforming those areas which are not protected into managed production zones, like terrestrial agricultural systems? Do we know how to carry out such a transformation (Walker *et al.* 2004)?

3.2.2 Evidence of past states of marine ecosystems

Evidence for large amplitude fluctuations in the state of marine populations and ecosystems has accumulated at nearly all accessible spatial and temporal scales. Recent records based on direct observations of marine systems provide the basis for understanding contemporary interannual fluctuations. Historic records (based on the fishing trade), diaries, and public records, provide information about the state of marine systems over several centuries (Quéro 1998; Jackson *et al.* 2001; Ojaveer and MacKenzie 2007; see Box 3.1). Further back in time, historic records are often lacking but palaeo-oceanographic studies have revealed patterns of fluctuations in marine systems over thousands of years (Lotze *et al.* 2006). In this section, we present some of the evidence for variations in the state of marine systems at time scales ranging from a few decades to millennia.

Fisheries statistics, including information on numbers and sizes of fish caught at different ages, are collected routinely for many stocks by sampling of commercial catches and special research surveys. Some of these time series, which may also include information on maturity, fecundity, liver weight, fat content, and other measures of condition, go back nearly a century (e.g. Fig. 3.1-upper panel). In contrast there are few consistent, geographically exten-

sive time series for other components of marine biota. The variability of global fish catches has been increasing over time (Halley and Stergiou 2005).

In the Mediterranean Sea, the bluefin tuna (*Thunnus thynnus*) population has been harvested with passive nets for several centuries and catches have been precisely recorded by trading companies. Ravier and Fromentin (2001) examined 54 individual time series of trap catches from the western Mediterranean and the Atlantic coast of Portugal, Spain, and Morocco. They were able to reconstruct a composite time series of the relative abundance of the bluefin tuna population for more than 3 centuries, which displays pseudo-cycles of 20 and 100 years, with amplitudes of a factor of 20 (Fig. 3.1-middle panel). The development of the bluefin tuna fishery in northern Europe shows that it was an important part of the ecosystem in the early 1900s (MacKenzie and Myers 2007).

Catch data for the Baltic Sea show fluctuations in cod (MacKenzie *et al.* 2007a), herring, eel, and whitefish fisheries as far back as the sixteenth century (Poulsen *et al.* 2007). Mid-nineteenth-century New England fishing logs provide geographically specific daily catch records which can be used to estimate the biomass of cod on the Scotian Shelf at that time (Rosenberg *et al.* 2005). The estimate of nineteenth-century cod biomass of 1.2 million t is very close to the estimated carrying capacity from twentieth-century data (Myers *et al.* 2001) and four times higher than the peak total biomass estimated in the 1980s (Mohn *et al.* 1998).

Major losses of marine mammal, sea turtle, and seabird populations through harvesting and incidental catch in fishing operations over the last 3 centuries have led to radical reorganization of many marine ecosystems (e.g. Jackson *et al.* 2001; Estes *et al.* 2007) resulting in fundamental changes in energy flow and utilization. These impacts and resulting changes in abundance levels have been chronicled in historical accounts and a diverse array of fishery-related records (Starkey *et al.* 2008).

Palaeo-oceanographic investigations show variations in marine ecosystems over longer time scales. Baumgartner *et al.* (1992) analysed multiple sediment cores sampled from anoxic sediment in the eastern north Pacific and were able to reconstruct the history of sardine and anchovy populations for nearly 2 millennia (Fig. 3.1-lower panel). They found large amplitude, multidecadal fluctuations in the abundance of both species. Fish bones from Mesolithic Stone Age excavations in Denmark record the catching of a number of warm water species during the Atlantic period (ca. 7000–3900 BC), when mean summer water temperature was 1.5–2.0°C higher than at present (Enghoff *et al.* 2007). Several of the species (anchovy-*Engraulis encrasicolus*, European sea bass-*Dicentrarchus labrax*) caught then have been increasing in abundance again during the warming period over the past 20 years (MacKenzie *et al.* 2007b). In the north Pacific, Finney (1998) reconstructed the history of sockeye salmon abundance in the Kodiak Islands for 300 years by measuring changes in sedimentary $\delta15N$. Again, the reconstructed history displays large changes in population abundance over time. When Finney *et al.* (2002) extended their analysis further back in time, they found that the variations over the last 2 millennia were of even greater amplitude.

For plankton we have two excellent time series which cover the California coast (1949-present: www.calcofi.org) and much of the North Atlantic (1946-present: www.sahfos.org). The latter uses the Continuous Plankton Recorder (CPR) on regular shipping routes to generate spatially and taxonomically detailed monthly time series of zooplankton (and some phytoplankton) (Planque and Reid 2002). The plankton community structure in the North Sea and adjacent areas has undergone large amplitude variations, with a general northward movement of mesozooplankton species (Beaugrand *et al.* 2002b), change in the balance of diatoms versus dinoflagellates (Leterme *et al.* 2006), and shifts in the phenology of major plankton species (Edwards and Richardson 2004).

Ecosystem changes in the past decades, in the North Sea and elsewhere, have been contemporary with

---

Box 3.1 Bassin d'Arcachon—historic records of 300 years of harvesting

The Bassin d'Arcachon is a tidal bay on the southern Bay of Biscay, French Atlantic coast, with an area of 155 km² at high tide. It is the principal area of oyster culture in Europe, with annual production of 18,000 t plus export of spat to many other areas. Fish have historically been plentiful in the Bassin, and in the eighteenth century about 500 people fished there during the summer. About half of the fishermen followed the fish out into the open sea in winter, when it became too cold for them in shallow water, returning in April. A fuller account of the fisheries and of the impacts of climate is provided by Quéro (1998).

Two authors, Le Masson du Parc (1727) and Duhamel du Monceau (1771) wrote accounts of the fisheries in the Bassin and identified the principal species in the catch (Box 3.1, Fig. 1), which included harbour porpoise (*Phocoena phocoena*), bramble shark (*Echinorhinus brucus*), and angel shark (*Squatina squatina*). The latter was fished with a special net and the fishermen roped the tails with string to a wooden buoy, to bring them back alive.

Harbour porpoise are now a rare species in France and have disappeared from the southern Bay of Biscay; however globally they are still common. Only six specimens of bramble shark were recorded in the Bay of Biscay during the twentieth century and the species may be threatened, but data are deficient. In the mid-1800s, the annual catch of angel shark at Arcachon was around 25 t per year, but in 1996 the catch was 144 kg from the Bassin and 147 kg from the sea. This species is listed as critically endangered.

About 18 species of skates and rays were caught at the beginning of commercial fishing by steam trawlers between 1869 and 1891. The maximum catch of skates and rays at Arcachon was 1881 t in 1922, which was 23.8% of the total landings. Between 1975 and 1989 the catches of skates and rays had dropped to less than 9 t, which was 0.45% of the total landings. Most of these species, such as the 'common' or blue skate (*Raia batis* L.) are now very rare. This is one of the largest (2–3 m), late maturing (>age 11) species of ray, which is now critically endangered, mainly due to fishing (Brander 1981).

continues

Box 3.1 *continued*

Box 3.1, Figure 1 Cover illustration of the new edition of Le Masson du Parc (1727).

Keith Brander

rapid evolution in fishing practices and major climatic variations (human induced or not). It is, therefore, still a challenge to attribute causes to the observed variations in populations or ecosystem structure. One way of gaining information on ecosystem fluctuations which are not influenced by intensive fishing and other anthropogenic drivers is by mining data on periods before fishing began.

Variability seems to be the rule rather than the exception, even in unexploited marine populations and ecosystems; the longer the time series, the greater are the variations (Vasseur and Yodzis 2004). Whether

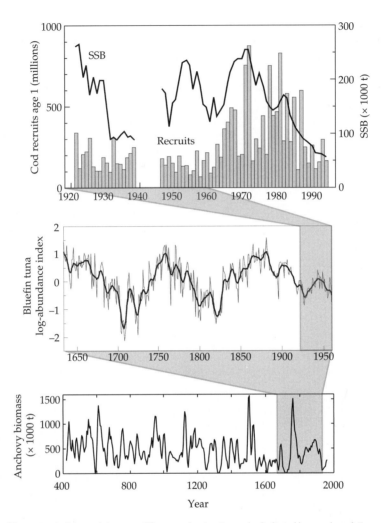

Figure 3.1 Evidence of fluctuations in fish populations over different overlapping time scales (indicated by grey shapes). Top panel-spawning stock biomass and recruitment of North Sea cod from contemporary sampling of fish catches and age structure for 1921 to 1993 (see Section 3.5.1). Middle panel-composite time series of relative abundance of bluefin tuna from the western Mediterranean and Atlantic for 1650 to 1960 (see details in text). Bottom panel-relative abundance of northern anchovy off California from 1700 BP (see details in text). (Baumgartner, personal communication.)

marine ecosystems are becoming less complex, and have lower mean trophic levels (Pauly *et al.* 1998) and simplified food webs is difficult to judge and certainly depend upon the timescale at which such observations are made. However, recent intensification of human impacts (Section 3.2.3) and the particular effects of fishing (Section 3.3) impose a change in the nature of the forces driving variations in marine ecosystems and the speed at which these variations may occur. Understanding the interplay between 'natural' modes of fluctuations and the effect of human distur-

bances will require intensified efforts in monitoring and modelling the state and dynamics of marine ecosystems.

3.2.3 Intensification of human impacts during the twentieth century: from local to global impacts

Human population growth has resulted in greatly intensified pressure on marine systems over the last century, affecting a broad spectrum of ecosystem services including provision of food, shoreline protection, and water quality. Over 40% of the human population now resides within 100 km of the coast; if current rates of population growth are maintained, the number of humans living near the coast will increase from 2.3 billion in 2000 to 3.1 billion in 2025 (cited in Duxbury and Dickinsen 2007). A recent global map of human impacts on marine ecosystems (Fig. 3.2) aggregates the effects of 17 anthropogenic drivers of ecological change, which can be grouped into pollution, climate change, fishing, shipping, and seabed structures. The areas of greatest impact are, not surprisingly, the busy shelf seas of north-east Europe and Japan, but even the Arctic and Antarctic are affected.

Overexploitation of marine resources has been identified as the dominant factor affecting biodiversity and abundance levels of marine organisms

(Millennium Ecosystem Assessment 2005; Lotze *et al.* 2006). Patterns of overall fish utilization and supply, including marine and freshwater capture fisheries and aquaculture, closely parallel the growth in the overall human population (FAO 2007; Fig. 3.3). Approximately 90% of the global capture fishery yield is from marine systems. The escalating demand for seafood resulted in nearly a tripling of global fleet capacity (gross registered tons-GRT) during the period 1970 to 1995 (http://www.fao.org/figis). Reducing fleet overcapacity remains the major impediment to controlling fishing mortality. Globally, 3% of the stocks for which information is available are classified as under-exploited, 20% are moderately exploited, 52% are considered fully exploited, 17% as overexploited, 7% are depleted, and 1% are listed as recovering (Fig. 3.3; FAO 2007). Excluding production by China, the capture production of fish, crustaceans, and molluscs

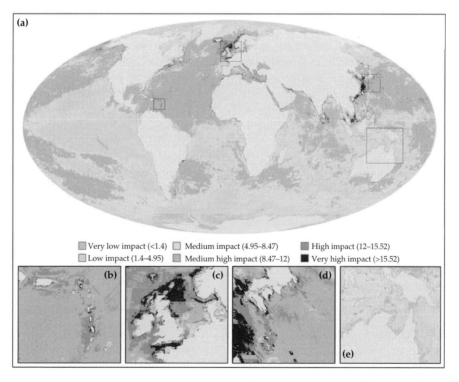

Figure 3.2 Global map from Halpern *et al.* 2008 reprinted with permission from AAAS showing (a) cumulative human impact on marine ecosystems. Insets show highly impacted regions including (b) the Eastern Caribbean, (c) the North Sea, (d) Japanese waters, (e) one of the least impacted regions, in northern Australia, and (e) the Torres Strait. The methodology uses expert judgement, standardization, and weighting to combine the 17 anthropogenic drivers. Terms describing degree of impact correspond to the 'per cent degraded' scheme of Lotze et al. (2006). (See Plate 8).

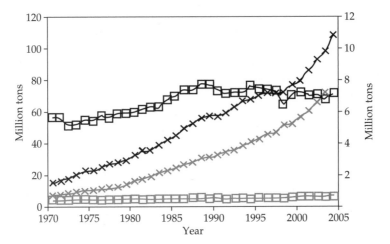

Figure 3.3 Trends in world capture fisheries (squares—left axis) plus aquaculture production (crosses—right axis) excluding problematic data from China. Black is marine and grey is fresh water. (Modified from Food and Agricultural Organization of the United Nations 2007).

has been declining annually by 233,000 t since 1989 (Brander 2007a).

Coastal development has also resulted in widespread loss of wetlands, mangrove forests, and rooted aquatic vegetation, which collectively serve as nursery areas for commercially and recreationally important species and as important buffers against storm damage in nearshore areas. Lotze *et al.* (2006) report declines of 50% or more of wetlands, seagrasses, and other submerged aquatic vegetation in estuarine and coastal areas. The Millenium Ecosystem Assessment (2005) cites habitat loss as a major determinant of loss of biodiversity in the coastal zone over the last century. It has been suggested that the impact of events such as Hurricane Katrina and the 2004 Asian tsunami was exacerbated by loss of protective vegetation (Lotze *et al.* 2006).

While enhanced environmental protection has resulted in recent declines in the incidence of some forms of pollution (e.g. heavy metal contamination), nutrient enrichment has continued unabated, resulting in eutrophication of coastal waters and growing incidence of anoxic and hypoxic conditions (Lotze *et al.* 2006). The effect of nutrient enrichment on coastal systems has been amplified by loss of the filtration capacity of aquatic vegetation and overexploitation of suspension feeders such as oysters.

The extraction of marine mineral resources represents a growing industry which includes sand, coral, gravel, and shell for aggregate; beach replenishment and cement manufacture; and magnesium, salt, sulphur, diamonds, tin, gold, and heavy minerals (Wiltshire 2001). The extraction of deep sea deposits such as manganese nodules and crusts and methane hydrates continues to attract interest. These extractive activities could affect critical marine habitats and marine organisms dependent on them. The magnitude of impact on marine ecosystems related to these activities is difficult to assess, but a substantial body of knowledge has built up concerning the impacts of oil exploration, extraction, and oil spills (e.g. Hjermann *et al.* 2007). The issue is of renewed concern as the Arctic Ocean is being opened for oil exploration and may have particularly vulnerable ecosystems (Thorne and Thomas 2008).

3.3 Fisheries-induced changes

Collectively, the many effects of fishing on populations not only reduce the population abundance, but also make them more sensitive to additional fishing pressure, additional mortality, and environmental variability (Ottersen *et al.* 2006; Anderson *et al.* 2008; Perry *et al.* 2010a; Planque *et al.* 2010). Environmental effects are generally observed on younger stages (typically eggs and larvae), while fishing primarily affects larger and older individuals. The combination of the two results in increased variability in abundance and greater risk of collapse. Fishing acts through well-recognized effects on population dynamics, but also causes subtler changes in the behavioural and geographic structure of populations. Some of the effects are irreversible or slowly reversible; in order to reverse the effects of fishing and to halt a negative trend of population abundance, individuals may need to get older, memories of preferential migratory pathways may have to be rebuilt, genetic diversity may have to be restored, and new habitats may need to be recolonized. All of these take time, and may slow down the recovery of highly depleted stocks. The sensitivity of fisheries and marine ecosystems to human impacts is discussed further in Section 3.4. The history of the fisheries at Greenland over the past century (Box 3.2) illustrates some of the consequences of rising levels of exploitation and their interaction with environmental changes.

3.3.1 Demographic change

The fundamental effects of fishing and other human impacts on marine population dynamics are associated with changes in the population age structure. Fishing changes age structure by removing individuals over a range of ages each year. Since fish of older ages will have been fished for more years than younger individuals, the effect is to skew the age structure from its natural form to one with relatively fewer older individuals. The effect of this truncation of the age structure, if recruitment remained constant, would be to reduce population biomass. However, some populations may compensate for the reduction in biomass when fishing begins by increasing their recruitment. Eventually, however, as fishing increases further, recruitment will begin to decline, raising the possibility of population collapse through recruitment overfishing.

To understand how truncation of the age structure can lead to overfishing, we first note that it reduces the total amount of reproduction in the lifetime of the average individual in the population. We know from discussions of human populations

that lifetime reproduction is an important indicator of whether individuals in the population will replace themselves in the next generation. If every couple has two offspring, which survive to reproduce there will be zero population growth, while if they produce more than two the population will increase, and vice versa. Fish in natural, unfished populations typically produce up to several million eggs per year of which only a tiny fraction need survive to reproductive age for replacement. What makes the prediction and management of marine fisheries, as well as the assessment of other human impacts on fish populations, extremely difficult is that we know little about how the survival from egg to reproductive age is governed, or what minimum number of eggs must be produced in a fish lifetime for the population to persist.

The dynamics underlying this qualitative description can be seen in a graphical representation of the equilibrium condition of a model of an age-structured population with density-dependent recruitment (Sissenwine and Shepherd 1987; Botsford 1997). The population is represented by an assumed density-dependent relationship between the number of eggs produced each year, and the number of young individuals (recruits) they produce (Fig. 3.4). The equilibrium level of recruitment of a natural, unfished population will be at the intersection of the egg-recruit relationship and a line through the origin with slope 1/LEP, where LEP is lifetime egg production. As the amount of fishing or other mortality increases, LEP declines, moving the equilibrium to the left. Recruitment will remain constant until it reaches the knee of the curve where it will begin to decline. The equilibrium recruitment will be zero when the slope of the straight line (i.e. 1/LEP) becomes greater than the slope of the egg-recruit relationship at the origin. The value of LEP at which this happens is the critical replacement threshold. Fishery biologists have examined the value of this threshold for a number of species for which the necessary egg-recruit data exist. They express the critical replacement threshold as a fraction of natural, unfished lifetime egg production (FLEP) and have found it to be in the range of 30–50% (the critical replacement threshold is also known as the 'spawning potential ratio' or 'eggs per recruit') (Goodyear 1993; Mace and Sissenwine 1993; Myers *et al.* 1999).

Box 3.2 Greenland—effects of fishing and climate change

The cod fishery gradually developed during the 1920s and was dominated in its early days by a foreign offshore hook and line fishery before bottom trawls became more prominent (Horsted 2000). After the Second World War, the fishery expanded rapidly in the West Greenland offshore areas and annual catches rose to about 460,000 t in the early 1960s and then fell steeply. Following a brief increase in catch at the end of the 1980s, the fishery collapsed completely in the 1990s with no signs of recovery since then. Fishing mortality increased to about 0.8 at the beginning of the 1960s and fluctuated around this level until the stock collapsed (Box 3.2, Fig. 1). Hovgård and Wieland (2008) estimated that the equilibrium fishing mortality at this time was about 0.14 and they concluded that the collapse of the stock was primarily due to high fishing mortality, given the low productivity of the stock (in terms of recruitment per unit of spawning stock biomass). Mesh-size and minimum landing size regulations were introduced in the late 1960s, but the major stock decline occurred while the fishery was only restricted by modest access limitations (Horsted 1991). Subsequently, Total Allowable Catch (TAC)

regulations were introduced in 1974, but did not provide effective protection because they were typically set above the recommended level and ineffectively enforced (Horsted 2000). The drop in fishing mortality which occurred in the 1980s was probably too late due to the combined effects of low ocean temperatures and low spawning stock in Greenland waters.

The fishery for northern shrimp (*Pandalus borealis*) began in inshore areas in 1935. In 1970, a multinational offshore bottom trawl fishery started to develop and landings increased gradually to ~155,000 to in 2005 (NAFO 2007). Annual effort in the West Greenland shrimp fishery increased from about 80,000 h in the late 1970s to more than 200,000 h in the early 1990s and remained relatively high thereafter (Kingsley 2007). Considerable by-catch of juvenile cod and redfish has occasionally been observed (Sünksen 2007), but no quantitative studies on the impact of the shrimp fishery on the recruitment of commercially important fish species or benthic organisms are available to date. Catch restrictions were first imposed in 1977 and the fishery has since been managed by TAC. Survey indices

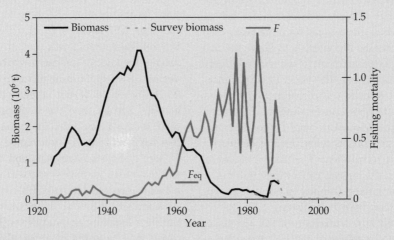

Box 3.2, Figure 1 Estimates of stock biomass (age 3+) and mean fishing mortality (1924–81: ages 5–12; 1982–9: ages 5–7) for Atlantic cod at West Greenland. F_{eq} the fishing mortality for equilibrium biomass per recruit is shown for the early 1960s (ages 5–12). (From Hovgård and Wieland 2008.)

continues

Box 3.2 *continued*

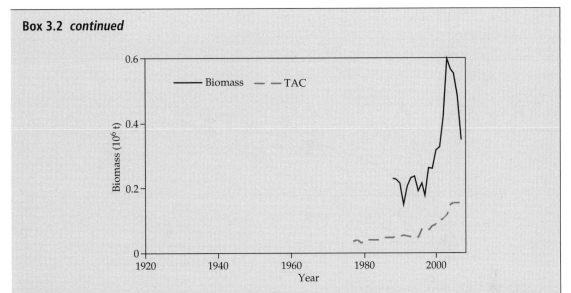

Box 3.2, Figure 2 Survey indices of stock biomass for northern shrimp (*Pandalus borealis*) and actual total allowable catch (TAC) for West Greenland.

showed a threefold increase in stock biomass of northern shrimp from 1997 to 2003 at relatively moderate TAC levels (Box 3.2, Fig. 2). Thereafter, the biomass indices from the survey began to decline, and concerns were raised that the current level of exploitation is no longer sustainable (NAFO 2007), in particular following the decline in recruitment since 2001 (Wieland *et al.* 2007). Sorting grids became mandatory in 2000,

which has likely contributed to an increase in biomass of juvenile Greenland halibut. In addition to high TACs, reasons for the decline in shrimp biomass may therefore also include predatory effects considering the strong negative relationship between Greenland halibut biomass and shrimp recruit survival found by Wieland *et al.* (2007).

Kai Wieland

While for many fisheries we can calculate FLEP from information on survival and fecundity, the critical replacement threshold is highly uncertain.

One useful conceptual consequence of these relationships for the question of human impacts on populations is that sustainability imposes an upper limit on the sum total of all sources of mortality on a fish population, and that total can be assessed in terms of the effects on LEP. This zero-sum characteristic of 'surplus' mortality necessitates that an increase in mortality from one source must be offset by decreases from other sources. The common view that single causes of population change can be considered separately is mistaken and results in erroneous views about management issues, for example, we should not have to reduce fishing

because pollution is really the cause of population decline, or because climate change is the cause of the decline (Schiermeier 2004). Such 'finger pointing' in policy debates misses the true nature of population dynamics. The dynamics of salmon populations (see Box 3.3) provide good examples of both the interactions between different factors and the 'fixed sum' characteristic of the mortality required to achieve sustainability.

Another consequence relates to the question of why it is important to study the effects of environmental variability on fish populations when we cannot do anything about it? The abundance or catch we see each year is a consequence of all of the survival factors. We are often interested in measuring the effects of remedial actions such as reducing

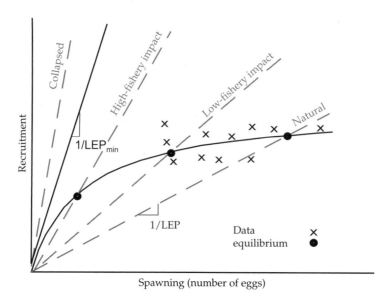

Figure 3.4 A simple schematic representation of how fishing reduces recruitment and may lead to overfishing: a plot of egg production versus the number of recruits that results from them, along with the lines through the origin with slope equal to 1/LEP under various conditions. The equilibrium is at the intersection of these. As fishing reduces LEP, the equilibrium moves to the left and recruitment declines, then eventually collapses. A fishery manager can observe the data points to the right ('x's), but does not know the egg-recruit relationship. Pressure to increase catches, profits, and employment pushes the equilibrium to the left, but the point at which the recruitment will collapse is unknown.

fishing or pollution, to determine their effectiveness through adaptive management. By studying the processes we aim to transform the environmental variability from noise to understandable signal and thus either factor them out or at least take into account their effects on populations and on management.

Specific fecundity (eggs produced per unit of female biomass) may change with the age and size of the fish (Marshall *et al.* 2006). Larger, older fish sometime have higher specific fecundity and their eggs and larvae may have higher survival rates due to having greater nutritional reserves (Berkeley *et al.* 2004; Trippel *et al.* 2005). Changes in age and size structure due to fishing can therefore have quite complex effects on reproductive output, and the calculation of LEP should take such demographic factors into account. However, when differential larval survival with female age was taken into account, the effects of reduced specific fecundity on population persistence and management were not substantial (O'Farrell and Botsford 2006). Higher nutritional reserves do not always translate into higher survival, as there is a trade-off between size

of the larvae (which reduces predation) and the energy provided by nutritional reserves (which reduces starvation; Fisher *et al.* 2007).

Removal of older or mature individuals from a population may alter the behavioural interactions among the remaining individuals. For example, older individuals consistently migrate towards the same spawning and feeding grounds. Transfer of knowledge among old and young individuals, through simple behavioural rules of schooling fish (e.g. Couzin *et al.* 2006), allows the population to maintain a 'conservatism' of migratory routes (Corten 2002). In contrast, the excessive removal of mature individuals and the consequent lack of guidance and transfer of knowledge may increase the percentage of the population that deviates from its established migratory routes. Searching for far and unexplored habitats (i.e. dispersal) does not always have a negative outcome on a population, as it opens up the possibility of colonizing new and perhaps more favourable habitats and to increase genetic diversity (Iles and Sinclair 1982). However, dispersal in fish is typically confined to early life stages, which are numerous and to a

Box 3.3 Coastal ecosystems and salmon in the north-east Pacific

Because salmon are anadromous, human impacts occur in freshwater ecosystems as well as marine ecosystems and provide greater opportunities to illustrate interactions among multiple impacts. Human impacts on salmon have conventionally been expressed as the four 'H's: harvest, hydropower (dams), hatcheries, and habitat. Impacts on habitat, and through hydropower, are greater in the southern part of the range of Pacific salmon, the west coast of the contiguous United States, than in Alaska, where the freshwater environment is relatively pristine.

Salmon were harvested by Native Americans long before the European presence and even then, society required an explanation for the tremendous variability in salmon runs (Taylor 1999; Finney et al. 2002). Catch records began being kept with the advent of salmon canneries in the late 1800s (Taylor 1999), and there is no doubt that there were more salmon then than now. The current status of Pacific salmon is that stocks of all five species are abundant and sustainably harvested in Alaska, but less abundant and less clearly sustainable in Canada and the contiguous United States. Reasons for this difference include: (1) a more pristine freshwater environment in Alaska and (2) different oceanographic conditions, with most Alaskan populations in the Alaska Coastal Current and the more southern populations in the California Current System (CCS). In the CCS, off the west coast of the contiguous United States, salmon are dominated by two species, coho salmon (*Oncorhynchus kisutch*) and chinook salmon (*Oncorhynchus tshawytsha*). Coho salmon fisheries management in the CCS began to reduce allowable catch because of low abundance in the 1990s. In addition, at that time nine Evolutionary Sustainable Units (ESUs) of chinook salmon and four ESUs of coho salmon were listed under the US Endangered Species Act (http://www.nmfs.noaa.gov/pr/species/esa/fish.htm).

Management of human impacts on Pacific salmon requires an assessment of the complex combination of the four Hs and recently appreciated unpredictable oceanographic influences, to determine their combined effect on sustainability. Because these species are semelparous, evaluation of the effects of LEP on population persistence (Fig. 3.4) is much simpler. Since reproduction occurs only once in a Pacific salmon's lifetime, twice the inverse of individual fecundity is the survival necessary to sustain the population. Since this is a single number, we can express lifetime mortality as the sum of the logarithms of mortality due to causes in each stage (Box 3.3, Fig. 1). This allows a clear view of how human and natural impacts combine in some years to produce a total mortality that reduces the reproduction in each individual's lifetime to less than its replacement value. The valuable insight from this figure as regards human impacts is that it is the sum of mortalities, not a single factor, that leads to populations being lower than the replacement level (i.e. unsustainable). Both the farmer withdrawing water from a river for irrigation of agriculture and the fisherman removing his catch contribute to mortality and the overall level of survival being less than sustainable. There is rarely a single cause of a decline in abundance, as is often proposed. Freshwater management and harvest levels that are sustainable under some ocean mortality conditions may not be sustainable under other ocean conditions. This is particularly problematic because ocean conditions are not predictable. For example, ESA recovery plans require management actions that are predicted to return populations to a recovered state, raising the question of what the assumed ocean mortality should be in those predictions?

We are gaining an increasing understanding of the ways in which ocean conditions affect salmon as we observe continuing unprecedented changes in environmental conditions. The US GLOBEC Northeast Pacific programme was founded in the

continues

Box 3.3 *continued*

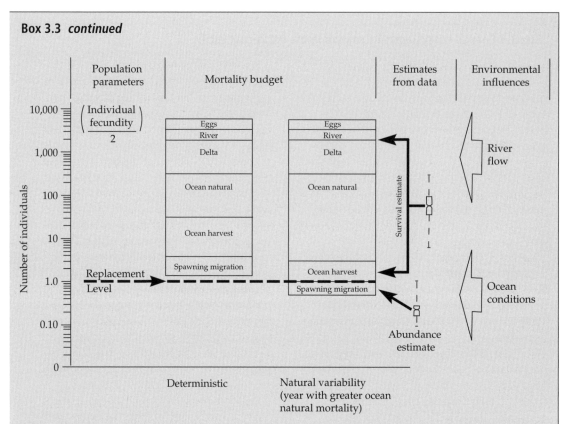

Box 3.3, Figure 1 The autumn run of Sacramento River chinook salmon as an example of how the sum of logarithms of mortality illustrates the way in which multiple factors, including water diversions in the Sacramento-San Joaquin Delta, harvest in the ocean, and variable survival due to ocean conditions, can lead to a lack of adequate individual replacement. Two examples of the mortality budget are shown, with the one on the right falling below the replacement level because of higher ocean mortality. The long-term estimates available, spawning abundance through spawning counts and survival estimates from coded wire tags, are indicated. Fishery catch is also known. Both ocean conditions and river flow are subject to environmental variability.

mid-1990s on the precepts that: (1) salmon populations covaried inversely between Alaska and the California Current, (2) salmon ocean survival was determined soon after ocean entry as juveniles, and (3) mesoscale physical variability was important to coastal ocean productivity and salmon survival. Interest was focused on the regime shift in ocean conditions in the mid-1970s, in which coho salmon in the CCS covaried inversely with coho in Alaska (i.e. decreasing in the CCS, increasing in Alaska), but chinook salmon did not covary (Botsford and Lawrence 2002). Coho and chinook salmon abundance were found to vary with interannual *El Niño* Southern Oscillation (ENSO) conditions in the CCS (Botsford and Lawrence 2002), and on decadal,

semi-basin scales, salmon abundance appeared to vary with the Pacific Decadal Oscillation (PDO; Mantua *et al.* 1997). Analysis of coded wire tag data from coho salmon for 1980–2004 indicates a lack of inverse covariability between the CCS and Alaska (Teo *et al.* (2009)).

During the last decade there have been big shifts in ocean conditions in the CCS, which have affected salmon populations and the public's awareness of the impacts of climate on the marine ecosystem. In 1997–8 there was an intense *El Niño* followed by a strong *La Niña* (Chavez *et al.* 2002; Schwing *et al.* 2002). The *La Niña* in 1999 led to a cooling along the coast, which appeared to signal a shift in the PDO to a different state (Bond et *al.*

continues

Box 3.3 *continued*

2003). These cool conditions lasted 4 years and were accompanied by a more productive CCS. The coastal zooplankton and pelagic fish communities shifted to being dominated by more northern species, and biomass anomalies of three cold water (i.e. northern) copepods were found to be positively correlated with the survival of coho salmon over the previous 14 years (Peterson and Schwing 2003).

These years were followed in 2005 by a warming event that was apparently driven by the unusually late initiation of sustained upwelling winds (Schwing *et al.* 2006; Barth *et al.* 2007). This led to lower primary and secondary productivity in the CCS during the spring and early summer, a shift in zooplankton community structure with less dominance by northern species, and changes in the nekton. A particularly remarkable example was the impact of reduced prey (krill) levels on a planktivorous bird, Cassin's auklet (*Ptychoramphus aleuticus*), which led to complete abandonment of their breeding site on the Farallon Islands (just west of San Francisco, California) for the first time in 35 years of observations (Sydeman *et al.* 2006). In 2006, there were again periods of low upwelling winds early in the season, and there was again no reproduction by Cassin's auklet.

In 2008, as this is being written, the projected return of spawners to the autumn run of chinook salmon on the Sacramento River (which enters the ocean at San Francisco) is the lowest on record, and the associated fishery has been closed for the first time in recorded history. These spawners are from the cohorts that would have entered the ocean in the springs of 2005 and 2006, the years of unusual upwelling winds and low reproduction by Cassin's auklet. The closing of the fishery has fueled extensive speculation by the public regarding the cause of the decline. These responses are mostly the single cause 'finger pointing' mentioned in the text. For example, a non-governmental conservation organization has teamed with fishermen in a critique of water diversion for agriculture in California's central valley as being 'one of the most significant and reversible' factors contributing to the decline of chinook salmon (NRDC 2008). The explanation by the National Marine Fisheries Service that the decline in the Sacramento River salmon run in 2008 is due to adverse ocean conditions in 2005 and 2006 is viewed with considerable public scepticism.

In summary, research on ocean influences on salmon has identified modes of variability that appear to be without historical precedent, but has not produced predictive results. However, it is providing a context which aids in understanding past (and present) variability. In some cases, just having the experience of dealing with interactions between population dynamics and environmental variability can aid in dealing with contentious issues such as the former range of ESA-listed species (Adams *et al.* 2007). It seems likely that the primary benefits of GLOBEC research will be more of this nature than of predicting the future.

Acknowledgement: Louis W. Botsford thanks H. Batchelder and W.T. Peterson for comments on the salmon box.

Louis W. Botsford

certain degree expendable, while mature individuals are scarce and precious. Increased random dispersal of adults is unlikely to be beneficial to the population.

Another important aspect of the dynamics associated with overfishing is that trajectories of decline may not be reversible or the rate of reversal may be much slower than expected from the population rates (recruitment, growth, and natural mortality) which prevailed before the collapse occurred (Hutchings 2000). Although stock-recruitment models predict higher growth rates and fast recovery times of population at low abundance (due to density-dependent effects), in nature the recovery of severely depleted stocks occurs at much slower rates or sometimes does not occur at all (e.g. Labrador and Newfoundland cod; Shelton *et al.* 2006). For some stocks that experienced protracted declines, that is, Pacific sardine in the California Current, Dungeness crab in central California, and the Eurasian perch in Lake Windermere in the United Kingdom, modelling results showed that an increase in individual growth rate may have locked populations into a lower stable state (Botsford 1981). The asymmetry between the rise and fall of fish stocks can be due to the more subtle and long-lasting changes that fishing induces on fish populations, particularly if such changes were an undetected contributory factor in the stock decline (Drinkwater 2002; Brander 2007b).

Finally, another effect of the truncation of age structure through fishing is the fundamental change in population dynamics on short time scales. For example, populations that are dominated by young individuals are more unstable because the buffering effect of strong year classes is confined to shorter periods, making the population more dependent on recruitment rather than on the existing stock (Hsieh et al. 2006). Furthermore, the effect of fishing on the spawning age structure can change the relative sensitivity of populations to fluctuations on different time scales (Botsford 1986; Bjornstad et al. 1999,

2004), and lead to cyclic behaviour (Botsford and Wickham 1978).

On a longer timescale, the indirect effects of fishing act as a selective pressure that counters those imposed by nature (Fig. 3.5). The selective pressure in exploited populations is generally to mature at younger ages and to grow slower in order to reproduce before being caught, and to escape the size-selective gears. In contrast, the selective pressure from the natural environment is to grow fast and large in order to escape size-selective mortality (Conover 2007; Edeline et al. 2007; Fig. 3.5). Which of these two contrasting forces prevails depends on their intensity. Clearly, under regimes dominated by high harvest rates the selective forces imposed by fishing become dominant. Consequently, age and size of maturity may get progressively smaller, causing a reduction in natality and possibly also survival during early life stages (Heino and Godø 2002; Olsen et al. 2004; Jørgensen et al. 2007).

3.3.2 Spatial dynamics

Spatial aspects of population dynamics of marine fish and invertebrate populations have seldom

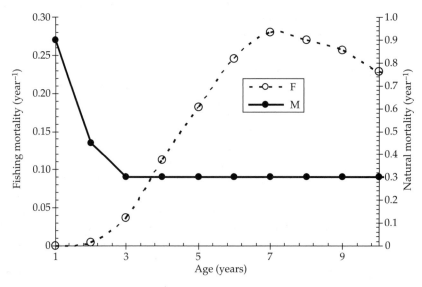

Figure 3.5 Fishing (F) and natural (M) mortality for different age classes of walleye pollock (*Theragra chalcogramma*) in the eastern Bering Sea (Ianelli *et al.* 2007). The contrasting trends of F and M illustrate the opposing selective pressure of natural and fishing mortality.

been accounted for in fishery management, but are beginning to receive more attention. With the rise in interest in spatial management through marine protected areas, spatial variability will be increasingly important. Most fish populations of interest in fisheries are metapopulations, that is, more or less distinct subpopulations linked by a dispersing larval stage, in addition to adult movement (Botsford *et al.* 1994).

The conceptual change in population persistence when one explicitly accounts for the distribution of a population over space is simply that the replacement concept must be expanded to include all possible replacement paths over space (Hastings and Botsford 2006). This means that it is the distribution of LEP over space that is important, and that linkages over space through larval dispersal and adult swimming are also important. Fishing probably changes the spatial distribution of LEP, even though that is currently rarely accounted for in management. Fishing could change dispersal paths, and it is quite likely that climate change will alter coastal circulation and thus exchange rates through larval transport. If that were to increase alongshore advection, as it could likely do in upwelling regions, it could have the effect of reducing local replacement and populations therefore failing to persist (Gaylord and Gaines 2000), an effect that would be more likely if the populations were heavily fished.

Fishing exerts spatial effects by reducing the geographic and genetic diversity of a population. For example, harvest typically peaks in feeding and spawning grounds and during times in which fish are concentrated and thus more easily captured. Particularly during spawning, fish may be isolated by genetic origin, which could result from behavioural interactions (Corten 2002) or imprinted mechanisms (Iles and Sinclair 1982) enabling them to conserve a memory for their spawning locations. Selective removal of these genetically homogeneous individuals will reduce the overall geographic and genetic diversity of the entire pool of stocks. The outcome of such loss is lower persistence. An example is the sockeye salmon in Bristol Bay whose resilience against major events of climate change has been attributed to the complex population structure of the entire pool of

stocks inhabiting the regions (Hilborn *et al.* 2003). Another example is the recent restructuring of walleye pollock spawning aggregations in the western Gulf of Alaska. The historically dominant spawning ground in Shelikof Strait (between Kodiak Island and the Alaska Peninsula) is no longer dominant (Ciannelli *et al.* 2007). Co-occurring with these geographic changes, the entire abundance of pollock in the region has reached an historical minimum, placing a threat on the viability of what once used to be a very active fishery. Shrinkage in the geographic extent of a stock is an early warning that the population is losing its genetic or social structure, and that it may be heading for a rapid decline. For example, the collapse of the north-west Atlantic cod (Newfoundland and Labrador Seas) was preceded by shrinkage of their spatial distribution (Atkinson *et al.* 1997; Warren 1997). The California sardine population has also been described to significantly shrink its distribution range during low abundance periods (Lluch-Belda *et al.* 1989; MacCall 1990; McFarlane *et al.* 2002).

3.3.3 Effects of fishing on species composition and biodiversity

Fishing increases the mortality rate and reduces the life expectancy of fish within the size and age range of the gear being used. Life expectancy (in years) is the inverse of the total annual mortality rate. In heavily exploited areas, such as the North Sea, the mortality due to fishing is many times greater than the mortality due to natural causes and life expectancy is correspondingly reduced (Fig. 3.6).

The relationship between life expectancy and LEP was explained in Section 3.3.1. Species which are vulnerable to fishing prior to becoming mature (because they are large and mature at a late age) are particularly sensitive to being overexploited, because many of them will be caught before reproduction begins. Fishing can be expected to remove such sensitive species first (whose recruitment trajectory has almost no harvestable surplus, i.e. close to the 'natural' sloping line in Fig. 3.4). This has been observed in many cases (Brander 1981; Quero

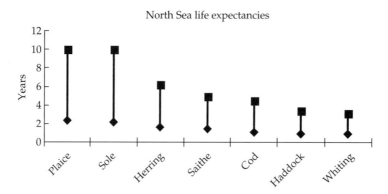

Figure 3.6 Change in life expectancy (1/total instantaneous mortality) at fishable age for a selection of North Sea species. The squares represent the unfished and the diamonds the fished life expectancy. (Data from table 1 of Brander 2003b.)

1998; Jackson *et al.* 2001; Lotze 2007), but often occurred during previous centuries, when records were poor (see Box 3.1). Such disappearances therefore often happen gradually and almost unnoticed. Dulvy *et al.* (2003) estimated that on average 53 years elapse between the last sighting of a marine species and the scientific acknowledgement of its disappearance.

The small number (18 to 21) of marine species which have become globally extinct over the past 300 years, compared to on land (829 species), probably reflects lack of monitoring as well as greater remoteness and resilience of marine ecosystems. The life history characteristics (e.g. high fecundity and highly dispersive early life stages) of many marine organisms may confer greater resilience relative to terrestrial species. It is estimated that more than 133 marine species (including 32 bony fish, 19 cartilaginous fish, 14 marine mammal, and 12 bird species) have undergone local, regional, or global extinction. Most of the extinctions (55%) were ascribed to exploitation and 37% to habitat loss (Dulvy *et al.* 2003).

A recent meta-analysis concluded that the oceans' capacity to provide food, maintain water quality, and recover from perturbation has been impaired through loss of biodiversity (Worm *et al.* 2006), but other studies of the relationship between biodiversity and ecosystem functioning and services produce a more nuanced picture (Balvanera *et al.* 2006). The relationship between biodiversity and resilience is not well established. The structure of ecosystems and in particular the organization of species interactions may be more important to system stability (May *et al.* 2008).

Size and species-selective harvesting can result in marked changes in relative species composition in exploited systems (see Box 3.4). Resulting changes in trophic structure may affect fundamental patterns of energy flow and utilization in marine ecosystems. Such changes may represent alternate stable states that may be difficult to reverse with direct intervention designed to restore previous community structures.

Some of the issues are discussed further in Section 3.4.1. In addition to the potential loss of species, the consequences of loss of local sub-populations have also become a source of concern, including effects on level of sustainable production, risk of progressive stock collapse as successive sub-populations are extirpated, loss of resilience to other pressures such as climate change, and compromised ability to recover from periods of stock decline (e.g. Hilborn *et al.* 2003; Heath *et al.* 2008a).

3.3.4 Damage to or loss of habitats

Efforts to understand the direct and indirect effects of fishing on marine ecosystems have assumed increasing importance as we move towards developing ecosystem-based approaches to fishery management.

Box 3.4 Georges Bank—fishery depletion and management

Georges Bank, a highly productive submarine plateau located off the New England coast, has supported important commercial fisheries for over 3 centuries (German 1987). The strong rotary tidal forces on the bank coupled with topographic rectification results in the establishment of an anticyclonic gyre during the stratified season, resulting in the retention of planktonic organisms (although advective processes remain important at all times of the year). Upwelling of nutrient-rich water onto the bank and mechanisms of cross-front exchange and mixing result in high levels of primary production on the bank, historically fueling high levels of fish production.

A pattern of sequential depletion of fishery resources and large-scale changes in the relative abundance of different ecosystem components as a result of species-selective harvesting strategies has dominated the dynamics of the Georges Bank ecosystem (Fogarty and Murawski 1998). These changes include the depletion of marine mammal populations in the eighteenth century (Waring *et al.* 2004; Clapham and Link 2006), collapse of major fisheries such as that for Atlantic halibut by the mid-nineteenth century, and a series of fishery declines initiated by the arrival of distant water fleets on Georges Bank in 1961 (Clark and Brown 1977). The rapid escalation in fishing effort by these fleets resulted in an initial increase in landings due to a 'fishing up' effect. Groundfish stocks subsequently declined under heavy exploitation and the sequence was repeated for small pelagic fish (principally Atlantic herring and mackerel) and 'other' fish stocks (including elasmobranchs, large pelagics) throughout the next 2 decades (Fogarty and Murawski 1998; Overholtz 2002).

With the exception of some gear restrictions, the fisheries on Georges Bank were effectively unregulated prior to the 1950s. With the advent of fishing activities by the distant water fleet and subsequent decimation of fishery resources, the need for direct controls on fishing was evident. The establishment of a quota management system in 1973 by the International Commission for Northwest Atlantic Fisheries (ICNAF; Hennemuth and Rockwell 1987) based on ecosystem principles provided the nucleus for recovery of depleted stocks (see also Barange *et al.*, Chapter 9, this volume). A quota-based management system was maintained following the establishment of extended jurisdiction in 1977 but was replaced by constraints on mesh size, legal size limits for fish, and seasonal area closures in 1982 under widespread violation and lack of effective enforcement of the quota limits. When these qualitative measures failed to adequately protect fishery resources, court-ordered restrictions (including the use of large-scale year-round closures and limits to days-at-sea) were added in 1994 (Murawski *et al.* 1997; Fogarty and Murawski 1998).

The direct and indirect effects of fishing andxregulatory actions are reflected in fishery-independent abundance estimates for major fish groups (Box 3.4, Fig. 1). As principal groundfish populations (primarily gadoid and flatfish stocks) declined under heavy exploitation, small elasmobranch (skates and spiny dogfish) populations initially increased, possibly reflecting a form of competitive release. However, a redirection of fishing effort on these species as groundfish continued to decline resulted in reduced elasmobranch populations by the mid-1990s. Sharp reductions in fishing pressure resulted in increases in herring and mackerel populations through the mid-1990s (Overholtz 2002). With the imposition of more restrictive management starting in 1994, gadoid and flatfish populations have also recovered although the magnitude of the response has varied substantially among species.

continues

Box 3.4 *continued*

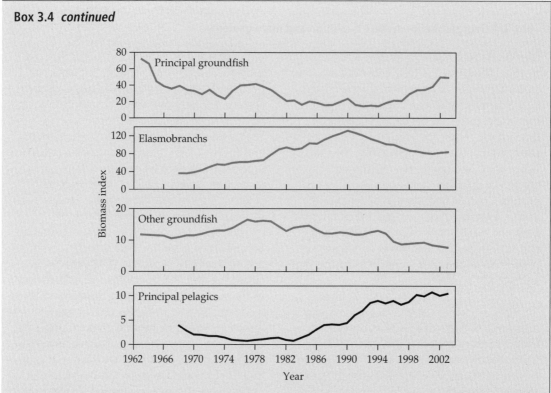

Box 3.4, Figure 1 Figure 1 Trends in biomass of major fish species groups on the north-east continental shelf of the United States based on fishery-independent research vessel surveys for principal groundfish, elasmobranchs, and other groundfish (smoothed mean kg/tow) and for principal pelagic fishes (predominately herring and mackerel; million tons) based on sequential population analysis. Georges Bank accounts for a dominant proportion of overall biomass of the demersal fish indices. (Figure courtesy of Northeast Fisheries Science Center, Woods Hole, Massachusetts, USA.)

Michael J. Fogarty

An emerging focus on habitat-related considerations is evident in the now extensive literature on this topic (Barange *et al.*, Chapter 9, this volume; Kaiser and deGroot 2000) and in legislative mandates such as the essential fish habitat provisions of the Sustainable Fisheries Act of 1996 in the United States (Benaka 1999).

Harvesting can affect habitat-related productivity in several ways ranging from reduction in structural complexity to effects on benthic production and prey availability (Hall and Harding 1997; Jennings and Kaiser 1998; Jennings *et al.* 2001; Kaiser *et al.* 2003). The former includes consideration of loss of shelter-providing structures and its effects on survivorship of sheltering species while the latter concerns patterns of

energy flow and availability to higher trophic levels. Sainsbury (1988, 1991) directly incorporated the effects of habitat availability in production models for fish assemblages on the north-west continental shelf of Australia. In his basin model, MacCall (1990) considered mechanisms of habitat selection and the implications for the development of spatially explicit harvesting strategies. Walters and Juanes (1993) developed models of predation risk in relation to habitat complexity and availability. Destructive fishing practices were shown to exert dramatic effects on productivity and biodiversity of coral reef systems (Saila *et al.* 1993), although this may not be immediate (Sano 2004). Hayes *et al.* (1996) explored the linkages between habitat, recruitment processes, and the implications

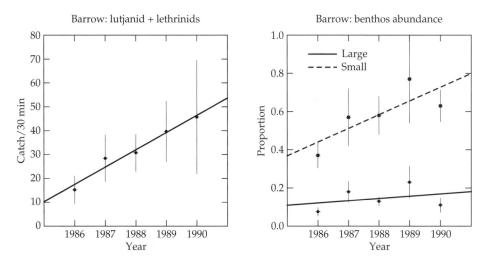

Figure 3.7 Abundance of two fish taxa (lutjanids and lethrinids) in a fishery closed area (left panel) and the proportion cover for two size classes (<20 cm and >20 cm height) of shelter-providing sponges (right panel) on the north-east continental shelf of Australia during 1986–90. (Courtesy K. Sainsbury.)

for management. Habitat-related considerations have now also been directly incorporated in models of marine-protected areas (e.g. Mangel 2000; Rodwell *et al.* 2003; Kaplan *et al.* 2009). A risk analysis framework was used by Fogarty 2005 to explore the effects of fishing-induced loss in habitat and carrying capacity on the shape of the production function and the recovery dynamics of overexploited resources.

Sainsbury (1988, 1991) provided the most extensive empirical evidence for the effects of habitat availability on fish abundance. Recovery of shelter-providing sponges was tracked following cessation of fishing in areas closed to fishing on the north-east continental shelf of Australia and linked to the abundance of commercially important fish species. Concomitant increases in abundance of two fish taxa and in the abundance (proportion bottom-area covered) of sponges were observed (Fig. 3.7). A habitat-impact model was found to be more consistent with the observed population trajectories than a standard production model (see also Sainsbury *et al.* 1997).

3.3.5 Consequences for ecosystem turnover rates and carrying capacity

Indicators derived from available catch and abundance time series, and various kinds of ecosystem models, have been used to examine the ecosystem effects of fishing (e.g. Rochet and Trenkel 2003; Cury *et al.* 2005; Shin *et al.* 2005). For example, there is much effort being directed at selecting indicators that can be used as a means for quantifying ecosystem changes, especially those induced by fishing (Garia and Staples 2000; Degnbol and Jarre 2004, Link 2005; O'Boyle *et al.* 2005). Trophic models have been constructed for the Benguela and Humboldt ecosystems for static periods before human intervention, and compared to models for several periods along the gradient of fishing pressures through time (Neira 2008; Watermeyer *et al.* 2008a,b; Neira *et al.*, in preparation). Dynamic trophic models of the southern Benguela, southern Humboldt, and South Catalan Sea were used to simulate the closure of fisheries on small pelagic fish and the collapse of small pelagic fish stocks (Shannon *et al.* 2009).

Fishing has been shown to have some severe effects on the structure and functioning of marine ecosystems by disturbing the trophic interactions underlying the food web. Fishing changes community structure, often with long-term consequences. For example, fishing reduces the mean size of individuals comprising stocks (e.g. Bianchi *et al.* 2000; Sibert *et al.* 2006), thereby resulting in communities that are dominated by species of small body size and which reproduce fairly rapidly. This phenomenon has been termed 'fishing down the marine food web' (Pauly *et al.* 1998). Collapses in small pelagic and other fisheries in recent years have also altered the mean trophic level of the catch, with large declines and/or fluctuations in recent decades in the northern Benguela (Cury *et al.* 2005), southern Humboldt (Neira 2008; Neira *et al.*, in preparation), and Mediterranean Sea ecosystems (e.g. Coll *et al.* 2006a).

Stocks of certain non-fished groups, usually those at low trophic levels and/or having high turnover rates, may proliferate under heavy fishing. In some instances, such as off Morocco, the negative effects of overfishing some species can be outweighed to a certain extent by the benefits from the proliferating species, in this case octopus (Gulland and Garcia 1984). However, in other cases fisheries may suffer, for example, where the proliferating species competes with other commercially important species for prey. Model simulations carried out by Shannon *et al.* (2009) suggest that in systems dominated by small pelagic fish, jellyfish may become more abundant when small pelagic stocks are heavily fished and thus when competition for zooplankton prey is lessened. It appears that jellyfish may alter the energy pathways of a food web (e.g. the Namibian case; Heymans *et al.* 2004). A model fitted to time series data available for the South Catalan Sea suggests that certain species not targeted by fisheries, such as benthopelagic fish, may have proliferated in response to declining stocks of competitors such as small pelagic fish and predators such as demersal fish, and a simultaneous increase in planktonic prey availability (Coll *et al.* 2006a; Shannon *et al.* 2009). The proliferation of species such as these may alter the food web substantially and divert the flow of energy to higher trophic levels, with detrimental consequences for production of fish species harvested by man. The Black Sea ecosystem has undergone a regime shift, reportedly due to overexploitation (resulting in trophic cascades through the food web) and eutrophication, which has led to the demise of small pelagic fish and

the proliferation of jellyfish (Daskalov 2002; Gücü 2002).

Thus, because marine ecosystems are complex webs of trophic and other interactions, fishing has effects on the ecosystem that are much wider than the target species themselves. For example, small planktivorous fish play a key role in the pelagic ecosystems of major upwelling regions (e.g. Benguela, Peru). Higher trophic levels are subject to bottom-up control by the abundance of planktivorous fish prey, while zooplankton are subject to top-down control, also by planktivorous fish but acting as predators. The control in such systems has been termed 'wasp-waist' (Rice 1995; Bakun 2006; Hunt and McKinnell 2006), and changes in abundance of small pelagic fish have large implications for species at both higher and lower trophic levels of the food web. Trophic models of several upwelling ecosystems at different (static) time periods suggest that flows to detritus may increase when the abundance of small pelagic fish decreases (Shannon *et al.* 2009), and that systems may shift from being driven by pelagically dominated processes to being dominated by benthic communities. Such appears to be the case off Namibia (van der Lingen *et al.* 2006b). Changes in the flows between benthic and pelagic components of food webs have been proposed as a means of assessing the impacts of fishing (Shannon *et al.* 2009).

We have touched on just a few of the proposed indicators for tracking ecosystem changes over time and particularly, fishing-induced changes. However, it is not a simple matter to disentangle fisheries-induced changes from environmentally induced changes, and in fact these two drivers act synergistically or antagonistically to produce the ecosystem changes we ultimately observe (van der Lingen *et al.* 2006b). For example, in the northern Benguela, anchovy and sardine stocks have collapsed while jellyfish *Chrysaora hysoscella* and *Aequorea aequorea* appear to have increased since the 1980s, reportedly due to sequential unfavourable environmental events, the effects of which were exacerbated by fishing more heavily than was sustainable at the low-stock biomass levels that prevailed (Bakun and Weeks 2004).

3.4 Sensitivity of marine ecosystems

3.4.1 Interactions between fishing and environment in driving observed ecosystem changes

Sensitivity is used here in a fairly broad sense to include both how direct the response to human impact is and also how resilient the marine ecosystem is to perturbation. Human impacts may alter the sensitivity of individuals, populations, or ecosystems and there may be interactions between several simultaneous factors (e.g. fishing and climate; Perry *et al.* 2010a; Planque *et al.* 2010). Some of the underlying population processes have already been introduced in Section 3.3.

Species richness, that is, the number of species present, is an important element of ecosystem structure, which in turn affects food-web complexity and ecosystem function. Latitudinal gradients in species richness are well documented across terrestrial, freshwater, and marine systems with fewer species present at higher latitudes, but the causes remain unclear (Townsend *et al.* 2000). Gradients in primary production, seasonality, specialization of predators, environmental harshness, and evolutionary age of the system have all been examined in the ecological literature, but none provide unambiguous explanations (Begon *et al.* 2005). In the ocean basins, species richness of fish communities has been noted to decline with poleward distance from the tropics in both hemispheres (Cheung *et al.* 2005), with covarying effects due to depth (Macpherson and Duarte 1994; Macpherson 2003). In the North Atlantic, there is a correlation between richness and temperature, such that for a given latitude both richness and temperature are greater in the eastern Atlantic than the western (Frank *et al.* 2007). Latitudinal species diversity gradients for groundfish on the north-west Atlantic continental shelf respond to interannual changes in bottom temperature, which is in turn linked to a dominant climate signal, the North Atlantic Oscillation (NAO) (Fisher *et al.* 2008). Species diversity may therefore be sensitive to climate change.

Semi-enclosed seas such as the Baltic and Mediterranean tend to be outliers in the latitudinal pattern of richness. The former is colder and younger in a geological and evolutionary sense than the North Atlantic at large, and exhibits lower species richness than expected for the latitude, while the Mediterranean is warm for its latitude and provides habitat for both temperate and subtropical species.

Production at successive levels in the food web can be considered as a function of primary production and represented as the product of trophic level and efficiency of transfer between trophic levels (Ryther 1969; Aebischer *et al.* 1990; Schwartzlose *et al.* 1999; Chavez *et al.* 2003). Regions which are poor in nutrients and the necessary trace elements required for phytoplankton growth, such as tropical gyres, can support only small fishery yields. In contrast, regions where oceanographic processes or major river inputs deliver large quantities of nutrient to the surface waters typically support substantial fisheries. Comparing across ecosystems in the north-west Atlantic, Frank *et al.* (2005) found a positive correlation between long-term average primary production and fishery yield. However, within individual ecosystems the relationship over time between primary and higher trophic level production was clearly positive in some, and clearly negative in others. Positive relationships between prey and predator time series are indicative of resource-driven or 'bottom-up' control of the food-web components, while inverse relationships indicate predation-driven or 'top-down' control. The latter has long been recognized in some lake systems and in nearshore or intertidal marine communities where species richness is low and the system is dominated by a small number of predators (Chapin *et al.* 1997). Recently, it has become clear that both bottom-up and top-down control may also pertain in open-shelf ecosystems, and that fluctuation between these states is an important signal of pressure due to fishing and/or climate changes. Frank *et al.* (2006) concluded that the species poor, lower primary production systems which predominate at higher latitudes and lower temperatures in the north-west Atlantic shelf region are fundamentally top-down controlled, while the lower latitude, species rich, higher primary production systems are fundamentally bottom-up controlled.

Ecosystems are rarely driven entirely by only one type of control (bottom-up, top-down, or wasp-waist); they are driven 'by a subtle and changing combination of control types that might depend on the ecosystem state, diversity and integrity' (Cury *et al.* 2003). Fishing, and climatic factors which affect the productivity of individual species, also have the capacity to shift a system between bottom-up and top-down control. For example, Hunt *et al.* (2002) showed a pattern of reversibly alternating control in the Bering Sea pelagic ecosystem between bottom-up in cold regimes and top-down control in warm regimes. Similarly, Litzow and Ciannelli (2007) describe changes in the trophic control of the ecosystem along the shelf south of the Alaska Peninsula (North Pacific). The system was initially bottom-up controlled in the early 1970s (as indicated by positive correlations between prey (mainly shrimp and capelin) and predators (mainly Pacific cod), shifted to top-down control during the late 1970s coincident with a rise in temperature, and then reverted to bottom-up control. The oscillation between control processes was coincident with a change in state of the system from one in which shrimp and capelin abundance, measured by catch per unit effort in surveys and commercial landings, was high and Pacific cod were scarce, to one in which cod abundance was ~50 times greater and shrimp and capelin had declined by a factor of ~100. The mechanism of the state change was hypothesized to be warming mediated shifts in the survival of Pacific cod larvae related to the timing of zooplankton production, and temperature dependent migration patterns of cod. As previously mentioned, the removal of larger, predatory fish by fishing leaves ecosystems that are often dominated by small pelagic fish-'fishing down the food web' (Pauly *et al.* 1998; Pauly and Palomares 2005)-and, given that small pelagic fish are more sensitive to environmental effects than larger fish at higher trophic levels, it is possible that bottom-up effects may become more pronounced as a result of fishing (Cury and Shannon 2004; Shannon *et al.* 2009).

The sensitivity of 'wasp-waist' controlled systems to fishing and environment has already been discussed in Section 3.3.5. Changes in the marine ecosystem of the North Sea can be used to illustrate many of the issues concerning the interaction between fishing and environment (see Box 3.5).

In contrast to the North Sea, the planktivorous fish guild at high latitudes in the North Atlantic (e.g. Barents Sea) is dominated by capelin and herring with few other fungible species present (for a definition of fungibility see Box 3.5). Capelin dynamics show more evidence of top-down control by piscivory, than bottom-up control by zooplankton (Hjermann *et al.* 2004). In the context of the latitudinal gradient in species richness and trophic control noted by Frank *et al.* (2006, 2007), the scope for species fungibility is presumably reduced in species-poor systems, making them more prone to top-down control. Thus, in the species-poor Baltic Sea, cod and sprat abundances show strong inverse correlation indicating top-down control of sprat by cod, with cod dynamics being impacted by a combination of fishing and climatic factors (Köster *et al.* 2003a,b; Alheit *et al.* 2005). Similarly, Worm *et al.* (2006) noted that the proportion of fisheries in a collapsed or extinct state (indicating top-down control by fishing), declines monotonically with species richness.

In summary, with the possible exception of the major upwelling systems, the natural state of warm low latitude, species-rich ecosystems containing many fungible species is for bottom-up control of production processes, with natural variability in primary production driving variability in the production of higher trophic levels. With increasing distance poleward and lower mean temperature, decreasing species richness and fungibility makes food webs more prone to top-down control in which the dynamics of low trophic levels are strongly regulated by predation. Bottom-up controlled systems are more resilient to the undesirable large-scale restructuring of food webs associated with the trophic cascades which may be precipitated by the disruption of predator-species dynamics in top-down controlled systems. In the short term, it appears therefore that high latitude, species-poor systems may be more vulnerable to reorganization by climate change and/or fishing. However, in the long term, we might hypothesize that the poleward expansion of species ranges might increase species richness in high latitude systems making them less prone to top-down control and more resistant to trophic cascades, although perhaps more susceptible to bottom-up

Box 3.5 North Sea-changes in pelagic versus demersal energy flow

Fishing activity and catches of species of commercial interest have been monitored and recorded in the North Sea for several centuries. Changes in population abundance can be inferred from catch rates. For example, fishing mortality on cod (*Gadus morhua*) in the North Sea was moderately high from 1920 to the late 1930s, dropped during the war years and then climbed steadily from the late 1950s to the end of the twentieth century (Box 3.5, Fig. 1). The spawning biomass fluctuated considerably during the twentieth century. From the late 1950s to 1970 spawning biomass doubled, probably due to improved planktonic conditions for early life survival (Beaugrand *et al.* 2003a) and in spite of a concurrent doubling of fishing mortality. The declining trend in spawning biomass since 1971 is due to a combination of fishing pressure and adverse changes in plankton production, composition, and phenology.

In addition to changes in biomass of individual fish species, such as cod, interannual changes also occurred at the level of fish communities. Heath (2005) and Jennings *et al.* (2002) highlighted the decline of the demersal piscivorous fish commu-

nity due to fishing. In the case of pelagic, planktivorous species, individual populations may have been impacted by fishing, but other functionally similar species expanded to fill the vacant niches, thus maintaining the planktivore role in the system. Differential responses of the pelagic and demersal communities resulted in multidecadal changes in the trophic structure of the North Sea ecosystem.

Several studies have claimed to show that the pelagic food web in the North Sea has underlying bottom-up control. Aebischer *et al.* (1990) reported parallel trends in phytoplankton, zooplankton and herring biomass, kittiwake breeding success, and weather indices as evidence of bottom-up forcing. However, it is clear that herring biomass cannot be considered purely as a product of trophic forcing. More likely, fishing combined with environmental (including trophic) factors were involved in the temporal changes which, coincidentally, were correlated with zooplankton and phytoplankton abundance. Similarly, Richardson and Schoeman (2004) cited spatially coherent temporal correlations between phytoplankton abundance and zooplankton herbivores, and between herbivorous and carnivorous zooplankton as evidence of bottom-up food-web control in the north-east Atlantic and North Sea in particular. Frederiksen *et al.* (2006) showed that off the Scottish east coast in the north-western North Sea, diatom, zooplankton, and sandeel abundance, and the breeding success of some seabirds, were all positively correlated implying bottom-up controlled. Reid *et al.* (2001a) noted parallel changes in food-web components including plankton groups which indicated periods of rapid restructuring of the North Sea taxa in the late 1980s. However, Hunt and McKinnell (2006) considered the complexity of diagnosing the nature of trophic forcing within ecosystems and concluded that in fact the necessary data are rarely available, and that correlations between time series of selected species may not of themselves be sufficient

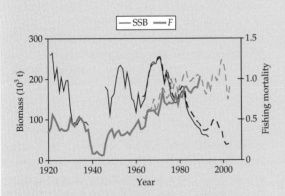

Box 3.5, Figure 1 Spawning stock biomass (black) and fishing mortality (blue) for North Sea cod. The series for 1920–90 (solid lines) are from Pope and Macer (1996). The series for 1963–2003 (dashed lines) are from the ICES Data Centre http://www.ices.dk/datacentre/ Stock Summary Database.

continues

Box 3.5 *continued*

evidence. The authors suggested that fisheries removals are probably decreasing the strength of the couplings between fished species and their prey, thus diminishing the likelihood or demonstrating bottom-up effects of climate change. A solution may be to consider time series of production by functional groups of species which encompass the total flux through the food web, rather than time series of biomass of individual species. Along these lines, Heath (2005) found significant positive correlations during the period 1973–99, between omnivorous zooplankton and planktivorous fish guild production, and between the productions of planktivorous and pelagic piscivorous fish guilds, giving added weight to the case for bottom-up control of the North Sea pelagic food web. Frank *et al.* (2007) found that, given the species richness in the North Sea, these correlations were consistent with the emergent pattern of bottom-up versus top-down control for shelf ecosystems around the North Atlantic.

A key factor in the bottom-up response of planktivorous fish production in the North Sea seems to be the extent of species redundancy, or the fungibility of species, that is, the degree to which species are interchangeable with others of the same functional type within the ecosystem. In the relatively species rich North Sea, Heath (2005) noted that sprat, herring, sandeels, and Norway pout appeared to be functionally interchangeable within the planktivorous fish guild, and that while the production of the guild as a whole was bottom-up driven by zooplankton production, the constituent species had undergone boom and bust phases in response to fishing and climatic factors which bore little or no relation to underlying plankton production. For example, as herring production declined during the 1970s partly through overfishing, sprat production expanded to fill the food-web niche vacated by herring, reversing in the 1980s as herring stocks recovered.

Species fungibility may be less effective in buffering the effects of fluctuations in individual species in the North Sea demersal food web. Several studies have noted that species richness and diversity of the groundfish community have declined over time, and the growth and maturity characteristics of species changed, in the most heavily fished parts of the North Sea (Greenstreet *et al.* 1999; Jennings *et al.* 1999; Greenstreet and Rogers 2006). Heath (2005) estimated that production of demersal piscivorous fish as a guild had decline since 1973 due to reductions in the major commercial species by fishing and lack of sufficient replacement by non-target species to maintain the guild-level production. Also, there was a highly significant negative time-series correlation between the consumption of macrobenthos by demersal fish and the production of macrobenthos carnivores, which was symptomatic of top-down control. Thus, within the North Sea, different branches of the food web displayed fundamentally different control mechanisms.

Michael Heath

effects of climate change. Indeed, recent evidence, that fish species diversity gradients respond to interannual variability in bottom temperature, suggests that the effect may be quite rapid (Fisher *et al.* 2008).

From a management point of view, identifying top-down controlled systems which may be vulnerable to reorganization by climate change and/or fishing, based on species richness, primary production characteristics, and temperature, would seem to be an important step to devising more holistic schemes for ecosystem harvesting. Another development indicated by such considerations is a move towards fisheries management targets based on functional groups, in addition to those based on target species, with the aim of conserving ecosystem stability as well as individual species (Hughes *et al.* 2005).

Trophic models fitted to time series of catch and abundance data have been used to explore the contributions of fishing, the environment, and internal trophic flow controls towards driving observed ecosystem fluctuations/changes. In the southern Benguela, (Shannon *et al.* 2004b, 2008), model simulations suggest that the environment was more important than fishing in changing the ecosystem dynamics in the second part of the twentieth century. These studies suggested that availability of mesozooplankton prey to anchovy and sardine, driven by environmental conditions, and the availability of anchovy and sardine as prey to their predators, could be key processes in this ecosystem, although the underlying mechanisms require further analyses and investigation. By comparison, a trophic model fitted to time series data for the southern Humboldt upwelling system (Neira 2008; Neira *et al.*, in preparation) found that fishing may explain nearly a third of the observed variability in the time series examined and that the dynamics of this ecosystem may have been further affected by a long-term change in primary production. In the case of the northern Humboldt Current, fishing accounted for only about one-fifth of observed variability in the time series used to fit a trophic model, and environmental forcing was important. Environmental forcing which acts in the traditional bottom-up way was found to be more important than environmental forcing acting directly on the interactions between meso- and macrozooplankton (prey) consumptions by anchovy and sardine, or between anchovy and common sardine consumption by their respective predators (Taylor *et al.* 2009). A comparable model of the South Catalan Sea (Coll *et al.* 2006a) suggested that fishing and the environment played equally important roles in driving the observed changes in that ecosystem. These studies highlight the importance of considering synergistic effects of fishing, internal flow controls (predator-prey interactions), and environmental forcing acting at different levels within the food web.

3.4.2 Coastal versus open ocean

Open ocean and coastal or inner-shelf ecosystems may differ in their responses to human activity. The exposure of coastal ecosystems to impacts generated by land-based human populations such as nutrient input, waste disposal, and engineering activity is greater. Coastal and open ocean ecosystems differ in energy inputs and finally in species richness.

In the open ocean, aggregates formed of dead phytoplankton cells and faecal pellets in the surface waters, referred to as marine snow, are progressively digested and respired by microbial activity during their sedimentation into the mesopelagic and bathypelagic zones (Lampitt and Antia 1997; Lampitt *et al.* 2001). As a consequence, the material reaching the sea floor is denuded of carbon and nutrients, and only a small fraction of annual primary production in the euphotic zone is available to support deep ocean benthic food webs. Deep sea taxa have evolved to survive in this impoverished environment by slow growth rates, late maturation, extreme longevity, and low fecundity compared to shallow sea taxa inhabiting equivalent temperature ranges. Species richness typically declines with depth below ~1,000 m in the deep ocean, and with increasing latitude (Macpherson and Duarte 1994; Rex *et al.* 2000; Macpherson 2003; Kendall and Haedrich 2006). This combination of life history characteristics and poverty of species makes deep sea fish and communities exceptionally vulnerable to fishing (Devine *et al.* 2006) and recovery times will be correspondingly long. Given the paucity of species, it is expected that the deep ocean benthopelagic food web would be controlled by top-down predation, and that the removal of fish from such a system would lead to growth in abundance of invertebrates and other taxa in the more heavily fished areas, but it is not known whether this is the case.

In contrast to the deep ocean, the sea floor in shallow-shelf seas and coastal regions receives high nutritional quality phytodetritus, and in areas where the seabed is within the photic zone filter-feeding benthos may graze directly on live phytoplankton. Whereas the sensitivity of deep sea benthic communities to human intervention is a product of their impoverished nutritional state, the sensitivity of shallow sea communities arises from their relatively highly enriched nutritional state. Eutrophication of shallow-shelf seas adjacent to major centres of human population by nitrogen and phosphorus from agricultural, industrial, and sewage discharges is an endemic phenomenon. The excess organic matter resulting from nutrient enrichment leads to hypoxia and anoxia especially in areas subject to water column stratification. Diaz (2001) reviewed global patterns in the incidence of oxygen deficiency and revealed a clear trend of increasing hypoxia. The early stages of eutrophication may increase food-web production and fishery yields, but as nutrient emissions increase the damaging effects of over-enrichment with organic matter become manifest first as seasonal benthic mortality events (e.g. Elmgren 1989; Diaz and Solow 1999; Nilsson and Rosenberg 2000; HELCOM 2002) and ultimately reduced species richness and declining fishery yields (e.g. as in the Black Sea; Mee 1992). The natural trophic state of shallow sea and coastal marine ecosystems is probably slightly net autotrophic (i.e. a small net sink for carbon dioxide and phosphorus), but systems disrupted by enrichment become net heterotrophic (i.e. net sources of carbon dioxide and phosphorus) due to the increasing predominance of organic degradation processes over primary production (Heath 1995).

3.4.3 Introduced/non-native species

Invasive species have exerted profound effects on aquatic ecosystems, resulting in fundamental restructuring through changes in habitat, predation impacts, and competition with native species. Aquatic nuisance species are 'organisms introduced into new habitats…that produce harmful impacts on aquatic natural resources in these ecosystems and on the human use of these resources (http://www.anstaskforce.gov/)'. Freshwater systems, such as the Great Lakes of North America, provide some of the best-known examples of the effects of introduced species, through direct mortality (e.g. lamprey effects on lake trout) or habitat disturbance (zebra mussels). A good source of information on introductions of non-indigenous marine species in the North Atlantic (including viruses, bacteria, fungi, plants, invertebrates, and

fish) and their impacts is provided by International Council for the Exploration of the Sea (ICES) (2007f). An up-to-date collection of papers on marine bioinvasions was edited by Pederson and Blakeslee (2008). Ballast water discharge, escape of aquaculture species, intentional or accidental release of aquarium species, live release of fishing bait, and the loss of live marine organisms transported for human food all potentially serve as vectors for the introduction of invasive marine species (e.g. Carlton 1989; Carlton and Geller 1993). Introduced species may undergo dramatic increases in abundance if their natural predators, diseases, and parasites are not present in the new environment.

In the Black Sea, a population explosion of the introduced gelatinous species *Mnemeopsis leidyi* occurred in the mid-1980s (Grishin *et al.* 1994), and it became a major competitor with the indigenous pelagic fish stocks for zooplankton production (Grishin *et al.* 1994; Shiganova 1998). This compounded the effects of overfishing (Daskalov 2003), leading to a trophic cascade which was manifest as extreme phytoplankton blooms (Yunev *et al.* 2002).

The introduced colonial tunicate *Didemnum* sp. has rapidly increased in abundance off the northeastern United States, in coastal waters, and on the continental shelf. *Didemnum* was first detected there in 1988 (Bullard *et al.* 2007) and is now a dominant species in nearshore and continental waters off New England. Since 2002, it has been found on gravel beds on Georges Bank (Fig. 3.8), which are an important substrate for settlement and growth of sea scallops and provide shelter for juvenile cod and haddock. The tunicate carpeting the gravel could impair these habitat functions. As of summer 2007, *Didemnum* was abundant in two gravel areas totaling over 230 km^2 (88 square miles) (Fig. 3.8). Its offshore spread may continue, or it may be limited by the mobile sands that border the gravel, or (to the north and east) by temperatures too cold to reproduce.

Introduction of lionfish *Pterois volitans*, a venomous coral reef species native to the Indo-Pacific region, has recently been documented off the east coast of the United States (Whitfield *et al.* 2002, 2006; Hare and Whitfield 2003). Adult lionfish have been observed

Figure 3.8 Invasive *Didemnum* (tunicate) mat on gravel substrate on Georges Bank. (From http://woodshole.er.usgs.gov/project-pages/stellwagen/didemnum. Photographed by Page Valentine and Dan Blackwood, U. S. Geological Survey.)

from southern Florida to Cape Hatteras, North Carolina. Juvenile lionfish have now been observed off North Carolina and Bermuda and as far north as New York (Fig. 3.9). Available information on juvenile and adult distributions suggests reproductive populations of lionfish exist off the south-eastern United States. Overwintering mortality of juveniles may account for the current lack of evidence for adults north of Cape Hatteras. However, under projected climate warming scenarios, the potential range of adult lionfish may extend northwards (Kimball *et al.* 2004). Potential fishery-related impacts of the lionfish introduction are currently unknown but this species may compete with economically important grouper species for habitat and food resources (Hare and Whitfield 2003). Grouper populations are currently overfished, possibly freeing resources for lionfish to exploit.

3.4.4 Human activities as triggers or amplifiers of abrupt ecosystem changes

Large-scale changes in marine ecosystems over relatively short time frames have been recorded in a number of systems. The interplay of harvesting and underlying ecosystem processes has been implicated in many of these cases. Lluch-Belda *et al.* (1989) coined the term 'regime shift' to describe alternating global patterns of sardine and anchovy abundance. Collie

et al. (2004) define a regime shift as a low-frequency, high-amplitude change in oceanic conditions that may propagate through several trophic levels; they distinguish between smooth, abrupt and discontinuous ecosystem responses. Smooth regime shifts occur when an ecosystem state variable exhibits a linear response to a forcing function, while abrupt and discontinuous regime shifts arise as a result of underlying non-linear system dynamics. The latter category gives rise to the possibility of alternate stable states in which forcing by oceanographic and/or anthropogenic factors results in rapid shifts between ecosystem configurations. The characteristically reddened spectrum of marine processes (Steele 1985) coupled with harvesting pressure can result in synergistic effects which increase the likelihood of persistent shifts in ecosystem states (Collie and Spencer 1994).

The interaction between temperature and harvesting pressure has been recognized as a contributing factor in the collapse of sardine populations on the west coast of the United States and the specification of sustainable harvesting levels for this species is now tied to changes in temperature (Jacobson *et al.* 1995). Hare and Mantua (2000) identified a regime shift in the North Pacific starting in 1976/7 that resulted in dramatic changes in Pacific salmon populations and other ecosystem components. Choi *et al.* (2004) noted that decadal-scale changes in temperature and stratification affecting benthic-pelagic coupling, combined with overharvesting of cod and other species led to a restructuring of the Scotian Shelf ecosystem to one dominated by small pelagic fish and characterized by sharply reduced productivity of the groundfish assemblage.

In systems dominated by top-down controls (see Section 3.4.1), rapid changes in ecosystem configuration can be triggered by human activities, resulting in indirect effects at two or more trophic levels removed from the level directly affected by anthropogenic forcing (e.g. Frank *et al.* 2005). This effect is termed a trophic cascade. Frank *et al.* (2007) further noted that ecosystem susceptibility to top-down control and resiliency to exploitation are related to species richness and to temperature. Top-down controlled systems are sensitive to overfishing and depletion of predator species (Christensen *et al.* 2003) and the resulting changes may not be reversible with

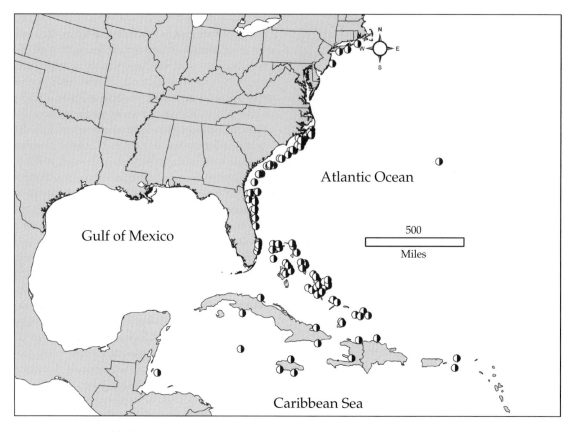

Figure 3.9 East coast of the United States and the Atlantic Ocean, showing locations of lionfish sightings from August 2000 to January 2005 (Whitfield *et al.* 2006).

the reduction or elimination of fishing mortality. Examples are the increase in shrimp production and fisheries in high-latitude ecosystem where cod has been depleted by fishing (Worm and Myers 2003; Frank *et al.* 2005). In the Black Sea, overfishing of marine predators provoked cascading changes, resulting in increases in abundance of planktivorous fish and jellyfish, and a corresponding decrease in zooplankton and increase in phytoplankton (Daskalov 2002) and decrease of the mean trophic level in the ecosystem (Daskalov 2000). Knowlton (1992) noted that in coral reef ecosystems, removal of herbivores through harvesting and/or an increase in eutrophication related to human activities can result in a shift to an algae-dominated system.

Species loss through environmental events may also precipitate trophic cascades in top-down control sys-

tems. For example, unusually warm conditions during summer 1999 caused mass-mortality of invertebrate species in the north-western Mediterranean (Cerrano *et al.* 2000; Perez *et al.* 2000) allowing other species, pre-adapted to the new conditions, to replace them and potentially precipitating food-web restructuring.

3.5 Summary and conclusions

We reviewed studies showing that human impacts on marine ecosystems due to fishing, habitat degradation, and other factors have intensified with time and that evidence of alteration of marine populations and ecosystems goes back many centuries, particularly in coastal areas. Awareness of such long-term changes due to both natural causes and human impacts has increased, as historical and archaeological studies have reconstructed information on the species composition, relative abundance, and age composition of exploited fish. Prior to the recording of fish catches and biological characteristics in order to track and control fisheries for sustainable harvesting, some detailed records were kept for purposes of taxation and trade. Greater attention to the effects of climate and lengthening time series provided by monitoring programmes now give us clear evidence of past changes in marine ecosystems, but the evidence comes mainly from the fish component of the ecosystems and from a few well-studied areas of the world oceans.

The accumulation of more complete and detailed information about what marine ecosystems used to be like, not only over the past century, but over many centuries of human impact, may in principle provide a clearer target for efforts at conservation and restoration. However the evidence of past variability, even in the absence of human impacts, shows that there is no static, historically pristine state to aim at; change is the norm. Overexploitation of fish stocks is the main cause of loss of biodiversity and decline in abundance of marine organisms. This has stimulated great public concern and political pressure to protect fish stocks and the ecosystems of which they are a part. It is imperative to act in order to prevent further degradation of marine habitats and loss of biodiversity and to restore sustainable fisheries.

We describe the effects of fishing on population age structure and resilience in some detail, using the concept of LEP to explain the dynamics of replacement. Because replacement depends on all sources of mortality it is their collective effect which determines whether exploitation is sustainable. Resilience and the rate at which stocks recover depend on details of behaviour, internal dynamics, alternate stable states,

metapopulation structure, and ecosystem changes. Population recovery may therefore take much longer than estimated from models based on pre-decline vital rates especially if they do not include relevant processes, such as those itemized above. Furthermore, since we know that future environmental conditions will be different from those in the past, because of climate change, this means that the recovered state of a population or ecosystem may be different from the past state.

Fishing causes damage to habitats and alters the trophic structure of ecosystems in a variety of ways, with effects which go far beyond the species targeted by the fisheries. Although most of the changes which have been observed have been detrimental (e.g. loss of production of desirable species), there could in principle also be desirable changes, resulting in cultivation of desirable species, or using spatial planning to achieve desired objectives in different areas. Turning principle into practice however requires much greater understanding of the processes governing such ecosystem changes than we presently possess.

The EAF is being developed to ensure sustainable yields of exploited species, while conserving stocks and maintaining the integrity of ecosystems and habitats on which they depend, but considerable difficulties remain in making this fully operational. One issue concerns objectives, for example the extent to which high food production from the sea is compatible with the maintenance (or restoration) of ecosystems and habitats. Conflicting objectives need to be resolved, trade-offs evaluated, priorities established, and effective control measures agreed between different interest groups and countries-quite a tall order, but one with which we will grow increasingly familiar as we come to terms with the need to take responsibility for management of a small planet.

PART II

Advances in understanding the structure and dynamics of marine ecosystems

Having set the scene in Part I, the following four chapters describe how our understanding of marine ecosystems has advanced in the 1990s and 2000s. Chapter 4 considers the Global Ocean Ecosystem Dynamics (GLOBEC) approach of using 'target species' in the main ecosystems studied globally. The target species have usually been ones whose ecological role is pivotal in the food web, either as prey or as predators. Their population dynamics, behaviour, and eco-physiology have been intensively studied in order to gain deeper insight into the functioning of the relevant ecosystem. The chapter documents the successes of this approach and also points out the remaining unanswered questions. This leads to Chapter 5 which describes our understanding of the dynamics of physical processes

mainly at fine to intermediate scales, and how these dynamics affect the food web and ecosystem. The physical forcing processes include tides, winds, stratification and retention features, and how these influence the dispersal and trophic dynamics of plankton and fish. New ecological theories have developed from this understanding and these are discussed in their relationships to the recruitment of fish populations.

Having discussed physical-biological interactions, Chapter 6 moves on to review new approaches, methods, and techniques developed during GLOBEC to study feeding, behaviour, reproduction of zooplankton, fish as well as top predators like mammals, both in the laboratory and aboard ship. It describes new optical and acoustical systems, animal tagging,

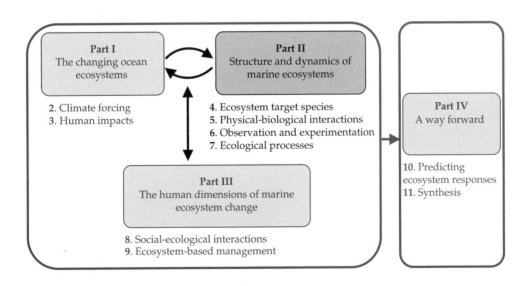

and telemetry and also methods used for retrospective data analysis relevant to describing past states of the ecosystem. It concludes by looking forward to discuss potential future observational and experimental approaches for marine ecosystem research. After Chapter 6, which covers the methodology used and developed in GLOBEC, Chapter 7 summarizes advances in understanding ecological processes at the middle and upper trophic levels of food webs. This is built upon the Joint Global Ocean Flux Study (JGOFS) legacy of primary production and microbial processes at the base of the food web. Ecosystem dynamics are discussed in relation to global change and some key advances in understanding ecological processes analysed. New emerging issues for future study are also highlighted.

CHAPTER 4

Dynamics of marine ecosystems: target species

Dian J. Gifford, Roger P. Harris, Stewart M. McKinnell, William T. Peterson, and Michael A. St. John

4.1 The GLOBEC approach: population dynamics of target species

GLOBEC (Global Ocean Ecosystem Dynamics) was initiated in 1991 as a multinational, multidisciplinary programme centred on the population dynamics of marine animal populations in the context of environmental variability and global change. GLOBEC's overall goal was and is to understand how global change affects the abundance, diversity, and productivity of marine populations. GLOBEC research themes include retrospective analysis, process studies, predictive and modelling capabilities, and feedback from marine ecosystem changes. Because of the central focus on population dynamics, species-level research is central to GLOBEC, and most field and modelling programmes have been directed at target species. In practice, most GLOBEC programmes have studied zooplankton and fish populations, with some consideration of other top predators such as seabirds and marine mammals.

4.2 Target organisms

Because of the daunting complexity of marine ecosystems, the concept of target organisms has been used as a means to focus studies and make comparisons between ecosystems centred on the population dynamics of the selected organisms. The designation of target organisms is species-centric by definition, and GLOBEC has thus taken a species-centric, rather than trophic-centric, approach to ecosystem dynamics (deYoung et al. 2004a). As a consequence, one of the challenges in the programme's synthesis phase has been to translate and integrate species-specific information into a whole-ecosystem context (e.g. Field et al. 2006; Steele et al. 2007).

4.3 What are the justifications for the target species approach?

Marine pelagic ecosystems are inherently complex due to the many interactions among the species that comprise them. At the level of individual organisms, interactions such as predator-prey relationships and competition act to create complex food web interactions. One way to reduce these diverse interactions to manageable proportions is to focus research efforts on a few species within an ecosystem. In so doing, the objective is to understand how biological interactions and/or physical processes affect the population dynamics of the species of special interest (e.g. deYoung et al. 2004a; Fig. 4.1). Target species are usually selected because of their ecological or economic significance, but they may also have conservation, cultural, or social significance. These categories need not be mutually exclusive.

Ecologically significant target species have usually been selected on the basis of two criteria: the species' suitability for studies using the comparative approach and the species' trophic role in the

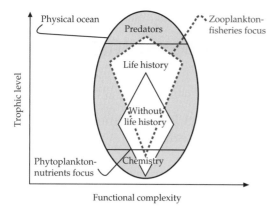

Figure 4.1 The GLOBEC species-centric approach is represented by the rhomboid (dashed line) contained between the two solid black lines. The trophic-centric approach encompasses the entire range of trophic and functional complexity contained within the spheroid. (After deYoung *et al.* 2004a, reprinted with permission from AAAS.)

ecosystem under study. The first criterion permits comparison of the population dynamics of a given species in different geographic regions. By selecting broadly distributed target species that play key roles in marine food webs, the comparative approach is used as a proxy for a natural experiment on how the species responds to regional differences in physical environmental forcing and/or different biological interactions such as predator-prey relationships. Completion of many local studies coupled with regional comparisons allows the population ecology of target species to be compared among different ecosystems. Concomitant use of spatial differences and similarities allows inferences to be made about temporal changes due to environmental variation, and by extension, global change. The second criterion requires that the target species have a quantifiable role in the trophic web of the ecosystem under study. In this case the target species may be an important consumer of other taxa, or a key component of the diet of another target species. Table 4.1 gives a summary of some of the target species adopted in the GLOBEC National and Regional programmes. *Calanus* and other large copepods were selected as target species in a number of GLOBEC programmes (Chile, China, Japan, South Africa, the United Kingdom, and the United States) because they are important components of the diets of other target

organisms, such as larval and juvenile fish and euphausiids. Similarly, euphausiids were targeted in the US Northeast Pacific GLOBEC programme because they are important prey items for juvenile salmonids and many other fish species including hake, the biomass dominant in the region (Batchelder *et al.* 2002a). Euphausiids are also critical prey for seabirds and baleen whales in this region. The multinational GLOBEC Southern Ocean programme also focused on euphausiids, along with several of their predators, including penguins and baleen whales (Hofmann *et al.* 2004a).

Target species may be economically significant by virtue of their contribution to local, regional, and national economies through subsistence, commercial enterprise, use by indigenous peoples, and tourism (e.g. bird-watching and whale-watching). Economically important GLOBEC target species include commercially harvested fish taxa (e.g. Atlantic cod, Pacific salmon, and small pelagic fishes, such as anchovy and sardine, characteristic of upwelling ecosystems), as well as seabirds and marine mammals. Target species of conservation significance may be listed, or under consideration for listing, as endangered. They may also be the subject of regional, national, or international conservation agreements (e.g. cod in some North Atlantic ecosystems and whales and penguins in the Southern Ocean). Target species of social or cultural significance have value to communities because of their historical, aesthetic, educational or recreational value, separate from their economic or conservation significance (e.g. salmon, cod and other fish species, seabirds, and whales).

4.4 The criteria that define target species

Ideally, target species are chosen because they play an important role in marine food webs. However, a target species may be selected solely because it is a crucial prey item for another species of interest. For example, Atlantic cod recruitment in a given region may depend on the species composition, stage structure, and abundance of copepod nauplii. Similarly, recruitment of some salmon species may depend on

Table 4.1 Examples of the range of target species and taxa studied in GLOBEC National and Regional programmes

Programme	Ecosystem studied	Zooplankton target species	Fish and apex predator target species
Chile	Humboldt Current	*Calanus chilensis*	*Engraulis ringens*
China	East China Sea	*Calanus sinicus*	*Engraulis japonicus*
	Yellow Sea continental shelves		
CLIOTOP	Open ocean pelagic ecosystems		Open ocean apex predators
ESSAS	*Sub-Arctic Seas:* Bering Sea, Barents Sea,	copepods, including *Calanus* spp.	Cod spp., Capelin
	Newfoundland-Labrador Shelf, Sea of Okhotsk,		Walleye pollock
	Greenland waters, Iceland waters, Oyashio Current		
France	Bay of Biscay		*Engraulis encrasicolus*
			Solea solea
			Dicentrarchus labrax
Germany	*Shelf seas:* Baltic Sea and North Sea	*Pseudocalanus* spp.,	Sprat
		Acartia spp.,	Herring
		Temora longicornis	
Japan	Kuroshio, Oyashio and their transition region	Copepods	Saury
		Euphausia spp.	Sardine
			Anchovy
			Walleye pollock
			Squid
			Salmon
Norway	Northern Norwegian shelf	*Calanus finmarchicus*	*Gadus morhua*
	Barents Sea		*Mallotus mallotus*
			Clupea harengus
			Melanogrammus aeglefinnus
Portugal	North-eastern Atlantic upwelling		*Sardina pilchardus*
			Trachurus trachurus
Southern Ocean GLOBEC	Southern Ocean	*Euphausia superba*	Cetaceans
			Penguins
			Birds
SPACC	Upwelling systems		Anchovy
			Sardines
Spain	Bay of Biscay		*Engraulis encrasicolous*
			Sardina pilchardus
Turkey	Black Sea	*Calanus exinus*	*Engraulis encrasicolous*
		Mnemiopsis leidyi	*Sprattus sprattus*
		Pleurobrachia pileus	
UK	Irminger Sea, North Atlantic	*Calanus finmarchicus*	
USA	*North-west Atlantic:* Georges Bank	*Calanus finmarchicus*	*Gadus morhua*
		Pseudocalanus spp.	*Melanogrammus aeglefinnus*
	North-east Pacific: northern California Current,	*Euphausia pacifica*	*Oncorhynchus kisutch*
	Coastal Gulf of Alaska	*Thysanoessa spinifera*	*Oncorhynchus tshawytscha*
		Calanus spp.	*Oncorhynchus gorbuscha*
		Neocalanus spp.	Cetaceans
			Birds
	Southern Ocean: Marguerite Bay, Western	*Euphausia superba*	Cetaceans
	Antarctic Peninsula		Birds

Box 4.1 Criteria for the selection of target species

- Likely to be impacted by hypothetical climate change scenarios
- Economically or ecologically important, either as a dominant member of the ecosystem or through interactions with other species
- Has larval planktonic stage, is holoplanktonic or pelagic (birds and mammals)
- Evidence that life history variability is linked to environmental variability
- Widely distributed, or having life history and/or ecological interactions representative of many

other species, thus providing opportunity for large-scale spatial comparisons
- Demonstrated evidence of long-term shifts in abundance
- Distribution associated with physical features and/or faunal boundaries
- Analogous species or taxa in other ecosystems
- Long-term data record of abundance and distribution, and (preferably) vital physiological rate measurements

the standing stock of large copepods and other zooplankton available to juvenile fish. Thus, if the primary scientific question concerns the early life history of a particular fish species, then the appropriate target species are fish larvae, juvenile fish, their predators, and their prey. On the other hand, *keystone species* have well defined and completely different roles in an ecosystem. Although the precise definition of a keystone species has been a matter of active debate (e.g. Power *et al.* 1996), in its simplest form, a keystone species has a disproportionate effect on its environment relative to its abundance: its removal from the ecosystem promotes either collapse or major structural changes in the system. The classic example of a keystone species is starfish from rocky intertidal communities (Paine 1969). Removal leads to overpopulation by, and competition between, the prey of the keystone species, mussels, and barnacles. It is more difficult to define a keystone species in pelagic ecosystems, because it is challenging to imagine how a prey species could be impacted to such an extent that the ecosystem structure is changed dramatically. Baleen whales in the Antarctic provide an example of a possible pelagic keystone predator. They are rare relative to their prey and their removal through harvesting has apparently altered the Antarctic ecosystem (Laws 1977, 1985). Because they are not rare and their impact is not disproportionate to their abundance,

gelatinous zooplankton predators are not keystone species as defined by Paine (1969). They may however function as *keystone modifiers* (Power *et al.* 1996) because their predatory activity can force striking changes in pelagic ecosytem structure, that is, in the Black Sea (Kideys 2002) and the coastal north-west Atlantic (Costello *et al.* 2006).

In GLOBEC studies, several criteria have been applied to the selection of target species (Box 4.1). GLOBEC target species are heavily weighted towards marine pelagic organisms, particularly zooplankton. However, vertebrates with largely (seabirds and seals) or wholly (whales) pelagic life histories have been selected in some ecosystems, as have anadromous fish such as salmon, whose life history is not entirely marine.

4.5 Where has the target species approach worked to allow comparisons of species and ecosystems among regions?

GLOBEC programmes have been successful in providing detailed illustrations of the spatial dynamics of plankton and fish larvae (e.g. Werner *et al.* 2001). Historically, these detailed simulations of spatial and temporal processes have focused on planktonic or early pelagic life stages of fisheries species over relatively short time scales (<1 year) and have largely ignored the benthic components of the ecosystem (but see Queiroga *et al.* 2006). The power of this approach is also apparent in recent publications dealing with physically complex basin and shelf ecosystems (e.g. Wiebe *et al.* 2001); Cowen *et al.* 2006; Spiers *et al.* 2006) where questions concerning dispersal, retention, and connectivity are addressed. More recently, these models have been extended to include older life history stages (e.g. Miller *et al.* 2002), providing a method for forecasting climate and fishery impacts over the entire life cycle of the target species (e.g. Megrey *et al.* 2007b).

Species-centric approaches have also been used to evaluate processes underlying fish movement including ideal free distribution, random walk, directional movement, and gradient tracking (Humston *et al.* 2004). Stock assessments provide an additional species-centric approach to forecasting ecosystem changes on future fish production. Historically, these projections were utilized to assess the short-term future of fish stocks in order to evaluate whether a stock was overfished or likely to become overfished. Such projection models have been adapted to evaluate the performance of different management strategies under variable environmental conditions (e.g. De la Mare 1996). This approach is consistent with deYoung *et al.*'s recommendation (2004a): 'rather than model the entire ecosystem we should focus on key target species and develop species-centric models'. Below, we review examples of several target taxa: large copepods typified by *Calanus* spp., krill (euphausiid species), salmonids, Atlantic cod, small pelagic fish, and apex predators (penguins and whales).

As applied in GLOBEC studies, the target species approach generally includes two major elements: data on the spatial and temporal distribution of the target species and measurements of vital physiological rates such as growth, ingestion, moulting, and mortality. When available, these elements are placed in the context of climatic, biological, and hydrographic time series. The complete suite, or portions, of the data may be used in models of individual target species (e.g. Werner *et al.* 2001) or of the ecosystem (e.g. Aydin *et al.* 2006; Steele *et al.* 2007).

4.5.1 Calanus and other copepods

The target species approach has been particularly successful for the copepod *Calanus finmarchicus*, which has been studied in a number of areas around the North Atlantic basin including Georges Bank, a highly productive retention area (Wiebe *et al.* 2002), the shelf seas of the United Kingdom (Bonnet *et al.* 2005), and the Irminger Sea (e.g. Heath *et al.* 2000, 2004). *C. finmarchicus* was also the focus of the Trans-Atlantic Study of *Calanus* (TASC) programme (Tande and Miller 1996). Target *Calanus* and related species tend to be spring/summer biomass dominants, particularly in temperate and boreal regions, and have been shown to undergo significant climate related biogeographic shifts (e.g. Beaugrand *et al.* 2002a; Chiba *et al.* 2004). Many GLOBEC programmes have focused on *Calanus* species and other large calanoid copepods that include *C. finmarchicus* and *C. helgolandicus* in the North Atlantic (Bonnet *et al.* 2005), *C. marshallae* and *C. pacificus* in the North Pacific (Strub *et al.* 2002), *C. sinicus* in the East China Sea (Lan *et al.* 2007), *C. chiliensis* in the Humboldt Current (e.g. Hidalgo and Escribano 2007), *C. agulhensis* in the Benguela Current (e.g. Richardson *et al.* 2003), and *Neocalanus* spp. in the coastal Gulf of Alaska (Strub *et al.* 2002) and the western subarctic Pacific (e.g. Tsuda *et al.* 2004). These GLOBEC programmes collected extensive data on the copepod's seasonal cycles of abundance and vital physiological rates.

The regions where *Calanus* was studied are characterized by variable hydrographic and climatic regimes, and the species' life history and population dynamics vary accordingly. For example, the North Atlantic *C. finmarchicus* population consists of a single generation at high latitudes, and multiple generations at lower latitudes (Heath *et al.* 2000). Different subpopulations of *C. finmarchicus* in the north-west Atlantic exhibit variable timing of entry/emergence into/from diapause (Johnson *et al.* 2008). Questions about *Calanus* and related species addressed by the GLOBEC regional programmes include the effects of hydrography and meteorology (e.g. stratification vertical mixing, transport, wind-induced washout) on the species' distribution and abundance. In the North Atlantic, there was emphasis on the location and dynamics of overwintering source populations (e.g. Heath *et al.* 2000, 2004). Production (mainly egg production), growth rates, and mortality rates (see Harris *et al.*, Chapter 6, this volume) were quantified in order to evaluate their influence on population dynamics and the persistence of local, regional, and basin-scale populations. The primary questions addressed by these studies focused on the copepod's population dynamics and their trophic role as prey for higher-order consumers. Synthesis of data from the combined North Atlantic studies revealed that the persistence of *Calanus* populations depends on invasion of overwintered stock in spring: both production and abundance appear to rely on proximity to an overwintering deep-water location (Heath *et al.* 2000).

4.5.2 Krill

Euphausiids are highly mobile organisms whose large size and patchy distribution in persistent (days-weeks) dense swarms present a potential food resource for predators (Ressler *et al.* 2005). The euphausiids *Euphausia pacifica*, *Thysanoessa spinifera*, and *T. inermis* were selected as target species in the US GLOBEC North Pacific programme. *E. superba* was the target euphausiid of the Southern Ocean GLOBEC programme (Hofmann *et al.* 2004a). *T. longicaudata* and *Meganyctiphanes norvegica* were target species of the UK GLOBEC programme in the North Atlantic. In all cases, euphausiids were selected for their key role in food web dynamics. In some parts of the Southern Ocean, *E. superba* is also a biomass dominant.

Given their importance in pelagic food webs and the lack of information on seasonal and interannual variations in their abundance and vital rates in most regions where they occur, euphausiids were an ideal target species. In the Southern Ocean, key questions centred on abundance and vital rate processes (Daly 2004) of *E. superba* during the winter season, a relatively little known portion of the life history. In this region a combination of bathymetry, sea ice, and top predator ecology leads to spatial differences in predation pressure on krill. Bathymetry and its influence on physical and biological processes appear to be key to understanding how top predators and their differing life history strategies

survive during austral winter and impact krill populations (Ribic *et al* 2008). In the northern California Current, euphausiid studies centred on collection of basic ecological data including the seasonal cycle of abundance in coastal (*T. spinifera*) and oceanic (*E. pacifica*) waters. Key questions addressed how these are modified by the coastal upwelling cycle. Seasonal and interannual variations in growth, moulting, and egg production rates were measured in relation to cycles of primary production and patterns of cross-shelf zonation as a function of upwelling strength. Relationships among krill biomass, biomass variations, recruitment and salmon growth, and survival were explored. In the Northeast Atlantic, the focus was on krill population dynamics and demography in the context of climate variability.

For inter-regional comparisons of euphausiid ecology, sufficient work has been completed on rates of molting, growth, reproduction, and seasonal cycles of production to allow comparisons of vital rates of *E. pacifica* and *T. spinifera* in the northern California Current (Feinberg and Peterson 2003; Gómez-Gutiérrez *et al.* 2005) and *E. pacifica* and *T. inermis* in the Gulf of Alaska (Pinchuk and Hopcroft 2005, 2006). Because vital rate measurement protocols in the GLOBEC Northeast Pacific programme were similar to those used in studies of euphausiids in the Southern Ocean programme and the Palmer Long-term Ecosystem Research programme, it is now possible to make a biogeographic comparison of *E. pacifica* and *E. superba*. The two *Thysanoessa* species can also be compared. However, such comparisons have not yet been conducted.

4.5.3 Salmon

Salmonid fish were target organisms in the Northeast Pacific programme where the major goal was to determine how variability in physical forcing and food web structure and dynamics influence juvenile salmon survival in the coastal ocean (Batchelder *et al.* 2005). The programme studied salmon in two regions, the northern California Current and the coastal Gulf of Alaska. Both have very different hydrographic regimes underlying biological production (upwelling versus downwelling; Ware and McFarlane 1989) and are potentially linked by large-scale climate patterns (Hare and Mantua 2000). The two regions are occupied by different salmon species, with trophic similarities among the members within region: piscivores (chinook and coho) in the south and planktivores (pink and chum) in the north. The fish exhibit differences in average marine survival between regions, with higher survival along the Alaskan coast. The juvenile life history stages of coho and chinook salmon were targeted in the northern California Current, and pink and chum salmon in the coastal Gulf of Alaska.

Studies in the northern California Current system confirmed that interannual variability in hydrography, chemistry, plankton, and the fish community was a consequence of whichever watermass predominated in any particular growing season. Coho and chinook salmon, which return to a natal stream to spawn, did not survive well under the influence of a relatively warm environment (Logerwell *et al.* 2003). In contrast to the northern California Current system where the summer and winter seasons alternate between upwelling and downwelling winds, the coastal Gulf of Alaska alternates between strong downwelling in winter and weak downwelling in summer with weak upwelling appearing occasionally. Studies done in the northern Gulf of Alaska provided a basis of comparison with the very different northern California Current and allowed hypothesis testing that would not otherwise have been possible. These comparisons led to the rejection of some important hypotheses (Grimes *et al.* 2007). For example, survival of coho salmon in the Alaska Current region can be an order of magnitude greater than what has been observed recently in the northern California Current (Shaul *et al.* 2007). The difference had been attributed to faster growth, large size, and greater accumulation of energy reserves by juvenile salmon in the ocean at northern latitudes. However, considerable variation was observed among sub-regions, and these expectations did not hold when regional comparisons were made (Trudel *et al.* 2006). The lack of consistent latitudinal trends was attributed to differences in local conditions and habitat utilization among species and life history types rather than a fundamental consequence of latitudinal clines in temperature and growing season (Trudel *et al.* 2007).

Some of the original hypotheses about linkages among Pacific salmon survival, climate, and the

ocean environment were not robust and did not withstand closer inspection (McKinnell 2008; Teo *et al.*, 2009). For example, correlation studies had suggested that energetic winter storms in the Gulf of Alaska could provide the physical basis for increased fish production (e.g. Beamish *et al.* 2004). It has been particularly difficult to find evidence to support these hypotheses when survival time series are analysed (Mueter *et al.* 2002; McKinnell 2008; Teo *et al.*, 2009). Specifically, survival of Fraser River sockeye salmon appears to have declined rather than increased under the influence of more severe winters in the Gulf of Alaska (McKinnell 2008). The most significant change in twentieth-century climate in the North Pacific (1976–7) did not generate similar survival trends in coho salmon and chinook salmon within the northern California Current system (Botsford *et al.* 2005). While coho salmon survival in the northern (southern) north-east Pacific tends to be better (worse) in years when sea surface temperature (SST) is warmer, there is little evidence that coast-wide interannual variation in SST generates a negative correlation in survival. This is probably because SST correlations along the coast are weakest during summer/autumn when the salmon occupy this coastal habitat (Mueter *et al.* 2002).

Regional comparisons have provided important clues to the geographic scale of ecosystem variability along the north-west American coastline (Grimes *et al.* 2007). For example, interannual patterns of coho salmon survival in southern British Columbia are shared with salmon originating from the US west coast (Teo *et al.*, 2009). This shared pattern is explained by the species' location with respect to two major hydrographic transition zones. The first, the British Columbia Bifurcation, separates the Alaska Current and the northern California Current. Its latitudinal position varies from year to year (e.g. Cummins and Freeland 2007). The second zone, of greater importance at lower latitudes, is the offshore-onshore transition between coastal zone upwelling and the more stable water of increased subtropical influence offshore. The dynamics of these features are determined largely by the timing and intensity of upwelling (downwelling) winds in summer (winter). Variation in each of these features contributes to the mix of water masses and species that occurs along the coastal shelf (Peterson and Keister 2003; Keister *et al.* 2005). Understanding these important linkages between atmosphere, ocean state, and salmon survival has proved to be of practical application in forecasting future returns (Logerwell *et al.* 2003; McKinnell 2007; Trudel 2007).

4.5.4 Small pelagic fish

Small pelagic fish include anchovy (Engraulidae) and sardine, herring and sprat (Clupeidae). Both families contain a number of genera and many species. Small pelagics are characteristically small schooling fish that broadcast pelagic eggs. They are widely distributed in oceanic and coastal waters, and shelf seas and are notably abundant in eastern boundary current regions characterized by seasonal upwelling. They also occur in temperate non-up-welling areas (van der Lingen *et al.* 2009). Shelf seas with significant stocks of small pelagics include the North Sea and the Baltic. Eastern boundary regions include the California, Humboldt, Benguela, Kuroshio-Oyashio, and Canary Currents. There are striking differences in the productivity of small pelagic fish species in the ecosystems in which they occur. A few species living in the relatively cool waters of eastern boundary currents and the coastal waters of Japan account for the majority of the global catch. A larger number of species live in tropical waters, but they are relatively less productive overall (Hunter and Alheit 1995).

Small pelagic fish were chosen as GLOBEC target species because of their economic and trophic importance. Economically, they comprise >25% of the global marine fish catch (Hunter and Alheit 1995). Trophically, small pelagics feed on plankton-based food webs, primarily on mesozooplankton (James 1988), and they are the prey of predatory fish and marine birds and mammals (van der Lingen *et al.* 2009). Because they have short lifespans with rapid turnover and great physiological plasticity, their populations are able to respond rapidly to environmental forcing and climate change.

Small pelagic fish occupy what is described as a 'wasp-waist' trophic position in which one or a few species are positioned between lower and upper trophic levels containing many species (e.g. Cury *et al.* 2000, 2004; Bakun 2005, 2006). Small pelagic planktivores located at the trophic 'waist', thus have the ability to control the flow of material and energy between the lower and upper trophic levels of the ecosystems in which they occur (Rice 1995). The life cycles of wasp-waist species include susceptible stages that allow their population abundances to vary greatly and independently of other components of the trophic web (Bakun 1996). Any physical or biological forcing from the top or bottom of the trophic web that affects the survival of larvae or other susceptible life history stages has the potential to be amplified and cascade up or down the trophic web (Bakun 2009).

Because of their very long history of exploitation, small pelagic fish have been studied intensively and long-term data are available for retrospective analysis (Alheit *et al.* 2009). Additional information about the variability of small pelagic fish populations prior to commercial fishing comes from a variety of sources, including historical records of artisanal fisheries, archaeological remains, and fish-scale remains in sediments (see Harris *et al.*, Chapter 6, this volume). Cumulatively, the different kinds of historical records show highly variable abundances of small pelagic fish prior to the onset of commercial fishing (Field *et al.* 2009). Individual populations are characterized by alternating episodes of high and low production, and stocks fluctuate dramatically over decadal time scales, with significant ecological and economic effects. The historical records indicate that a plethora of different scenarios for ocean ecosystem variability occurred in the past (Field *et al.* 2009).

The precipitous collapse of the Peruvian anchovetta fishery during the 1972 *El Niño* is the contemporary paradigm for environmentally forced fluctuation in a small pelagic fish population. Recent examples include northward distributional shifts of a number of small pelagic populations, including invasion of the North Sea by anchovy and sardine since the 1990s (Edwards *et al.* 2007) and increase of sprat populations in the Baltic in the late 1980s (Alheit *et al.* 2005). Decadal-scale alternations of dominance between anchovy and sardine have been observed in a number of regions (Bakun and Broad 2003) and are associated with environmental forcing (McCall 2009). Changes in geographic distribution are typically associated with changes in population dynamics as well as changes in spawning habitat (Bakun 2009).

A number of GLOBEC programmes focused on small pelagic fish: ECO-UP (Ecosystèmes d'Upwelling) in the Canary, Benguela and Humboldt Currents, BENEFIT (Benguela Environment and Fisheries Interactions and Training Programme), ENVIFISH (Environmental Conditions and Fluctuations in Recruitment and Distribution of Small Pelagic Fish Stocks), and IDYLE (Interactions and Spatial Dynamics of Renewable Resources in Upwelling Ecosystems) in the Benguela Current, NATFISH (Natural Variability of a Coastal Upwelling System and Small Pelagic Fish Stocks) in the Northwest African upwelling region, and SARDYN (Sardine Dynamics and stock structure) in the North-east Atlantic (http://www.globec.org). SPACC (Small Pelagics and Climate Change), a multinational GLOBEC regional programme, pursued the broad objective of identifying the environmental forces that control populations of small pelagic fish, taking a comparative approach to examine the mechanisms underlying fish production in different ecosystems. The overarching question was how does the coupled ocean-atmosphere system produce observed ecosystem differences in the production of small pelagic fish? SPACC compared physical environmental variability, zooplankton population dynamics, and fish population dynamics among ecosystems by means of retrospective analyses and process studies designed to examine linkages between population dynamics and ocean climate. Specific SPACC studies focused on productivity, resource availability, and reproductive habitat dynamics of small pelagics (Hunter and Alheit 1995). Studies were done in eastern boundary currents, including the Benguela, Humboldt, Kuroshio-Oyashio, Canary and California Currents, and shelf seas (e.g. Baltic, Mediterranean shelf).

Anchovy and sardine were the small pelagic taxa studied most intensively in GLOBEC programmes, and comparisons between these two groups are thus the most comprehensive. Overall, GLOBEC findings confirmed and expanded the results of previous studies. Anchovy and sardine are trophically distinct, exhibiting different feeding morphologies, behaviour, physiology, and habitat utilization (Barange *et al.* 2005). Both taxa are omnivores, deriving most of their nutrition from zooplankton, and the details of diet change with life history stage. Anchovy generally feed on larger particles and are smaller bodied and less migratory than sardine. Sardine tend to be longer lived. Anchovy tend to be distributed more coastally in upwelling regions, while sardine are more oceanic. The habitat utilized by both species expands and contracts with changes in population size, and most stocks appear to occupy predictable refugia at small population size (Checkley *et al.* 2009). In terms of hydrographic factors, both species exhibit similar temperature and salinity tolerances.

It is generally accepted that small pelagic fish exert top-down control on their mesozooplankton prey in upwelling ecosystems (Cury *et al.* 2000). Both high- and low-frequency variability affect the population dynamics of small pelagics. Not surprisingly, climate and fishing pressure are major bottom-up and top-down drivers of stock fluctuations. Climate warming has forced poleward distributional shifts in small pelagic fish, and other potential changes are predicted due to alteration in winds, hydrography, currents, stratification, acidification, and phenology (Shannon *et al.* 2009). Changes in the standing stocks of small pelagics may force significant changes at the top and bottom of the trophic web. From a trophic perspective, climate change most likely affects small pelagics indirectly by impacting their plankton food supply (van der Lingen *et al.* 2009). There may also be consequences for top predators. Seabirds respond strongly to changes in pelagic fish abundances, and both changes in diet and collapses of seabird populations have accompanied decreases in small pelagic species (Crawford and Shelton 1978).

Because overfishing of wasp-waist species has the potential to amplify natural fluctuations in their stocks, it has been suggested that overexploitation of small pelagic fish in concert with climate-driven stock fluctuations may provoke dramatic and possibly irreversible changes in ecosystems. Details of the interactive effects of environmental forcing and fishing on small pelagic fish and the ecosystems in which they occur remain to be fully resolved, but there are several cases where the story appears to be unambiguous. For example, the pelagic ecosystem of the northern Benguela has changed dramatically since the 1980s (Bakun and Weeks 2004). A series of unfavourable environmental events was exacerbated by heavy, unsustainable fishing on low stock biomasses, forcing the collapse of anchovy and sardine stocks off Namibia, accompanied by a proliferation of gelatinous zooplankton (Boyer *et al.* 2001). A similar regime shift occurred in the Black Sea in which a combination of overfishing, eutrophication, and an opportunistic invasive ctenophore species provoked trophic cascades down the food web and a collapse of small pelagic fish stocks (Kideys 2002) (see Moloney *et al.*, Chapter 7, this volume). In both of these cases, the fundamental structure of the trophic web changed from control at the wasp-waist to more bottom-up forcing leading to replacement of small pelagic fish by opportunistic gelatinous zooplankton species that may not transfer material and energy to higher trophic levels.

4.5.5 Cod

Atlantic cod (*Gadus morhua*) is a demersal gadid fish widely distributed around the coasts and continental shelves of the North Atlantic (Kurlansky 1997). A number of individual stocks are recognized for management purposes, but these do not necessarily reflect discrete ecological or population units. Cod spp. have historically been the focus of economically important fisheries throughout its range, and a number of these fisheries, including those of the North Sea, Georges Bank/Gulf of Maine, Canadian Grand Banks, and central Baltic Sea notably collapsed during recent decades and show no indication of recovery, even following cessation of harvesting (Sinclair and Page 1995). Many other stocks remain at risk, but others (e.g. Arcto-Norwegian cod) have been sustained. The decline of economically important cod fisheries is attributed to a coalescence of factors associated with overexploitation and environmental change due to climate warming (e.g. Beaugrand *et al.* 2003b; Beaugrand 2004; Alheit *et al.* 2005). Because of its broad geographic distribution and economic importance, cod was a logical target organism for GLOBEC studies,

being the focus of the Cod and Climate Change programme in the North Atlantic (Brander 1993).

The pelagic early life history stages of cod were target species of several GLOBEC programmes, including those mounted in the North Sea, central Baltic Sea, Norwegian shelf/Barents Sea, Icelandic shelf, and Georges Bank. The GLOBEC approach to cod centred on the assumption that recruitment to the adult stock of cod is determined by density- and habitat-dependent variability in survival during the egg and larval stages. Variability in egg and larval mortality and survival is independent of density and results from interaction of a complex of biological and physical factors operating over a range of temporal and spatial scales. Studies have generally included field sampling of larval distribution and abundance, prey distribution and abundance, and evaluation of larval condition and growth rates. Distribution is affected by advection (e.g. Lough and Bolz 1989; Vikebø *et al.* 2005), turbulence (e.g. Lough *et al.* 2006), and larval photoreception and behaviour (see Section 6.6). Abundance is a product of growth and survival. Growth rates are limited by temperature, food, and turbulence (e.g. Fiksen *et al.* 1998). The latter may exert both positive and negative effects on feeding and growth. Survival is affected by larval condition and physical forcing, including storms resulting in washout of eggs and larvae (Lough *et al.* 2006). During GLOBEC, cod larvae were studied perhaps most intensively in the Norwegian shelf/Barents Sea and Georges Bank ecosystems. These gadid stocks are among the best studied with respect to the influence of biological and physical processes on eggs, larvae, and juvenile fish.

Arcto-Norwegian cod spawn along the coast of central and northern Norway and have a long route (600–1200 km) of pelagic drift within the Norwegian Coastal Current to the Barents Sea. The long drift is significant because it encompasses a critical period for the formation of year-class strength. The vertical distribution of larvae during the drift period affects where they end up in the Barents Sea as well as their access to prey and thus, their biomass. There is a strong relationship between early larval survival and recruitment and temperature through direct effects on feeding, metabolism, and growth, and indirectly as a proxy for other climate parameters, particularly the advection of zooplankton-rich waters from the Norwegian Sea onto adjacent continental shelves. By far the most dramatic effect is the influence of temperature on feeding rates, metabolic rates, and thereby growth (Otterlei *et al.* 1999). The *in situ* light field and wind-induced turbulence also affect behavioural responses, and hence growth of larvae (e.g. Sundby and Fossum 1990; Fiksen *et al.* 1998). Temperature and salinity affect production at lower trophic levels directly: *C. finmarchicus*, is the primary prey item for Arcto-Norwegian cod larvae and juveniles. The recruitment-temperature relation of cod larvae is a proxy for prey abundance due to advection of relatively warm waters carrying *C. finmarchicus* from its centre of production in the Norwegian Atlantic Current to larval cod habitat in shelf and coastal regions of Norway and the Barents Sea (Sundby 2000).

On Georges Bank, cod spawn in late winter/early spring on the north-eastern area of the Bank, and are advected around the Bank to the south and west by a partially retentive clockwise circulation until they settle as juveniles on the Banks' southern flank (Lough and Bolz 1989). During the Georges Bank GLOBEC programme, high rates of growth and survival of cod larvae were associated with low salinity, a proxy for the influx of fresh copepod-rich water from the Scotian Shelf (Mountain *et al.* 2003). Egg and larval mortality varied among years and was related to salinity, with higher mortality associated with high salinity. A comparison with hatch date distributions from late stage eggs (Mountain *et al.* 2003) suggested that individuals hatched early in the season exhibited better survival than those which hatched later. Tidally induced mixing is a major source of turbulence on Georges Bank and turbulence can affect both contact rates between larvae and prey as well as post-capture handling of prey. Modelling studies suggest that this effect is most pronounced in late winter/early spring (Werner *et al.* 1996, 2001). In contrast, under the more stratified conditions characteristic of late spring, prey aggregations near the pycnocline are sufficient to promote rapid growth in the absence of turbulence-enhanced contact rates. On Georges Bank, loss of eggs and larvae to offshore Slope Water was long suspected to be a source of recruitment variability. Studies using coupled biological/physical models

have demonstrated that eggs and larvae lost from the Bank are located in surface water, with the degree of loss influenced by the magnitude, timing, and direction of wind stress. Water is retained on the Bank at depths >20 m and shoalward of the 70 m isobath during winter and early spring (Werner *et al.* 1993, 1996, 2001; Page *et al.* 1999).

Many of the major synthetic findings about cod, particularly in the context of global change, arise from retrospective analyses of long time series of fisheries, climate, and plankton data. The central Baltic Sea and the North Sea are two representative cod ecosystems that have been well studied in this context. Simultaneous dramatic changes in ecosystem structure in the North Sea and the central Baltic Sea occurred in the late 1980s, associated with climate-mediated changes in copepod size and abundance. In addition to the effects of overfishing since the late 1960s, fluctuations in the abundance and species composition of plankton have forced long-term changes in cod recruitment in the North Sea (Beaugrand *et al.* 2003b). North Atlantic-wide decreased survival of young cod is associated with changes in prey species composition and size driven by warming (Beaugrand 2004). In the North Sea, the large copepod *C. finmarchicus* was replaced by the smaller *C. helgolandicus*, changing the prey field available to larval cod. Central Baltic cod stocks were depleted by a similar regime shift in which warming forced changes in phytoplankton and zooplankton communities, and cod were replaced by sprat. The dramatic change in the early 1980s has been attributed to a combination of recruitment failure and overfishing (Køster *et al.* 2006). Changes in sea surface temperature in the central Baltic have been linked to heat fluxes associated with the positive phase of the North Atlantic Oscillation (Alheit *et al.* 2005). The late 1980s regime shift probably contributed to the lack of stock recovery.

4.5.6 Apex predators: penguins, whales, and seabirds

In the Northeast Pacific and the GLOBEC Southern Ocean programmes large predators located at the top of the trophic web were chosen as target species. These included humpback whales and a number of seabird species in the northern California Current System and Adelie penguins, other seabirds, and humpback and minke whales in waters of the West Antarctic Peninsula. In the context of apex predators, both programmes focused on the unifying theme of predator and prey aggregations in 'biological hot spots' facilitated by the interaction of bathymetry with hydrographic processes driving production from the bottom-up.

4.5.6.1 Northern California Current System

The northern California Current System is an upwelling eastern boundary current characterized by complex hydrography and bathymetry. It is part of a larger eastern boundary current that extends along the west coast of North America from the Strait of Juan de Fuca to the tip of Baja California (Hickey 1998). Decadal-scale climate-mediated shifts in the ecosystem structure of this system are well documented (e.g. Roemmich and McGowan 1995; Peterson and Schwing 2003), and enhanced productivity associated with coastal upwelling is known to provide predictable summer foraging for whales (Fiedler et al. 1998; Gill 2002) and seabirds. In GLOBEC studies of the northern California Current System, seabird distributions were associated with mesoscale features including upwelling-derived frontal features, low surface salinity, chlorophyll concentration (a proxy for plankton production), and zooplankton prey abundance (Ainley et al. 2005). Overall, seabirds preferred waters where food was concentrated, specifically near the inshore boundary of the upwelling jet where chlorophyll, micronekton, and zooplankton densities were high. Distance to the inshore upwelling front was among the most important variables explaining seabird abundance patterns. The GLOBEC northern California Current study also examined the ecological influence of Heceta Bank, a major bathymetric feature. Here, upwelling, the interaction of flow with bathymetry, and local recirculation combine to generate enhanced seasonal productivity that attracts foraging birds and cetaceans.

The humpback whale (*Megaptera novaeangliae*) was the most abundant large cetacean in the study region during late spring and summer. Humpbacks arrive in the northern California Current system in spring after spending the winter off Mexico (Calambokidis et al. 2000). During spring, whale occurrence was correlated primarily with warm sea surface temperature over the continental slope, water column depth, distance to the surface expression of the upwelling front and prey abundance. During late summer when humpbacks concentrated over Heceta Bank and off Cape Blanco, sea surface salinity was the most important variable, followed by latitude and depth. The major prey of whales were also concentrated in these areas at this time, suggesting that interactions between bathymetry and flow were important to both the whales and their prey (Tynan et al. 2005). Tynan et al. (2005) proposed that turbulence forced by topography on the continental slope affects foraging whales through its effects on prey distribution. By late summer humpback whales appeared to be responding to a cascade of trophic events enhanced by the flow-topography interactions and the broad upwelling signature over the bank. Maximum abundances of other target species such as juvenile salmon (Brodeur et al. 2004) and large euphausiids (Ressler et al. 2005), also occurred at the same time.

4.5.6.2 Southern Ocean

The Western Antarctic Peninsula supports large standing stocks of krill and apex predators. The distribution, production, and survival of krill in this area are strongly driven by physical forcing, with effects translated up the trophic web to apex predators. The GLOBEC Southern Ocean programme addressed the central hypothesis that the Antarctic Circumpolar Current facilitates transport of warm, nutrient-rich Circumpolar Deep Water onto Antarctic continental shelves of the Western Antarctic Peninsula via deep across-shelf troughs, keeping surface water above freezing and enhancing winter production and prey availability for top predators (Hofmann et al. 2002, 2004a). In waters of the Western Antarctic Peninsula, the combination of bathymetry, sea ice, and predator ecology leads to spatial differences in predation pressure on krill. Bathymetry, and its influence on physical and biological processes, appears to be key to understanding how top predators survive during austral winter and impact krill populations (Chapman et al. 2004; Friedlaender et al. 2006; Ribic et al. 2008).

The Adelie penguin (*Pygoscelis adeliae*), a target organism of the Southern Ocean programme, has been called 'a bellwether of climate change' (Ainley 2003). The species has been studied extensively as an indicator of environmental change in Southern Ocean food webs. There are about 2.5 million Adelie penguins in the Antarctic, and they exploit the pack ice zone more effectively than other seabird species. Adelie populations have changed dramatically in recent decades, coincident with climate warming, specifically associated with changes in sea ice. Adelies consume krill and other less densely distributed zooplankton taxa (Ribic *et al.* 2008). Because krill are the dominant prey of Adelie penguins on the Antarctic Peninsula (Ainley 2003), understanding penguin distribution and foraging ecology is particularly important in GLOBEC studies. During early austral winter, groups of Adelie penguins were observed hauling out on islands just north of Marguerite Bay, suggesting that the penguins may be foraging nearshore and hauling out on islands prior to the establishment of sea ice, thereby shifting their distribution offshore in winter. Thus, predation pressure by penguins and other ice-associated birds during austral winter may be strongly modified by the timing and extent of the development of pack ice (Chapman *et al.* 2004).

Late in the winter the distributions of seabirds other than penguins (gulls, several petrel species, and fulmars) were also associated with sea-ice characteristics (e.g. sea-ice concentration and sea-ice type) rather than the water column environment. Delays in sea-ice development may allow seabirds and other apex predators access to biologically important areas such as the Inner Shelf Water for a longer period of time, thereby increasing predation pressure on zooplankton and fish (Chapman *et al.* 2004). Seabird distributions were associated with water-mass structure and variability in bottom depth during austral spring. Some species had higher densities in Inner Shelf Water, while others were positively associated with variability in bottom depth in autumn.

Humpback (*M. novaeangliae*) and minke (*Balaenoptera acutorstrata)* whales were target cetacean species in the Southern Ocean programme. Both whale species were associated with coastal habitat, particularly fjords where topography is likely to concentrate prey. Their association with zooplankton prey and topographic features suggests that the whales are able to locate physical features that enhance prey aggregation (Friedlaender *et al.* 2006). Whale distributions were strongly associated with zooplankton concentrations in the upper and mid-water column (Friedlaender *et al.* 2006), as well as with sea-ice cover, which varied among years (Thiele *et al.* 2004). There was a lower limit of zooplankton concentration below which the whale-zooplankton relationship was not significant, and a positive association of whales with bathymetric slope (Friedlaender *et al.* 2006). When present, the sea-ice edge provided additional feeding habitat for minkes in winter when they followed ice-covered areas along the entire shelf edge, and for humpbacks during late summer and autumn. Foraging sites for these species are located primarily in coastal areas and are used in all years, but when ice margins are present, the locations attract larger numbers of whales (Thiele *et al.* 2004).

4.6 Where is the target species approach not appropriate, or where does it require alteration?

The target species approach is unlikely to work in tropical and subtropical environments due to high species diversity and lack of dominance. Ideally, in planning tropical ocean studies, a pragmatic approach would be taken: one would select species with broad, perhaps circumglobal, distributions, species that can be sampled quantitatively and relatively easily, species with identifiable life history stages, species that are abundant and present most or all of the time, species that are tractable for experimental work, and species that exhibit behaviours or attributes likely to be impacted by changes in the physical or biological environment.

4.7 Outstanding questions about target species

GLOBEC has focused on understanding the population dynamics of target species in the context of physical/biological interactions and environmental variation. This entails collecting a suite of information about target species over the species' entire annual cycle: distribution and abundance in space and time, measurement of vital physiological rates including feeding, reproduction, growth and mortality, as well as data on physical, chemical, and climatic forcing in the target study region. No GLOBEC programme managed to collect all of this information. Even the data-rich and temporally intensive Georges Bank programme covered only the first half of the annual cycle. While reasonably comprehensive (but integrated in time and space) data are available for most managed fish species, collection of zooplankton data is more problematic. This is especially true of larger organisms that have complex behaviour (e.g. krill swarming) or are relatively rare and/or difficult to collect and/or preserve (e.g. jellyfish). Apex predators are not only numerically rare relative to their prey, they are usually very large, cannot be collected or manipulated in experiments, and some, such as whales, spend much of their time out of the sight of scientists and out of the range of remotely deployed sensors. These are not egregious deficiencies of the various GLOBEC programmes, they simply reflect the realities of oceanographic research. The result is that there remain many pieces of basic information that have not been collected about virtually all GLOBEC target species.

Nevertheless, several areas emerge where outstanding questions and issues remain. For all target taxa, from copepods to whales, we lack information on complete *in situ* prey fields, how these vary in space and time, and how they are affected by environmental variability, and hence global change. In terms of vital physiological rates, many measurements of 'predation' rates have been made including grazing rates of mesozooplankton and feeding rates of higher order pelagic predators. However, with the exception of small pelagic fish, there has been little consideration of the impact of predation on prey populations. Mortality rates due to harvesting are reasonably well known for managed fish species, but *in situ* mortality rate estimates for marine zooplankton taxa are rare (but see Ohman and Wood 1996; Ohman and Hirche 2001; Eiane *et al.* 2002). In *Calanus* and related copepod species, the triggers that cue onset into and egress from diapause remain poorly known (e.g. Johnson *et al.* 2008). No single environmental factor (photoperiod, temperature, and chlorophyll) consistently explains entry/emergences dates in all regions. In many GLOBEC study regions, a question remains about the role of small copepods over the entire annual cycle rather than just the portion of the year when large copepods dominate mesozooplankton biomass. The UK Marine Productivity programme did study the regional and temporal variation of small numerically dominant *Oithona* spp. and concluded that their nauplii are likely to serve as food source for other copepod species prior to spring bloom (Castellani *et al.* 2007). Outstanding questions remain about the forcing of populations and food webs by bottom-up versus top-down processes. The latter is currently a focal point of GLOBEC synthesis studies.

Dynamics of marine ecosystems: integration through models of physical-biological interactions

Brad deYoung, Francisco E. Werner, Harold Batchelder, François Carlotti, Øyvind Fiksen, Eileen E. Hofmann, Suam Kim, Michio J. Kishi, and Hidekatsu Yamazaki

5.1 Introduction

Our understanding of biological and physical processes in the ocean has developed substantially over the past decade, as has our ability to sample and model the dynamics of coupled physical-biological processes at scales ranging from individual planktonic organisms to the shelf scale to the basin scale. Global Ocean Ecosystem Dynamics (GLOBEC) has made a major contribution to these advances. The study of marine ecosystems requires the integration of environmental drivers, for example, physical and biogeochemical, and the biological responses to them (Fig. 5.1). Just as an ecosystem is made up of a complex food web of organisms (Moloney *et al.*, Chapter 7, this volume), environmental variables must also be considered as an interconnected web of drivers.

New understanding of the coupling of biology and physics in the ocean typically begins with observations (Gifford *et al.*, Chapter 4, this volume). It is through understanding fundamental biophysical processes that we will be able to extrapolate, generalize, and ultimately to predict responses and future states of marine ecosystems. At the core of the GLOBEC approach has been the development and implementation of general physical-biological models of ecosystem dynamics based on processes that affect individual organisms (GLOBEC 1993b).

Building upon numerical models of the physical circulation, relatively sophisticated coupled bio-physical models that include growth, development through life history, and feeding and predation are now widely implemented. These models have enabled hypothesis testing and have begun to be used to forecast future states of the ocean ecosystems (Ito *et al.*, Chapter 10, this volume).

In studying the structure and function of marine ecosystems, GLOBEC has sought to determine how physical forcing and changes in the global environment affect abundances and production of organisms in the world's oceans. The primary approach has been a bottom-up investigation of the role of the physical environment as a major driver affecting marine animal populations. While the growth and survival of individuals may depend strongly upon a single variable, such as temperature, their ultimate recruitment to a population is the result of an integrated set of processes and ocean characteristics such as transport, stratification, and turbulence, modulated by light, biogeochemistry, and trophodynamics (Fig. 5.1). Fluid motions strongly influence the distribution and transport of the plankton, including the early life stages of upper trophic level animals, such as fish. In turn, the ability of marine organisms to find and capture food, grow, and survive is controlled by their distribution relative to their prey and potential predators.

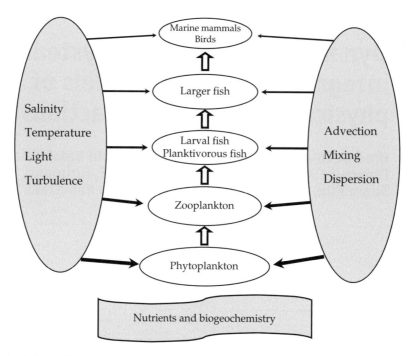

Figure 5.1 Oceanic and climate effects on marine organisms and populations. Local properties such as temperature, salinity, and light can affect the organisms' behaviour, physiology, and growth rates. At larger space- and time-scales the populations' connectivity and biogeography are affected by the dynamics of the coupled ocean-climate system. (Adapted from Sundby 2006.)

Predictions of the impacts of global change rely on understanding how the physical environment, directly and indirectly, affects animal abundance and its variability in marine ecosystems.

Exploration of the interactions between observations and integration through models has led to mechanistic understanding of key marine ecosystem processes. Coupled physical-biological models have linked individual organisms to local, regional and larger-scale physical processes, and in turn linked the biology and local and regional physics to basin-scale changes in global climate. Spatial variability and the connection between the physical environment and organisms are particularly important in the ocean (Steele *et al.* 1993) and the modelling undertaken in GLOBEC programmes has contributed significantly to our understanding. GLOBEC studies have relied on an interdisciplinary partnership between physical and biological oceanographers and development of such links has been a major achievement of the programme. GLOBEC

studies have successfully considered ecosystem responses to large-scale physical fluctuations such as the North Atlantic Oscillation (NAO) or the *El Niño* Southern Oscillation (ENSO; Lehodey *et al.* 2006), to variability at intermediate scales such as along tidal fronts (Franks 1997), and at smaller (turbulent) scales (Yamazaki *et al.* 2002).

As an approach to disentangling the complexity of marine ecosystems, a common and successful approach within GLOBEC has been to concentrate the biological resolution at the trophic level of the species of interest, and to decrease the resolution of the trophic representation and connections above and below the target species (Brander *et al.*, Chapter 3, this volume). This rhomboid approach (deYoung *et al.* 2004a) applies both to observational and to modelling studies (Gifford *et al.*, Chapter 4; Harris *et al.*, Chapter 6, both this volume). Keeping the focus on a key trophic level ensures that both sampling and modelling resources are directed towards the central processes of importance.

The challenge that emerges from the rhomboid approach is to develop models and observation systems that span and link trophic levels with different degrees of resolution and uncertainties, and embedding these formulations in large-scale representations of physics and biogeochemistry. It is unlikely, given the complexity of the processes and the breadth of scales, that a single model will simulate all possible ocean ecosystem states. The challenges are formidable given the wide range of spatial and temporal scales covered by the biological and abiotic subcomponents (Fig. 5.2). Scales span over 10 orders of magnitude from millimetres and seconds at the level of individual organisms to thousands of kilometres and decades and millennia when considering variability and evolutionary development in populations at basin and climate scales.

Processes at all scales can influence the variability of marine organisms and populations. GLOBEC studies have successfully described the effects of physical forcing on selected phases of target organism life cycles and/or on subsets of the population

(e.g. Runge *et al.* 2004). With increasing demands for a more ecological approach to marine fisheries and environmental management in general, there is a growing need to understand and predict changes in marine ecosystems. Physical oceanographic models have reached an advanced stage and have been successful in embedding biogeochemical formulations; however, extending these models up the food web to include both zooplankton and fish remains a challenge (Werner *et al.* 2007a). The difficulty arises in part because organisms at higher trophic levels are longer-lived, thereby integrating variability in abundance and distribution at basin and decadal scales. Additionally, organisms at higher trophic levels have complex life histories and behaviours compared to microbes, further complicating their coupling to lower trophic levels and the physical system. Recent full life-cycle models that link physical and population-level variability over several years and multiple cohorts and trophic levels have furthered the capability to simulate marine populations over space and time (e.g. Megrey *et al.* 2007b; Travers *et al.* 2007).

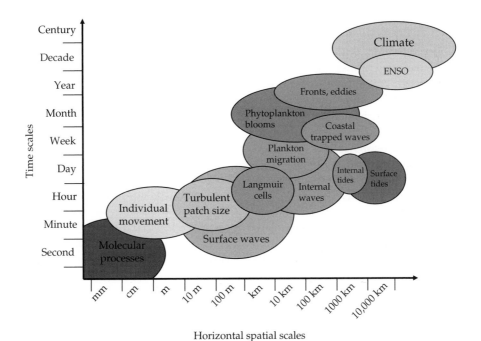

Figure 5.2 Schematic drawing showing the dominant space- and time-scales in the marine environment for physical processes and biological populations (cf. Haury *et al.* 1978 with kind permission from Springer Science and Business Media.)

The connections between variations in hydrodynamics or physical forcing and marine populations overlap across different spatial and temporal scales (Haury *et al.* 1978). At the smallest scales, those of the individual, the ability of organisms to encounter their prey is affected by turbulent processes (Fig. 5.2). At intermediate scales, from days to seasons, productivity, and perhaps the type of populations (community species composition), are influenced by the presence or absence of mesoscale features such as fronts or the strength of upwelling structures (Fig. 5.2). At longer time scales, variability arising from interannual fluctuations such as *El Niño*, or the NAO may affect the basin-scale variation of populations, on the order of thousands of kilometres. At the largest scales, variations resulting from global warming or the changes in the ocean's thermohaline circulation (THC) and evolutionary adaptations come into play (Ito *et al.*, Chapter 10, this volume).

Although the discussion and presentation of the processes and scales is discrete in this space-time representation (Fig. 5.2), the scales are not independent (Hofmann and Powell 1998). Changes at one scale are generally coupled through many different physical and biological processes to those at neighbouring scales. Storms affect mixing, which in turn affects stratification, nutrient availability, and primary production. Ensuing trophic interactions cascade upwards and affect secondary producers (zooplankton), larval fish, and ultimately fish stocks. Variability in physical dynamics in one region can affect characteristics of organisms (e.g. number, size, and condition) available to the next region along a transport path. In this way, variability can be introduced at any point along the trajectory from fish spawning grounds to nursery areas, with poor recruitment resulting from a single adverse event (e.g. flushing out of a nursery area due to a storm), or from a combination of factors along the way (e.g. changes in temperature, the presence of predators, etc.). The variability in a recruiting population will also depend on when a particular event occurred and what the distribution of larval fish characteristics may have been, for example, small-scale turbulence becomes less important as organisms develop from early life stages and become more mobile, making variability in local turbulence significant for small larvae but less so for larger larvae.

5.1.1 Relevant space- and time-scales

It has long been known that the spectra of plankton are red, meaning that they exhibit greater variability at larger spatial scales (Platt and Denman 1975). Spectra of phytoplankton are quite similar to those observed for temperature, suggesting physical control of their variability (Steele and Henderson 1992a). Thus the mesoscale spatial structure observed in physical fields is also reflected in the phytoplankton. The relation between the spatial patchiness of plankton (Steele 1974) and physical processes is still debated (Martin 2003). New field experiments and numerical modelling experiments such as those of Garçon *et al.* (2001) are beginning to clarify the influence of physical processes and scales on phytoplankton production in the ocean. Interestingly, zooplankton are organized into many more and smaller-scale patches than phytoplankton. It is presumed that the different scales of patchiness in phytoplankton and zooplankton are because the latter have greater swimming ability and can aggregate in response to available food or favourable environmental conditions. Conversely, the response times of zooplankton (e.g. change in biomass levels) to environmental changes are substantially slower than the response times of phytoplankton, so the distributions of zooplankton might reflect earlier conditions that have changed (e.g. past history).

Specific features in the ocean influence the concentrations of phytoplankton, zooplankton, and fish larvae. Mesoscale eddies influence the distribution of oceanic phytoplankton (Flierl and McGillicuddy 2002; Vaillancourt *et al.* 2003) and zooplankton (Riandey *et al.* 2005). Anticyclonic eddies in the northern hemisphere sustain new production by transporting nutrients into the surface mixed layer (McGillicuddy *et al.* 2001; Woodward and Rees 2001). Internal waves can also lead to localized abundances of zooplankton and phytoplankton. Surface slicks associated with internal waves and Langmuir cells are regularly seen on the ocean. Lennert-Cody and Franks (1999) used a simple model to show that we should expect to see the maximum concentration over the trough of the wave and the changes in concentration will depend linearly on the wave amplitude and the swimming efficiency of the organism. They suggest that the maximum increase in concentration should be less than twice the background level. Satellite imagery has been used to show that similar effects can be seen associated with the low-frequency Rossby waves (Cipollini *et al.* 2001; Uz *et al.* 2001). These baroclinic Rossby waves have length scales of hundreds to thousands of kilometres. As they propagate westwards, they cause vertical displacements of 10–100 m in the thermocline. Satellite images show a strong relationship between surface chlorophyll concentrations and sea level. This has been referred to as the Rossby rototiller (Siegel 2001). It remains to be determined how much of the observed surface chlorophyll signature is from new primary production as opposed to concentrating chlorophyll nearer the ocean's surface, where it can be detected by satellite remote sensing.

There are many reported correlations between time series of annual indices of biological constituents and climate in the literature but these often turn out to be ephemeral. For example, the annual average abundance of the copepod *Calanus finmarchicus* in the North Sea was highly correlated with the NAO index between 1960 and 1996, but subsequently the relationship broke down (Planque and Reid 1998). The long life cycles of target organisms, their migratory behaviour, and the decadal scales of environmental changes define the ocean basins as a natural spatial scale for such populations. Large-scale circulation features often define the geographical distributions of species, and changes in physical oceanic conditions at basin scales will affect the organisms growth and survival directly, for example, transport of larvae or prey (Sinclair 1988), changes in temperature affecting vital rates, or indirectly by changes in nutrient/food supply or changes in mixing/stratification (Lehodey 2001; Drinkwater 2002; Chavez *et al.* 2003).

Continental shelf and marginal sea ecosystems are affected by basin-scale forcing, and on decadal

scales cannot be studied in isolation (Robinson and Brink 1998). Modulations in the circulation and feeding environments of marginal seas can result in increased connectivity of distinct, previously isolated populations (Cowen *et al.* 2000), or they can affect growth and possibly population recruitment through changes in the feeding environment (Barber and Chavez 1986). For example, variations in populations of several North Sea fisheries target species have been related to the supply of an important copepod prey species, *C. finmarchicus*. Evidence from about 40 years of regular Continuous Plankton Recorder sampling reveals that the spatial variability of *C. finmarchicus* appears to be related to the North Atlantic circulation, to the NAO, and to climate change (Beaugrand *et al.* 2002b). In fact, decadal variability itself should not be considered as a fixed pattern but also as a modulation of climate change as illustrated by the recent compilation of data for the physical ocean (Levitus *et al.* 2001).

Increasing the biological information included in models has led to better understanding of the connections between the physical circulation and the patterns of biological distribution and dispersal. For example, the additional effects of pelagic-phase mortality and settlement substrate selectivity on possible dispersal outcomes show that physical dispersal alone overestimates the population distribution (e.g. see Cowen *et al.* (2000, 2006); Paris *et al.* (2005), for case studies in the Gulf of Mexico and Caribbean Basin). Thus, while ocean current trajectories might indicate the potential outcome of larval transport, they fail to account for the true probability of successful downstream transport because larval concentrations are reduced by several orders of magnitude by along-trajectory diffusion and mortality.

We have also observed during GLOBEC that not all variability in marine ecosystems is slow; some changes in aquatic ecosystems can also be sudden, giving rise to the consideration of threshold responses (Scheffer and Carpenter 2003). Marine ecosystems can exhibit relatively sudden and dramatic changes in form and function, often referred to as regime shifts (Hare and Mantua 2000), with switches between contrasting, and otherwise persistent states, occurring abruptly (deYoung *et al.* 2008; Overland *et al.* 2008). While such shifts have been noticed in all the major ocean basins, the dynamics underlying the observed changes are not always identifiable (deYoung *et al.* 2004b), but are likely driven by natural forcing, human activities, or perhaps by a combination of the two. Three key drivers of oceanic regime shifts are abiotic processes (e.g. changes in ocean stratification), biotic processes (e.g. internal food web dynamics), and changes to structural habitat (e.g. bottom type). These drivers can include natural and anthropogenic components that operate simultaneously and whose influences are difficult to separate (e.g. climate). Abiotic factors, such as global warming or large-scale oscillations in the atmosphere and ocean, are generally the most easily identified. Biotic drivers can include restructuring of food webs resulting from overfishing and internal population dynamics of key species, an example of the latter being the alternation of sardine and anchovy populations in upwelling systems (Schwartzlose *et al.* 1999). The loss of structural habitat can be the result of natural abiotic events such as hurricanes or of anthropogenic effects such as dynamite fishing in coral reefs or the introduction of exotic species. These drivers generally act together so their separation, as is the case for the space- and time-scale diagram (Fig. 5.2), is an idealization for purposes of developing conceptual models.

5.2 Processes affecting individuals

The oceanic environment directly influences the growth rate of organisms, through temperature-dependent metabolic processes, but also indirectly by affecting the abundance or distribution of prey (feeding environment) upon which larger organisms depend. The effect of the ocean on individuals depends on the organisms' size and life stage, for example, ocean currents strongly influence the dispersal of planktonic organisms (which typically can only control their vertical position), while the distribution of larger animals such as adult tuna is determined by both their horizontal (directional) and vertical swimming capabilities. While it is rare that a single driver determines the observed ecosystem response, the approach of isolating a single process is helpful in building a mechanistic understanding of the relation between the physical driver and the organisms' response.

Abiotic factors affect the growth of organisms, their position in the water column, their survival, and their dispersal or retention. Our discussion will focus on a few key abiotic factors including light, turbulence, and temperature (Fig. 5.1).

5.2.1 Light

Light is essential for primary production in the ocean and is also an important factor influencing feeding and feeding cycles. At mid to high latitudes, the seasonally varying diurnal light cycles influence the position in the water column of organisms that migrate vertically searching for prey or to avoid predators, as well as the available feeding hours for visual predators (Suthers and Sundby 1996). Light and temperature also influence the water column's vertical density structure that in turn influences both the circulation and the vertical position of many organisms and plankton. The vertical distribution and life cycles of many marine organisms are affected by the seasonal and diurnal cycles of light and temperature.

Light influences marine food webs through bottom-up effects on photosynthesis and productivity. Light also has a top-down effect in influencing the visual foraging efficiency for fish. Aksnes *et al.* (2004) showed that for light-limited fjord ecosystems, the abundance of zooplanktivorous fish is proportional to the vertical extent of the visual feeding habitat, which they related to the inverse of the light absorbance coefficient. They suggest that zooplankton habitat is defined by the product of the depth and the light-extension coefficient and that these, and perhaps other light-related factors, could serve as predictors of fish and zooplankton. Aksnes (2007) showed that variations in the Secchi depth accounted for 76–85% of the variation in combined fish biomass. Sørnes and Aksnes (2006) showed that the abundance of mid-water fishes in a Norwegian fjord is inversely related to light absorbance.

5.2.2 Encounter rates and turbulence

Most of the important processes that influence plankton, including growth, reproduction, and mortality, are regulated by the rate at which organisms encounter food (Kiørboe 2008). Turbulent motions in the ocean contribute to encounters between predators and prey by either increasing the turbulent encounter rate between predators and prey or by simply concentrating predators and prey by moving them from more to less turbulent waters. Discussions of the understanding and issues around turbulent encounter and feeding in the ocean are summarized in Browman (1996) and Dower *et al.* (1997).

The importance of turbulence was first raised by Rothschild and Osborn (1988) who suggested that turbulence could critically influence the encounter rates between planktonic predators and their prey. Their work subsequently stimulated many field studies, laboratory experiments (see Harris *et al.*, Chapter 6, this volume), and theoretical and model studies (e.g. Sundby and Fossum 1990; MacKenzie *et al.* 1994; Werner *et al.* 1996). MacKenzie *et al.* 1994 suggested that the response should be dome-shaped (see Fig. 5.3), that is, as the level of turbulence increases there should be an increase in the encounter rate and also in the feeding rate. Beyond some critical turbulence level, however, there should be a decline in the feeding rate as the probability of successful feeding decreases as turbulence increases. In

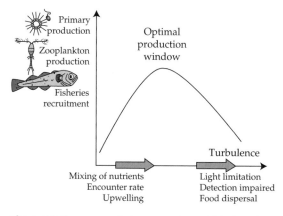

Figure 5.3 The response of primary, secondary, and fisheries production rises with turbulence but then declines beyond a certain turbulent level creating a dome-shaped response curve (Visser and Stips 2002).

spite of many attempts, the dome-shaped relationship has not yet been established in the field, most likely because turbulence can have both a negative and positive effect and so the observations are mixed (MacKenzie and Kiørboe 2000). Other important variables that should be considered include the concentration of the organisms, their sizes, their visual and swimming capabilities, search strategies, and the type and intensity of the turbulence. Because the perceptive distance of predators is generally greater than that of the prey, increasing turbulence increases the encounter rate with predators faster than for prey, implying that turbulence affects marine organisms.

Numerical modelling has been used to explore the effect of turbulence on the encounter rate and feeding of fish larvae (Werner *et al.* 1996). In their model, these authors included advection, feeding, and effects of turbulence on encounter rates and capture success. They found that regions of enhanced larval growth and survival, resulting from the enhancement of contact rates and effective prey concentrations by turbulence within the tidal bottom boundary layer, coincided with hydrodynamically retentive subsurface regions of Georges Bank, on the east coast of the United States, defined in earlier studies (e.g. Werner *et al.* 1993). Near-bed turbulence played a significant role in increasing contact rates, but high values close to the bottom decreased the larvae's capture success and thus the smaller number of survivors close to the bottom. Fiksen and MacKenzie (2002) developed a detailed mechanistic model that they applied to cod larvae on Georges Bank to show that cod are only food limited in deeper waters of the bank. In their result, there is only a negative influence of seabed turbulence on feeding since the light levels are so low.

The search volume geometry, that is, how organisms look for prey, was recognized early on as an important factor of encounter rate (Browman and Skiftesvik 1996) determining when the prey fall within the visual field of the predator which then has the opportunity to attack. The theoretical benefit of turbulence is greatly reduced when more realistic formulations for the search volume shape are employed (Galbraith *et al.* 2004). Galbraith *et al.* (2004) also demonstrated that behavioural choices of the predator, such as movement durations, could lead to an improved energetic benefit of feeding. But in their model, the prey field is fixed and consequently the influence of turbulence on the larval encounter rate is underestimated (Mariani *et al.* 2007).

The concepts developed by Rothschild and Osborn (1988) were applied to fish larvae, but these have now been expanded to zooplankton (Kiørboe and Saiz 1995; Wiggert *et al.* 2008) and phytoplankton (Kiørboe 2008). These organisms encompass a wider range of scale, from 1 cm to 1 μm or less, so the effects of turbulence should be most significant for organisms with perceptive distances close to the Kolmogorov mixing length scale (mm to cm) and less important for smaller and larger organisms (Kiørboe and Saiz 1995). Zooplankton can swim independently of local flow when swimming speed is equal to or greater than root-mean-square turbulent velocities (Rothschild and Osborn 1988) and turbulent intensity can exceed the swimming capability of some species. Haury *et al.* (1990) showed that zooplankton vertical distribution could be selectively altered depending on their swimming ability. Mackas *et al.* (1993) reported that some copepods consistently appear in mixing layers, but some other species are observed just below mixing layers. They hypothesized that while some copepods 'like' turbulent regimes, others prefer quiet water. Visser *et al.* (2009) determined how fast a zooplankter should swim to optimize its fitness in balancing the turbulent benefits of increased encounter rate with prey and with predators as turbulence levels increase.

The conventional view of turbulence as a purely diffusive process is still valid when considering only the net effects of turbulence, that is, spatially and temporally averaged quantities. However, the local structure of turbulence may be more important for micro-organisms than these ensemble-averaged quantities. Direct numerical simulations show that turbulent flows have coherent structures (Pécseli and Trulsen 2007) because the flow must satisfy the governing equations. The dominant pattern is composed of elongated thin tubes (Vincent and Meneguzzi 1991). Yamazaki (1993) suggested that such an organized structure may play a significant role in a 'micro-cosmos' ecosystem, and proposed that organized structures help plankton find mates and detect prey/predators. This is a difficult hypothesis to test because one must understand the details of turbulent flow structures as well as the behaviour of organisms.

The potential for organisms to directly influence the physical ocean has been proposed in several studies, for example, the idea of shading by phytoplankton influencing the thermal structure in the tropical oceans (Manizza *et al.* 2005). The possibility that zooplankton could influence mixing in the ocean has been considered (Huntley and Zhou 2004) and recent estimates suggest that zooplankton could generate about one-third of the total mixing of the world's oceans (Dewar *et al.* 2006). Thus, zooplankton would provide as much energy for mixing as is now estimated for internal waves and tides. Experimental studies of turbulence and zooplankton abundance appear to show that zooplankton can generate an observable amount of turbulence (Kunze *et al.* 2006). However, Visser (2007a) argues that most zooplankton are simply too small to really generate any mixing and that their energy goes directly into heat.

5.2.3 Temperature

Perhaps the most central oceanic environmental factor is temperature, which directly influences the organisms' vital rates (e.g. Hirst and Bunker 2003; Folkvord 2005). Huntley and Lopez (1992) offered a synthesis of the temperature control on growth. Indeed, models have been successful in relating growth of marine organisms as a function of temperature alone (Campana and Hurley 1989). Such success does not imply that temperature is the only factor influencing growth but, rather, that in the absence of food limitation it can be a proxy for other variables influencing growth.

Temperature exerts strong physiological and metabolic controls on marine organisms, so it is not surprising that it has a strong influence on both their distribution and characteristics. Metabolic rates of zooplankton depend on temperature, generally rising in a linear fashion with inflection points leading to declines at higher temperatures (Heinle 1969). Temperature-specific physiological rates might be subject to adaptation to changing temperature conditions in zooplankton species, as shown for fish (Pörtner and Knust 2007). It is possible that zooplankton and fish larvae can use vertical migratory behaviour to 'regulate' their temperature somewhat, by moving from warm surface water to cooler water in or below the thermocline, as is done by *Calanus sinicus*, which utilizes the 'cold pool' in the Yellow Sea (Wang *et al.* 2003). Early studies showed that maximum fitness, in a purely physiological sense at least, is achieved by remaining in the warmest surface water (Orcutt and Porter 1983) but other factors such as predator avoidance have been suggested to be more important than temperature (Loose and Dawidowicz 1994). In copepods, adult size generally increases with decreasing temperature (McLaren 1963), which may reflect a physiological adaptation, limiting food supply, or some influence of differential mortality (Runge and Myers 1986). Timing to maturity is strongly temperature-dependent with increasing temperature generally leading to increased growth and development rates and shorter time to maturity (in the absence of food limitation). The relationship among body size, temperature, and growth in copepods was used by Ikeda (1985) to reveal latitudinal variations in the metabolic rates and standing stock biomass both of which show peaks near the equator and at high latitudes, above 40° latitude.

In a study of North Atlantic cod stocks, Brander (1995) showed that weight-at-age of cod is strongly temperature-dependent and that mature cod from the Labrador shelf are substantially smaller than those found at the southern limit of their range on Georges Bank where water temperatures are higher. Brander (1995) found that the weight-at-age can vary by a factor of two or more over the distributional range. Interestingly, the size of cod eggs, which is also highly correlated with temperature, decreases with increasing temperature (Miller *et al.* 1995). Studies defining the relation between temperature and growth and body size have important biological and practical implications. Biologically it means that changing temperatures and habitat can not only influence distribution but also growth, survival, and productivity. On the applied side, it implies that stock assessments of fish, which generally are based on weight and not number, must consider the influence of temperature on growth,

weight-at-age, and the geographic implications of variable body size. One challenge in the consideration of weight-at-age is to keep maturation processes and growth processes separate (Aksnes *et al.* 2006).

The relation between preferred habitat and the organism depends on predation, the organism's life history, available food and habitat, temperature and oceanic conditions, and physiology and metabolism. Most factors covary in ways that have not yet been adequately quantified. Work on feeding and growth (Peck *et al.* 2005) reveals how the metabolism of juvenile haddock increases with increasing temperature. The cost of feeding in haddock is about 3% of the energy of the food consumed. Such results are important to understand the field observations that reveal correlations between temperature, recruitment, and survival to the fishery (Campana and Hurley 1989).

Spatially explicit individual-based models (IBMs; see Section 5.4), which include variable temperature distributions, have been used to explore environmental controls on marine organisms (T.J. Miller 2007). Hinckley *et al.* (1996) showed that the size distribution of a zooplankton population was sensitive to the variability in temperature experienced by individuals transported through a region. Temperature affected growth rate, which in turn contributed to differences in horizontal dispersal that arise through vertical behaviour. Heath and Gallego (1998) used simulated temperature distributions (taken from a circulation model) as a proxy for the feeding environment to determine individual growth rates of larval haddock. They found that the model-derived spawning locations resulting in the highest larval growth rates coincided with the observed preferred spawning locations.

5.2.4 Effects of global warming

The ecological impacts of a changing climate are already evident in terrestrial and marine ecosystems, with clear responses from the species to the community levels (e.g. Harley *et al.* 2006). The effects of a warmer climate are hypothesized to result in important changes that can impact marine populations not just by affecting organisms directly as indicated above, but also through changes in environmental conditions such as changes in wind patterns and intensities, strength of ocean currents, increased stratification in the surface layers, loss of sea ice, and impacts on levels of turbulent mixing (Fréon *et al.* 2009; also Ito *et al.*, Chapter 10, this volume).

Drinkwater (2005) reviewed projections based upon established environmental relationships and physiological studies, for distributions of cod in the North Atlantic. He suggested that cod distributions would shift northwards, decline in the southern reaches of their range such as in the North Sea and Georges Bank and move northwards to the Labrador Sea, off Greenland, and occupy larger areas of the Barents Sea. Changes in temperature are also suggested to influence annual migrations of certain species. Salmon, which undergo large anticlockwise migrations around the North Pacific (Pearcy 1992), have shown temperature preferences that suggest possible changes in which some species, such as sockeye salmon (*Oncorhynchus nerka*), could be either excluded from the Pacific Ocean or severely restricted by the overall area that would support their growth (Welch *et al.* 1998). At lower latitudes, Loukos *et al.* (2003) found that changes in primary production (Bopp *et al.* 2001) and temperature would yield significant large-scale changes of skipjack tuna (*Katsuwonus pelamis*) habitat in the equatorial Pacific. Somewhat like an *El Niño* situation, warming conditions east of the dateline would extend the favourable habitat for tuna from the present zones that are more concentrated in the western Pacific.

Beaugrand *et al.* (2002b) demonstrated, using Continuous Plankton Recorder (CPR) data, that strong biogeographical shifts in the north-east Atlantic in copepod assemblages have occurred over the past few decades. The CPR data show a northward extension of more than 10° in latitude of warm-water species and a decrease in the number of colder-water species. Different research groups have proposed to use three-dimensional, ecosystem-biogeochemical models to predict further biogeographical shifts of planktonic communities to anticipate their impact on ecosystem dynamics and marine aquaculture (deYoung *et al.* 2008).

Changes in currents and temperature at large scales are also likely to affect dispersal of marine populations, especially the early life stages. One example is the potential changes in larval drift and development of Arcto-Norwegian cod (*Gadus*

morhua) that may arise for a reduction in the strength of the global THC. Using a regional model forced by a global climate model, Vikebø *et al.* (2007b) found a reduction in the THC relative to the present-day circulation for a warming climate. The impact of the change in circulation and ocean temperature on the cod resulted in southward and westward shifts in the distribution from the Barents Sea onto shelf regions, a reduction in the predicted individual growth of the pelagic juveniles, and an increase in the number of larvae and pelagic juveniles that advected towards regions where they are unable to survive. Our ability to link existing coupled biological-physical models to climate models capable of examining climate change scenarios will be critical in assessing potential direct and indirect impacts of climate change on population connectivity.

5.3 Framing theories and hypotheses

Over the years many different hypotheses have been proposed to explain the influence of the environment in regulating survival of larvae in the ocean. Although these ideas were developed for and have motivated fisheries research, primarily on fish larvae, the general questions of how oceanic characteristics influence survival of marine organisms also apply to zooplankton. The factors that influence survival include temperature, circulation, prey, predators, and growth rate (Houde 2008). The development of hypotheses explaining how these factors influence survival has guided much of the research since the work of Hjort (1914). Hjort developed two central hypotheses which have since been expanded upon by others (Fig. 5.4) as new data have been obtained, and new models have been developed. Hjort suggested in his 'Critical Period' hypothesis that the survival of fish larvae was determined early in the larval development, just after the yolk absorption, when the larvae must find suitable amounts of planktonic prey (Houde 2008). His second hypothesis was that dispersal of early life stages, called 'Aberrant Drift', would strongly influence larval survival. These two basic ideas underlie many of the other hypotheses that have been developed since and have formed the basis for much GLOBEC research.

Since feeding is central to growth, it has long been expected that key life stages dependency on food

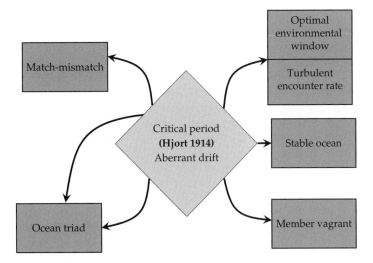

Figure 5.4 The ideas of Hjort (1914) have stimulated the development of many different hypotheses to explain the influence of the environment on the recruitment of fish larvae.

supply would be a central constraint on survival. Folkvord (2005), for example, showed that larval cod always grow at temperature limited rates, suggesting no food limitation. In contrast, Beaugrand *et al.* (2003a) found that a plankton index can explain a good deal of the recruitment success in the North Sea. Hirst and Bunker (2003) found that for adult copepods the degree of food limitation increases with temperature but found that juveniles are very close to food saturation at all temperatures. Many fish larvae feed primarily on zooplankton, particularly eggs and nauplii, and the 'match-mismatch' hypothesis suggests that the timing of larval appearance should be in phase with the dynamics of phytoplankton and zooplankton (Cushing 1975). Thus, years where the zooplankton and larvae are temporally matched result in better recruitment to the population.

The 'migration triangle' (Harden-Jones 1968) and 'member-vagrant' (Sinclair 1988; Sinclair and Iles 1989) hypotheses include spatial considerations. In the member-vagrant hypothesis, circulation controls the dispersion of larvae away from a favourable environment, with population density playing a role in population expansion at high abundances; this hypothesis may be most relevant for banks on the continental shelf where closed, retentive circulations are a persistent oceanographic feature (Page *et al.* 1999). In contrast, the migration triangle hypothesis suggests that

the circulation transports larvae from the spawning grounds to favourable nursery areas, with the surviving juveniles and adults returning to particular feeding grounds. Work on plaice by deVeen (1978) supports a model in which the fish spawn on the same ground each year and the larvae migrate to the north-east growing in the nursery ground and then rejoining the adult stock. Another stock that fits this model of spatial migration linked to ontogenetic development is the cod stock that spawns each spring in the Lofoten area of northern Norway. Larvae then drift with the coastal current into the Barents Sea where juveniles develop there on nursery grounds. Growing juveniles later rejoin the adult stock that migrates through this same region returning to form a spawning aggregation each spring (Vikebø *et al.* 2005).

The 'stable ocean' hypothesis of Lasker (1978) was developed based on observations in the California Current System (CCS), considering the role of vertical stratification in concentrating prey for fish larvae at the pycnocline. Lasker and colleagues worked on northern anchovy *Engraulis mordax* and found that the larvae survived well on a single species of dinoflagellate for which the initial capture efficiency of the anchovy larvae was rather low. They proposed that the prey concentration had to be very high for successful larval development. Rearing studies of the larvae at sea showed that a stable layer of prey was needed for successful larval growth. Retrospective historical analysis showed that the best year classes were during periods of weak winds, the worst during periods of strong upwelling (Peterman and Bradford 1987). Cury and Roy (1989) suggested that recruitment in upwelling ecosystems is dome-shaped with the greatest success occurring in moderate wind conditions with a balance between advective loss and foraging success associated with enhanced turbulent-regulated encounter. They called this the 'Optimal Environmental Window'.

The 'Encounter Rate' hypothesis (Rothschild and Osborn 1988) invokes a physical mechanism, in this case turbulence, for bringing prey and predators together. If larvae are chronically food-limited, a subject of some debate (see Leggett and deBlois 1994), then physical factors that enhance the encounter of larvae and their prey could influence survival of larvae. While early models focused on the turbulence impacts on encounter alone, work by MacKenzie *et al.* (1994) extended the analysis to include the entire feeding process, from encounter, through pursuit, to attack and the feeding behaviour of the larval fish. There are limited observations of any effect at sea (Sundby 1997) most of which include only indirect measurements of turbulence.

For upwelling systems, Bakun (1996) suggested an 'ocean triad' in which there are three characteristics that must come together for optimal productivity. First, there must be ocean enrichment from the upward mixing of deeper waters that generally have higher concentrations of nutrients. Upwelling, ocean eddies, and tidal mixing can inject deep nutrients into surface waters (Bakun 2006a). Second, there must be concentration of food organisms, and or other particles, at frontal zones to permit effective and efficient feeding. In upwelling regions this can be the layer of surface water that moves offshore as the upwelling front develops. Finally, the third component is the retention of larvae within the high food regions (see Boxes 5.1 and 5.2).

Box 5.1 Vertical swimming behaviour and retention

The selection of vertical location by larvae influences both the tactical or short-term balance between growth and mortality and the longer-term consequences, which are more strategic, connected to the large-scale circulation. Small movements in the vertical can lead to significant changes in environmental conditions, because of the strong vertical gradients of properties such as temperature, light, predator-prey concentrations, and horizontal and vertical currents. As an example of the impact of the vertical shear, particles just 10 m apart on the spawning grounds of the northeast Arctic cod end up hundreds of kilometres apart after 100 days (Vikebø *et al.* 2005).

In considering the influence of retention, we have to examine the overall fitness of effects (Box 5.1, Fig. 1). In considering the influence of dispersion, fitness will depend on the water column structure, the drift trajectory, and the ultimate settling area (Fiksen *et al.* 2007). It is the consequences for the organism that matter: the influence on growth and mortality, the potential for contacts with predators and prey, and the opportunities for growth, survival, and reproduction. We do not always see all these effects as we observe changing distribution patterns but they act as the underlying drivers of a population's response to environmental forcing.

In dynamic environments such as upwelling areas it is unclear how zooplankton and other larvae manage to stay in favourable areas. Observations show that zooplankton do well in these very dynamic and advective regions, so there must be mechanisms that allow them to remain and to grow in these regions. Batchelder *et al.* (2002b) explored the influence of diel vertical migration (DVM) on the distribution and retention of zooplankton off the Oregon coast using a two-dimensional coastal upwelling model, an individual-based copepod model including growth, development, reproduction

continues

Box 5.1 *continued*

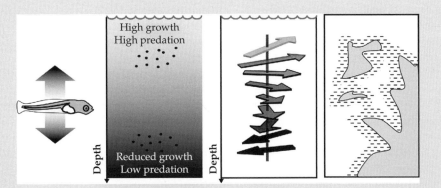

Box 5.1, Figure 1 This figure illustrates the local and strategic trade-offs in larval habitat selection. To the left are the classical behavioural elements of the pelagic environment, setting up a local trade-off between growth and survival. To the right, the large-scale drift consequences of local depth selection. The panel on the right illustrates the terminal settlement area when the drift phase is over. This area may also influence the fitness and should be included in the evaluation of the success of the behavioural strategy. (From Fiksen *et al.* 2007.)

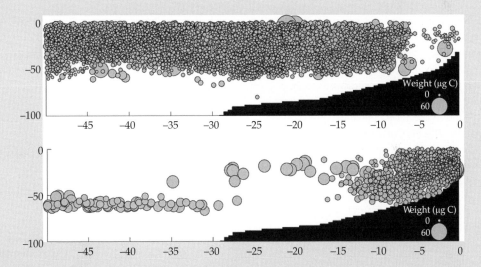

Box 5.1, Figure 2 Two different model simulation results for a coastal upwelling situation. The upper panel shows the copepod distributions after 40 days for a simulation in which the copepods do not undergo diel vertical migration. In the lower panel the copepods undergo vertical migrations where the speed is dependent on light, food concentration and the individual's weight, and hunger. The size of the bubble is related to the weight of the individual. (From Batchelder *et al.* 2002b with kind permission from Elsevier).

and starvation mortality, and an NPZ model to represent prey fields. Non-migratory copepods disappear from the nearshore ecosystem, as the surface dwellers are flushed offshore, and deep dwellers starve and die. Conversely, in simulations where individual copepods underwent DVM, they interact with the sheared flow and are retained

nearshore when the amplitude of DVM is great enough to put them into near-bottom onshore flow during the day (Box 5.1, Fig. 2).

Upwelling and downwelling structures alternate as winds shift and it is then reasonable to consider how such variability might influence larval dispersal, and how behaviour, such as DVM, might

continues

Box 5.1 *continued*

influence dispersal. Shanks and Brink (2005) explored this question in relation to the cross-shelf transport of bivalve larvae. Although cross-shelf currents were 10 to 100 times faster than swimming speeds, the larvae remained within 5 km of the shore. The bivalve larvae do not behave as passive particles. One possible mechanism that enables nearshore trapping of plankton is for individuals to swim against vertical currents. At nearshore convergence zones, larvae or zooplankton that swim upwards can become trapped and concentrated by the flow field. In upwelling conditions, larvae can move downwards and become entrained into the onshore deep flow, and concentrated near the shore. The details of how larvae make these behavioural choices, and how they sense their oceanographic surroundings, are not known. Some larvae do get advected to unfavorable habitats and are lost, yet enough 'choose' the right behaviours that keep them in the preferred habitat and enable long-term population persistence (Metaxas 2001).

Box 5.2 Coastal upwelling systems

The southern Benguela upwelling system, off the west coast of southern Africa, is one of the most dynamic and well studied coastal regions in the oceans. The upwelling regime extends from Cape Point on the southern end northwards to St. Helena Bay. Anchovy which spawn and migrate through this region have been intensively studied for decades. The Benguela upwelling is both highly productive and variable. For example, estimates of anchovy recruitment vary from 30 to 500 billion individuals (Lett *et al.* 2006). The transport of spawning products from Agulhas Bank to the spawning grounds to the north is influenced by the strength of the coastal current, the strength of the upwelling currents, and the biological process of spawning (Box 5.2, Fig. 1). Hydrodynamic models of this complex current system have been an important tool in exploring the synchronization between the physical processes of the currents and the biological processes of spawning and drift.

The processes of the Benguela system have stimulated ideas around biophysical dynamics in upwelling regimes. For example, Bakun's ocean triad (1996) spans a wide range of physical and biological processes. Many different enrichment processes are possible including upwelling, wind-mixing, and tidal rectification. Concentration can take place via convergence front formation or water column stabil-

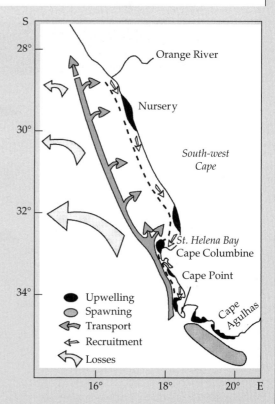

Box 5.2, Figure 1 Schematic of anchovy eggs and larvae dynamics in the southern Benguela. The eggs and larvae are transported from spawning areas in the western Agulhas Bank to nursery areas located on the west coast (Mullon *et al.* 2002).

continues

Box 5.2 *continued*

ity. Processes favouring retention can include many different aspects of the circulation that on the shelf are also typically connected to the bottom topography. Lett *et al.* (2006) applied a Lagrangian tracking model to test Bakun's triad model (Box 5.2 Fig. 2). They built maps of enrichment and retention for the southern Benguela upwelling ecosystem presenting the results in relation to what is known about the reproductive strategies of the two pelagic clupeoid species that are abundant there, anchovy (*Engraulis*

encrasicolus) and sardine (*Sardinops sagax*). The resulting maps show good agreement between areas of enrichment and retention and observed oceanographic and biological features (Box 5.2 Fig. 3). The results do not show strong overlap between enrichment and retention and because the particles in their model were purely passive they were unable to explore the influence of concentration. These results offer insight into the role of physical processes of transport in the Benguela upwelling system.

Box 5.2, Figure 2 Modelled particle tracks (white dots) in the Benguela current system transported over a 6-week period; the land mass is in red and the depth contours are the remaining shaded areas. (Courtesy of Christian Mullon). (See Plate 10).

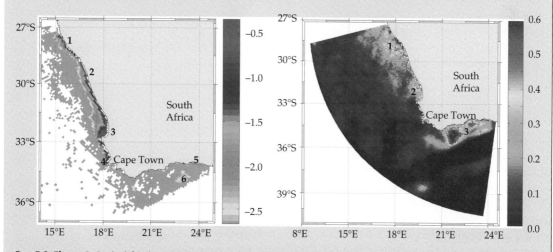

Box 5.2, Figure 3 On the *left* is the enrichment intensity obtained through simulated particle upwelling-regions 1–4 are the most enriched off the west coast. On the *right* is the map of the simulated pattern of retention. Values correspond to the proportion of particles retained averaged over the period 1992–9 and depth. Only the areas 1 and 2 of retention match the regions of enrichment of the left-hand panel. Although recruitment for both anchovy and sardine is considered to occur predominantly off the west coast; the high retention predicted for region 3 may not result in observed successful recruitment. (From Lett *et al.*, 2006). (See Plate 11).

continues

Box 5.2 *continued*

Mullon *et al.* (2002) used this modelling approach to explore the evolutionary implications of spawning in the Benguela upwelling system. They used the model simulations to index the environmental factors important for anchovy recruitment. They were able to establish a hierarchy of factors, temporal, spatial and environmental, that influence anchovy recruitment. While none of the results perfectly matches the observations of recruitment they do reveal different aspects of the influence of offshore advection and cold water.

Although these hypotheses strongly influenced the development of fisheries oceanography over many decades, they cannot in isolation provide a framework to extend our understanding of the problem of fisheries recruitment. The processes that influence ocean survival, of both fish and zooplankton, cut across life stages and operate over a wide range of spatial and temporal scales. We now know that factors influencing recruitment are different for different species and that they vary in time as well (Houde 2008). It is unlikely that a single process dominates, particularly for an extended period of time or over a range of locations. Growing understanding of the complexity of the physical environment, both in scale and connected extent, and the wide range of biophysical processes that influence growth and survival, has led to scepticism that any single process, representing a simplified aspect of a very complex system, is likely to provide, on its own, convincing explanatory power (Houde 2008). In his review of the critical research needs, Houde (2008) identified additional fundamental work required to advance our understanding including longer and better time series, improved coupled models, and greater consideration of all life stages.

Spatially explicit models presently being employed carry embedded within them many of the concepts discussed above. Indeed, these models offer the potential for simulations to test the characteristics of these different processes within a single system. The hypotheses discussed here are still used but more for purposes of discussion rather than explicit testing through any directed programme. Although biophysical models are greatly simplified relative to the marine ecosystems that they are meant to represent, they are still much more complex than the conceptual models discussed above. With the development of more sophisticated biophysical models we might expect more opportunities to test existing hypotheses and the development of new ones as the models are used as exploratory simulation tools (T.J. Miller 2007).

In wind-forced coastal upwelling regions, it is a particular challenge to determine the pathways of marine production from nutrients, through phytoplankton to zooplankton and top predators (Bakun 1990). There is evidence that there are inter-annual and decadal changes in wind-forcing that lead to changes in upwelling intensity. Investigations on the west coast of North America have linked reductions of upwelling intensity with declines in primary productivity, potentially related to lower biomass and fisheries yields for pelagic fisheries (Ware and Thomson 1991).

5.4 Modelling approaches used in GLOBEC studies

Many of the coupled numerical model/observational systems in use today (Proctor and Howarth 2008) and now central to the study of marine ecosystems were outlined in the early 1990s at the outset of the GLOBEC programme (GLOBEC 1993b). The vision developed at that time was to achieve an understanding of marine ecosystems that would enable predictive capability. The observational components would integrate *in situ* and remote physical, biological, and chemical measurements. The modelling components were envisioned as being multi-scale, with nested and fully coupled approaches linking coastal, shelf-break, and oceanic regions. Data assimilation methods and Observing System Simulation Experiments (OSSEs) were identified as central to the observing-modelling system.

Work during the past decade within GLOBEC and related programmes have built on scientific and technological breakthroughs that made possible research at the frontiers of interdisciplinary modelling. Enabled by increased computational speeds and capabilities, three important achievements have included: (1) the development of hydrodynamic models that captured motions from turbulence to mesoscales (allowing the consideration of physical processes affecting individuals and populations within a single formulation), (2) the coupling of hydrodynamic and biological models, and (3) the increased structure and complexity of biological models building on laboratory and field experiments (see Harris *et al.*, Chapter 6, this volume) providing information on target species, their behaviours, vital rates, and trophic interactions (see reviews in Werner *et al.* 2001a; Hofmann and Friedrichs 2002; T.J. Miller 2007). In the sections that follow, we highlight modelling advances that helped realize aspects of the coupled observing-modelling systems discussed in the GLOBEC (1993b) Numerical modeling Report of the first meeting of an international GLOBEC working group.

5.4.1 Circulation models

The advent and establishment of realistic coastal circulation models (cf. Haidvogel and Beckmann 1999), including unstructured and nested grids (Greenberg *et al.* 2007), non-hydrostatic models (Fringer *et al.* 2006), and large-eddy simulations (Hughes *et al.* 2000) have enabled the quantitative study of key physical processes in varying degrees of approximation. Similar to developments in the basin-scale modelling, public domain 'community' shelf-circulation models, such as the Regional Ocean Modelling System (ROMS; Haidvogel *et al.* 2008) and the Princeton Ocean Model (POM; Ezer *et al.* 2002), are now more available with supportive user communities and substantial Web sharing of model resources making it easier to obtain and configure shelf-circulation models for specific sites.

An important development in circulation models has been the implementation of generalized vertical coordinate systems, which allows more effective transition across the deep/coastal ocean boundary (e.g. Chassignet *et al.* 2006). Along with this has been the development of robust procedures for one-way nesting of models with differing spatial resolution that allows coupling of local circulation to larger scale (e.g. regional) circulation processes. In addition, there have been many other developments including well-developed sub-models for the evolution of coupled biological and geochemical tracers (e.g. Rothstein *et al.* 2006a); robust procedures for one-way nesting of models with differing spatial windows and resolution (Curchitser *et al.* 2005; see Figs. 5.5 and 5.6); efficient algorithms or multivariate data assimilation of physical variables resulting in useful nowcast/forecast systems in limited area domains (e.g. Robinson and Lermusiaux 2002); multiple options for representing mixing and turbulence making it possible to examine the sensitivity of different solutions to these formulations when implementing any particular circulation model (Umlauf and Burchard 2005); and preoperational prediction systems for global, regional, and local areas (Proctor and Howarth 2003). In coming years, with the establishment of sustained observing systems, further progress will include the refinement of operational forecasts and analysis of systems at basin scales, more robust techniques for downscaling climate scenario modelling to regional and local scales (Hashioka and Yamanaka 2007), and the emergence of powerful alternatives for multi-scale ocean modelling based upon the availability of novel approaches and techniques for interdisciplinary modelling and data assimilation (e.g. Gregg *et al.* 2008).

5.4.2 Linking 'simple' individual-based models and hydrodynamics

Dispersal studies of organisms in the ocean began over a century ago (Hjort 1914) with the goal of determining how dispersal during early life history stages influenced recruitment. Improved

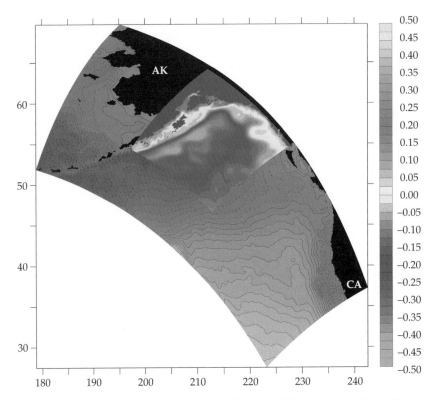

Figure 5.5 The grid for the ROMS model structure for the northeast Pacific showing the high resolution sub-grid set inside a coarser resolution large-scale domain. (Courtesy of A. Herman). The colour scale shows the anomaly of sea surface height in metres. (See Plate 9).

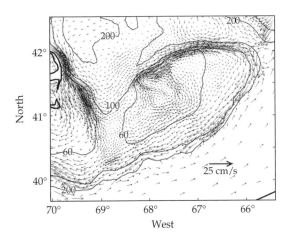

Figure 5.6 Surface circulation for Georges Bank. The circulation is strongly influenced by the bottom topography leading to retention of larvae and plankton on the Bank. (Werner *et al.* 1993.)

measurements and numerical modelling capabilities have made it possible to track the drift of particles in a realistic manner (North *et al.* 2009).

Nonetheless, the accurate determination of when a group of particles (organisms) arrives at a particular location or where those particles originated from remain a modelling challenge that is further compounded when organism behaviour is included (Batchelder 2006; Christensen *et al.* 2007). Modelling an organism's trajectory through the ecosystem is essential to understanding the spatially dependent biological and physical factors that influence growth and survival (Fach *et al.* 2006). For systems dominated by strong tidal flows or well-defined geostrophic flows, model solutions provide good approximations of these trajectories. For systems with weak and variable currents, neither models nor observations may be able to clearly separate advection from variability that could lead to dispersal. While we understand many of the physical processes that lead to this particle dispersion, it remains a challenge to properly represent it in models.

In the early 1990s, advances in hydrodynamic models and the ability to quantify the Lagrangian properties of the circulation though the use of IBMs, allowed GLOBEC and related research programmes to answer ecological and site-specific questions about retention-dispersal (Levin 2006; T.J. Miller 2007). The simplest studies ignored biotic factors such as feeding and predation; but included imposed traits such as swimming behaviours. Topics that were successfully investigated were the space-time pathways of larval fish from spawning grounds to nursery areas (Bartsch *et al.* 1989; Hermann *et al.* 1996b; Vikebø, *et al.* 2007a), the retention and transport of fish and invertebrate larvae on submarine banks (Werner *et al.* 1993; Tremblay *et al.* 1994; Fogarty and Botsford 2007), the identification of fish-spawning locations (Heath and Gallego 1998; Stegmann *et al.* 1999), the retention of copepods on shelf regions (Batchelder *et al.* 2002b; also see Box 5.2), and the exchanges and connectivity of copepod populations between deep ocean basins and shelf regions (Hannah *et al.* 1998; Miller *et al.* 1998; Speirs *et al.* 2005).

5.4.3 Modelling swimming and behaviour

Many marine organisms, including many plank-tonic forms, are not entirely passive and exhibit various behaviours that include active swimming, directed vertical migration, and vertical migration via changes in buoyancy. Behaviour can arise in response to environmental cues, or it can be ontogenetic (i.e. as a function of larval develop-ment), with eventual schooling behaviours emerg-ing as the organisms reach juvenile or adult stages. Many studies have established the importance of vertical behaviour in population dispersal, reten-tion, settlement, and connectivity. Larvae located at different depths will be subjected to different currents and thus their Lagrangian trajectories will be different (e.g. Batchelder *et al.* 2002b) and their trajectories are also influenced by their pelagic-phase duration (Tremblay *et al.* 1994). Similarly, horizontal swimming rates have been shown to significantly affect distributions of individual organisms (e.g. Werner *et al.* 1993). Behaviours can be complex and their explicit consideration essential in many cases. The question from the modelling standpoint is not whether behaviours

should be included, but rather how they should be included (North *et al.* 2009).

The ability of plankton and larval fish to swim in the horizontal is negligible relative to the strength of the horizontal currents, although in certain cases they may decrease dispersal through directed swim-ming (Leis 2006). In contrast, their swimming abili-ties in the vertical are sufficient to enable control of their depth in the water column, and affect their dispersal as well as their growth and survival. Selection of vertical habitat for larvae drifting in currents was found by Fiksen *et al.* (2007) to influ-ence the local short-term trade-off between growth and mortality and the more strategic and long-term consequences related to the large-scale circulation regime. If ending up in particular habitats depends on the swimming ability of settling juveniles, it can be expected that organisms prioritize growth and survival along their drift trajectory.

Behaviour can be implemented in IBMs by sim-ple rules, for example, seeking to maximize growth or to minimize mortality risk. However, these sim-plifications are probably not realistic as other trade-offs need to be considered that may vary with stage, age, feeding history, and over generations when longer time scales are considered (Carlotti and Wolf 1998). Additionally, variability in the environment can be large enough that it may override these life history-based assumptions. Alternatives to impos-ing 'simple' and likely unrealistic behaviours are to consider modifications of behaviour through adap-tation (e.g. Giske *et al.* 2003). Dynamic program-ming methods allow organisms to 'find' optimal habitats by balancing risks of predation, growth, and advective loss. Examples include the adapted random walk (Huse 2001) and the individual-based neural network genetic algorithm (ING; Giske *et al.* 1998) that allow adaptive behaviours to emerge in populations in complex environments. The ING method provides a way to implement behaviour in individual-based models. Strand *et al.* (2002) pro-vide an example whereby an IBM that uses artificial evolution is used to predict behaviour and life his-tory traits on the basis of environmental data and organism physiology. In this approach, evolution-ary adaptation is based on a genetic algorithm that searches for improved solutions to the traits of habitat choice, energy allocation, and spawning strategy. Behaviours emerging from model studies can complement results from field and laboratory efforts and allow for predictions to be attempted.

Swimming is a crucial behaviour of organisms since it provides the ability to select position, either through vertical and horizontal movements. Such movement allows shifting from unfavourable to more favourable conditions, presumably leading to an improvement in survival or growth, or to increased local retention (see Box 5.2). Although we have only limited knowledge of why an organism chooses to move, or the cues it uses, we do have much greater biological understanding of the factors influencing swimming. Swimming is often a means to find prey and while increased swimming can increase the potential for encounters with prey it does also increase the risk of an encounter with a predator (Visser 2007b). Thus in considering behaviour, such as swimming, one must consider the overall impact on fitness.

Swimming is included in individual and population models through small-scale interactions between predator and prey (Visser 2007b) or through the escape of predators (Kiørboe and Saiz 1995). The simplest parameterization for swimming behaviour assumes a linear response (Gerritsen and Strickler 1977) or random-walk swimming (Yamazaki et al. 1991). IBM models offer a large range of sophisticated representation of swimming behaviour related to internal and external factors (Carlotti and Wolf 1998) and allow trade-offs between specific types of rules (e.g. temperature response) and life history theory (e.g. ontogenetic development), as suggested by Tyler and Rose (1994). The distribution patterns that result from these models of swimming are too crude because swimming rates are related not only to the weight or size of organisms, but to temperature, salinity, or light as well.

5.4.4 Individual-based models

Size/age-structured models and IBMs offer advantages in their inherent flexibility to include ontogenetic detail of the target organism, as well as mechanistic descriptions of feeding and swimming behaviours; see reviews by Carlotti et al. (2000) and Werner et al. (2001a, b). Additionally, developments

in spatially explicit IBMs allow inclusion of detailed parameterizations of biological processes required for quantitative estimates of dispersal/retention of zooplankton and early life stages of fish (e.g. Batchelder et al. 2002b; T.J. Miller 2007; North et al. 2009). A strength of IBMs is that properties of ecological systems can be derived by considering the collective properties of individuals that constitute them, that is, IBMs are able to account for inter-individual variability and rare individuals, or rare circumstances affecting a few individuals, which contribute strongly to determining population strength, or variance (e.g. of growth rates) within populations (see Boxes 5.1 and 5.2). The differences among individuals result in unique life histories, which when considered as a whole give rise to growth and size distributions that provide a measure of the state of the population. IBM and size/age-structured formulations require substantial information about organisms throughout their life history and are perhaps best suited for species-specific studies where the biological traits of the target species are well known. Moreover, density-dependent processes (e.g. cannibalism) can be difficult to include in IBMs because of the difficulty of simulating realistic densities (Batchelder et al. 2002). If population, or ecosystem-level statements are sought using IBMs (Spiers et al. 2005; Thorpe et al. 2007), the large numbers of individuals required by the simulations can render the ensuing long-term simulations computationally costly. The use of 'super-individuals' (Scheffer et al. 1995; Carlotti and Wolf 1998) or probability distributions based on variability of physiological processes (e.g. Bochenek et al. 2001) provide approaches for extending results from IBMs to cohort and population scales.

Consideration of longer time scales and multiple cohorts also raises the need to quantify the ability of modelled organisms to be adaptive in the sense that individual traits will be subject to evolution by natural selection. This would enable models to include an important force in relation to ecosystem functioning and biodiversity, namely the continuous adaptation of gene pools to changing biotic and abiotic forces. Community and ecosystem dynamics thus become an emergent property of the models solely dependent on the description of individual traits and the physical environment. The trait-based approach is therefore not limited by the need to pre-define functional groups as is the case in the majority of contemporary ecosystem models.

Present practices of predefined model structures need to evolve into a framework where food webs, diversity, and biogeochemical cycles emerge from basic principles at the level of individuals (e.g. Loreau *et al.* 2001; Bruggeman and Kooijman 2007). Ecological mechanics (e.g. encounter, mortality, and behaviour) must be coupled with evolutionary models of how individuals, functional types, or even communities are shaped by natural selection (Giske *et al.* 1998) so that properties will emerge from simulations that will allow investigation and prediction of how environmental change will alter ecosystem functioning. The present limitations of rigid ecosystem models are unrealistic, with proposed alternative approaches including trait-based models or forms of adaptive systems theory (e.g. Norberg 2004; Bruggeman and Kooijman 2007).

At present, the biological research community is just beginning to utilize this mode of modelling, with adaptive IBMs being one example. So far this approach has been used to study animal distributions (Giske *et al.* 2001), behavioural decisions (Strand *et al.* 2002; Giske *et al.* 2003), and life history strategies (Fiksen 2000). There has been less focus on speciation, emergence of morphologies and life history strategies, and population dynamics in systems with interacting species. In a specific example of fish populations, Fiksen *et al.* (2007) suggest a procedure that frames larval behaviour and fish-spawning strategies in the tradition of evolutionary ecology, both conceptually and formally.

Recommended IBM practices (Grimm and Railsback 2005) should emphasize: emergent properties from basic, transparent, and mechanistic assumptions on growth, mortality, behavioural abilities, and drift processes; behavioural strategies or rules that show variability between individuals; and selection processes that incorporate fitness consequences along the trajectory and at the settlement location. Models that simulate evolution based on genetic algorithms represent one efficient tool for such investigations. Such models can tackle sufficient complexity and are capable of evaluating consequences of larval behaviour in flow fields and integrating effects across time scales. Recommended

modelling practices are those that use transparent and mechanistic processes in growth, mortality, behavioural abilities, and drift; apply behavioural strategies or rules that allow true variability between individuals; and include an evolutionary selection procedure to assess fitness consequences along drift trajectories and at the settlement location. This procedure enables studies on how organisms can adapt to environmental change through natural selection.

5.4.5 Ecosystem models

Few models explicitly consider the trophic links from nutrients and primary producers, through the food web to fish, birds, and mammals. The need to consider multiple interactions across trophic levels that include vastly different reproduction rates and lifespans (see Harris *et al.*, Chapter 6, this volume) and the different time scales of important processes at different trophic levels makes modelling the entire marine ecosystem a challenge. The limited observations of the interactions between species, and the complexity of life stages and life histories, abundances, behaviours, and spatial distributions result in multiple approaches to modelling marine ecosystems and make development of generic models unlikely. A consequence is that numerical representations across trophic levels differ substantially, and include mass balance or budget models, size- or age-structured models, and IBMs, making the links between different trophic levels or different groups of organisms difficult and uncertain (e.g. see Fig. 5.7).

Mass-balance models consider the dynamics of material or energy flow (see Box 5.3). To render the problem tractable, species are usually combined into groups that are assumed to function similarly, with some loss of detail including species richness, and simplified biological traits such as age- or size-structure, and behaviour (Anderson 2005). Examples of lower trophic level mass-balance models focusing on the nutrient-phytoplankton-zooplankton-detritus components (NPZD) models are those of Fasham (1993) and Le Quéré *et al.* (2005). The NPZD models have been coupled to ocean circulation models and used to study nutrient cycles, characteristics of phytoplankton blooms, biogeochemistry, and the carbon budget in oceanic and coastal regions (e.g. Doney

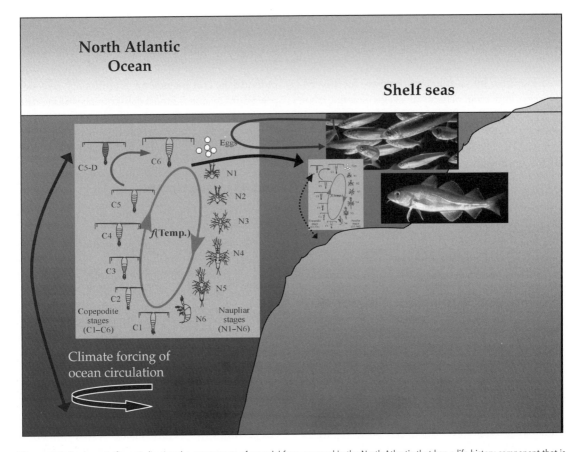

Figure 5.7 A schematic figure indicating the components of a model for a copepod in the North Atlantic that has a life history component that is present both in the open ocean and on the shelf. The copepod is influenced by the large-scale ocean circulation, the circulation on the shelf, and organisms on the shelf and in the open ocean that are predators and prey. The model structure needs to account for these different aspects of the organism's life history.

Box 5.3 NEMURO fish modelling

With the overall goal of describing key elements of the North Pacific ecosystem, a biomass-based model was developed by the PICES Model Task Team as an initial step in identifying and quantifying the relationship between climate change and ecosystem dynamics (Batchelder and Kashiwai 2007). The approach built on a common multi-compartment model that describes the energy and mass flux from inorganic nutrients through higher trophic levels and recycling of those nutrients through multiple trophic pathways. Building on previous NPZD formulations (Fasham 1993), a model was constructed for nitrogen-limited ocean marine ecosystems, for example, the North Pacific, with state variables representing functional groups characterizing North Pacific phytoplankton and zooplankton species, but at the same time, attempting to keep the model formulation ecologically as 'simple' as possible (Box 5.3, Fig. 1). The resulting lower trophic level marine ecosystem model was named NEMURO (North

continues

Box 5.3 *continued*

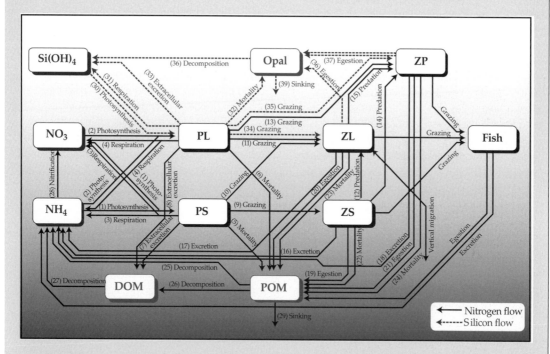

Box 5.3, Figure 1 The NEMURO model simulates the predator-prey interactions and biogeochemical cycling of phytoplankton, zooplankton, nutrients, and detritus (Kishi *et al.* 2007b). The state variables are nitrate (NO_3), ammonium (NH_4), small phytoplankton biomass (PS), large phytoplankton biomass (PL), small zooplankton biomass (ZS), large zooplankton biomass (ZL), predatory zooplankton biomass (ZP), particulate organic nitrogen (PON), dissolved organic nitrogen (DON), particulate silica (Opal), and silicic acid concentration ($Si(OH)_4$). The fish model simulates the daily growth and mortality of the target fish, and is coupled to NEMURO via fish consumption dependent on zooplankton, and fish excretion and egestion contributing to the nitrogen cycle. The coupled models were called NEMURO. FISH (NEMURO For Including Saury and Herring) (In Megrey *et al.* 2007b with kind permission from Elsevier). The NEMURO and fish bioenergetics models can be solved simultaneously (coupled) or separately (uncoupled), allowing for investigation of the feedbacks between herring dynamics and their prey.

Pacific Ecosystem Model Understanding Regional Oceanography; see Kishi *et al.* 2007a,b). Extensions of NEMURO including fish as higher trophic level (HTL) state variables, were achieved with the development of NEMURO.FISH (NEMURO For Including Saury and Herring; Megrey *et al.* 2007a,b). Pacific herring (*Clupea harengus pallasi*) and saury (*Cololabis saira*) were the two target fish considered.

   A special issue in Ecological Modelling (Kishi *et al.* 2007b) details the approach taken to produce integrated and quantitative statements about biophysical components of the North Pacific marine ecosystem. Included are case studies and in-depth modelling studies focused on the North

Pacific that address oceanic biogeochemistry, regional and seasonal variability of phytoplankton and zooplankton dynamics, 40–50 year retrospective time histories of phytoplankton and zooplankton, model formulations that include links to higher trophic levels, as well as studies that take initial steps towards estimating future states of marine ecosystems subjected to global warming.

   The approach of using a single, common model as an initial approximation of the North Pacific ecosystem showed strengths and weaknesses. A benefit of a single NEMURO and NEMURO.FISH formulation was that results across studies were interpreted without the effects introduced by

continues

Box 5.3 *continued*

different choices of parameter values or model formulations. Such differences in formulations or parameter sets can arise when models are used in specific geographic locations, or to study particular events, without consideration of the formulations' or parameters' validity over larger domains or longer time scales. The development of the saury and herring components of NEMURO.FISH was also done with a common framework so that interspecific comparisons were possible. An example is the study of Megrey *et al.* (2007a) in which a single

formulation approach allowed for comparisons across two fish species and two oceanic regions in an attempt to determine if the response to a common environmental/climate forcing would manifest itself equivalently in both fish species. Using a common formulation to establish baseline studies in models like NEMURO and NEMURO.FISH was necessary to be able to determine variability arising from spatial or temporal variations, climate forcing, or environmental variability independent of the model's structure.

Box 5.4 Connectivity in the Southern Ocean

The Antarctic Circumpolar Current (ACC) is one of the most dramatic ocean current systems on the planet. This current carries more water than any other current system in the world's oceans, over 130 Sverdrups (1 Sverdrup = 10^6 m^3 s^{-1}), more than 100 times the transport of all the rivers in the world combined. The circumpolar nature of the ACC (Box 5.4, Fig. 1) provides the potential for connectivity between all regions of the Southern Ocean. Moreover, the water masses that are formed in the Southern Ocean and are transported by the ACC are integral components of the global ocean thermohaline circulation, which result in the Southern Ocean having strong connectivity to this larger circulation. The production and melting of sea ice in the Southern Ocean provides one of the largest seasonal signals experienced by any marine ecosystem. Sea ice is important in regulating air-sea exchanges, upper ocean structure, mixing and light levels, and habitat structure. Variability in the sea ice environment directly influences physiological process and recruitment success of many key Southern Ocean species and also regulates the very short period, 2 to 3 months, when primary production occurs. These unique aspects of the Southern Ocean have

large effects on the structure and function of its marine ecosystems.

Antarctic krill (*Euphausia superba*) is a key component of the Southern Ocean food web and provides a primary prey item for many higher trophic level predators including such predators as penguins, seals, whales, and seabirds. Murphy *et al.* (2007) estimate that Antarctic krill represent about 70% of the prey consumed by top predators in the Scotia Sea. It is because of its important role in trophic connectivity that Antarctic krill was selected as a target species (Gifford *et al.*, Chapter 4, this volume) for GLOBEC studies. While krill are clearly key, other grazers such as salps dominate in certain areas of the Southern Ocean (Pakhomov *et al.* 2002) and at times can replace Antarctic krill (Loeb *et al.* 1997). Other krill species (e.g. *E. crystalorophias)* dominate in areas of permanent pack ice (Brierley and Thomas 2002) and in some areas and seasons copepods provide the major link between primary producers and higher predators (Atkinson *et al.* 2001). Thus, an emerging view of Southern Ocean food webs is one of heterogeneity and variability. Connectivity of the Southern Ocean is key to understanding the heterogeneous nature of food-web structure in

continues

Box 5.4 *continued*

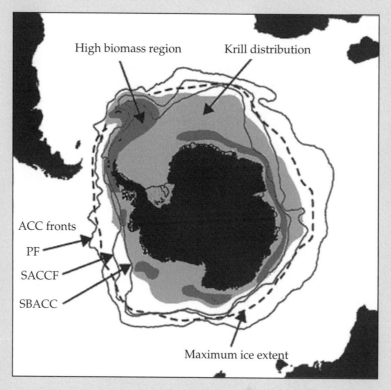

Box 5.4, Figure 1 The circumpolar distribution of krill from many years of field sampling (Hofmann and Murphy 2004). Also shown are the mean positions of the fronts associated with the Antarctic Circumpolar Current (ACC). They are the Polar Front (PF), the Southern Antarctic Circumpolar Current Front (SACCF), and the Southern Boundary Antarctic Circumpolar Current (SBACC).

the region. For example, Antarctic krill are found throughout the Southern Ocean (Marr 1962), but their distribution is non-uniform, with higher densities found along the western Antarctic Peninsula and in the Scotia Sea regions (Atkinson *et al.* 2008). The maintenance of this non-uniform distribution in a system that is connected by a circumpolar current and the implications of this for Southern Ocean ecosystem structure and function are central to understanding how this system will respond to climate change.

The dependence of the South Georgia ecosystem in the Scotia Sea on upstream sources of Antarctic krill (Hofmann and Murphy 2004) provides a clear example of the importance of

connectivity via the circulation in the Southern Ocean (Box 5.4, Figs. 2 and 3). Observations of krill biomass and currents near South Georgia in the Scotia Sea showed temporal and spatial variability in abundance and biomass flux (Murphy *et al.* 2004a) that was related to variability of the Southern ACC Front. These studies showed that horizontal flux of krill via the ACC is a major contributor to krill abundance at South Georgia and hence to maintenance of top predator populations. The source of this krill biomass has been inferred from Lagrangian modelling studies that show that krill transported to South Georgia likely originate in sea ice covered regions of the southern Scotia Sea a few

continues

Box 5.4 *continued*

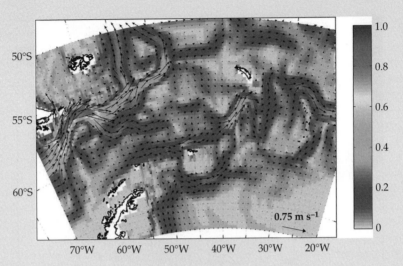

Box 5.4, Figure 2 Six-month average of the depth weighted mean velocity fields for the upper 182 m of the circulation model OCCAM in the Scotia Sea for October 1999–March 2000. The colours show the magnitude, arrows (not shown at every point for clarity) have been capped at 0.75 ms⁻¹. (Reprinted from Murphy *et al.* 2004 with kind permission from Elsevier). (See Plate 12).

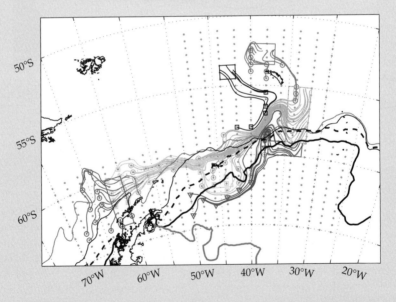

Box 5.4, Figure 3 Lagrangian tracks of particles related at the beginning of October 1999 that passed through the high krill biomass areas during January 2000. Based on releases in a 0.5° latitude by 2° longitude grid on the depth-weighted mean velocity field over the upper 182 m. Different colours are related to different regions of high biomass and the associated particles tracks. The September to January ice-edge is shown-September 1999 with a thin grey line; October 1999 with a solid black line; November 1999 with a broken black line; December 1999 with a thick black line; and January 2000 with a thick grey line. (Reprinted from Murphy *et al.* 2004 with kind permission from Elsevier). (See Plate 13).

continues

Box 5.4 *continued*

months prior and are then transported north across the Scotia Sea to South Georgia during the summer (Murphy *et al.* 2004a). The existence of many potential krill source regions along the western Antarctic Peninsula and in the Weddell Sea (Fach *et al.* 2006) suggests that connectivity pathways between South Georgia and the larger Southern Ocean are complex and variable. Circum-Antarctic Lagrangian modelling studies (Thorpe *et al.* 2007) further strengthened the ideas of connectivity of krill populations at larger scales and over longer times.

Sea ice is important to survival and subsequent recruitment of krill larvae and modelling studies have shown that sea ice is critical for successful transport of younger krill stages (Fach *et al.* 2006). The modelling work of Murphy and Reid

(2001) showed enhanced inputs of krill to South Georgia across the Scotia Sea during cold periods. These inputs do not take place every year but seem to sustain the local system until the next event. Lagrangian modelling studies using results from a global ocean circulation model suggested that the krill flux transport and pathways respond to larger-scale climate signals (Thorpe *et al.* 2004). The current warming of the western Antarctic Peninsula and Scotia Sea region may reduce the habitat suitable for Antarctic krill and favour other macro- and mesozooplankton species (Whitehouse *et al.* 2008). Thus, the continued existence of the present South Georgia marine ecosystem may depend on shifting source regions and different transport pathways.

et al. 2004). In coastal oceans and shelf seas, the NPZD pelagic model has been coupled with benthic components and related to terrestrial/riverine inputs (Fennel *et al.* 2006; Radach and Moll 2006).

Coupled hybrid modelling methods that represent higher trophic levels (e.g. fish) using Lagrangian-IBMs, but represent lower trophic levels (e.g. prey) using Eulerian-based NPZD models are becoming more commonly used. Hybrid approaches have largely focused on coupling of lower trophic (NPZD) levels to larval fish (e.g. Hermann *et al.* 2001) with realistic, spatially explicit models that include the full life cycle of higher trophic levels, such as fish populations only recently being implemented (e.g. Megrey *et al.* 2007a,b; Lehodey *et al.* 2006). The hybrid Lagrangian-Eulerian copepod model of Batchelder *et al.* (2002b) was unusual in including all vital rates (behaviour, growth, reproduction, and mortality) so that the life cycle of the Lagrangian element (the copepod) was closed and multiple generations could be tracked through space and time. Bridging the gaps between the different numerical representations of different trophic levels remains as one of the challenges of hybrid methods.

Other important ecosystem modelling approaches are size-spectral (Zhou and Huntley 1997; Maury *et al.* 2007) and food web models (Plagányi 2007; also see Moloney *et al.* Chapter 7 for additional discussion). To a certain degree these methods sacrifice mechanistic and biological detail, but they provide important diagnostics based on mass and energy balance considerations and include a much broader range of trophic levels. Size-spectral models assume that fundamental size-based processes that determine the use and transfer of energy in communities respond to changes in temperature and primary production in consistent and predictable ways, based on empirical observation. Given temperatures and rates of primary production, size-based methods allow biomass and production of consumer communities to be calculated. There is a need for using groups of different sensors able to measure density of organisms from unicellular organisms to large vertebrates, over a large size spectrum in connection with model building and validation. Trophic levels have been for a long time observed independently and new instrumentation allows simultaneous observation of phytoplankton, zooplankton (crustaceans as well as jellyfish), and fish. Increasing observations of size-structure based on new tools such as optical plankton counter, acoustic and, video systems including image analysis (Culverhouse *et al.* 2006) are consistent with the development of zooplankton size-structured models (see Moloney *et al.* Chapter 7 for additional discussion).

5.4.6 Modelling population connectivity

In studies of marine ecosystems there are trade-offs between the desire to represent ecological complexity and the limits of data and understanding which forces simplification. The spatial representation of higher trophic levels presents additional challenges as active behaviour, for example, directed swimming, large-scale migrations, schooling and other aggregations, become important and thus the populations' distributions depend to a lesser degree on hydrodynamics alone. Nevertheless, an important achievement has been the development and use of models that include more of the necessary components of the marine ecosystem to address the questions being raised. As a community, and as a result of the GLOBEC approach, we are now asking questions about the large-scale connectivity of marine populations (e.g. Wiebe *et al.* (2007); also see Boxes 5.3 and 5.4) and the influence of lower trophic productivity on predators, and vice versa, requiring more sophisticated and complete representations of marine ecosystems.

Population connectivity refers to the exchange of individuals among geographically separated subpopulations that comprise a metapopulation (Cowen *et al.* 2007). Connectivity studies seek to identify spatial patterns related to population linkages. As such, models must be spatially explicit and resolve the scales of source and sink populations. The use of IBMs and improved physical circulation models described in previous sections has been central to successful model investigations of spatial connectivity, retention, and dispersal (Box 5.3). Pathways of larval connectivity that are conducive to self-recruitment are essential for population persistence (Gaines *et al.* 2007). Life history traits may have evolved to exploit hydrographic regimes that improve the odds of larvae returning close to the parent population (Strathmann *et al.* 2002). Physical

processes can also influence biological processes such as the timing of reproduction, larval transport and behaviour, and the timing of settlement (Cowen *et al.* 2000). Induced behaviour (e.g. diel vertical migration (DVM), foraging, and predator avoidance) and other individual properties (i.e. larval stage duration, age/stage-dependent vertical migration) have been shown to influence population connectivity as much as currents do and to reduce dispersal (Tremblay *et al.* 1994).

It is interesting to consider how adults (parents) make trade-offs between offspring numbers and different aspects of offspring quality. For instance, for cod spawning in the southern regions of the Norwegian coast for their eggs and larvae to mature in warmer water, the migration distances to nursery grounds are large and the costs of migration are high both in time and energy. The actual spawning distance can therefore be seen as a life history strategy of energy allocation between growth, egg production and migration costs, driven by oceanographic (or ecological) benefits to offspring. These issues can now be tackled by exploration with an optimality model, using drift trajectories of larvae to determine the benefits in terms of temperature exposure (Jørgensen *et al.* 2008; Opdal *et al.* 2008).

From the large number of physical and biological processes that relate to the dispersal and recruitment of marine organisms, general points can be extracted which help define the connectivity problem. First, the temporal and spatial correlation scales over continental shelves may be relatively short (on the order of days and kilometres). Second, the relative contributions of key processes will likely be site-specific and depend on coastal geometry, proximity to estuaries or deep ocean boundary currents, seasonal stratification, and wind-forcing. Third, individual physical processes contain variable length and time scales. Thus, hydrodynamic transport and dispersion modulated further by biological properties, is fundamentally a multi-scale process. Highly resolved flow fields are needed in order to embed behavioural models within hydrodynamic models to examine processes involving biophysical interactions (Werner *et al.* 2007b).

There are many pelagic organisms found both in the deep ocean and on the continental shelves.

Some of these species are also found in more than one ocean basin or are found in multiple gyre systems within a particular ocean basin (Boxes 5.4 and 5.5). Understanding how circulation connects these gyres and connects the deep basins with adjacent shelves remains one of the central challenges for oceanographers (Wiebe *et al.* 2009). In the context of the ecosystem, we are interested in whether sub-populations of species exist, or whether there is substantial genetic exchange among distant oceanic or oceanic and shelf populations.

5.5 Data assimilation, integration with field observation, and skill assessment

We have made progress in developing complex numerical ecosystem structures, and are now working to address questions that have been raised by the new modelling and observations. Work is now directed to connect the various components of the models, for example, properly representing trophic interactions such as predator-prey interactions. Modelling systems now being considered will include more trophic levels, from the bottom to the top, so-called end-to-end models (see Moloney *et al.* Chapter 7, this volume):

- Multi-scale circulation models of the ocean and atmosphere;
- Coupled sub-models (e.g. ecosystem dynamics), of differing structures and spanning different trophic levels that build on process-based studies that provide the functional forms of rate processes and the parameter coefficients needed to represent transitions (e.g. energy/mass flow) among ecological state variables, and allow for adaptive traits;
- Observational networks to provide data for initial and boundary conditions, forcing functions, etc.; and
- Advanced methods of data assimilation to optimally merge the models and data.

The last component is important because of inherent limitations in model accuracy, and limits on how much data can be routinely collected in the ocean. Data assimilation is needed to provide the best-combined estimate of the oceanic state taking into account uncertainties in both the models and the data.

5.5.1 Data assimilation in physical-biological models

Coupled physical-biological models offer a framework for examining the processes (environmental forcing, internal individual or population processes, etc.) that determine observed structure in marine population distributions. However, the success or skill of these coupled models depends on the ability to construct a valid simulation of the natural system. The 'forward problem' initializes a coupled model with observations, steps forward in time, and then compares the predicted state with the next set of observations. Successful model solutions, with small (often subjective) discrepancies between observations and predictions, can be used as a basis for diagnosis of the processes controlling the observed patterns. Unfortunately, the forward approach does not always work well, owing to limitations in the models, in the observations, or in both. Inverse methods provide an alternative approach that is particularly useful for determining the model inputs (e.g. parameters and forcing functions) that minimize the misfit between observations and predictions, thereby producing an optimal solution from which the underlying dynamics can be examined. At the heart of this problem lies the topic of data assimilation, which is the systematic use of data to constrain a mathematical model (Hofmann and Friedrichs 2002; US GLOBEC 2007).

In the 1970s, ocean general circulation models (OGCMs) were implemented and enhanced our understanding of ocean circulation processes. Initial applications of these models focused on simulation of the large-scale structure of ocean currents with clear successes, but also with limitations. Data assimilation offered an approach to improve OGCMs (e.g. Wunsch 1996) and with recent increases in near real-time physical data from both satellites and *in situ* observations (e.g. ARGO; Freeland and Cummins 2005), it is now possible to use data assimilative OGCMs for global ocean state estimation (Stammer *et al.* 2002). Advances in coastal ocean models and observational infrastructure have led to realistic data assimilative models in the coastal ocean (Robinson and Lermusiaux 2002).

Implementing data assimilation in coupled physical-biological models has been challenging because of the lack of adequate data (Hofmann and Friedrichs 2002). Historically, biological and chemical data were obtained almost exclusively by ship surveys, and thus were limited in both space and time. However, advances in satellite and mooring instrumentation, made it feasible to begin the development of data assimilative coupled physical-biological models. As a result, the last 2 decades have seen dramatic increases in the quantity and types of data that are used in such models, and the development of robust and varied approaches for assimilating these data (e.g. Gregg *et al.* 2008).

Initial results have been encouraging as data assimilation approaches have shown promise for improving model capabilities, for example, reducing model-data misfit by recovering optimal parameter sets using multiple types of data (Friedrichs *et al.* 2007). Data assimilation analyses can demonstrate whether a given model structure is consistent with a specific set of observations, and if not, it can isolate model assumptions that have been violated. Thus, although the assimilation of data into a model cannot generally overcome inappropriate model dynamics and structure, it can serve to guide model reformulation and identify previously overlooked processes.

5.5.2 Models and experimental field design

As physical and biological models progressively matured, they have served as platforms for hypothesis testing and guiding sampling protocols and observatory/observing system design. For example, numerical simulations based on high frequency, quasi-synoptic *in situ* measurements of physics and larval distributions have proven effective in testing larval transport hypotheses (Helbig and Pepin 1998a,b). Implementation of real-time, *in situ* observing systems will be necessary to provide the data to define the oceanographic environment. However, the proper deployment and design of such observing systems and observatories is not trivial as attested by the range of physical and biological processes present in natural systems. In turn, the effectiveness of the sampling strategies is determined by the accuracy with which the observations can be used to reconstruct the state of the natural system being studied. Given the

limited opportunities for evaluation of sampling strategies against objective criteria with purely observational means, numerical models offer a framework for investigation of these issues (Walstad and McGillicuddy 2000), and Observational System Simulation Experiments (OSSEs) will be valuable for designing future field experiments and long-term observation programmes for physics and biology.

5.5.3 OSSEs

The aim of OSSEs is to model observation systems to quantify their sampling properties and optimize their design, that is, to simulate unknown ocean properties to better measure and discover them. OSSEs can be utilized for multiple purposes including: (1) to guide the design of an observation system and its components, (2) to optimize the use of observational resources, (3) to assess the impact of existing or future data streams (e.g. for nowcasting and forecasting of requisite accuracies), (4) to understand the interactions of system components and improve system performance, (5) to evaluate and validate system performance using quantitative error estimates, and (6) to compare data assimilation methods (GLOBEC 1994).

One example is that of McGillicuddy *et al.* (2001) who used OSSEs to assess quantitatively the synopticity of surveys of Georges Bank. Model simulations were used with realistic spatio-temporal fluctuations of adult females of the planktonic copepod *C. finmarchicus*. Simulations were constructed with two types of assimilation procedures used to dynamically interpolate between surveys. Assuming the simulations were representations of the real ocean, the model fields were subsampled in space and time along a typical cruise track. Maps constructed from objective analyses of the simulated data were compared with 'reality' as represented in the original simulations. Results indicated a total error of ~50%, comprised mostly of incomplete spatial sampling. It was found that adjustment of the station positions for displacement by the mean flow reduced the error by nearly half. The present advanced stage of modelling and data assimilation systems suggests that the continued development and implementation of OSSEs to design observing systems for use in quantitative physical-biological studies will become more routine.

5.5.4 Skill assessment of physical-biological models

In the past decade, large interdisciplinary oceanographic programmes (e.g. JGOFS, GLOBEC, and now IMBER) have included model prediction as a specific research goal and activity. However, more work is needed before this becomes realistic and achievable. Until high-resolution biological and chemical data are available over large regions of the ocean, and until a better understanding of the dynamics of marine systems is attained, data assimilation in coupled physical-biological models will likely be used more for model improvement and parameter estimation than for operational prediction. While simulation of ocean

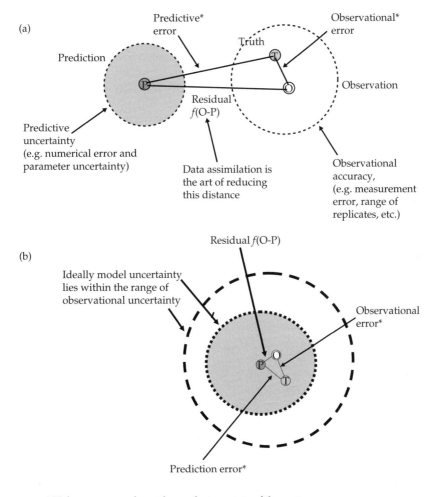

Figure 5.8 Schematic diagram of the relationships between model prediction (P), observations (O), and the true state of the system (T). Both P and O are assumed to have a halo of uncertainty. Figure 5.8a shows the case for a model with no skill and Fig. 5.8b shows the case for the ideal model, with inner circle representing model uncertainty and outer circle representing observational uncertainty. (Reprinted from Stow *et al.* 2009 with kind permission from Elsevier).

physics has approached a high level of sophistica-tion, the complexity of the biological processes, and to a lesser extent the chemical processes, presents an expansion of state variables and their interaction. As a result, it is quite common to find either complex physics coupled to reduced-com-plexity biology or complex biology coupled to simplified physics (Lynch *et al.* 2008). Of concern also is the fact that biological observations (e.g. phytoplankton chlorophyll-*a*) are often not well matched to modelled state variables (e.g. phy-toplankton carbon or nitrogen), requiring conver-sions that are inadequately characterized and likely non-uniform in space and time. This intro-duces additional uncertainty that is difficult to isolate from inadequate process representation or incorrect parameterizations. Hence there is nota-ble diversity in what is possible in 'replicating observations', and more importantly, in assimilat-ing them into simulations and creating forecast systems.

The definition of model 'skill' depends on context-specific factors such as the goals of the modelling exercise and the spatio-temporal scales of importance (Stow *et al.* 2009). Because truth cannot be measured, observations are used as surrogates and rather than asking how well does the model rep-resent the truth, the question becomes how well does the model fit the data? Model predictions and obser-vations each have uncertainties and the true (and unknown) state of the system is assumed to lie within the observational uncertainty (Fig 5.8a). We can say that a model has skill when the observational and predictive uncertainties overlap, eventually reaching a state of complete overlap (Fig. 5.8b). Skill assess-ment requires a set of quantitative metrics and proce-dures for comparing model output with observational data in a manner appropriate to the particular appli-cation. The definition of skill and the determination of the skill level necessary for a given application are context-specific, and no single metric is likely to reveal all aspects of model skill; the use of several metrics can provide a more thorough appraisal (Stow *et al.* 2009). The routine application and presentation of rigorous skill assessment metrics will also serve the broader interests of the modelling community, ultimately resulting in improved forecasting abilities as well as helping assess our limitations. Lynch *et al.*

(2008) and references therein discuss the needs in developing quantitative skill assessment methods in physical-biological models, including aspects of mis-fits, errors, and other metrics faced when assessing model skill.

5.6 Understanding recruitment: the role and challenges of modelling

For most of the past century, fisheries science has focused on the general problem of recruitment, that is, predicting the number of young at birth that will survive to a size or age when they become accessi-ble to commercial harvest (Hilborn and Walters 1992). The desire to understand controls on recruit-ment has been driven in large part by the needs of fisheries management to have a forecast of fish abundance (Rose and Cowan 2003). Prediction of recruitment to marine fish populations and its inter-annual variability requires understanding the causes of variation in growth, mortality, and sur-vival during the larval and early life stages (Houde 1987, 1996).

Recruitment, which is the successful passage of an individual from egg to a size where it becomes commercially accessible, is determined by the com-bined effects of the many factors influencing the organism at each stage in its development. Mortality and growth are connected processes that determine age-specific survival, for example, slow growth due to reduced feeding lengthens the period when an individual is small and vulnerable to predation, and therefore increases mortality. As such, low tem-perature, low prey availability, or a combination of the two would have the same effect (Houde 1989). The path to success for a larva is to grow quickly so that it can better (1) avoid or escape predation, (2) find food, and (3) take control of its environment and cease being a passive planktonic organism (Leggett and DeBlois 1994).

As with body size (Section 5.2), recruitment in many fish is strongly correlated with temperature (for Atlantic cod, cf. Planque and Frédou 1997 or Ottersen and Loeng 2000). For Atlantic cod, the relation between recruitment and temperature has itself been found to depend on temperature, for example, to be non-linear. In areas of the Atlantic where the bottom temperature is relatively low (Iceland, West Greenland, and the Barents Sea) recruitment increases with warm water temperature anomalies. In areas where the bottom temperature is already relatively warm (Georges Bank, the North Sea, and the Celtic and Irish Seas) recruitment tends to decrease with warm water temperature anomalies. This results in a dome-shaped relationship between bottom temperature and cod survival and recruitment. These results suggest a direct link with temperature associated with preferred thresholds but the mechanism regulating this relation has not yet been determined. Drinkwater (2005) used this result, combined with climate forecasts of temperature change, to predict the response of cod stocks in the Atlantic to climate change.

One of the major problems with spawner-recruit data is that the observed high degree of variability can confound interpretation of the relationship between the two (Walters and Ludwig 1981). Over the years, the modelling approach has shifted from focusing on the factors that influence the mortality rate to determining the characteristics of the survivors (T. J. Miller 2007). With better understanding of the causes of the variation in larval growth and mortality, coupled physical-biological models can play an important role in establishing the relationship between spawners and recruitment, that is, models may be able to help determine the contribution of

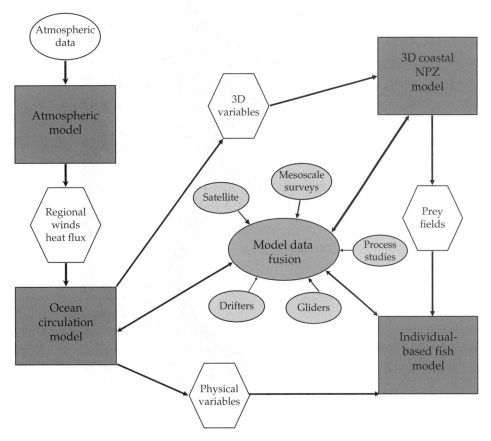

Figure 5.9 Coastal and regional models, both physical and biophysical, are often connected to many other models. In this schematic, the atmospheric model drives a large-scale ocean circulation model which provides boundary condition information for the local circulation model and for the fish model. The observational data that are used to drive the ocean circulation models are treated in a fusion process before being fed into the models (adapted from Mantua *et al.* 2002).

larval stage variability to overall recruitment variability. Modelling studies are now central to most large-scale fisheries oceanography programmes, for example, FOCI in the Gulf of Alaska (Megrey *et al.* 2002), Baltic Sea studies (Hinrichsen *et al.* 2001), Georges Bank (Lough *et al.* 2005), the IDYLE programme off South Africa (Mullon *et al.* 2003), and many others. These studies have addressed recruitment regulation dependent upon feeding and growth, transportation of eggs and larvae, and predation (T.J. Miller 2007).

Population dynamics and population regulation are the result of the complex mix of different processes occurring in different life stages (Houde 2008). In some species it is the juvenile stages that are considered critical to understanding density-dependence and recruitment variability (Cowan *et al.* 2000), and recent analyses have documented density-dependent mortality in the juvenile stage, suggesting that density-dependent adult growth may be more widespread than previously thought (Lorenzen and Enberg 2002). Therefore, improving only our knowledge of larval stage growth and mortality (even with density-dependence) may not help in predicting overall recruitment if significant density-dependent effects occur after the larval stage (Runge *et al.* 2005).

Movement of juvenile and adult fish depends more on biological factors and individually driven decisions and less on physics. In addition, juvenile and adult fish generally spend less time feeding than larvae, and tend to eat larger prey that are themselves less controlled by physical transport. Indeed, many fish species eventually become piscivorous, where their prey becomes larval and juvenile fish. Thus, the key role played by physical-biological models for larval stages of generating transport, temperature, and prey fields will be lessened in the simulation of juvenile and adult fish.

Approaches to and expectations of recruitment studies have shifted from a focus on just prediction and forecasting to development of constant and intensive monitoring of early life stages and the variables that affect recruitment (Houde 2008). Such monitoring efforts are not practical except for a limited number of species. Today the goal is to expand our understanding of the causes of recruitment variability rather than 'solving the problem' (Houde 2008). A series of linked models, each temporally and

spatially scaled to a particular life stage, would enable full life simulations over multiple generations; the physical-biological modelling being one component in this chain (Fig. 5.9). These models should be capable of predicting trends in growth and mortality rates and possible density-dependent effects.

5.7 How has improved understanding of the physical environment improved ocean management?

The tools developed for studying marine ecosystems and the understanding gained through the studies are often suggested to be relevant to improved ocean management (see Barange *et al.*, Chapter 9, this volume). New knowledge includes improved understanding of the influence of the ocean environment on population processes of commercially harvested resources, and protected and endangered species of concern, and better understanding of marine ecosystem structure and processes that provides the link among the organisms and between the organisms and the physical environment. The challenge for ocean management is how to use our new skills and knowledge.

Many different models have been used to investigate the effects of physical forcing and fishing on marine ecosystems. However, effective management of living marine resources still requires more reliable predictions of the combined effects of environment and fishing, especially in the context of the adoption of an ecosystem approach to fisheries and the ongoing concerns about the long-term effects of climate change. Climate and fishing affect different components and attributes of the food web in different ways, for example, light on phytoplankton growth, temperature on zooplankton spatial distribution (Beaugrand *et al.* 2002b, 2008), currents on recruitment success of small pelagic fish (Mullon *et al.* 2003), fishing on the size distribution of fish, and oxygen concentration on tuna distribution (Cayré and Marsac 1993). Fishing and the environment also have indirect effects on organisms when direct effects propagate through the food web. Therefore, to assess the direct and indirect effects of climate and fishing on ecosystem dynamics, models must represent the key linkages among ecosystem components from the bottom to the top of the food web.

The growing knowledge of the complexity of ecosystems, which has yielded surprises such as the cascading effects of fishing (Frank *et al.* 2007) and sudden regime shifts in the ocean (Francis *et al.* 1998; Steele 2004; King 2005; Beaugrand *et al.* 2008), has led to reconsideration of the narrow single-species approach to managing fish resources in the ocean (Pikitch *et al.* 2004). Over the past few decades, there has been a shift first towards a fisheries oceanography approach, in which the environment is considered important to recruitment, and later to an approach referred to as the ecosystem approach to fisheries (Botsford *et al.* 1997). Ecosystem approaches to fisheries have been part of national (USCOP 2004) and international (Sinclair *et al.* 2002; FAO 2003; Jamieson and Zhang 2005) management plans, and some discussions and guidance have been developed for broader approaches that consider all human activities in relation to the marine environment (ICES 2005). Despite the recommendations to move towards the ecosystem approach, relatively few ecosystem-based management plans have been implemented. The Southern Ocean is one place where ecosystem considerations have been well developed. The Commission for the Conservation of Antarctic Marine Living Resources (CCAMLR) management plan has a very explicitly defined ecosystem approach (Constable *et al.* 2000). It is now accepted that including interspecific interactions and physical environmental influences would contribute to greater sustainability by reducing uncertainty. Such an approach requires both process understanding and numerical descriptions of key interactions and the coupling between the physical and ecological environment. An example is the multispecies virtual population analysis of predation by multiple predators on herring and mackerel on the north-east US shelf (Tyrrell *et al.* 2008).

A key gap in expanding the ecosystem representation in fisheries models is the representation of the species in a system. Expanding the number of species in a model is both a modelling and an observational problem. Some species are represented as functional groups, where the single group represents all the species present. Beaugrand *et al.* (2003a) and Reid *et al.* (2003b) showed that seasonal differences and shifts in dominance of two similar North Atlantic copepods (*C. helgolandicus* and *C. finmarchi-*

cus) impact recruitment of Atlantic cod in the North Sea, thus showing the importance if the differences in detailed life history between two very similar copepods. Increased biological resolution can also mean greater detail in the age- and stage-structure of particular species. Explicit consideration of zooplankton ontogeny will be required to properly represent the preferred prey of the early life stages of fish. Ito *et al.* (2007) in their modelling study of Pacific saury, noted that their growth was overestimated compared to observed values. Field observations revealed that large sardine populations grazing on zooplankton reduced the zooplankton available as prey for saury. It was not until sardine predatory effects on zooplankton were invoked that modelled and observed values of saury growth were plausibly reconciled. Ito *et al.* (2007) concluded that the competitive effect of sardines on saury growth can be an important characteristic of ecosystem and necessary to address in future studies of saury populations.

Improvements in hydrodynamic models have dramatically enhanced understanding of space-time connectivity in marine populations, which is important to the development of management strategies for living marine resources. For instance, the design of marine protected areas (MPAs) or the exploration of future scenarios requires a quantitative description of population connectivity. An understanding of the dispersal and mixing of populations will contribute to the explanation of variability in fisheries (Fogarty and Botsford 2007). Realistic descriptions and models of habitat, hydrodynamics, larval transport pathways, and adult growth and survival will provide a mechanistic understanding of how local populations may be interconnected (Cowen *et al.* 2007). Validated, spatially explicit models will also be useful in designing and assessing MPAs in that they will provide the degree to which populations are connected and provide estimates of the exchange between adjacent and distant areas.

MPAs provide an approach that potentially may improve fisheries, by enhancing the rebuilding overharvested stocks, protecting essential fish habitat, and reducing risk of stock collapses or extinctions (Gell and Roberts 2003). The closures associated with an MPA may be partial or complete. The closure of a particular area ensures that there is an unfished por- tion of the population thereby limiting, or at the very least reducing, the harvest rate and the potential for fishing down the population (Botsford *et al.* 1997). The closure thus acts as an insurance buffer against overfishing and as a buffer in the face of uncertainty about the stock status as used in fishery management plans. The population connectivity of marine popu- lations, primarily through larval dispersal but also through migration, is an important consideration in the design and function of closed areas because the reserve has the potential to enhance production out- side of its border. For individuals that are relatively immobile, such as benthic invertebrates, the reserve gives them an opportunity to live longer and grow larger, thereby greatly increasing their fecundity.

For many years the ecological and oceanographic perspective was that marine populations were quite 'open', and that larvae were plentiful and widely dispersed (Caley *et al.* 1996). Genetic studies sug- gested that populations were fairly homogeneous over large spatial scales (Doherty *et al.* 1995) and modelling studies of passive larvae showed that they are carried great distances (Roberts 1997). More recent work on the ecology (Cowan *et al.* 2000) and genetics (Palumbi 2004) of marine population con- nectivity suggests that larval retention near local populations may be more important for maintain- ing population structure and persistence than was previously believed. There has been significant progress in linking dispersal and connectivity in systems such as coral reefs given their smaller spa- tial scales and apparently relatively closed popula- tion structure (Cowan *et al.* 2000).

5.8 Directions for future work

There are many exciting challenges ahead in deter- mining the linkages between the biological and the physical ocean. Rapid increases in our ability to make detailed measurements, coincident with advances in numerical modelling, data assimila- tion approaches, etc., promise additional advances in coupled physical-biological modelling over the next decades. Coupled physical-biological models are now available that can provide information on the physical and feeding environment for marine

organisms. In the long term, the existing coupled models that are generally limited to a single trophic level will need to be melded with models for other organisms at differing trophic levels. Additionally, multiple 'species' at a single trophic level should be modelled to investigate potential climate-driven changes in dominant species abun- dances, such as those that occurred during the mid-1970s regime shift in the North Pacific (King 2005). We need more creative approaches to link- ing models that operate on different biological, spatial, and temporal scales.

Our approach to sampling the ocean has under- gone a shift as we move towards operational data collection systems, now being developed for both the coastal and for the open ocean (e.g. ARGO-Freeland and Cummins 2005). Such programmes will yield vast amounts of data that will be used in modelling and will also lead to new observations of how organ- isms respond to the changing ocean environment. The operational use of these data will become one of the key applications of the marine science that we have reviewed in this chapter and one of the benefits of the many years of development of the techniques deployed and the understanding achieved. Such operational use will pose further opportunities as well as challenges to the marine science community, with demands for greater model fidelity, better approaches to estimating uncertainty, and more robust sampling techniques.

In scientific studies addressing questions of bio- logical and physical coupling, we can expect that a focus on target species will continue to be neces- sary, with key advances needed in the representa- tion of interactions between species both at the same trophic level and between trophic levels. There is much to be learned using existing approaches but it will also be necessary to develop new strategies to explore the coupling across trophic levels. Not only do we want to know how changes in primary and secondary production influence productivity of top predators, but we need the abil- ity to model such interactions so that we can fore- cast how changing ocean conditions will lead to changes in overall ocean productivity.

Models are now better able to explicitly include the resolution needed to tackle ecological scale pro- blems. Until the early 1990s, the available options

Box 5.5 Large-scale circulation and zooplankton in the North Atlantic

One of the most studied zooplanktonic organisms in the ocean is the copepod *Calanus finmarchicus* (Harris *et al.* 2000). This particular species has received a lot of attention because of its importance in the marine ecosystem of the North Atlantic. Many research programmes have focused on *Calanus*, including several national GLOBEC programmes (see Gifford *et al.*, Chapter 4, this volume), but two in particular stand out, the Trans-Atlantic Study of *Calanus* (TASC) and the UK Marine Productivity (MarProd) programme. While the original intent of TASC was a full basin-scale study of *Calanus*, as implemented, the programme had a much greater

focus on the north-east Atlantic. The MarProd programme, composed mostly of scientists from the United Kingdom, ran from 2000 to 2005 and sought to identify the dominant spatial and temporal scales of the physical oceanographic features and zooplankton population dynamics with a particular focus on the Irminger Sea, between Iceland and Greenland. The new data collected during MarProd were used to initialize, guide, and validate spatially explicit models of zooplankton and their food and predators.

In the early 1990s, Backhaus *et al.* (1994) offered a hypothesis to explain the observed

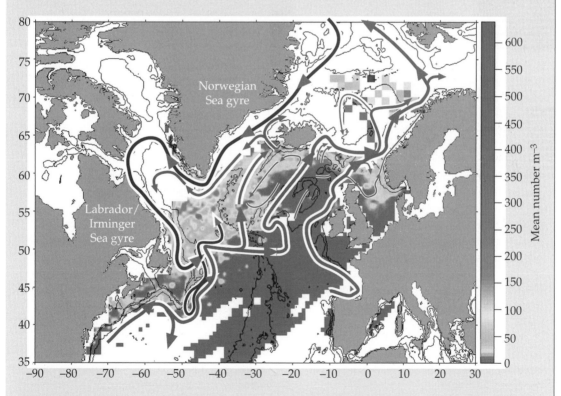

Box 5.5, Figure 1 The abundance of *Calanus finmarchicus* in the surface waters of the North Atlantic. The arrows and lines indicate the major features of the sub-polar gyre. The thickness of the line approximates the strength of the current and the colour indicates whether it is a locally warm (red) or cold (blue) current system. The peak concentrations of *Calanus* occur in the eastern and western cores of the gyres. (From deYoung *et al.* 2004 reprinted with permission from AAAS). (See Plate 14).

continues

Box 5.5 *continued*

spatial patterns and seasonal evolution of
C. finmarchicus in northern European waters. In
reviewing available data, both on *Calanus* and the
circulation of shelf and shelf-edge waters, they
proposed a linkage between the deep ocean
circulation, the *Calanus* populations found there,
and the copepod's abundance and seasonal
circulation on the shelf. This suggestion stimulated
consideration of the coupling between shelf
waters and oceanic waters and led to recognition
of the importance of the open ocean for the life
history and abundance of *Calanus* (Box 5.5, Fig.
1). This work suggested that there is a repopula-
tion of the North Sea by *C. finmarchicus* copepodites that overwinter in the deep water off the
continental slope between Scotland and Norway.
Topographic steering of currents brings the
overwintering *Calanus* to the entrances to the
North Sea. This closure of the life cycle, coupling
the shallow North Sea and the deep ocean,
requires a complex life history in which *Calanus*
overwinter at depths greater than 600 m in the
Norwegian Sea Deep Water, and are returned to
the North Sea in the spring by a northward
counter-flowing coastal shelf jet.

Careful analysis and integration of the data
collected through TASC and MarProd demonstrated that the bulk of the *Calanus* biomass is in
the deep ocean. Maps of the overwintering
biomass, when *Calanus* are in their resting phase,
show that they are present in all the three
primary gyres of the North Atlantic. The juxtaposition of this distribution with the schematic
diagram of the horizontal circulation clearly raises
many questions. How are the populations
maintained within each gyre? What is the degree
of coupling between the gyres? And how does
the coupling change with the documented
interannual and decadal variability exhibited in
the North Atlantic?

Bryant *et al.* (1998) and Tittensor *et al.* (2003)
explored some of the regional aspects of the
coupling of *Calanus* with the circulation in the
North Atlantic. It was not until the work of Speirs
et al. (2005) that a single model spanning the
North Atlantic was able to address the question of
the larger-scale coupling of population dynamics
and circulation. They were able to build upon the
comparative ecological integration of Heath *et al.*
(2004) who brought together Optical Plankton
Counter and net plankton data to describe the
overwintering distribution and depth of
C. finmarchicus in the North Atlantic (Box 5.5, Fig.
1). Speirs *et al.* (2006) used their model to
demonstrate a high-level of connectivity throughout the North Atlantic domain but were also able
to use it to show dependency of mortality on
temperature and to explore different hypotheses
for diapause. While their model only has a
two-layered ocean, with a surface and a deep
layer, they were the first to demonstrate the
degree of connectivity of a single organism
spanning an ocean basin.

Our perspective on the oceanographic dynamics
of *Calanus* has changed because of the work done
since the early 1990s as part of GLOBEC. We have
a much clearer understanding of the importance of
the coupling between the shelf and the deep ocean
and a recognition that coupling across the ocean
basin likely influences zooplankton population
dynamics. Research during this decade has shifted
our perspective on shelf ecosystem dynamics. We
now understand that the shelf is connected to the
deep ocean, not just for a supply of nutrients, but
also for organisms, such as zooplankton and
phytoplankton, that can be carried across the shelf
break, in both directions. Until this work, we saw
the deep ocean more as a sink for material carried
off the shelf than as a source for organisms
important to shelf ecosystem dynamics.

Box 5.6 Modelling tuna

Model studies of skipjack tuna (*Katsuwonus pelamis*) in the Pacific Ocean (Lehodey 2001) have been motivated by their economic importance and the need to better understand the population dynamics of this fish that moves across the ocean basin. The modelling approach taken by Lehodey and colleagues has tuna as the central target

species (Box 5.6, Fig. 1). The most detailed representation of biology is required for the tuna that are represented by a two-dimensional (horizontal) spatially resolved, age-structured cohort model. The resolution of biological detail reduces with distance down the trophic scale, the food of tuna being represented by a simulated

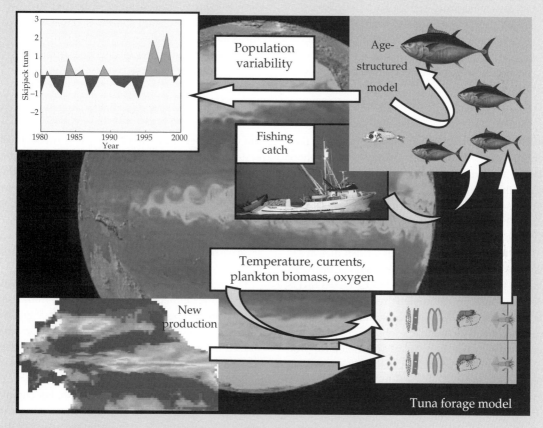

Box 5.6, Figure 1 The lower left image shows a snapshot of the simulated time-varying distribution of NO₃-based primary production (referred to as 'new production') integrated over the euphotic zone in the Pacific, from a coupled physical-biogeochemical model (Chai *et al.* 2003 with kind permission of Springer Science and Business Media). These data together with simulated temperature and velocity data drive the tuna population model (insert *top right*). The basin-scale population of skipjack is represented by a (horizontally) spatially resolved age-structured cohort model, which tracks the numbers of individuals-at-age over space and time (From Lehodey 2001 with kind permission from Elsevier). The relative spatial distribution of recruitment, and age-specific migration rates are the only two aspects of the tuna model that are coupled to the underlying physical-biogeochemical system. Coupling is achieved through the dispersal pattern of eggs and larvae, and dynamic linkage to an intermediate model of the prey of tuna (referred to as 'forage', lower right panel). Coupling to the tuna model is achieved by spatial scaling of the overall mortality rate of forage biomass. (See Plate 15.)

continues

Box 5.6 *continued*

bulk-biomass variable representing all prey species. The prey model in turn is driven by an underlying biogeochemical model of the Pacific Ocean coupled to a three-dimensional, ocean circulation model that captures *El Niño* events and the regime shift in the late 1970s (Chai *et al.* 2003).

The strengths of this model are that it simulates skipjack dynamics over its entire geographic range in the Pacific Ocean, and establishes an interface between the existing, statistically based, fishery-stock assessment models and biogeochemical modelling efforts motivated primarily by carbon cycle and climate issues. The food web has been successfully simplified to retain an element of

space-time dynamics without an overhead of taxonomic resolution. A limitation of the tuna model is that despite being age-structured to capture the essential details of annual cohort variability, it does not fully close the life cycle, and there are no dynamic density-dependent processes in the representation of life history. Thus, the age-structured model is strongly driven by external information from the statistical assessment model. Despite this, preliminary studies using the same approach have also been applied at longer time scales to show potential distributional shifts of skipjack due to anthropogenic global warming (Loukos *et al.* 2003).

---

were often set by computer capacity or model limitations. Most studies still decouple atmospheric forcing from the ocean, imposing it as an external driver through specification of heat and momentum fluxes across the ocean's surface (although see Ito *et al.*, Chapter 10, this volume for recent advances). Also, turbulence is rarely modelled explicitly, but rather it is expressed in terms of diffusivities or viscosities; and important features such as coastlines, bottom topography, boundary layers, mixed layers, etc., are only approximately resolved. However, despite these idealizations, the scales of the models are now much closer to the observed spatial and temporal scales of the physical and biological processes being considered. The challenge in studying marine ecosystems, either through modelling or observations, is to bring together those components of the system that are known, or at least our best knowledge of them, and develop a conceptual framework that enables and leads to the extraction of new results and understanding.

We must devote more attention to the influence of changing biology as organisms grow. We have learned much about how the influence of the ocean environment changes as organisms move from egg and larval stages to adult phases with more complex structures and behaviour. The complexity of life cycles and of marine ecosystem structure forces

difficult choices in how we study and model these systems. Some selectivity is necessary, for example, target species, as the changing relation between different life stages of organisms and the physical and biotic environment requires explicit consideration of life cycle complexity (e.g. behaviours, size-dependent predation, and prey selectivity); otherwise we might miss important processes that control dynamics of populations.

The impact of human activity on the oceans, from coastal development to climate change, from fishing to agricultural run-off, has brought a greater sense of urgency to the need to understand and indeed to predict future ocean states. For fisheries management, we need better forecasts of fish abundance to ensure the long-term viability of fish stocks in the face of growing demand for fish. For climate change, we need to understand how warming and changing levels of CO_2 in the atmosphere will influence the ocean and the services that we extract from it (Ito *et al.*, Chapter 10, this volume). Biophysical models coupled with future scenarios of climate conditions are crucial for assessing and forecasting what might occur in the coming decades. Modelling approaches developed during GLOBEC offer much potential as tools to guide ocean management and policy development now and in future that includes multiple environmental challenges.

CHAPTER 6

Dynamics of marine ecosystems: observation and experimentation

Roger P. Harris, Lawrence J. Buckley, Robert G. Campbell, Sanae Chiba, Daniel P. Costa, Tommy D. Dickey, Xabier Irigoien, Thomas Kiørboe, Christian Möllmann, Mark D. Ohman, Jeffrey A. Runge, Enric Saiz, and Peter H. Wiebe

6.1 Sampling and technological advances in support of GLOBEC science

Global Ocean Ecosystem Dynamics (GLOBEC) research has used a nested set of observations, experiments, and models in space and time to address the question of how climate change may affect marine populations. One of the great challenges has been to use observations effectively that span roughly 10 orders of magnitude spatially and temporally to understand variability in physical and biological environments. A major achievement has been the fostering of a coupled modelling and observational programme in a number of well-selected ecosystems globally. The advances in inter-disciplinary observation and experimentation over the past decade, reviewed in this chapter, have led to significant progress in understanding the structure and functioning of ocean ecosystems. The concept of target organisms (Gifford et al., Chapter 4, this volume) has been central to this approach and is linked to advances in the understanding of individual organism behaviours and population processes. New sampling and observation systems have been developed, particularly acoustic and optical. Shipboard, laboratory, and in situ process studies have been linked with new approaches to understanding trophic complexity. New approaches, applying techniques to retrospective studies, have contributed to understanding past ecosystem states

and widespread use of a comparative approach has revealed new insights into the role of target species and the dynamics of marine ecosystems globally.

In this chapter, observational (field and laboratory) approaches and how they have helped broader advances in the field are presented. The unique challenges of sampling marine ecosystems and quantifying key ecosystem process are considered, and the significant developments are highlighted. How the methods adopted worked in different regions and on different species are evaluated. New methods for observation and experimentation represent one of the significant contributions to the legacy of GLOBEC.

While the focus is on methodology there is a clear contribution to advances in understanding the structure and dynamics of marine ecosystems. The material reviewed relates in particular to physical-biological processes (De Young et al., Chapter 5, this volume) and food web processes (Moloney et al., Chapter 7, this volume).

6.2 New approaches to the trophic complexity of marine ecosystems

Research on new methods to understand trophic links has not resulted in major advances during GLOBEC. However, a number of issues that will need to be addressed in the future have been clearly identified during the programme. These include the identification of digested and visually unrecognizable remains in gut contents, measuring real food concentrations in the field, the effects of food quality and mechanisms like intraguild predation.

Sampling trophic complexity has been a challenge at two main levels: (1) understanding who eats whom, and (2) understanding the effect of eating different prey both for the predator and prey populations.

Understanding who eats whom has posed a surprising number of methodological challenges. Methods can be classified into two main groups (Bamstedt *et al.* 2000), quantification of gut contents (the usual method for fish and using gut fluorescence for zooplankton) on the one hand, and incubation methods on the other (zooplankton). A third alternative is markers such as stable isotopes (e.g. Bode *et al.* 2003) or lipids (Dalsgaard *et al.* 2003) that, although not providing detailed information on the diet, do provide an integrative view of the average diet of the organism.

Methods based on gut contents offer the advantage of allowing feeding to be estimated under natural conditions, without biases resulting from incubation conditions. Therefore the estimates obtained with these types of methods are more likely to be closer to the real situation. However, methods based on gut contents have the limitation of not being able to estimate feeding rates for food items that do not resist digestion or leave unidentifiable remains in the gut. For fish this is the case with, for example, ciliates and other soft-bodied organisms that are likely to be important for the first larval stages but that cannot be reliably identified in gut contents (Fukami *et al.* 1999). The same situation occurs for zooplankton where ciliates have been identified as an important food source (Calbet and Saiz 2005) but cannot be identified directly in the gut in contrast to chlorophyll *a* by fluorescence. However, considering the methodological problems involved, and that incubations are not realistic for fish above a certain size, it is clear that examination of gut contents remains a key technique to understand trophic interactions. In this sense a very promising tool is the application of molecular methods to identify digested items in the gut (e.g. Nejstgaard *et al.* 2003); Blankenship and Yayanos (2005); Sheppard and Harwood (2005); Durbin *et al.* (2008); and see Box 6.1). The application of microarray techniques that allow gut contents to be tested for the presence of a large number of potential prey items should make molecular approaches one of the main tools for understanding trophic links in future (King *et al.* 2008). However, at the moment this field is limited by the low number of primers and sequences available for marine organisms (in particular for small organisms and invertebrates).

A further problem in estimating what are the actual *in situ* ingestion rates of zooplankton that became evident during the GLOBEC programme is how to define food concentration and what proportion of the *in situ* food field is really available. Even for something easy to measure such as chlorophyll the presence of thin layers (McManus *et al.* 2003) results in available concentrations that vary dramatically within the water column. It is therefore important to evaluate the vertical position of zooplankton and fish larvae in relation to the food distribution to estimate the real ingestion rates. Furthermore, it is also important to establish which fraction of the apparent food field is really available to the predator, either because of size (too small) or other reasons limiting availability. This has been an obvious problem when defining food preferences for fish, where selectivity indices may be more dependent on the mesh size of the net and the layers sampled to evaluate the prey field (zooplankton) than on the real fish behaviour. Estimating the prey distribution with traditional methods involves an enormous amount of work (analyses of net samples for different size ranges and different layers that is beyond the practical reach of most studies. However, a very promising approach based on image analysis (*in situ* and in the lab) and automatic recognition (Benfield *et al.* 2007 and see Box 6.2) has been developing during the last decades. This new technology is opening a completely new approach to characterizing the spatial distribution of pelagic organisms, offering the possibility of a much better estimate of the three-dimensional distribution of both predators and prey (see Section 6.5).

Box 6.1 Molecular techniques to establish trophic links

A promising new strategy for assessing feeding in small invertebrates is the use of molecular methods to detect prey-specific nucleic acid molecules as biomarkers of trophic interactions (Sheppard and Harwood 2005). Various different assays have been developed, but the general strategy of these methods is to purify DNA from stomach contents followed by detection and possible quantification using Polymerase Chain Reaction (PCR) amplification-based methods targeting gene fragments associated with prey organisms. Increasingly, this approach is being utilized to tease apart food webs, establish tropic links, and to estimate *in situ* feeding rates. Molecular approaches provide a means by which stomach content analyses can be conducted directly on field-caught animals without the potential of bias from incubation-based methodologies (Nejstgaard *et al.* 2003, 2007). A distinct advantage of a DNA-based molecular approach compared to gut fluorescence and direct microscopic observation is the ability to detect non-pigmented and macerated prey. Two general approaches have been used. The first is to use end-point PCR to qualitatively identify prey species in the guts of predators, while the second is to use real-time quantitative PCR (qPCR) to quantify the amount of DNA of a prey species in the stomach of a predator.

In the first approach PCR amplification primers with different specificities for prey have been used for amplification of genetic markers including species-specific (Bucklin *et al.* 1998; Nejstgaard *et al.* 2003; Vestheim *et al.* 2005), group-specific (Jarman *et al.* 2006), and universal (Blankenship and Yayanos 2005). Species-specific primers only amplify gene fragments associated with a prey species of interest and require a priori knowledge of the marker gene sequence of the prey species to design primers. In contrast, universal primers for a particular genetic marker amplify DNA of all of the different prey species. These amplification

products may be separated through the use of clone libraries or other DNA profiling techniques (Troedsson *et al.* 2008a) and sequenced, enabling prey species to be identified (Blankenship and Yayanos 2005). Because predator DNA is always abundant compared to prey DNA, PCR amplification using universal primers is typically biased towards amplification of predator DNA and rarer prey sequences from the stomachs may fail to amplify. Different approaches have been utilized to attempt to overcome this problem. Blankenship and Yayanos (2005) attempted to minimize it by dissecting the stomachs to reduce the amount of predator DNA in their DNA purifications. They were able to further reduce the amount of predator DNA amplified by digesting with a restriction enzyme that cut only predator DNA within the target PCR amplicon prior to amplification with universal primers (Blankenship and Yayanos 2005).

qPCR offers a method for quantifying the amount of prey DNA present in the stomachs of predators and provides a basis for determining predator feeding rates. This approach allows quantification of the starting amount of DNA template and is based on the detection of a fluorescent reporter molecule that increases exponentially as PCR amplicons accumulate with each cycle of amplification. Fluorescence is measured at each cycle and an amplification plot is generated from the fluorescence data for standards and samples. Samples with higher amounts of target DNA exhibit increases in fluorescence after fewer number of amplification cycles than samples with less target DNA.

During the past decade qPCR has begun to be applied widely in ecological studies including the quantification of algal species in marine planktonic and sediment environments and for investigations of protist parasites and pathogens of marine metazoans (Frischer *et al.* 2006; Handy *et al.* 2006; Lyons *et al.* 2006). PCR-based assays are

continues

Box 6.1 *continued*

now becoming routine in marine ecology studies, especially to detect and quantify free-living organisms. Typically, standard curves are prepared using a dilution series of organism numbers from which DNA is extracted so prey abundance can be expressed as organism concentrations. However, quantitative estimates of target prey or parasite species in predator or host organisms presents a unique set of methodological challenges including the development of efficient quantitative DNA extraction and purification protocols, minimization of DNA digestion and degradation, minimization of PCR artefacts associated with the detection of the target organism in the environment of a host organism, and importantly, the use of appropriate quantitative calibration standards.

The first application of qPCR to measure feeding rates of marine organisms was in a laboratory study of the appendicularian *Oikopleura dioica* feeding on several different algal species (Troedsson *et al*. 2008b). In this study algal DNA in the guts and filtering apparatus was quantified using species-specific primers targeting the 18S gene. In these experiments ingestion rates were measured over short time intervals and digestion of DNA was not apparent. To calculate ingestion rates and filtering apparatus trapping rates, standard curves were prepared from cloned genes as well as cultured algal cells that the animals were fed so prey abundance could be expressed both in units of gene copy numbers and cell concentration.

More recently, a similar approach was applied to investigate feeding of different copepod species in laboratory and field studies (Nejstgaard *et al*. 2007). Results were compared directly with feeding rates derived from parallel studies utilizing gut pigment methodology (laboratory studies) and rates based on direct microscopic analysis of simultaneously conducted bottle incubation experiments (field studies) (Nejstgaard *et al*. 2001a,b). Both laboratory and field studies demonstrated robust quantitative relationships between gut DNA content and independently obtained gut content or feeding rate estimates for the specific prey. However, when absolute

estimates of prey algae recovered from copepods based on DNA were compared to independent estimates of ingested algae, they suggested that algal consumption was underestimated by the DNA-based qPCR assays. In these studies it was hypothesized that the underestimation by qPCR was due to digestion of prey DNA either after consumption or during post-handling steps associated with DNA purification.

A more general application of this method would be to combine information on gut contents in the field over time with a measure of how rapidly food disappears from the guts either through digestion or evacuation (Durbin and Campbell 2007). Ingestion rate is calculated from the equation: $C = 24\ SR$, where C is consumption over a 24 h period, S the mean gut content over 24 h and, R the exponential digestion rate. If there is strong diel migration, or indications of diel feeding periodicity in non-migrating copepods, ingestion is calculated for each time interval and then summed over 24 h (Durbin *et al*. 1995).

Methods have been developed to apply this approach in measuring copepod ingestion of multicellular organisms (nauplii), (Durbin and Campbell 2007); specifically predation by *Calanus finmarchicus* in the Georges Bank/Gulf of Maine region. In this region there is a limited number of potential prey species (Durbin and Casas 2006) making it possible to design species-specific primers for the mtCOI gene for each prey. Laboratory experiments were carried out with *Acartia tonsa* N1 and N2 as prey and adult female *Centropages typicus* as predator. The relationship between *A. tonsa* mtCOI gene copy numbers copepod^{-1} for stages N2-C1 copepod carbon was similar across stages indicating that copy number could be used as a measure of copepod biomass. *A. tonsa* DNA was detectable in the guts of the predators for as long as 3 h. Exponential rates of decline in prey DNA from the stomachs of the predators are similar to those measured for gut pigments.

Conversion of the copy number ingestion rates to numbers of each nauplius stage ingested is complicated by the fact that copy numbers change with

continues

Box 6.1 *continued*

stage. In order to apportion these copy number ingestion rates amongst different stages it was suggested that estimates of the clearance rates of each stage by the predator determined in laboratory experiments be used together with the abundance of each stage in the field, to calculate the relative proportions of the copy number of each stage ingested. The actual numbers of each stage ingested in the field are calculated from these proportions and the *in situ* measurements of total mtCOI copies of each prey species ingested. At present this work is still in the development stage and there is a need to actually calibrate it in the laboratory against more traditional methods. One disadvantage is that this method cannot be used to measure cannibalism, which may be significant in some copepods (e.g. Bonnet *et al.* (2004); Ohman *et al.* 2004).

There are clear advantages and disadvantages of DNA-based methods compared to many of the classical approaches for investigation of zooplankton trophic interactions. The primary advantage of the DNA-based methods is the ability to obtain species-specific information of the trophic interac-

tion, both qualitatively as well as quantitatively. However, due to digestion problems, the technique is at best semi-quantitative at this point, although there are some promising assays aiming at profiling digestion to obtain absolute quantification. Further, for organisms with complex trophic interactions, DNA-based techniques are still laborious and expensive. There are a number of promising high throughput sequencing as well as profiling techniques available today, but they are still relatively expensive. Therefore, we believe that the strength of the DNA-based techniques will be in the combination with classical approaches mainly because DNA-based techniques offer much better resolution of specific trophic interactions when such resolution is necessary in the data analysis. However, biotechnological companies are making rapid advances in user-friendly, high-throughput and affordable assays, and we predict that many of the methods reviewed here will be standard in most studies only a few years from now.

**Edward G. Durbin, Ann Bucklin,
Jens C. Nejstgaard, and Marc E. Frischer**

The other general method used to estimate trophic relations of zoo- and phytoplankton involves incubations. Incubations also present a large number of yet unsolved problems. Other than their intrinsic limitations such as wall effects and enclosure (see Section 6.5) the main problem of incubations is the enhancement of trophic cascades that bias feeding estimations. The classic example is that of copepods, ciliates, and phytoplankton (Nejstgaard *et al.* 2001a,b). The copepod concentrations needed to estimate feeding rates on phytoplankton are often high enough to strongly reduce the ciliate population in the container and hence their feeding pressure on phytoplankton, resulting in an increase of phytoplankton in the incubation bottle instead of a decrease. Different methods have been proposed to eliminate this type of error (Nejstgaard *et al.* 2001a,b) but in general all involve much more laborious procedures multiplying the number of samples to be analysed. In this sense, future improvements may be

expected through the use of image analysis systems allowing the enumeration of organisms in a rapid and effective way (see Box 6.1). A further problem of incubations is their limited capacity to estimate predation on prey present at low concentrations. As an example, laboratory experiments have shown that copepods are able to ingest a number of large organisms such as meroplanktonic larvae (Kang *et al.* 2000; Lopez-Urrutia *et al.* 2004) and copepod eggs and nauplii (Bonnet *et al.* 2004). The ingestion of such large organisms in the field may be occasional but important in terms of contribution to diet because of their large size. However, standard incubation experiments under field conditions do not allow the quantification of such trophic links. Again, the most promising technique in the future seems to be based on the identification of prey in the gut through molecular techniques (Box 6.1).

A more generalized problem in obtaining a generic understanding of who eats whom is the lack

Box 6.2 RAPID visualization of zooplankton predator/prey distributions

Collecting, identifying, and counting zooplankton has traditionally been a time-consuming, labour-intensive process. For these reasons, there is normally quite a long lag between sample collection and the visualization and analysis of distributions of biological taxa and corresponding hydrographic properties. The advent of imaging systems capable of recording the contents of defined volumes of water holds great promise for advancing our ability to describe the three-dimensional spatial distributions of zooplankton. This is particularly true for taxa that either avoid conventional nets and pumps, or for fragile organisms that are not well preserved during net sampling.

While imaging systems deliver massive volumes of data about zooplankton, exploiting their true potential is hampered by an inadequate ability to process this torrent of information. The development of exciting new software tools that semi-automatically and automatically process image data to extract zooplankton target information has begun to reveal the enormous power of *in situ* imaging systems. The same software tools can also be applied to digitized images of preserved or live plankton in order to accelerate their processing.

Automated image classification begins with an image data set from one of the many innovative *in situ* and laboratory systems developed to study plankton. See Benfield *et al.* (2007) for a review of currently operational systems. The first step is to isolate valid plankton targets from the background of each image. This process is termed segmentation and requires algorithms to locate the external boundary of each object. A somewhat larger bounding box is usually applied to the target to ensure that subtle features such as antennae or tentacles, that may not have been part of the perimeter of the object, are included. Segmentation may also include a screening process for focus detection to ensure that only targets imaged within the in-focus depth of field of the camera are included in the analysis.

Once a valid, in-focus object has been isolated from the background image, most software packages employ a step called feature extraction. Features are characteristics of the plankton image and its metadata that contain taxonomically useful information. These may include many of the morphological features that conventional taxonomy employs to distinguish different taxa as well as length to width ratios, circumference, and other allometric measurements. More often, features include a diverse suite of optical characteristics of the image. These optical features can include things such as the range of greyscale and colour levels in an image, brightness, texture, color, specularity, contrast gradients, and a host of optical characteristics that the human eye may or may not perceive. The completion of the feature extraction stage results in a collection of images of plankton, each isolated from their parent image, and associated with a series of characteristic features.

Computer-based classifiers require training before they can attempt to identify the contents of a series of unknown images. This requires that an individual or group constructs a training data set. A training set consists of taxonomic categories that each contains a series of representative images of each taxon. There must be a training category corresponding to each taxon that the researcher wishes the software to attempt to distinguish. Each category should include a large number of representative images of that taxon. A training set consisting of the best examples of each taxon will not be useful because such images may not share features that are common with the majority of images in the unknown sample. If organisms of the same taxa are imaged in two or more typical orientations (e.g. copepods in a lateral and dorsal view), then it may be necessary to dedicate a category to each orientation.

Once a training set has been assembled, it can be used to build a classifier. Classifiers are mathematical algorithms that learn to classify unidentified plankton images by constructing

continues

Box 6.2 *continued*

decision mechanisms. These decision mechanisms use relationships between features associated with the images in a training set and the label provided by a human expert to classify images of unknown identity. Misclassification errors must be quantified with a confusion matrix. It should be noted that automated classification systems are at best, expected to perform as well as a human expert. Experts do make mistakes (Culverhouse *et al.* 2003). When mistakes are incorporated into the training set, boundaries between features associated with each taxon will be less well defined and accuracy will likely suffer. Ideally an expert system should be capable of learning from misclassification errors to improve overall classification accuracy.

There are currently several examples of software tools that incorporate most or all of the above activities to enable computerized classification. These include ZOOIMAGE (www.sciviews.org), ZooProcess and Plankton Identifier (www.zooscan.com), Visual Plankton (www.whoi.edu/instruments/vpr), and SIPPER software (http://figment.csee.usf.edu/~shallow/sipper/papers/SipperSoftwareManual.pdf). At present, each of these packages is primarily designed to function with a single instrument type, although ZOOIMAGE has the capability of functioning with both scanner-based instruments such as ZOOSCAN (Grosjean *et al.* 2004) and the FlowCAM (Sieracki *et al.* 1998).

Considerable progress is being made using computers to conduct the labour-intensive classification of plankton samples. Accuracies of 70–80% or better have been demonstrated for 10–20 class problems. A notable success has been the demonstration of an accuracy of 88% for a 22-class phytoplankton problem with individual class accuracies ranging from 69 to 99% (Sosik and Olsen 2007). Based on work with the Video Plankton Recorder (VPR) and other systems such as SIPPER, it appears likely that similar performance is achievable for mesozooplankton using support vector machine (SVM) and other classification algorithms.

Once we are able to employ computers to advance past our current image bottleneck, the oceanographic and plankton ecology communities will be able to tap into the wealth of information that imaging systems can provide about planktonic predators and their prey. Far too often the time lag between data collection and interpretation is unacceptably long. When computers are able to do the hard work while oceanographers are at sea collecting the data, we will be able to observe plankton ecology on time scales that permit real-time responses to interesting predator-prey interactions.

An example of the type of interactions that could be visualized while at sea is provided by Global Ocean Ecosystem Dynamics (GLOBEC) data collected with the VPR in Wilkinson Basin, Gulf of Maine. During 1998 and 1999, there were dramatic changes in the abundances of diapausing *Calanus finmarchicus* and their invertebrate predators (see Box 6.2, Fig. 1). For example, in 1998, relatively few *C. finmarchicus* were present while physonect siphonophores were very abundant. In contrast, *C. finmarchicus* were very abundant during 1999 while siphonophores were relatively sparse. This relationship may have been partially a consequence of predation pressure by siphonophores on *C. finmarchicus* because their regions of high densities were inversely distributed in 1998 (see Box 6.2, Fig. 1) while there was no obvious spatial relationship in 1999.

Although these spatial patterns illustrated in Box 6.2, Fig. 1 were not obtained using semi-automated techniques, we have begun using a new interactive sorting tool called Plankton Interactive Classification Tool (PICT) to rapidly separate VPR images into their constituent taxa. PICT was developed by the University of Massachusetts Computer Vision Laboratory as part of a project to develop flexible software tools to classify plankton images. PICT combines segmentation and feature extraction with a classifier to semi-automatically sort images into taxonomic categories. As images are placed into these by a human operator, all unknown images that share features with those

continues

Box 6.2 *continued*

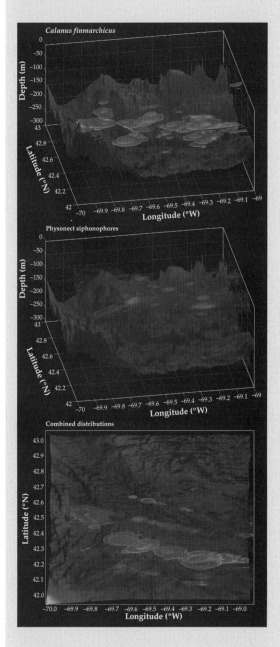

classified objects are then allocated to appropriate categories. In this manner, a training set can be rapidly assembled for construction of a classifier.

There is clearly a need for flexible software classification tools that bring automated image classification capabilities to the broad constituency of users who employ a diverse suite of imaging systems. Many current research imaging systems do not have associated image classification software. Research on Automated Plankton IDentification (RAPID) is a global initiative designed to bring zooplankton ecologists, engineers, software and hardware developers together to develop new tools that seamlessly work with imaging systems to locate, extract, learn, classify, and count zooplankton in near-real-time. Associated with the work of Scientific Committee on Oceanic Research (SCOR) Working Group WG130 (http://www.scor-wg130.net/index.cfm), RAPID is committed to developing practical and flexible software tools for the oceanographic and plankton ecology communities.

Box 6.2, Figure 1 Spatial distributions of diapausing *Calanus finmarchicus* (top panel) and physonect siphonophores (middle panel) in Wilkinson Basin during December 1999 as measured with a towed Video Plankton Recorder (VPR). Observations of each taxon were converted to abundances and interpolated in three-dimension using GLOBEC EasyKrig 3.0 software. Isosurfaces in this visualization correspond to the highest densities. For *C. finmarchicus* these densities were 100–300 individuals per m³ and for siphonophores (1–4 colonies m⁻³). Combined distributions in plan view are illustrated in the bottom panel, which demonstrates that the patches of *C. finmarchicus* are generally absent from regions of high siphonophore densities (Christian Briseno). (See Plate 20).

Mark Benfield

of a theoretical basis allowing the design of robust experiments to test the predictions of theory. Most studies have been opportunistic experiments where the trophic links in one area for a target species are measured. Field experiments are usually observational rather than aimed at testing a hypothesis. The extrapolation of such results and incorporation in a general synthesis is difficult without a theoretical basis. However, there is a developing body of theoretical research on encounter rates between predator and prey (Visser and Kiørboe 2006; Visser 2007b) providing such a theoretical basis (see Section 6.5). There is a need for future studies to link field studies with the predictions of theory.

Understanding the effects of the different trophic links between predator and prey populations also offers a number of challenges. On the one hand there is the effect of food composition on the physiology of the organisms. Both the stoichiometric composition of the food (Jones and Flynn 2005) and the presence of toxins through physiology or feeding inhibition (Miralto *et al.* 1999; Selander *et al.* 2006) have been shown in the laboratory to have the potential to influence population dynamics. The relevance of such effects in the field is still unclear (Irigoien *et al.* 2002; Irigoien *et al.* 2005; Pierson *et al.* 2005) and needs further field research combining high-resolution sampling with food composition and diet studies. One further problem related to this issue is how to incorporate the complexity of the food composition, where the effects can be not only species but also strain dependent (Wichard *et al.* 2008), into models. A possibility emerging from a better understanding of biodiversity (Irigoien *et al.* 2004) could be to use a statistical approach. At low food concentrations food limitation is more likely to have an effect than food composition. At medium phytoplankton concentrations diversity is high and zooplankton may be expected to find an appropriate diet in terms of composition through selection. It is at high phytoplankton concentrations when diversity is low that food composition could result in a decrease of zooplankton production below the values expected from food concentration alone. Because phytoplankton diversity is related to phytoplankton concentration (Irigoien *et al.* 2004) an index of probability of reduced production could be developed for high concentrations. This would not be based on the idea that a diverse diet is necessarily better than a diet based on a single food source, but on the idea that as diversity decreases the probability of food choice being toxic or nutritionally deficient increases. This approach could simplify the inclusion of food quality in models because experiments would not need to test all species and strains but rather estimate the number of times where a single diet decreases production in comparison with a mixed diet.

Predator effects may also have important impacts on prey populations. Top-down control and trophic cascades are increasingly being demonstrated for the marine environment (e.g. Worm and Myers 2003; Frank *et al.* 2005). However, the predatory effect of fish on zooplankton populations has rarely been convincingly demonstrated (Möllmann and Köster 2002). Other mechanisms such as intraguild predation and cannibalism (Polis *et al.* 1989; Köster and Möllmann 2000; Ohman and Hirche 2001) can have an important effect on the population dynamics of the species as well. Intraguild predation (competitors that eat each other) releases pressure on the basic prey and favours dominance when resources are abundant. An example of such an interaction is that of copepods, ciliates, and phytoplankton (Gismervik and Andersen 1997). Although phytoplankton are their main food source, by having high feeding rates on ciliates copepods might actually release pressure on phytoplankton, therefore affecting carbon fluxes. Similar mechanisms may occur with predation by copepods on eggs and nauplii that could determine the success of cohorts (Ohman and Hirche 2001) or the succession of species (Irigoien and Harris 2006). Similarly planktivorous fish have been shown to limit their own mortality rates by preying on early-life stages of their predators (Köster and Möllmann 2000).

GLOBEC associated programmes have highlighted such mechanisms, but our understanding of the relevance of those interactions is still very limited. Recent studies strongly suggest that predation is more likely to influence population dynamics in the field than food composition (Pierson *et al.* 2005). Therefore trophic complexity should be considered in a wider view, and for an organism one should consider both prey and predators, including the often ignored predators on the early stages. To have such a wider view of the trophic links of organisms we urgently need new approaches allowing better estimates of *in situ* trophic links in the field.

6.3 Sampling and observation systems

GLOBEC programmes have developed and exploited new sampling and *in situ* optical, video, and acoustical methods, together with satellite tracking of individual organisms. These observational advances have provided comprehensive measurements of ecosystem properties on time scales from minutes to years and on space scales from less than millimetres to the global.

At the beginning and during the GLOBEC field programmes, considerable effort was expended to develop and exploit a number of new sampling tools and techniques as well as to use mature technologies to sample the zooplankton target species and their predators and prey, and to measure important environmental variables concurrently. Programmes studying marine ecosystems worldwide recognized that existing sampling technologies were not sufficient to meet the objectives. As a result a meeting was held 'to discuss the existing capabilities and potential developments in acoustical and optical technology, methodology, and instrumentation for measuring spatial and temporal distributions and assessing the behaviour of animals in the sea' (US GLOBEC 1991; see also GLOBEC 1993).

The resultant field work consisted of broad-scale and process surveys of the study sites at designated station locations, along the transect lines between stations, or in the vicinity of drogues or dye patches used to track water parcels. Fixed location moorings with a combination of physical and biological instrumentation and near-surface drogues have been used to provide continuous data to fill in the time gaps between survey and process cruises. In addition, in some studies, tags on large mammals and sea birds were used to acquire environmental data as well as information about animal location and behaviour. The data from tags, buoys, and drogues were frequently transmitted to land via satellite telemetry. The special issue, Southern Ocean GLOBEC (Hofmann *et al.* 2004c) provides examples and illustrations of these methodologies applied at the US GLOBEC study sites.

Sensors that operate on quasi-continuous spatial and temporal scales were viewed as essential if GLOBEC was to link small-scale process measurements to population parameters in the quest to understand the dynamics of zooplankton target species populations. Although seawater transmits visible light poorly because it is absorbed, scattered, and reflected more than in air, increasingly powerful video and camera chip technology made it possible to develop a number of new optically based sensors for zooplankton research. Transmission of sound in the moderate- to high-frequency range (38 to 1,000 kHz) is much greater and suitable for studies of zooplankton, which can be detected 10s to 100s of metres from the sound source. Thus, it was recognized that the integration of acoustical and optical technology would be highly beneficial and that the technologies were both complementary and synergistic in their potential utility. The need for synoptic sampling of both the biological and physical characteristics of the water column was stressed at the outset of the programme. The following is a summary of some of the technological developments and their application. There are two aspects to be considered: the sensors themselves and their modes of deployment.

6.3.1 Optical systems

Three optical sensor systems were principally used in the GLOBEC field programmes to study the distribution and abundance of zooplankton and fish eggs and larvae: the Video Plankton Recorder (VPR), the Continuous Underway Fish Egg Sampler (CUFES), and the Optical Plankton Counter (OPC) (Table 6.1). Other optical systems were also used to study phytoplankton distributions based principally on fluorometry and light attenuation (e.g. Barth *et al.* 2005), and they will not be discussed further here.

VPR: The VPR is a high-magnification underwater video microscope that images and identifies plankton and seston undisturbed in their natural orientations and that can quantify their abundances at sea in real time (Davis *et al.* 1992, 2005). The original VPR had four analog video cameras and a strobe light; each camera imaged concentrically located volumes of water ranging from less

Table 6.1 Regional GLOBEC programmes that used optical or acoustical senor systems and the net systems deployed in the field work. GLOBEC programmes that did not employ optical or acoustical sensors are not included in this table.

GLOBEC Programs ≥>		Georges Bank, Gulf of Maine	NEP California Current	NEP Alaska	SO GLOBEC	Arabian Sea	Canada GLOBEC	ICOS	TASC	Mare Cognitum	UK GLOBEC Marine productivity	Baltic Sea German GLOBEC
Instrumentation												
Acoustical systems	HTI	X	X	X	X							
	BioSonics	X										
	Simrad		X		X				X	X		
	TAPS		X	X								
	ADCP				X	X						
Imaging systems	OPC		X		X		X	X			X	
	VPR	X			X			X				X
	CUFES		X									
Net systems	MOCNESS	X	X	X	X	X			X	X		
	BIONESS						X		X			
	MultiNet								X			x?
	ARIES										X	
	Ocean Sampler										X	
	WP2								X	X		
	Bongo net	X	X			X	X		X			X
	CPR						X		X			
	Other (RingNets)	X	X				X		X	X	X	
	CalVET		X									

Note: NEP, North-East Pacific; SO, Southern Ocean; ICOS, Investigation of *Calanus finmarchicus* migrations between Oceanic and Shelf seas; TASC, Trans-Atlantic Study of *Calanus finmarchicus*.

than 1 to 1,000 ml, but it was subsequently modified to a one or two-camera system. The systems typically imaged a volume of about 5.1 ml at 60 Hz (3×10^{-4} m^3 s^{-1}). An image processing system was also developed that was capable of digitizing each video field in real time and scanning the fields for targets using user-defined search criteria such as brightness, focus, and size (Davis *et al.* 1996; Tang *et al.* 1998; Davis *et al.* 2004; Hu and Davis 2006; see also Box 6.2). The targets are identified using a zooplankton identification programme to provide near-real-time maps of the zooplankton distributions. Targets that meet the criteria are sorted into different taxonomic categories, enumerated, and measured together with the location, time, and depth at which they were observed. The software can also be used to post-process data from internally recording VPRs that are deployed autonomously. The VPR has typically been deployed as the primary zooplankton sensor along with environmental sensors in a V-fin vehicle that is undulated from the surface to some depth (100 m or greater). Recently, the VPRII has been developed that substantially improves the original version through use of a high-resolution digital camera, an automatically undulating towfish capable of tow speeds up to 12 knots on a trackline offset from the wake of the ship, and an improved software interface for automatic identification and display of plankton taxa together with hydrographic data (Davis *et al.* 2005).

A number of VPR-based systems were used in GLOBEC programmes. The original version and modified versions were used in surveys of Georges Bank and the Gulf of Maine (Benfield *et al.* 1996; Gallager *et al.* 1996; Norrbin *et al.* 1996; Ashjian *et al.* 2001; Davis *et al.* 2004). A one-camera system was used on the Bio-Optical Multifrequency Acoustical and Physical Environmental Recorder (BIO-MAPER-II) vehicle (described below, see also Box 6.3) to map the vertical and horizontal structure of zooplankton and nekton in the deep basins of the Gulf of Maine (Benfield *et al.* 2003; Lavery *et al.* 2007). A VPR was mounted on a 1 m^2 Multiple Opening/Closing Net and Environmental Sampling System (MOCNESS) net system to map the fine-scale distributions of larval cod prey items (Broughton and Lough 2006; Lough and Broughton 2007) on Georges

Bank. In the Southern Ocean GLOBEC Programme a two-camera system mounted on BIOMAPER-II was used on four survey cruises (Ashjian *et al.* 2008) to map the distribution of larval krill and other zooplankton. In addition, euphausiid furcilia populations living under sea ice were quantified using a stereo VPR mounted on a Remotely Controlled Vehicle (ROV; Gallager *et al.* 2001). In the Baltic Sea GLOBEC programme, a VPR was used by Schmidt *et al.* (2003) to examine the distribution of *Pseudocalanus*.

CUFES: Image resolution constraints inherent in the use of standard video formats have driven the development of optical systems that utilize higher-resolution formats. Development of the CUFES (see Box 6.4) utilizes a line-scanning digital camera to quantify the abundance of fish eggs (Checkley *et al.* 1997, 1999). Mounted on shipboard, seawater from a surface intake is channelled through a fish egg concentrator and viewport. A digital camera creates images of the water that are recorded on a microcomputer- based image processor. Near-real-time estimates of egg abundances are possible with this system. CUFES has been used by a number of countries involved in GLOBEC Small Pelagic Fish and Climate Change (SPACC) projects (Hunter and Alheit 1997; Checkley *et al.* 1999, 2000).

OPC: The OPC, a non-image-forming device, has been used widely. The OPC was developed during the mid-1980s (Herman 1988) and was redesigned in the 1990s (Herman 1992). This instrument measures changes in the intensity of a light beam that occurs when a particle crosses the beam. Light intensity attenuation caused by the passage of a particle across the light sheet is detected and counted, and the magnitude of the change in light intensity is used to determine the size of the particle. The detectable size range is nominally between 250 um and 14 mm. A more sophisticated version of the OPC, the Laser OPC (LOPC) was developed to provide higher sampling frequency and improved information about the shapes of particles as they pass through the laser sheet beam (Herman *et al.* 1998).

As part of the US GLOBEC Northeast Pacific Study, an OPC and a fluorometer mounted on a vertically undulating SeaSoar were used to survey zooplankton and phytoplankton in the California Current between 42.5 and 44.7°N in spring and

Box 6.3 BIOMAPER-II

Sampling of plankton communities historically has been a costly, labour-intensive activity, due in large part to the effort needed for sorting and identifying organisms collected by nets, pumping systems, or water bottles. Thus, in the planning phases of the Global Ocean Ecosystem Dynamics (GLOBEC) programme, more efficient, higher-resolution samplers were designed, tested, and deployed in the field sampling at many of the study sites. Video and acoustic technologies employed have demonstrated the capability for cost-efficient plankton sampling and identification. One such system is the Bio-Optical Multifrequency Acoustical and Physical Environmental Recorder, or BIOMAPER-II. This is a towed system capable of conducting quantitative surveys of the spatial distribution of coastal and oceanic plankton/nekton (Wiebe *et al.* 2002).

BIOMAPER-II consists of a multi-frequency sonar, a Video Plankton Recorder (VPR-Davis *et al.* 1992) system (Davis *et al.* 2005) and an environmental sensor package (CTD, fluorometer, transmissometer). The latter sensor set is used to describe the hydrographic and environmental characteristics of the water column that then can be related to plankton distributions and abundances.

The acoustic system collects backscatter data from a total of 10 echo sounders (5 pairs of transducers with center frequencies of 43 kHz, 120 kHz, 200 kHz, 420 kHz, and 1 MHz), half of which are mounted on the top of the towbody looking upward, while the other half look downward. This arrangement enables acoustic scattering data to be collected for much of the water column.

These acoustic frequencies were chosen to bracket the transition from the Rayleigh to geometric scattering regions for zooplankton and micronekton in the range of 1 to 200 mm. The software enables data aquisition on five frequencies with each pair of transducers. The range of the 0.5 m depth strata allocated for each transducer is dependent on frequency with the lowest

frequencies given the longest range and highest frequency the shortest range (i.e. 43 kHz = 200 m, 120 kHz = 200 m, 200 kHz = 149 m, 420 kHz = 100 m, 1,000 kHz = 35 m). Echo integration is normally conducted at 12 s intervals to provide volume backscattering data at all five frequencies. Split-beam data are normally collected at the four lower frequencies, which enables individual targets to be identified and target strength (TS) determined.

Acoustic data from the up- and down-looking transducers are processed in real time and combined to provide a vertically continuous acoustic record extending from the surface to at least 200 m, and at most 350 m, depending on the position of the BIOMAPER-II along its undulating towyo path.

The VPR is an underwater video microscope that images and identifies and counts plankton and seston in the size range 0.5–25 mm, often in real time. The VPR video data augments the high-resolution acoustical backscatter data. The two systems together allow high-resolution data to be obtained on zooplankton in the water column. The range-gated acoustical data provides distributional data at a higher horizontal resolution than is possible with an independent VPR, while the video data provides high-resolution taxa-specific abundance patterns along the towpath and allows for direct identification, enumeration, and sizing of objects in acoustic scattering layers, so that the VPR data can be used to calibrate the acoustical data.

BIOMAPER-II in combination with a Multiple Opening/Closing Net and Environmental Sampling System (MOCNESS) was used on a series of five US GLOBEC cruises in the Gulf of Maine in a project to examine the overwintering stock of *Calanus finmarchicus*. The high-frequency volume back-scattering data provided the most complete coverage of the Gulf of Maine basins on the cruises. Although the backscattering data did not reflect the distribution of the zooplankton and micronekton biomass directly, patterns in the

continues

Box 6.3 *continued*

Box 6.3, Figure 1 Deployment of BIOMAPER-II at sea (Peter Wiebe).

acoustics data can augment the interpretation of the net tow data taken concurrently. There were clear day/night shifts in some of the profiles of volume backscattering indicating diel vertical migration. During the day, depths below 100 m generally had higher backscattering and surface values were lower. The reverse generally occurred at night. Ground truthing the acoustics data to provide biologically meaningful information has been a significant aspect of the work (see papers by Chu and Wiebe (2003); Warren *et al.* (2003); Benfield *et al.* (2007); Lavery *et al.* 2007).

BIOMAPER-II was also used on the four Southern Ocean GLOBEC broad-scale surveys on the Western Antarctic continental shelf region in the Marguerite Bay environs. In this work, krill distribution and abundance were determined on two austral fall cruises and two winter cruises when pack ice covered the entire survey region. Acoustic volume backscattering was used as an index of the overall biomass of zooplankton. Distinct spatial and seasonal patterns were observed that coincided with advective features (Lawson *et al.* 2004). The general pattern of backscattering across most of the survey area involved low backscattering in the surface mixed layer, moderate backscattering in the pycnocline, a midwater zone that typically had faint scattering, and when the bottom was within range of the transducers, a well-developed bottom scattering layer extending 40 to 100 m above the bottom. More sophisticated methods that capitalize on the full multi-frequency data set were developed. These distinguished the scattering of krill from that of other zooplankton taxa, delineating krill aggregations in the acoustic record, and then estimated krill length, abundance, and biomass in each acoustically identified aggregation (Lawson *et al.* 2008a,b). The distribution of krill was characterized by many small aggregations closely spaced relative to one another, punctuated by much fewer aggregations of very large size that accounted for the majority of overall biomass in the region. The greatest number of aggregations was found at depths less than 100 m, but aggregation biomass was usually greatest at deeper depths. There was little association between the characteristics of individual aggregations and the mean length of krill estimated acoustically, and thus little evidence for any size- or age-related changes in aggregative behavior.

Peter H. Wiebe

Box 6.4 Continuous, Underway Fish Egg Sampler

The Continuous, Underway Fish Egg Sampler (CUFES) was developed during the 1990s to improve sampling of the typically highly patchy distributions of pelagic fish eggs, and was first applied to test the hypothesis that spawning by Atlantic menhaden (*Brevoortia tyrannus*) occurs during storms along the western wall of the Gulf Stream. Over the past decade the CUFES has been used in many regions, principally to study the distribution of eggs of small pelagic fishes such as anchovy and sardine. The CUFES has now become the standard sampling tool for mapping spawning habitat used by participants in the Small Pelagic Fishes and Climate Change (SPACC) regional

programme of Global Ocean Ecosystem Dynamics (GLOBEC), which ensures compatibility for inter-ecosystem comparisons.

A CUFES system consists of a high-volume (ca. 0.5 m³ min⁻¹), submersible pump either fixed rigidly to the ship's hull or pumping through the hull via a sea-chest; a sample concentrator; and a mechanical sample collector. Water is pumped from pump depth (around 3 m for the external configuration or 6 m for the through-hull configuration) to the concentrator, where particles retained by a 500 μm mesh (or occasionally smaller) are concentrated in a reduced flow. This flow is then directed to the mechanical sample collector, which allows for

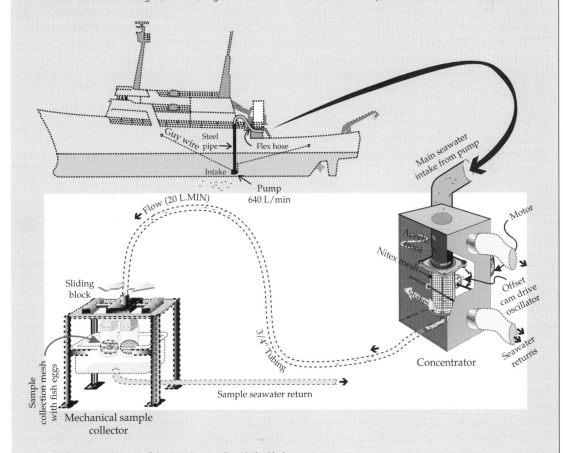

Box 6.4, Figure 1 Schematic of the CUFES system (David Checkley).

continues

Box 6.4 *continued*

sequential collection of samples which are generally examined immediately after collection and hence provide near-real-time information on egg abundance. Because the pump can be used while the vessel is both on-station and underway, the CUFES collects many more samples and provides much higher spatial resolution than is possible using standard, on-station ichthyoplankton samplers such as a California Cooperative Oceanic Fisheries Investigations Vertical Egg Tow (CalVET) net. In contrast to the CalVET net which collects a vertically integrated sample at a single location, however, the CUFES only samples at a fixed depth and thus collects horizontally integrated samples at that depth when used while underway, or point-samples from that depth when used while on-station.

The major disadvantage of the CUFES is its inability to sample the entire egg vertical distribution range, but because the eggs of pelagic fishes are typically positively buoyant and abundant near the surface, near-surface sampling allows inference about the areal abundance (i.e. the number of eggs under 1 m² of sea surface) and distribution of pelagic fish eggs. Such inference depends on a statistically significant relationship between egg concentration at CUFES pump depth and areal egg concentration, which is the case in most published studies using CUFES and demonstrates the efficacy of CUFES as a sampler of pelagic eggs.

CUFES samples are used to characterize spawning habitat in terms of space and time, and environmental data (temperature, salinity, fluorescence, etc.) collected concurrently with CUFES samples have been used to characterize spawning habitat in terms of hydrography, such characterizations being subsequently used to develop models of spawning habitat. The high spatial resolution of CUFES-derived data has been used to examine the fine-scale spatial structure of egg patches and to optimize sampling design, including the location and spacing of transects and stations for net collections. CUFES has also been incorporated into the Daily Egg Production Method (DEPM) of estimating spawner biomass, with on-board egg counts being used to determine when full water column samples should be taken with a CalVET net. However this adaptive sample allocation is critically dependent on a thorough understanding of the relationship between the abundance of eggs at the CUFES pump depth and vertically integrated egg abundance, and how this may change under different oceanographic conditions. This has stimulated significant research effort aimed at deriving realistic egg vertical distribution models.

Future development involves automation of the counting and staging of eggs of target fish species in CUFES through the use of progressive-scan cameras and line scan video (see Box 6.2). Such automation would reduce cost and provide real-time data on egg abundance and distributions under all conditions. More details on the CUFES are available at http://cufes.ucsd.edu.

Carl van der Lingen and David M. Checkley

autumn 2000 (Zhou and Zhu 2002). An OPC mounted on a MOCNESS was used in Southern Ocean GLOBEC on autumn and winter process and survey cruises (Zhou *et al.* 2004) to examine size spectra changes between these two seasons and to estimate zooplankton growth and mortality. The OPC has also been used in GLOBEC-related projects in the northern North Atlantic. In the investigation of *Calanus finmarchicus* migrations between oceanic and shelf seas off north-west Europe (Heath 1999b), the Autosampling and Recording Instrumental Environmental Sampler (ARIES; Dunn *et al.* 1993) was equipped with a Mark II OPC and used in conjunction with the serial plankton collector on ARIES to assess the vertical distribution of *Calanus* and other zooplankton in the Faroe-Shetland Channel (Heath *et al.* 1999b, 2000). The NERC Marine Productivity programme (UK GLOBEC) also used ARIES equipped with an OPC to study the dynamics of zooplankton in the Iceland Basin and Irminger Sea on four cruises during 2001 and 2002 (Heath *et al.* 2008b). Sampling extended to depths greater than 2,500 m. The OPC has also been incorporated into the along-track surface sampling CUFES system used in the SPACC surveys.

6.3.2 Acoustic systems

High-frequency acoustics played a significant role in a number of the GLOBEC field programmes in part due to the rapid pace of technological development of high-speed microprocessors, accessory electronic components, and concomitant software that made a new generation of acoustical instruments possible in the 1990s. Several different acoustic systems have been used. Essentially all of the acoustic sensors were made by commercial companies either as standard off-the-shelf units or as special units configured to meet programmatic requirements. They included single-frequency systems (dual-beam [BioSonics Inc.], split-beam [Hydroacoustic Technologies Inc.-HTI, Simrad Inc.], multiple-beam [Acoustic Doppler Current Profilers, ADCP-Teledyne RDI Inc]) and dual or multiple frequency systems (BioSonics Inc.; HTI Inc.; Tracor Acoustic Profiling System-TAPS, Tracor Inc.; Simrad Inc.). Some were hull mounted systems (Simrad-typically EK500 with 38, 120, and 200 kHz; ADCP-typically 153 kHz), while others were deployed over the side in towed vehicles or profiling systems as described below.

There are two fundamental measurements relevant to the acoustic detection of zooplankton: volume backscattering (integration of the energy return from all individuals in a given ensonified volume, i.e. echo integration) and target strength (TS-echo strength from an individual; Foote and Stanton 2000). Depending upon the construction of the echo sounder and transducers, either or both of these measurements can be obtained. With a single-frequency single-beam transducer, only volume backscattering can be determined directly. A given return cannot be used to discriminate individual size, although statistical procedures have been developed to provide estimates of the animal assemblage size distribution using the data from single-beam transducers (Clay 1983; Stanton 1985a,b). With a series of single-beam transducers operating at different frequencies (e.g. TAPS) and a frequency-dependent theoretical model(s) of back-

scatter from individual animals, it is possible to estimate animal size distribution in addition to volume backscattering (Greenlaw and Johnson 1983; Holliday and Pieper 1989, 1995). Both dual-beam (Ehrenberg 1974) and split-beam (Ehrenberg 1979) systems provide a direct means of determining individual TS. With a dual-beam system, the TS of an animal that is detected with both beams can be estimated directly, but not its angular position. In contrast, a split-beam system, with a receiving transducer array divided into four quadrants, provides both TS and angular position (Chu and Wiebe 2003).

No single echo sounder or acoustic data processing methodology was used in all of the GLOBEC study areas (Table 6.1). Several echo sounders were used in surveys of Georges Bank and the Gulf of Maine: a 420 and 1,000 kHz BioSonics system was mounted in a towed V-fin, the 'Greene Bomber', for early GLOBEC work on the Bank (Wiebe and Greene 1994; Wiebe *et al.* 1996); a more advanced digital BioSonics system was used in the towed vehicle 'BIOMAPER' to survey the bank until lost at sea (Wiebe *et al.* 1997); its replacement, BIOMAPER-II, was used principally to survey the Gulf of Maine and carried an HTI multiple frequency sonar with pairs of up- or down-looking split-beam transducers operating at 43, 120, 200, 420, and 1,000 kHz (Wiebe *et al.* 2002, see Box 6.3 for more details). The Greene Bomber, outfitted with an HTI sonar operating at 120 and 420 kHz, was used to complete the bank surveys.

A variety of echo sounders were used to survey zooplankton in the Southern Ocean GLOBEC programme. BIOMAPER-II was used on all four of the US survey cruises to map zooplankton distributions and krill patchiness (Lawson et al. 2004, 2008a,b). Independently, another 120/420 kHz HTI towed system and a hull mounted ADCP were used on the Southern Ocean process cruises to survey krill and study their larval development and patch dynamics (Daley 2004; Zhou et al. 2004). A Simrad EK 500 (38/120/200 kHz) was also used to observe krill layers while they were being sampled by a MOCNESS to study net avoidance behaviour (Wiebe et al. 2004).

In the north-east Pacific California Current studies, a four-frequency (38, 120, 200, and 420 kHz) HTI system with the transducers oriented in a down-looking configuration in a fixed-depth towbody (15 m) was used to measure the fine-scale backscattering from zooplankton (Sutor *et al.* 2005). This same system was used by Ressler *et al.* (2005) to qualitatively map the distribution of euphausiids along the Oregon and California coast using the difference in backscattering at 38 and 120 kHz to identify euphausiid aggregations. A similar approach was used by Swartzman *et al.* (2005) to study euphausiid distributions along the Pacific Coast using a SIMRAD EK-500 split-beam echo sounder operating at 38 and 120 kHz. In the Gulf of Alaska, a TAPS acoustic system was deployed on moorings on the Seward line.

In the northern North Atlantic, there were two major programmes that utilized acoustics as an integral part of their sampling programmes. The Trans-Atlantic Study of *Calanus* (TASC), which focused on experimentation, modelling, and field sampling of *Calanus finmarchicus*, employed an EK 500 operating at 38 and 120 kHz (Kaartvedt *et al.* 1996; Dale *et al.* 1999, 2001; Bagoien *et al.* 2001). During Mare Cognitum, a regional GLOBEC programme in the Nordic Seas (Greenland, Iceland, and Norwegian Seas;Fernö *et al.* 1997) extensive surveys of zooplankton and fish stocks (principally herring and cod) were conducted in the 1990s also using a 38/120 kHz EK 500 (Kaartvedt *et al.* 1996; Torgersen *et al.* 1997; Misund *et al.* 1998; Melle *et al.* 2004).

In the Arabian Sea, the two GLOBEC cruises used a hull mounted 153 kHz ADCP as the principal acoustic instrument to study diel migration of zooplankton and mesopelagic fish and their spatial distribution (Luo *et al.* 2000; Hitchcock *et al.* 2002). A 12 kHz echo sounder was also used on an ancillary basis (Luo *et al.* 2000).

6.3.3 Conventional zooplankton collection systems

Although imaging and acoustic systems such as those described above provide substantially increased sampling frequency and ease of analysis thus allowing biological and physical gradients in the ocean to be examined at high resolution, they do not eliminate the need to collect animals for species

and stage identification, rate process experimentation (e.g. feeding, growth and development, egg production-see Section 6.5), or biomass determination. Thus, net systems and/or pumping systems were used in all of the GLOBEC programmes to collect animals to determine their spatial distributions and for rate process measurements (Table 6.1). In addition, the net collections were used to ground-truth or calibrate the optical or acoustic measurements, or for inter-comparison purposes.

For quantitative depth-specific sampling, the principal sampling systems used were the MOCNESS (Wiebe *et al.* 1985), Bedford Institute of Oceanography Net and Environmental Sampling (BIONESS; Sameoto *et al.* 1980), Multi-net (Weikert and John 1981), ARIES (Dunn *et al.* 1993), and Ocean (Dunn *et al.* 1989) samplers (Table 6.1). A number of other non-opening/closing nets were also used either for quantitative vertically integrated or oblique sampling, or for collecting animals for experimental purposes. The Bongo net (McGowan and Brown 1966; Posgay and Marak 1980) was used in many of the Pacific and western North Atlantic studies and the WP-2 net was principally used in northern North Atlantic work (Table 6.1). The Continuous Plankton Recorder (CPR, Hardy 1926) was used in the Canadian and UK GLOBEC programmes. A variety of other ring-nets with varying mesh sizes were also used to sample the zooplankton. Also in the north-west Pacific surface tows were made with a Nordic 264 rope trawl with a mouth opening of approximately 30 × 18 m to capture juvenile salmon (Brodeur *et al.* 2004).

Field comparisons between optical/acoustical sensors and net collections: Inter-comparison of optical and acoustical data with net tow collections is an essential part of the process of assuring that the data from a given instrument can be related to data produced by the other instruments and under what conditions they may be valid. A number of studies have been conducted with the sensors typically used during the GLOBEC programmes. The VPR has been the subject of two inter-comparisons with MOCNESS (Benfield *et al.* 1996; Broughton and Lough 2006). The OPC has been the subject of significantly more calibration work (Heath 1999b; Zhou and Tande 2002; Nogueria *et al.* 2004). High-frequency acoustics data produced by the Biosonics and HTI systems has also been used in a number of calibration studies using net collections (e.g. Wiebe *et al.* 1996; Lawson *et al.* 2004; Ressler *et al.* 2005; Sutor *et al.* 2005). In most of these comparisons, taxon-specific abundance and size data were used with appropriate acoustic backscattering models (Lavery *et al.* 2007) to predict the volume backscattering. The predictions were then compared to the observed backscattering. This approach has also been used once to examine the biological interpretation of mean volume backscattering strength of ADCP data (Fielding *et al.* 2004). The VPR data combined with MOCNESS data have also been used to interpret acoustic backscattering data (Benfield *et al.* 1998, 2003; Lavery *et al.* 2007; Lawson *et al.* 2008). A clear message that comes from virtually all of these inter-comparisons is that each sensor or sampler has distinct built-in biases and a high degree of caution and in many cases ground-truthing is required in using them to make inferences about the quantitative distribution and abundance of zooplankton.

6.3.4 Animal tags and telemetry

While a number of GLOBEC programmes have included studies of top predators, most of this work employed traditional ship-based survey methods where the distribution of top predators was correlated with oceanographic features (Chapman *et al.* 2004; Ainley *et al.* 2005; Bluhm *et al.* 2007; Ribic *et al.* 2008). This approach has been critical to developing an understanding of trophic relationships and the importance of biophysical forcing of their distribution. This work has shown that apex predators occur in areas where oceanographic features such as currents, frontal systems, thermal layers, sea mounts, and continental shelf breaks increase the availability of prey (Hui 1979; Haney 1986; Ainley and DeMaster 1990; van Franecker 1992; Hunt 1997; Tynan 1998). All these features and processes are thought to impact predator distributions by physically forcing prey aggregations and, thus, creating areas where foraging efficiency can be increased (Ainley and Jacobs 1981; Croxall *et al.* 1985; van Franecker 1992; Veit *et al.* 1993). Indeed, for many predators, regions of highly localized productivity may be essential for reproduction

and survival (Haney 1986; Costa *et al.* 1989; Fraser *et al.* 1989; Hunt *et al.* 1992; Veit *et al.* 1993; Croll *et al.* 1998, Croll *et al.* 2005).

However, the survey approach has limitations as the associations are limited to population-level studies where the distribution of animals is correlated with oceanography. Although these studies have been and continue to be quite informative, they do not provide insights into the strategies employed by individual animals, nor can they provide information on the spatial or temporal course of these interactions. Advances in satellite telemetry, electronic tags, and remote sensing methods are providing new tools that allow us to follow the movements and behaviour of individual animals. These studies provide insights into the links between predators, prey, and the oceanic environment (Boustany *et al.* 2002; Block 2005; Crocker *et al.* 2006; Shaffer *et al.* 2006; Biuw *et al.* 2007). These new tools make it possible to extend our understanding beyond linkages of prey and predator distributions with environmental features (Biuw *et al.* 2007). The key to understanding the processes that lead to high predator abundance is the identification of the specific foraging behaviours associated with different environmental conditions (Guinet *et al.* 1997).

These new tools or electronic tags have provided field biologists with a new form of 'biotechnology' that allows the study of complex behaviour and physiology in freely ranging animals (Costa and Sinervo 2004). This technology has produced data loggers small enough to be attached to animals while they freely go about their activities (Block 2005; Shaffer and Costa 2006). Information on the movement patterns, depth utilization, and/or diving behaviour are obtained when the tags are recovered (archival tags) or when transmitted via satellite. Archival and satellite linked tags have made possible the study of ocean basin-scale movements, oceanographic preferences, and behaviours of many pelagic species (Delong *et al.* 1992; McConnell *et al.* 1992a,b; Costa 1993; Block *et al.* 1998; Klimley *et al.* 1998; Lutcavage *et al.* 1999; Block *et al.* 2001; Gunn and Block 2001; Boustany *et al.* 2002; Metcalfe 2006). Further advances in data compression have made it possible to get significantly more information through the limitations of the ARGOS system, including detailed oceanographic and behavioural information (Fedak *et al.*

2001). As these new techniques and tools became available they have been incorporated into GLOBEC programmes such as Southern Ocean GLOBEC (Burns *et al.* 2004; Burns *et al.* 2008; Costa *et al.* 2008) and the CLimate Impacts On TOp Predators (CLIOTOP) programme. These new tools used in conjunction with established survey methods are providing an understanding of the distribution of oceanic organisms in relationship to their changing physical and biological environments. There are a variety of new devices that can be used to track fish and other marine organisms such as miniature Global Positioning System (GPS) devices (Rikardsen *et al.* 2007) and GPS tags that can be deployed on marine organisms that frequently come to the surface (www.wildlifecomputers.com). There are a number of programmes that are using these new technologies to gain an understanding of the large-scale movements of marine organisms, such as the Tagging of Pacific Pelagics programme (www.topp.org; Block *et al.* 2003), the Pacific Ocean Tracking Project (www.postcoml.org), and the Ocean Tracking Network (http://oceantrackingnetwork.org).

A comparison of the advantages and disadvantages of the two approaches of studying top marine predators can be seen in Table 6.2. While *in situ* environmental data can be collected using both methods, electronic tags allow us to follow the animals wherever they go. In contrast, survey data are limited to areas where the observation platform can go. In some cases this can lead to a significant bias in our understanding of the distribution of a species. For example, prior to the deployment of electronic tags, northern elephant seals were thought to range just offshore along the west coast of North America (Fig. 6.1a), whereas, tracking data showed that they travel across the entire north-eastern Pacific Ocean (Fig. 6.1b). Tagging data provide a time series that can last from months to in some cases years, and provide behavioural information that can be used to identify behaviours and associated habitats. Depending on the type of tag deployed data acquired can range from a simple surface track (Fig. 6.2a), to a surface track with a dive profile (Fig. 6.2b) or a surface track and dive profile with associated environmental data (Fig. 6.2c; temperature, salinity, and/or light level). Such behavioural data are important to identify differences in the movement patterns and habitat utili-

Table 6.2 Comparison of survey and tagging methods to determine the distribution of marine animals.

Measure of animal distribution and abundance

Survey	*Electronic Tags*
Advantages:	Advantages:
Can sample; hard to study species	Long time series
Environmental data	Animal behaviour
Physical environment	Dive pattern
CTD, chlorophyll	Animal movements
	Home range
	Habitat utilization
	Environmental data
	Physical environment
	CTD, chlorophyll
Disadvantages:	Disadvantages:
Snapshot	Must be able to tag animal
Only know about area surveyed	No direct measure of abundance
Biased measure of range	
Sample bias	
Animal behaviour	

zation of different species. For example, some species may travel over considerable distances (southern elephant seals), while others may remain within a smaller home range (Weddell seals, Fig. 6.3). Such differences in behaviour would not be apparent with traditional survey methods. However, tagging data have some significant limitations. Foremost among these is that data can only be collected from animals that can be tagged and that there is as yet no way to derive estimates of animal abundance. For these reasons, Southern Ocean GLOBEC employed both approaches and is in the process of integrating these complementary approaches to obtain a more complete picture of the ecology of top predators (Burns *et al.* 2004; Ribic *et al.* 2008).

Archival tags: Archival tags are data logging tags that record data as a time series from sensors that measure depth (pressure), water temperature, salinity, animal body temperature, and light level. The major limitation of archival tags is that they must be recovered in order to obtain the data. However, judicious choice of animals or use on exploited species where a reward is offered has provided a wealth of information on the foraging behaviour and habitat utilization of a large group of marine organisms (Block 2005; Shaffer and Costa 2006;

Shaffer *et al.* 2006). Archival tags have provided tracks covering up to 3.6 years (Block *et al.* 2001).

Movement patterns can be derived with archival tags by examining changes in light level to establish local apparent noon. In turn, longitude and day length can be estimated from time of sunrise and sunset to determine latitude (Ekstrom 2004). These locations can be further corrected using sea surface tempera-

tures (SST; Teo *et al.* 2004; Shaffer *et al.* 2006). Salmon researchers have also been using depth and temperature archival tags to discern more about the behaviour and movement of salmonids in relationship to their environment. The data intensity of these devices allows studies of both fine- and large-scale behavioural patterns, migratory routes, and physiology, all in relation to the environment (Boehlert 1997).

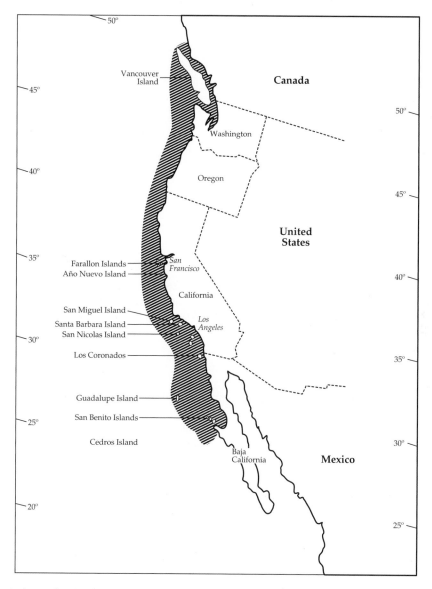

Figure 6.1 (a) Distribution of northern elephant seals as determined using boat and plane based surveys. (continues). Reproduced with permission of Ecological Society of America, from Foraging ecology of northern elephant seals, Le Boeuf, B. J., Crocker, D. E., Costa, D. P., *et al.* Ecological Monographs, **70**, 2000; permission conveyed through Copyright Clearance Center, Inc.).

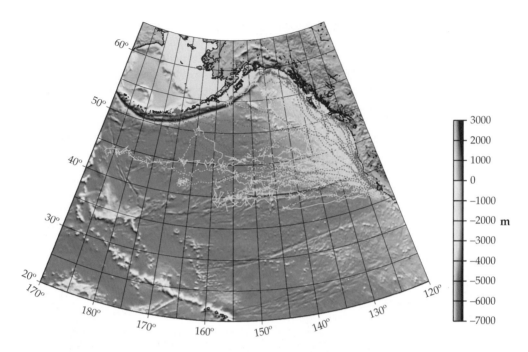

Figure 6.1 (continued) (b) Distribution of northern elephant seals determined using satellite telemetry. Reproduced with permission of Ecological Society of America, from Foraging ecology of northern elephant seals, Le Boeuf, B. J., Crocker, D. E., Costa, D.P. *et al.* Ecological Monographs, **70**, 2000; permission conveyed through Copyright Clearance Center, Inc.). (See Plate 16).

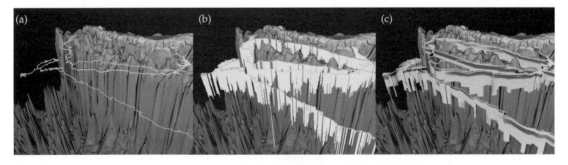

Figure 6.2 Track of southern elephant seals in the western Antarctic Peninsula obtained using the SMRU CTD-SRDL 9,000 tag. (a) Surface track. (b) Shows the surface track and with diving behaviour. (c) The temperature and salinity profiles obtained from animal dives. (From Costa, Goebel, and McDonald, unpublished data.) (See Plate 17).

Argos satellite tags: Satellite tags provide at sea locations and have the advantage that the data can be recovered remotely without the need to recover the tag. Satellite-linked data recorders have expanded our understanding of the fine-scale movements of marine birds (Weimerskirch *et al.* 1993, 2000; Burns and Kooyman 2001), sea turtles, (Renaud and Carpenter 1994; Polovina *et al.* 2000), sharks (Eckert *et al.* 2002; Weng *et al.* 2005), and marine mammals

(McConnell *et al.* 1992a,b; Le Boeuf *et al.* 2000; Shaffer and Costa 2006). Since the antenna on the satellite transmitter must be out of the water to communicate with an orbiting satellite, the technology has mainly been used on air-breathing vertebrates that surface regularly. For large fish and other animals that remain continuously submerged, the ability to transmit at the surface is not possible. For these organisms, a pop-up satellite archival tag (PSAT) has been

Figure 6.3 Differences in the movement patterns of southern elephant seals (yellow), crabeater seals (red), and Weddell seals (green) along the Antarctic Peninsula. The tracks cover the same time period during 2007. (From Costa, Goebel and McDonald, unpublished data). (See Plate 18).

developed (Block *et al.* 1998; Lutcavage *et al.* 1999; Block *et al.* 2001; Boustany *et al.* 2002). Pop-up satellite tags combine data storage tags with satellite transmitters. The pop-up satellite device communicates with the ARGOS satellites that serve to both up-link data and calculate an end-point location. Importantly, the tags are fisheries independent in that they do not require recapture of the fish for data acquisition.

GPS tags: Development of a GPS tag has increased the precision of animal movement data to within 10 m compared to the 1–10 km currently possible with ARGOS satellite tags. Such precision is allowing measurements of animal movements relative to the mesoscale features and will provide higher-resolution locations for the physical oceanographic data collected by the animals. However, standard navigational GPS units require many seconds or even minutes of exposure to GPS satellites to calculate positions and the on-board calculations required consume considerable power. A GPS system that can obtain GPS satellite

information in less than a second and can transmit the location information within the narrow bandwidth of the ARGOS system has now been developed. The Fastloc system uses a novel intermediate solution that couples brief satellite reception with limited on-board processing to reduce the memory required to store or transmit the location.

Marine animals as oceanographers: An exciting, recent development from observing diving predators such as marine mammals, fish, and birds has been the realization that electronic tag-bearing animals can be employed as autonomous ocean profilers to provide environmental data in diverse ocean regions (Costa 1993). A significant advantage of such oceanographic data is that they are collected at a scale and resolution that matches the animals' behaviour (Fig. 6.2). As more environmental information is gathered and delivered from the tagged animals, new insights will be obtained about their individual behaviours, as well as how they respond to environmental variability on daily, seasonal, and interannual time scales. Animal-collected oceanic data can complement traditional methods for assimilation into oceanographic models. The feasibility of marine animals as autonomous ocean profilers has been proven by deployments of temperature and salinity tags on a variety of marine species, such as marine mammals (e.g. Boehlert *et al.* 2001; Hooker and Boyd 2003; Campagna *et al.* 2006; Biuw *et al.* 2007; Costa *et al.* 2008), seabirds (e.g. Weimerskirch *et al.* 1995; Charrassin *et al.* 2002), turtles (McMahon *et al.* 2005), and fish (Weng *et al.* 2005). While the acquisition of such environmental data has been ongoing, only recently have these data begun to be used to address specific oceanographic questions (Charrassin *et al.* 2002; Costa *et al.* 2008).

The most advanced oceanographic tag is the Sea Mammal Research Unit 9000 CTD~SRDL (Satellite Relay Data Logger; www.smru.st-andrews.ac.uk). In addition to collecting data on the animal's location and diving behaviour it collects conductivity, temperature, and depth (CTD) profiles. The tag looks for the deepest dive for a 1- or 2-hour interval. Every time a deeper dive is detected for that 1–2-hour interval, the tag begins rapidly sampling (2 Hz) CTD from the bottom of the dive to the surface. These high-resolution data are then summarized into a set of 20 depth points with corresponding temperatures and conductivities. These 20 depth points include 10 pre-defined depths and 10 inflection points chosen via a 'broken stick' selection algorithm. These data are then held in a buffer for transmission via ARGOS. Given the limitations of the ARGOS system, all

records cannot be transmitted; therefore a pseudo-random method is used to transmit an unbiased sample of stored records. If the SRDLs are recovered, all data collected for transmission, whether or not they were successfully relayed, can be recovered. An example of the kind of coverage provided by these tags can be seen in Figure 6.4.

6.4 Advances in shipboard, laboratory, and *in situ* process studies

GLOBEC process studies have required new experimental approaches to investigating specific mechanisms which are thought to link ecosystem responses with environmental variability. Innovative methods to understand key components of the population dynamics of target species, both zooplankton and fish, have been used, focusing particularly on reproduction, growth and mortality, and between-species interactions. An extensive programme of laboratory experimentation on zooplankton and fish maintained under controlled conditions has been fostered. These experiments have focused on determining vital rates, such as feeding, growth, and reproduction of target species and this information has been especially valuable for model parameterization.

The GLOBEC focus on the influence of global change on marine animal populations required investigation of processes controlling abundance and productivity and how these processes are affected by environmental variability. The abundance

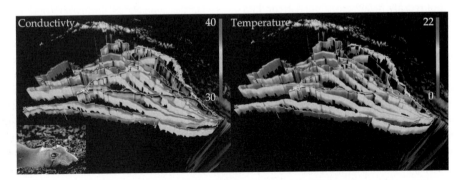

Figure 6.4 Left: conductivity and right: temperature profiles obtained from seven female elephant seals migrating across the North Pacific Ocean. The different coloured lines refer to the tracks of individual seals and the 'curtain' effect shows the depth over which the CTD data were obtained. The coloured bars are the scale for conductivity (mS/cm) and temperature (°C). Inset lower left: female elephant seal with CTD tag on her head. (From Costa, unpublished data). (See Plate 19).

of a pelagic population distributed in some defined volume of the ocean can be expressed as:

1. $dn/dt = (b - d)n - \varepsilon n + \iota n_b,$

where n is the number of individuals per unit volume, b and d are the instantaneous population birth and mortality rates, and ε and ι are the emigration and immigration rates and n_b is the abundance of individuals in the surrounding water (Aksnes et al. 1997). In a population of planktonic copepods, primary target organisms in GLOBEC studies (see Gifford et al., Chapter 4, this volume), the change in abundance is a function of the recruitment rate (R_i) into each life stage, i, the hatching or moulting rate, M_i, into the next stage and the instantaneous mortality rate, d_i, excluding advective terms, as follows (Aksnes et al. 1997):

2. $dn_i/dt = R_i(t) - M_i(t) - d_i(t)n_i(t).$

The variable, M_i, is determined by the stage-specific development rates and the initial input into the population, R_1, can be estimated from the measurement of egg production rates. Understanding change in productivity also involves processes influencing change in mass (typically as carbon or nitrogen), requiring investigation of influences on rates of growth and feeding in planktonic populations (e.g. Omori and Ikeda 1984).

In the following sections, we describe approaches used to measure and investigate processes determining birth, growth, feeding, and mortality rates in the GLOBEC-related studies. A common feature across many of these approaches is the ability to capture and observe living zooplankton in controlled settings, either in shipboard or shore laboratories or in mesocosms. These approaches contributed to and continue to be a source of quantitative understanding and parameterization of the rate processes determining population dynamics and production for incorporation into coupled physical-biological models that simulate effects of climate forcing on the secondary and higher levels of the pelagic ecosystem (e.g. deYoung et al. 2004a; Runge et al. 2005).

6.4.1 Zooplankton reproduction

6.4.1.1 Reproduction of planktonic copepods
Climate forcing may directly impact R_1, the rate of production of new individuals, through the bottom-up influence of ambient temperature and food supply. In the 2 decades leading up to the start of the GLOBEC programmes, laboratory studies of the effect of temperature, food concentration, and food quality on rates of egg production of planktonic copepods (e.g. Corkett and McLaren 1969; Runge 1985b; Kleppel 1992; reviewed for calanoid copepods in Mauchline 1998) had already indicated that these environmental variables could have a strong influence on copepod birth rates. New techniques for estimating female-specific egg production rates (eggs per female per day) of broadcast-spawning planktonic copepods, involving incubation of females immediately after capture (e.g. Dagg 1978; Durbin et al. 1983; Runge 1985b), expanded capabilities for measurement of in situ egg production rates of both broadcast spawners and copepods bearing eggs until hatching, for which egg production can also be estimated by the egg ratio method (Edmondson 1960; Aksnes et al. 1997).

The measurement and understanding of factors determining variability of copepod egg reproductive rates increased enormously during the GLOBEC years. While not all these studies were conducted with GLOBEC support, clearly research from GLOBEC national programmes both stimulated and contributed significantly to advancement of knowledge of zooplankton reproduction. Further assessment (e.g. Laabir et al. 1995; Saiz et al. 1997) and refinements to the incubation method were made (see methodology and procedures review in Runge and Roff 2000) and the method was applied to a wide diversity of copepod species in habitats ranging from the tropics to the polar seas (e.g. Plourde et al. 2001; Durbin et al. 2003; Hopcroft et al. 2005; Napp et al. 2005; Stenevik et al. 2007; Peters et al. 2007). A review of studies of copepod egg production rates collected between 1977 and 1999 showed significant Michaelis-Menten type relationships between egg production rate and chlorophyll a for broadcast spawners but not for egg-bearing species (Bunker and Hirst 2004). Increasing temperature was found to have little effect on broadcast spawners and more so in egg-bearing species (which must wait for temperature-dependent hatching before producing a new clutch of eggs). Considerable focus in North Atlantic and north-east Pacific GLOBEC programmes was on target species in the genera *Calanus*, *Neocalanus*, and *Pseudocalanus*. In the north-west Atlantic in spring and early summer, the egg production rates of *Calanus finmarchicus* followed a hyperbolic relationship with chlorophyll *a*, but with a critical concentration corresponding to an average chlorophyll concentration in the upper 50 m of 1.6–1.8 µg l^{-1}, above which food does not generally limit egg production (e.g. Campbell and Head 2000; Runge *et al.* 2006). This is considerably lower than the critical concentration indicated by the general equations in Bunker and Hirst (2004). In stratified summer waters, however, the relationship with chlorophyll *a* breaks down (e.g. Runge and Plourde 1996; Jonasdottir *et al.* 2005), suggesting different trophic connections to microzooplankton (e.g. Ohman and Runge 1994), and there is clearly high variability in the relationship between *Calanus* species egg production and chlorophyll in the North Atlantic (e.g. Gislason 2005) as well as the upwelling coastal regions of the north-east Pacific (e.g. Peterson *et al.* 2002). Moreover, there may be genotypic variability in the use of internal lipid stores to fuel egg production between populations across basins (cf. Runge *et al.* 2006 and Mayor *et al.* 2006).

A number of GLOBEC-related studies have investigated the composition and nutritional quality of food and the connection to microzooplankton as prey, providing insight into sources of variability in the chlorophyll-egg production relationship. In the laboratory, studies combined culture techniques for maintaining copepods in laboratory settings with sophisticated manipulation and analysis of nutritional composition, including concentrations of essential amino acids (e.g. Helland *et al.* 2003), lipids (e.g. Shin *et al.* 2003; Peters *et al.* 2007), and nitrogen (e.g. Augustin and Boersma 2006).

New methods for estimating egg production rate by assessment of the state of gonadal maturity in copepod females were also developed (e.g. Niehoff and Hirche 1996; Niehoff 2007). By calibrating the state of gonadal maturity in a female population with incubation measurement of egg production in the same population, the female-specific egg production rate can be determined from preserved net samples (Niehoff and Runge 2003). The product of the female-specific rate with the corresponding female abundance (individuals m^{-2}) yields the population egg production rate (eggs m^{-2} day^{-1}). This variable has been used to estimate mortality rate of the eggs and early naupliar stages of *Calanus* species (e.g. Ohman *et al.* 2002; Hirst *et al.* 2007) and can be applied broadly to measure the production of copepod eggs and naupliar stages as prey for early life stages of fish.

6.4.1.2 Hatching success

The measurement, on-board ship or in shore-based laboratories (methodology discussed in Runge and Roff (2000) and Pierson *et al.* (2005); see also Buttino *et al.* (2004) for application of fluorescent probes), of the fraction of eggs spawned that successfully hatch into nauplii took on important significance for understanding the control of copepod populations after the suggestion that feeding by female copepods on diatom diets in the laboratory can result in production of deformed nauplii or complete inhibition of egg hatching (Poulet *et al.* 1995; Ban *et al.* 1997; Ianora *et al.* 2003). Causative agents have been found to be volatile unsaturated aldehydes transformed from diatom derived polyunsaturated fatty acids (Pohnert *et al.* 2002); the ability to form reactive aldehydes varies among diatom species and even among isolates of the same species. This mechanism in diatoms for inhibition of embryogenesis and nauplius development has been argued to be a plant-herbivore interaction that prevents copepod grazers from efficiently utilizing spring diatom blooms and suppresses copepod recruitment (Ianora *et al.* 2004). In a global comparative study, hatching success across copepod species and ocean habitats was found to be high most of the time, even during diatom blooms (Irigoien *et al.* 2002). There were, however, occasional periods of low hatching success, and viability of early naupliar stages was not observed. Shipboard experiments to observe feeding behaviour, reproduction, and hatching success of *Calanus pacificus* in a coastal Pacific ocean habitat revealed that female copepods are capable of avoiding ingestion of harmful diatoms when feeding on a natural mixture of phytoplankton and microzooplankton prey (Leising *et al.* 2005), but that harmful effects of low hatching success and naupliar viability can occur during unicellular diatom blooms when there are few other food choices (Pierson *et al.* 2005).

6.4.1.3 Euphausid reproduction

The euphausids, *Euphausia pacifica* and *Thysanoessa inermis*, were target organisms (see Gifford *et al.*, Chapter 4, this volume) in the US GLOBEC Northeast Pacific programme and *Euphausia superba* was a target organism in the Southern Ocean. Incubation methods have been employed to observe spawning characteristics and egg production of these species. Findings indicate species-specific responses in spawning strategies to differences in food supply in specific environments. Pinchuk and Hopcroft (2006) observed a spawning period during April and May, coinciding with the spring bloom, for *T. inermis* and from early July through October, coinciding with development of seasonal stratification, *for E. pacifica* in the Gulf of Alaska. They found a strong, hyperbolic relationship between brood size of *E. pacifica* and chlorophyll *a*, in which the average chlorophyll concentration at which brood size was maximal was approximately 0.7 µg chlorophyll *a* l^{-1}. In contrast, the absence of a significant relationship between brood size of *T. inermis* and chlorophyll *a* concentrations indicates either reliance on stored lipid reserves or feeding on prey other than phytoplankton. In laboratory observations of spawning behaviour, Feinberg *et al.* (2007) found high variability in egg production characteristics among individuals of *E. pacifica*, suggesting that this species has a very plastic reproductive strategy in different environments. Quetin and Ross (2001) used direct observations of *E. superba* egg production in their analysis of the role of spring sea ice retreat and the extent of spring sea ice in determining the intensity and timing of reproduction and subsequent recruitment into populations along the western Antarctic Peninsula.

6.4.2 Growth and development rates

During the GLOBEC programmes both laboratory and field studies were employed to investigate the growth and development rates of target zooplankton species for the ecosystems under investigation.

6.4.2.1 Laboratory studies

Studies of growth and development rates in the laboratory were undertaken to assess growth in the field (e.g. Campbell *et al.* 2001b), to aid in investigations of mortality (Eiane *et al.* 2002; Durbin *et al.* 2003; Eiane and Ohman 2004; Ohman *et al.* 2004), to calibrate new techniques for estimating growth rates *in situ* (Wagner *et al.* 2001), for use in population and biophysical coupled models (e.g. Lynch *et al.* 1998; Miller *et al.* 1998; Stegert *et al.* 2007), and to allow for estimates of secondary production rates through data integration techniques. A laboratory study of *Calanus finmarchicus* growth and development rates by Campbell *et al.* (2001a) was a key component of the US GLOBEC Georges Bank programme. It was identified early on that these data were sorely needed, especially for the construction of biophysical coupled models that would be used to guide the development of the programme and future studies. The work was designed to investigate the growth and development rates of this target species under a range of temperatures and food concentrations that spanned the environmental conditions encountered on Georges Bank. There were several major findings from the study: (1) Maximum stage-specific development rates as a function of temperature were fully described by a series of Belehrádek functions. (2) The effect of food limitation on development and growth rate was determined. (3) Food requirements for growth were greater than those for development. (4) Growth rates were not equivalent for all stages; it was unwise to estimate secondary production rates of the population from egg production rates alone.

Another series of experiments was undertaken to investigate growth and development rates of *Calanus helgolandicus* under the auspices of the European Trans-Atlantic Study of *Calanus finmarchicus* (TASC) initiative (Harris *et al.* 2000; Rey *et al.* 2001; Rey-Rassat *et al.* 2002b). These studies focused on the effects of food quality on the growth and development rate of naupliar stages and on food concentration with respect to copepodite stages. Rey *et al.* (2001) found that naupliar growth and development rates were different when grown on different algal species and that factors influencing development were different than those for growth.

In a related experiment, Rey-Rassat *et al.* (2002b) described a new method for estimating growth rates in laboratory studies based on the initial weight of each stage that better described the growth within a stage compared to earlier studies. Also, these authors found that food requirements for growth were greater than those for development for copepodite stages, the same conclusion reached by Campbell *et al.* (2001a) for *C. finmarchicus*.

6.4.2.2 Incubations with natural populations

Over the course of the GLOBEC programmes, two main methods were used to determine moulting and/or growth rates for naturally occurring populations of copepods. The first was the artificial cohort method and variations thereof. This method was first proposed by Kimmerer and McKinnon (1987) and involves construction of artificial cohorts from naturally occurring populations by sequential sieving of the catch and a following incubation under ambient environmental conditions. Moulting rates can then be determined from the change in the stage frequency distribution between the initial sample and final sample collected after the incubation period (e.g. Liu and Hopcroft 2006a) or from a series of samples collected over time (Campbell *et al.* 2001b; see papers for details of methods). The main criticism of the method is that non-uniform age distributions within a stage can bias estimates of development rate from moulting rate. Growth rates can be estimated from knowledge of initial and final stage distributions, stage-weights, and incubation time (Liu and Hopcroft 2006a, 2007b). The artificial cohort technique is useful when numerous stages/species are present and it is not practical to sort for single stage incubations, but care must be taken when interpreting results. A second approach is the direct measurement of moulting/growth from incubations (e.g. Renz *et al.* 2007, 2008). This method has the advantage that a direct measurement of growth can be determined from initial and final weight measurements (e.g. Campbell *et al.* 2001b), although it has the same potential bias for estimating development rate as the artificial cohort technique (Hirst *et al.* 2005). To estimate the growth and moulting rates of euphausiids, incubations with individual animals were the method of choice (e.g. Daly 2004; Pakhomov *et al.* 2004; Ross

et al. 2004; Pinchuk and Hopcroft 2007). In these experiments moulting rates were determined in the same manner as for the copepod experiments, and growth rate from the incremental length increase between the euphausiid and its moult, and length:weight relationships.

Incubation studies require a substantial effort to obtain even a very few measurements, but they have been the cornerstone for understanding the variability in growth and development processes of target zooplankton species. They have provided knowledge on the relationships between temperature and food on the growth and development of naturally occurring populations that would otherwise be unattainable (e.g. Liu and Hopcroft 2006a). Comparisons with laboratory measurements and ambient and enriched incubation treatments have provided important insights into the role that food limitation may play in limiting secondary production rates (e.g. Campbell *et al.* 2001a). Although rate measurements from laboratory experiments are often used in biophysical coupled models, the field measurements are necessary for ground-truthing.

6.4.2.3 In situ methods

Several new techniques for estimating growth rates *in situ* have been under investigation for some time (see Runge and Roff 2000). One of the more promising techniques uses nucleic acid ratios, specifically total RNA:DNA ratios measured with a microplate fluorescent assay technique (e.g. Wagner *et al.* 1998, 2001). The obvious advantage of using this technique is the ability to obtain estimates of growth rates of individuals in naturally occurring populations without having to worry about the potential bias of 'bottle effects' associated with incubation techniques. However, it was found that the RNA:DNA ratios were sensitive to temperature, food, and stage of development of the species of interest and therefore, extensive laboratory calibration was required before the approach could be applied to field populations. The technique was used successfully to demonstrate the importance of food limitation on growth rates of *Calanus finmarchicus* on Georges Bank and the Gulf of Maine (Campbell *et al.* 2001b; Durbin *et al.* 2003) and was also shown to be a very good predictor of egg production rates for this same species (Durbin *et al.* 2003). Another approach, employing measurement of aminoacyl-tRNA synthetase enzyme activity level, has been shown to be a significant index of somatic growth of copepodid stages in laboratory experiments, but has yet to be worked out as reliable measure of growth rates in the sea (Yebra *et al.* 2005).

6.4.3 Feeding studies

In the GLOBEC programmes, studies of ingestion of key species of copepods and euphausiids were undertaken in order to understand their feeding behaviour under natural conditions, including selectivity, ingestion rates, and daily food requirements. Generally, two approaches were used: the gut pigment method (Irigoien *et al.* 1998; Pakhomov *et al.* 2004) and bottle incubations (Irigoien *et al.* 1998, 2000; Meyer-Harms *et al.* 1999; Harris *et al.* 2000; Liu *et al.* 2005; Dagg *et al.* 2006). The gut pigment method has best been used to estimate grazing impacts on phytoplankton or as a complement to the bottle incubation approach, but it is not adequate to estimate total food intake because of the importance of non-pigmented microzooplankton in the diets of mesozooplankton. It does however have the advantage of being an *in situ* method and by definition eliminates the question of 'bottle effects'. The method had been criticized in the past because of a belief that pigment degradation in the gut was variable and could destroy up to 90% of the pigment resulting in substantial errors in estimates of chlorophyll ingestion. However, it has recently been demonstrated that the 'disappearance' of pigment both by degradation and evacuation processes is accounted for during gut evacuation rate measurements and therefore the method is valid as long as concurrent estimates of pigment disappearance are determined (see Durbin and Campbell 2007).

The bottle incubation method was the gold standard approach for measuring feeding rates in the GLOBEC studies and probably will continue to be for the foreseeable future. This approach allows for the estimate of total ingestion including both phytoplankton and microzooplankton food sources. Phytoplankton ingestion rates were determined through changes in chlorophyll (Liu *et al.* 2005; Dagg *et al.* 2006), pigment composition by High Performance Liquid Chromatography (HPLC; Irigoien *et al.* 1998; Meyer-Harms *et al.* 1999), cell

counts by automated (flow cytometer, FlowCAM), or microscopic counting methods (Liu *et al.* 2005; see papers for methods). Ingestion of microzooplankton was estimated by the Utermöhl microscopic approach (Irigoien *et al.* 1998; Liu *et al.* 2005). In general, it was found that the target mesozooplankton species fed selectively, sometimes preferring certain phytoplankton groups (Meyer-Harms *et al.* 1999) or microzooplankton (Liu *et al.* 2005). However, microzooplankton were not an important food source when they were not abundant (Irigoien *et al.* 1998). In addition, the method allowed for the determination of predictive relationships between food concentration and ingestion (Dagg *et al.* 2006) and for the estimation of food requirements and seasonal changes in total ingestion (Irigoien *et al.* 1998; Liu *et al.* 2005). The method does not, however, adequately estimate the ingestion of large rare food sources such as small metazoans or particle aggregates. This will most likely come from the future development of new techniques that can quantify rates *in situ* through genetic analysis of stomach contents (e.g. Nejstgaard *et al.* 2003; see Box 6.1).

6.4.4 Zooplankton mortality

The comprehensive GLOBEC field studies made it possible to address zooplankton mortality, an important process affecting zooplankton behaviour, population growth rates (see equation (1) above), abundances, and spatial distributions. Unlike many other rate processes, however, mortality rates relevant to natural populations cannot be measured in incubations in the laboratory or shipboard because it is impractical to include all sources, or even the dominant sources, of mortality within a container. Mesocosm experiments have proven useful for assessing background mortality rates when most predators are excluded (e.g. Hygum *et al.* 2000). Direct measurement of mortality *in situ* has not been done because of the challenges associated with tracking individual zooplankton, although recent technological advances may open the door to this possibility (Steig and Greene 2006). However, several important developments in the realm of indirect means to estimate zooplankton mortality have occurred during the GLOBEC era.

One approach utilizes general life history principles to solve for steady-state mortality rates over a broad range of environmental temperatures and copepod body sizes (Hirst and Kiørboe 2002). These authors observed a negative size-dependence of development rates of planktonic copepods, which is generally consistent with allometric scaling of other biological processes (e.g. Peters 1983). Combining this observation with average egg production rates, Hirst and Kiørboe (2002) solved for the average mortality over a generation. For broad-cast-spawning species, they suggested that average mortality rates of copepods living in the epipelagic zone could be described by body size together with ambient temperature. For egg-sac-bearing copepods, average mortality rates were not related to body size but varied with environmental temperature. Another approach based on allometric principles is the use of plankton biomass size spectra to infer rates of growth and mortality (Edvardsen *et al.* 2002; Zhou *et al.* 2004). In the absence of immigration, emigration, and patchiness, the biomass spectrum is defined primarily by growth, which leads to propagation from smaller- to larger-size classes, and mortality, which reduces abundance within a size class. This approach assumes that all organisms of the same size grow and die at the same rate. Commonly, OPCs (see Section 6.3.1) have been used to assess the biovolume size spectrum, assuming that all particles sensed are living zooplankton, which is not the case in all ocean regions (e.g. Heath *et al.* 1999b; Checkley *et al.* 2008). Both steady-state and non-steady-state applications of biovolume spectra have been reported (e.g. Zhou 2006).

By definition, methods that assume equilibria, such as some allometric methods or the use of Production:Biomass ratios to approximate average lifespan mortality, cannot resolve the time-dependent variations that affect seasonal and interannual variations in populations. Averaged over a growing season or a year mortality may balance birth, but it is the variability in both that determines the timing of population variations and the temporal variability of abundance and secondary production. During the GLOBEC years, different inverse methods have been developed and refined to solve for time-dependent rates in stage-structured populations (Wood 1994; Aksnes and Ohman 1996; Caswell

2001; Li *et al.* 2006). Such inverse methods utilize the observed abundances and stage structure of a field population, usually together with independent estimates of development rates, to estimate mortality rates that would be consistent with the observed stage structure. These inverse methods are commonly described as either *horizontal* life table methods, referring to changes in demographic structure of a population followed sequentially over time, or *vertical* methods, referring to the static stage structure of a population measured at a single point in time. Although both horizontal and vertical methods were under development prior to GLOBEC, they advanced and were applied more extensively during the GLOBEC years. Some of the comprehensive GLOBEC field studies provided unusual opportunities where all essential measurements needed to make these estimates (including ocean circulation, egg production rates, stage-specific abundances and vertical distributions, measurements of food concentration and temperature) were available.

Of the inverse horizontal methods, the Population Surface Method (Wood 1994) was successfully applied to subpopulations of *Calanus* in two Norwegian fjords that were geographically close (ca. 20 km apart), but had markedly different predation regimes. One fjord (Sørfjorden) was dominated by zooplanktivorous fish, while the other (Lurefjorden) had few fish and high population densities of carnivorous zooplankton (Eiane *et al.* 2002). Eiane and co-authors found pronounced differences in the stage-specific patterns of mortality in the two fjords, apparently a consequence of different size/stage preferences of the two groups of predators. McCaffrey (2000) showed that even if the final abundances of adults were the same in the two fjords, the observed differences in stage-specific mortality significantly alter rates of secondary production. A delay-difference method was used to investigate the time-dependent mortality of *Calanus finmarchicus* at Weathership M in the central Norwegian Sea (Ohman and Hirche 2001). These authors suggested that the spring onset of population growth of *C. finmarchicus* may be affected as much by reductions in mortality rate as by increased birth rate. They uncovered a density-dependent mortality relationship for *C. finmarchicus* in the open

Norwegian Sea, apparently caused by cannibalism on eggs, a mechanism demonstrated in the laboratory by Bonnet *et al.* (2004). Density-dependent mortality of eggs and early nauplii was subsequently observed for the same species on Georges Bank in the north-west Atlantic (Ohman *et al.* 2004; Ohman *et al.* 2008). Heath *et al.* (2008b) suggested that cannibalism is perhaps a factor explaining egg mortality of *C. finmarchicus* in the Irminger Sea. Density-dependent mortality of zooplankton has been shown to be a key stabilizing mechanism for plankton predator-prey interactions (Steele and Henderson 1992b). Hirst *et al.* (2007) found a positive correlation between egg mortality and abundance of adult female *Calanus helgolandicus*, but in this case suggested that there were not sufficient females present to account for the observed egg mortality.

Another GLOBEC-related development of inverse methods was the use of an adjoint method to solve for mortality rates of late naupliar and copepodid stages of *Calanus finmarchicus* on Georges Bank (Li *et al.* 2006). In this approach an explicit model of the climatological mean circulation on Georges Bank (Naimie *et al.* 2001) was combined with observed stage structure (Durbin and Casas 2006) and temperature- and food-dependent development rates (Campbell *et al.* 2001b) to estimate the expected moulting fluxes of successive developmental stages, together with mortality rates. The approach resulted in time- and space-dependent mortality rates in such a way that their effects could be compared with the corresponding fluxes from both advection and diffusion (Li *et al.* 2006).

Applications of vertical life table (VLT) methods have proven illuminating in a number of field situations where following the sequential development of a population through time is impractical. The method requires that observed ratios of different developmental stages, as well as their rates of development, are constant for a period of time at least equivalent to the duration of each stage pair and that there be no rapidly passing cohort that causes stage structure to change quickly (Aksnes and Ohman 1996). Such methods were used to compare mortality rates of *Pseudocalanus* spp. and *Calanus finmarchicus* co-occurring on Georges Bank (Ohman *et al.* 2002). This study revealed that the high fecundity of broadcast-spawning *Calanus* is compensated by very high early stage mortality, while the low-fecundity *Pseudocalanus* has correspondingly low egg mortality (Fig. 6.5). Understanding such trade-offs, which have also been modelled theoretically (Kiørboe and Sabatini 1994), is key to forecasting differential responses of different species to changes in climate forcing.

In a study in the California Current, VLT methods revealed that upwelling regions of elevated food supply can also be regions of elevated mortality of *Calanus pacificus* (Ohman and Hsieh 2008), suggesting that there are trade-offs between regions of enhanced food supply and enhanced predation risk.

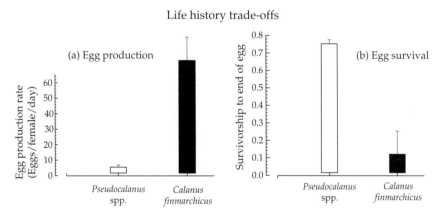

Figure 6.5 Trade-offs between egg production and egg survival for the broadcast-spawning copepod *Calanus finmarchicus* and the egg-carrying *Pseudocalanus* spp. co-occurring on Georges Bank, north-west Atlantic. (From Ohman *et al.* 2002 with kind permission of Springer Science and Business Media.).

In the Irminger Sea, a modified VLT method suggested that nauplius 3 and 4 *Calanus finmarchicus* may have elevated mortalities below a threshold chlorophyll concentration (<0.6 mg m^{-3}, Heath *et al.* 2008b), and that spatial differences in mortality may be key to explaining spatial patterns in recruitment. South-west of Iceland, deep overwintering C5 *C. finmarchicus* show remarkably low mortality rates, which increase in April and June when animals enter near-surface waters (Gislason *et al.* 2007). This study suggests that *C. finmarchicus* successfully minimizes predator encounter through deep dormancy. Another development in the GLOBEC era was the establishment of guidelines for reasonable bounds of mortality rates (Dam and Tang 2001), which have led to reinterpretation of some earlier results.

During GLOBEC, a new perspective emerged on some of the causal agents of zooplankton mortality. The presence of large numbers of resuspended benthically derived hydroids of the genus *Clytia* was rediscovered on Georges Bank (Madin *et al.* 1996; Concelman *et al.* 2001) and their potential to ingest copepod eggs, nauplii, and fish larvae was established (Madin *et al.* 1996). These hydroids are especially abundant in the shallow bank crest region (Concelman *et al.* 2001) where they are resuspended by vigorous tidal shear and associated with mortality of *Calanus finmarchicus* eggs and nauplii (Ohman *et al.* 2008). In the California Current, mass mortality of euphausiids was found to be linked to infestations by a parasitic ciliate (Gomez-Guteirrez *et al.* 2003). Cannibalism by *Calanus* on its own eggs and early nauplii is now recognized to be widespread (Bonnet *et al.* 2004; Ohman *et al.* 2008), and likely occurs in other abundant species of broadcast-spawning copepods (e.g. Landry 1978) and perhaps other types of zooplankton as well.

In addition to the development and application of inverse and allometric methods to field situations, some of the new conceptual insights about mortality that have developed during the GLOBEC era include:

- The importance of timing of mortality in affecting seasonal population dynamics.
- The existence of spatial differences in rates and stage-specific patterns of mortality that lead to unequal risks in different localities.

- The role of density-dependent mortality, operating through cannibalism on early life history stages, as a mechanism altering population growth and ecosystem dynamics.

The focus of much of this work has been on copepods, the biomass dominant in the mesozooplankton in the upper pelagic ocean. General trends and parameters describing growth and reproduction relationships to food and temperature have been analysed and described (e.g. Hirst and Bunker 2003, 2004). Despite its limitations, chlorophyll *a* concentration, which is widely measured and can be estimated remotely, has been shown to have utility as a proxy of food availability.

These results form the foundation for quantitative analysis and parameterization of population rate processes, which in GLOBEC programmes has especially advanced development of life cycle population models of copepod species in the genus *Calanus* and *Pseudocalanus* (e.g. Stegert *et al.* 2007) and coupling to physical circulation models (e.g. Speirs *et al.* 2006; Moll and Stegert 2007).

6.4.5 Estimating growth of fish larvae

Growth and development of marine fish larvae are linked. Both proceed at rates that are dependent on temperature, food availability, and quality, among other environmental and intrinsic factors. There are several approaches to estimating growth rate in marine fish larvae, falling into three general categories: (1) serial sampling of a cohort; (2) otolith microstructure analysis; and (3) biochemical approaches. While the latter two approaches were employed widely in the GLOBEC programme, results from the most widely used of the biochemical approaches, RNA:DNA ratio analysis, are considered here.

Serial sampling of a cohort, while routinely used with cultured larvae and often used to calibrate other methods, is limited in the field due to difficulties in identifying and following individual cohorts, since most marine fish species spawn over several weeks to months. In most instances otolith microstructure analysis is used to age the larvae and to identify cohorts.

Since the 1970s it has been recognized that most marine fish deposit daily rings in their otoliths

(Brothers *et al.* 1976). Otolith microstructure analysis has provided a wealth of information on larval age, growth rate, hatch size, and environmental history (Campana 2005). Estimates of larval age and birth date were critical to estimating mortality rates from field surveys in several of the GLOBEC field programmes (Mountain *et al.* 2008). Growth rate integrated over the life of a larva can be estimated from size-at-age. Based on the assumption that otolith diameter is proportional to larval length, growth history can be back calculated from ring diameter and growth rate over periods as short as a day can be estimated from ring width.

Numerous biochemical and molecular approaches to estimate larval growth rates have been employed with varying degrees of success and acceptance (Ferron and Leggett 1994). The underlying concept is that the concentration or activity of certain constituents, such as enzymes, lipids, hormones and nucleic acids, vary in proportion to food availability and growth rate. The challenge is to identify a constituent that varies reproducibly on the appropriate timescale and to rigorously test and calibrate the method with larvae of known environmental history and growth rate. Bulk ribonucleic acid (RNA) concentration has been related to growth rate in a wide variety of organisms ranging from elephants to viruses. The three classes of RNA, ribosomal, messenger, and transfer, are key components of the molecular machinery for protein synthesis and are regulated in response to the availability of nutrients and the need for protein synthesis. For purposes of estimation of growth rate or nutritional condition in fish larvae, RNA is usually normalized to DNA content, although dry weight and protein have also been used to account for the effect of size (Buckley *et al.* 1999). DNA is the carrier of genetic information and DNA content per cell is usually considered constant. The RNA:DNA ratio is an index of the protein synthetic machinery per cell.

Larval RNA:DNA ratio responds to changes in feeding conditions within about a day or two depending upon water temperature (Buckley *et al.* 1999). This time frame is appropriate to the persistence of many features of the physical and biotic environment important to growth and survival of fish larvae. The relations among larval RNA:DNA ratio, water temperature, and growth rate have been calibrated for a range of species of interest to the GLOBEC programme, including Atlantic cod and haddock. Also, GLOBEC facilitated the comparison of results among species and it now appears that there may be a single relationship among RNA:DNA ratio, water temperature, and growth rate in temperate marine fish larvae that can be used to estimate growth of species for which no species-specific calibration is available (Buckley *et al.* 2008). This development should greatly increase the utility of the approach.

While RNA:DNA ratio analysis requires special handling including sorting at sea and storage at low temperatures, large numbers of larvae can be processed and the analysis can be completed at sea if necessary. Over 10,000 individual cod and haddock larvae and early juveniles were analysed for RNA, DNA, and protein content as part of the Georges Bank programme. This unprecedented sampling effort revealed seasonal and interannual trends in recent growth rate that were related to photoperiod, water temperature, and food availability (Fig. 6.6) (Buckley and Durbin 2006; Buckley *et al.* 2006). While the relationship between photoperiod and growth rate were similar among years, the response to temperature varied among years with distinct maxima in growth rate observed near 6 to 7°C in some years (Fig. 6.6). In other years, when prey was abundant, no temperature optimum was observed. At times growth of larvae was food limited (Fig. 6.8). Although estimates of starvation mortality of young cod and haddock larvae were usually low (<2% day^{-1}), starvation mortality of larvae was particularly high (5 and 9% day^{-1} respectively) in 1995 when their copepod prey were scarce.

Results from the German GLOBEC programme revealed similar seasonal trends in recent growth of sprat larvae in the Baltic Sea (Petereit *et al.* 2008). Maximum growth of sprat larvae occurred between 7 and 8°C shortly before the photoperiod maximum in June. Again, trends in growth rate were related to food availability.

6.5 Zooplankton individual behaviours and population processes

Innovative new approaches have enabled GLOBEC researchers to investigate small-scale behaviour of

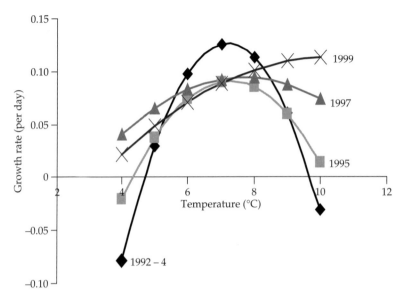

Figure 6.6 Growth rate of larval cod versus water temperature on Georges Bank. (From Buckley *et al.* 2004, 2006.)

both zooplankton and fish and to survey the diversity in behavioural responses. These studies have tackled the effects of physical (e.g. turbulence and light) and biological variables (e.g. behaviour and size) on predator-prey encounter rates, capture success, and feeding efficiency, eventually determining competitive interactions among organisms.

One approach to achieving the GLOBEC aim of predicting zooplankton population processes in the ocean is to make inferences from studies of the behaviour of individuals. This is not a common approach. While a respectable number of studies of individual behaviour in zooplankton have been conducted during the past, few have attempted extrapolations to population processes in more than very general terms. The dynamics of a population, that is, its variation in numbers, age-composition, birth and death rates, and vertical and horizontal distributions are the result of events happening at the level of the individuals. Thus, essential features of the dynamics of zooplankton populations can be understood only in the context of individual behaviours, and descriptions of individual behaviours may, in turn, be used to predict properties of the population. In this section we first briefly summarize past studies of

zooplankton individual behaviours and then, through a few examples, demonstrate how this reductionist mechanistic-behavioural approach to population dynamics may be applied to marine zooplankton populations.

6.5.1 Brief review of behavioural studies

To illustrate the evolution in research focus and the improvement in knowledge over the last 30 years, we will review five major topics where most contributions on zooplankton behaviour fall. We restrict ourselves to larval fish and copepods, with emphasis on the latter, and emphasize studies that directly observe behaviour, but note that it has become increasingly common to couple pure observational studies with 'black-box'-type incubation experiments and modelling.

6.5.1.1 *Feeding behaviour*

The development of high-speed cinematography and copepod tethering techniques in the early 1980s allowed the very detailed observation of copepod feeding at the smallest scales (e.g. Alcaraz *et al.* 1980; Koehl and Strickler 1981; Paffenhöfer *et al.* 1982; Strickler 1982, 1984; Price *et al.* 1983; Price and

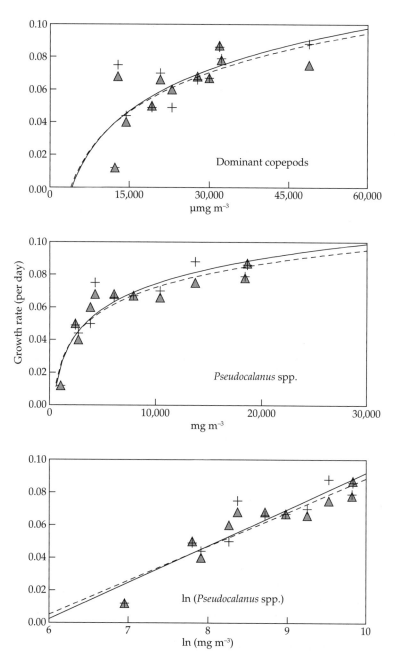

Figure 6.7 Relationship between potential prey biomass and growth rate for 7mm cod larvae on Georges Bank. (Reprinted from Buckley and Durbin 2006 with kind permission from Elsevier).

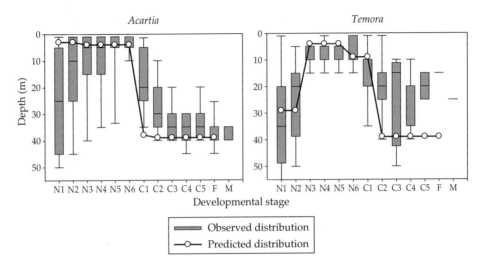

Figure 6.8 Observed and predicted vertical distributions of different developmental stages of two copepod species in a Swedish fjord. Box plot shows median residence depth, the 25 and 75% (boxes), and the 5 and 95% (whiskers) fractiles of the observed depth distributions. The open symbols and lines show the median depth predicted from a habitat optimization model based on mechanistic insights into individual behaviours. The plots also demonstrate the general trend found in other species, of small stages residing near the surface, and larger stages deeper down. (Modified from Titelman and Fiksen 2004.) (Modified from Kiorboe 2006. Copyright 2006 by the American Society of Limnology and Oceanography).

Paffenhöfer 1984). Issues like prey perception (chemo- versus mechanoreception), the generation of feeding currents to enhance prey encounter, and the mechanisms of prey selection and actual capture were thoroughly studied (Paffenhöfer and Lewis 1989, 1990; Yen *et al.* 1991). At somewhat larger scales (millimetres to centimetres and seconds to hours), and with the easy access to video, since the late 1980s a large body of research has been conducted with free-swimming animals. These studies have revealed important aspects of the diversity of copepod foraging strategies (ambush, cruising, and suspension feeding), swimming behaviours and time budgets in response to environmental variables (e.g. Jonsson and Tiselius 1990; Tiselius and Jonsson 1990; Saiz 1994), often combined with incubation experiments to test mechanistic models (Svensen and Kiørboe 2000; Henriksen *et al.* 2007). Regarding visual predators like fish larvae, a similar development has taken place (observational studies: e.g. Munk and Kiørboe 1985; Munk 1992; MacKenzie and Kiørboe 1995b, 2000; Hunt von Herbing and Gallager 2000; Hunt von Herbing *et al.* 2001; modelling studies: Fiksen and Mackenzie 2002; Galbraith *et al.* 2004).

6.5.1.2 *Effect of turbulence on feeding*
The impact of turbulence on zooplankton feeding is a topic that stems from the papers by Strickler (1985) and especially Rothschild and Osborn (1988), who showed theoretically that small-scale turbulence can increase encounter rates between particles, with implications for predator-prey relationships and ecosystem processes as well. Regarding feeding, subsequent studies conducted on copepods and fish larvae have shown that the enhancement of encounter rates due to turbulence (Marrasé *et al.* 1990; MacKenzie and Kiørboe 1995a) interacts strongly with predator behaviour resulting in a dome-shaped relationship which appears to be species-specific (MacKenzie and Kiørboe 1995b; Saiz and Kiørboe 1995; Saiz *et al.* 2003). Due to technical difficulties, only a few studies have actually observed the behaviour of zooplankters under turbulent conditions, either free-swimming (Saiz and Alcaraz 1992; Saiz 1994; MacKenzie and Kiørboe 1995a; MacKenzie and Kiørboe 2000; Seuront *et al.* 2004b) or tethered (Costello *et al.* 1990; Marrasé *et al.* 1990; Hwang *et al.* 1994). In this regard, incubation experiments (Saiz *et al.* 1992; Saiz and Kiørboe 1995; Caparroy *et al.* 1998), and modelling exercises

(MacKenzie *et al.* 1994; Kiørboe and Saiz 1995) have been a required complement.

6.5.1.3 *Predator escape behaviour*
Although some early studies demonstrated the different ability of copepods to perceive fluid disturbances and showed how water mixing could affect it (Singarajah 1969, 1975; Haury and Kenyon 1980), it was not until the 1990s and later that this topic was thoroughly addressed. Two major lines of research have been followed. One line has focused on the morphological adaptation and the neurophysiological performance of mechanoreceptors in copepods (e.g. Yen and Nicoll 1990; Yen *et al.* 1992; Bundy and Paffenhöfer 1993; Weatherby and Lenz 2000; Fields *et al.* 2002; Fields and Yen 2002; Fields and Weissburg 2004), and on the determination of the hydrodynamic signals that trigger the escape response in front of an approaching predator (e.g. Yen and Fields 1992; Kiørboe *et al.* 1999; Doall *et al.* 2002). The second line of research focuses on the perceptive performance and escape ability of copepods (e.g. Buskey and Hartline 2003; Titelman and Kiørboe 2003), and how, from an ecological point of view, the swimming and perceptive performance of copepods may affect predation risk through the effects of enhanced encounter (due to higher motility, higher hydrodynamical conspicuousness) and escape success (which integrates perceptive performance and escape ability) (e.g. Viitasalo *et al.* 1998; Broglio *et al.* 2001; Titelman 2001; Waggett and Buskey 2006).

6.5.1.4 *Motile behaviour*
There have been two major topics in the study of motility behaviour in copepods at the individual level. One of them relates to the ability of copepods to form swarms or aggregations in relation to microstructures in the water column (e.g. Cassie 1959; Owen 1989; Ambler 2002), and has focused on behavioural processes that allow copepods to find and stay in patches (triggered by either physical or chemical gradients, food patches, etc.; e.g. Tiselius 1992; Saiz *et al.* 1993; Buskey *et al.* 1996; Lougee *et al.* 2002; Bochdansky and Bollens 2004; Woodson *et al.* 2005). The second line of research has described swimming patterns and examined their ecological implications (e.g. Buskey 1984;

Buskey *et al.* 1993; van Duren and Videler 1995; van Duren and Videler 1996; Mazzocchi and Paffenhofer 1999; Gallager *et al.* 2004). Several studies have assessed the adaptive value of motile behaviour by means of modelling, for example, effects on encounter rate, optimal foraging behaviour, risk of predation, etc. (Bundy *et al.* 1993; Tiselius *et al.* 1993; Visser and Thygesen 2003; Seuront *et al.* 2004a; Visser and Kiørboe 2006). At a different level, changes in motile behaviour of zooplankton result in patterns of diel vertical migration and swarming behaviour with important demographic consequences in populations (e.g. Batchelder *et al.* 2002a; Zhou and Dorland 2004).

6.5.1.5 *Mate-finding behaviour*
Early studies demonstrated remote mate detection by chemical signals in copepods (Katona 1973; Uchima and Hirano 1988; Yen 1988; and several earlier studies), and later studies provided the details (Doall *et al.* 1998; Strickler 1998; Tsuda and Miller 1998) necessary for modelling mating signals and mate finding behaviour (Yen *et al.* 1998; Bagoien and Kiørboe 2005a,b). Although the number of actual species studied is still limited, very different mechanisms and strategies have been demonstrated (chemical trails: Tsuda and Miller 1998; Weissburg *et al.* 1998; Bagoien and Kiørboe 2005a,b; pheromone clouds: Kiørboe *et al.* 2005; hydromechanical cues: Strickler 1998; Bagoien and Kiørboe 2005b). Despite these differences, a general size-dependent pattern in mate finding capacities has emerged (Kiørboe 2006, 2007).

6.5.2 Individual behaviours and population properties

We consider three examples below, which, respectively, demonstrate how vertical distribution patterns, mortality rates, and population densities can be predicted from a mechanistic description of individual behaviours.

6.5.2.1 *Vertical distribution*
The ocean is stratified, typically such that the temperature and food availability for zooplankton is high in the well-illuminated surface layer, whereas

predation risk from visual predators is low at depth and varies diurnally at the surface. Consequently, many zooplankters undertake diurnal vertical migration, such that they always reside at a depth where the ratio of gain to risk, or some other measure of fitness, is maximized. Vertical migration is well documented and is a classical example of how individual behaviours dictate the distribution of the population. Several authors have elaborated on the general idea of fitness optimization with respect to vertical distribution in zooplankton attempting to predict vertical distributions from behaviour (e.g. Aksnes and Giske 1990; Ohman 1990; De Robertis 2002). The study by Titelman and Fiksen (2004) is particularly illuminating in the present context as it combines detailed mechanistic descriptions of predator encounter risk from visual (fish) and tactile (zooplankton) predators as functions of individual prey behaviours, individual predator avoidance capability, and temperature-dependent growth rates in a habitat optimization model to predict the ontogenetic vertical distribution pattern of various copepods. The general prediction from this exercise is that nauplii and small copepods will reside near the surface, while later developmental stages and larger copepods should reside deeper, consistent with observations (Fig. 6.8).

Similar considerations of the trade-off between feeding opportunities and predation risk would predict that zooplankters should reside shallower in the water column when feeding conditions are poor and deeper when they are better. Such variation in feeding opportunity may be a simple function of food concentration, but may also be mediated by variation in small-scale turbulence. If turbulence enhances the predator-prey contact rate, as suggested by some laboratory studies (see above) elevated levels of turbulence during wind events should lead to a deeper optimal zooplankton residence depth. Although other hypotheses lead to a similar prediction (Pringle 2007), observations of vertical distributions of copepods (Mackas *et al.* 1993; Lagadeuc *et al.* 1997; Incze *et al.* 2001; Visser *et al.* 2001) and fish larvae (Heath *et al.* 1988; Reiss *et al.* 2002) consistently show that these zooplankters reside deeper in the water column during wind events than during calm weather. This prediction is robust, because even in cases where turbulence has a negative effect on feeding,

deeper residence should be preferred in turbulent environments because turbulent intensities typically decline with depth.

6.5.2.2 *Motility and mortality*

Most zooplankters move, either by passive sinking or active swimming, and/or they produce feeding currents. There are gains and risks associated with moving. Specifically, moving enhances the chance of encountering food and mates, but moving also elevates the risk of meeting predators and has energetic costs. The optimal motility is that which maximizes gains over risks, in whatever units are relevant for the situation considered. Here we examine the case of mortalities in mate-searching pelagic copepods. Because it is typically the male that has to find the female, rather than vice versa, males often swim faster and with more directional persistence than females. This implies a higher mortality in males than in females and leads to female-biased adult sex ratios in field populations. The male should swim at the speed which optimizes the number of females he will encounter during his adult life. That speed may depend on the feeding strategy of the male. Some adult males do not feed at all (common among calanoid copepods), others cruise through the water while feeding and thus may feed and search for females simultaneously (most copepods of the superfamily Centropagoidea), while others again are ambush feeders and, thus at any point in time *either* feed *or* search for females (common among Oithonid copepods). Analytical predictions of the swimming velocity that optimizes the trade-offs between mate encounters, predation mortality, and energetics as well as empirical evidence suggest that the optimal swimming velocities of males with these different feeding strategies are dramatically different: ambush feeders swim at very high velocities when they swim, and at orders of magnitude faster than the females; non-feeding and cruise-feeding males swim quite slowly and at speeds that are within a factor of 2 of those of the females (Kiørboe 2008; Table 6.3). Simple models allow one to estimate the ratio of male to female mortalities (or average longevities) from differences in energetics (feeding or not) and swimming speed and, in turn, to predict adult sex ratios in field populations (Kiørboe 2008). The correspondence between observed and

Table 6.3 Observed swimming speeds and predicted male:female sex ratios in pelagic copepods with various male feeding strategies.

Male feeding strategy	Relative swimming speed (body lengths s^{-1})		Observed male:female swimming speed	Predicted male:female sex ratio
	Male	Female		
Ambush (*Oithona davisae*)	27	1.5	18	≥0.1
Non-feeding (*Pseudocalanus elongatus*)	4.4	2.5	1.8	≥0.25
Cruise-feeding (*various Centropagoidea*)	4.0–7.4	1.9–3.3	1.2–2.4	≥0.4–0.8

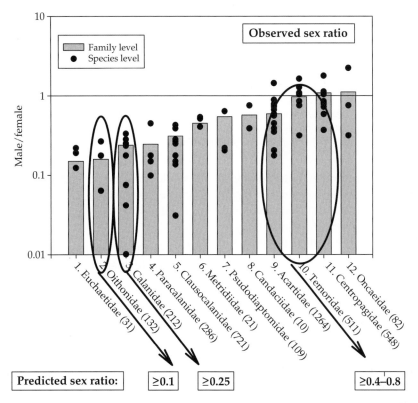

Figure 6.9 Observed and predicted sex ratios of copepod field populations. Sex ratios were predicted from observations and models of individual behaviours as described in the text and detailed in Kiørboe (2008). Field observed sex ratios were based on more than 4,000 samples that were compiled by Hirst and Kiørboe (2002) and taken from Kiørboe (2006). (Modified from Kiørboe 2006. Copyright 2006 by the American Society of Limnology and Oceanography).

predicted sex ratios is sufficient to provide yet another example of how important properties of the population can be predicted from observations of individual behaviours (Fig. 6.9).

6.5.2.3 Mate-finding and population dynamics
In organisms with sexual reproduction there must be a minimum critical population size below which

mate encounters are too rare to allow population maintenance and the population will go extinct. Similarly, at low population densities, population growth may vary in proportion to population density, leading to negative density dependence (Allee effect). These population phenomena may in particular be relevant to small zooplankters that live in a big three-dimensional world where finding a

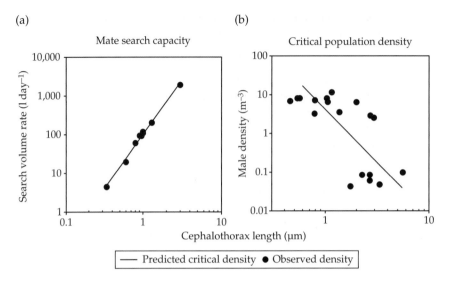

Figure 6.10 (a) Mate search capability of pelagic male copepods expressed as volume searched for female per day. (Modified from Kiørboe 2006. Copyright 2006 by the American Society of Limnology and Oceanography). (b) Predicted critical density of adult copepod males required to maintain a population (line) compared to observed seasonal minimum densities of adult males in the eastern North Sea (closed symbols). (Modified from Kiørboe 2006. Copyright 2006 by the American Society of Limnology and Oceanography).

mate is a challenge. Zooplankton have developed ways to enhance mate encounters: they may aggregate at certain depths for mating (Tsuda and Miller 1998), they may form mating swarms (Ambler 2002), and they may advertise their presence and position to potential mates using hydromechanical or chemical signals (Katona 1973; Strickler 1998). In pelagic copepods the latter appears to be the most widespread means of enhancing mate encounters, and mate search capacities are substantial, yet finite (Fig. 6.10). Critical population densities can be predicted from estimates of mate search capacities and mortality rates (Gerritsen 1980; Kiørboe 2006; Choi and Kimmerer 2008) and appear to fit seasonal minimum densities pretty well (Fig. 6.10). It also follows from these considerations that there may be negative density-dependent population regulation at low but typical densities of pelagic copepods, which can explain why winter population densities appear to have such a strong impact on population densities during the subsequent summer, several generations later (Kiørboe 2006). Critical population densities and negative density dependence may in particular be pronounced in populations with very biased sex ratios, as seen for many pelagic copepods (see above). In fact, the relative scarcity of males in some populations may

lead to fertilization limitation and substantially reduced population growth, even when food is plentiful (Kiørboe 2008).

6.6 Methods applied to retrospective studies on past ecosystem states

GLOBEC programmes have used retrospective analyses and time series studies to identify and understand the characteristic, natural, modes of historical forcing and marine ecosystem variability over a range of temporal scales. Early studies detected global synchrony of fish abundance and its significant correlation with various climatic indices in the twentieth century, while studies using palaeo-climate proxies such as fish scales have revealed centennial-scale variations. Many recent studies have conducted community-level or functional-level analyses on historically collected zooplankton samples rather than comparing biomass alone. The aim has been to elucidate mechanisms responsible for observed variations in abundance. These approaches have revealed biogeographical shifts and changes in phenology at lower trophic levels induced by climate and physical forcing, and subsequent match-mismatch with higher trophic levels. Multivariate analyses to investigate spatio-temporal variation of plankton communities have been applied, and inter-calibration of sampling gear has been necessary to deal with historical samples collected using different methods.

It is widely known that the catch of commercially important fish, such as sardine, salmon, herrings etc., many of them target species (see Gifford *et al.*, Chapter 4, this volume), fluctuates on a multi-decadal scale. Retrospective studies in the early years of GLOBEC detected significant correlations between regional ecosystem changes and large-scale climatic forcing indicated by the various climatic indices, for example North Atlantic Oscillation

(NAO), Pacific Decadal Oscillation (PDO), Southern Oscillation Index (SOI), etc. (see Drinkwater *et al.*, Chapter 2, this volume). The Kawasaki diagrams (Kawasaki 1991) exhibited global synchrony at a multi-decadal scale of variations in the abundance of common fish species, providing evidence of the influence of large-scale climatic forcing on regional ecosystems. Analysis of fish scales in anoxic sediments has revealed the fluctuation of fish abundance over the past 2 millennia far before commercial exploitation started (see Box 6.5 for the details). All these facts demonstrate obvious climate-ecosystem links, but what mechanisms lie behind these links?

Box 6.5 Fish debris indicate many modes of variability

As GLOBEC-related studies have the overall objective of understanding the mechanisms of physical and biological change, it is worthwhile to review palaeo studies when considering recent observations of ecosystem change. Fishery catches in the twentieth century suggest several paradigms of ecosystem variability. There are alternations in the catch of sardines and anchovies in various regions of the world as well as opposing variations in abundance between salmon catch in the Gulf of Alaska and salmon and pelagic fish in the California Current and these fluctuations appear to have a 50–60-year periodicity (Lluch-Belda *et al.* 1989; Kawasaki 1991; Mantua *et al.* 1997; Chavez *et al.* 2003). However, the palaeo-archives from sedimentary records, such as fish scales, nitrogen isotope signatures in Alaskan lakes, and other palaeoclimate proxies, indicate that the modes of variability observed in the twentieth century only exemplify a small portion of the total range of past variability. Seasonal variations in sediment flux to hypoxic sediments results in the presence of laminae or annual varves in the bottom sediments. The lack of oxygen both inhibits benthic organisms from disturbing the sediment layers as well as preserves the remains of fish scales, which sink to the seafloor.

Box 6.5, Figure 1 Fish scales typical of an upwelling community, mostly sardines and anchovies from sediment core off Namibia (diameter of fish scales around 5–10 mm). (From Struck *et al.* 2002.)

continues

Box 6.5 *continued*

Within Santa Barbara Basin in the California Current, there can be a persistent high abundance, or complete absence, of sardine scales in sediment layers corresponding to nearly 100 years duration (Baumgartner *et al.* 1992). Assuming that fish scale deposition off central California, which is near the center of the population ranges, generally reflects small fish pelagic populations in the California Current, then the 50–60 year periodicity is not likely to be a good predictor of future changes.

There are also co-occurring periods of high- and low-scale abundance for sardines and anchovies for decades at a time, both within the California Current (Baumgartner *et al.* 1992) and within the Humboldt Current (Valdés *et al.* 2008). Thus, while sardines and anchovies within a given region vary out of phase at times (e.g. the twentieth century), this pattern is not consistent through time and offers little chance of predictability.

Furthermore, centennial-scale periods of high sardine-scale fluxes in the California Current coincide with low inferred salmon returns to the Gulf of Alaska, whereas the two species co-varied in abundance during the twentieth century (Finney *et al.* 2002). Also, sardines did not appear to co-vary off California and Japan during the nineentieth century (Field *et al.*, in press), as they

did during the twentieth century. Therefore, the alternations between different regions observed in the twentieth century may not be typical of all modes of climate variability.

Thus, many of the patterns of variability in fish populations, as well as other palaeoclimate records of sea surface temperature (SST) in different regions of the Pacific (Gedalof *et al.* 2002; D'Arrigo *et al.* 2005), indicate that the Pacific Decadal Oscillation (PDO)-like interdecadal variability observed in the twentieth century is not consistently observed in the palaeo records. Work in progress off Peru and in the Benguela Current is also revealing modes of variability that have not been observed in the twentieth century. Inferred SST from the alkenone unsaturation index, in the Benguela Current showed an abrupt shift nearly 1,000 years ago. The percent of anchovy scales is apparently higher during warmer SST periods, a pattern not observed in other boundary currents.

Research programmes that test the paradigms and establish the mechanisms of change offer hope for interpreting complex ocean processes and histories. Rigorous testing of existing paradigms and hypotheses is essential for understanding and predicting future marine ecosystem changes.

David Field

To answer this question, and to construct future ecosystem change scenarios, GLOBEC has both developed and applied various analytical methods for historically collected samples and data.

6.6.1 Community- and functional-level analysis

Long-term zooplankton collections, which were historically sampled during the mid to late twentieth century mainly as a part of fisheries surveys, are to be found in institutes worldwide. While early analyses of these considered the annual average of total plankton biomass, most recent studies have applied community-level and functional-level analyses. Classification of zooplankton

species based on their geographical distribution range has revealed biogeographical shifts of zooplankton communities induced by climate and physical forcing. The Fourth IPCC Report pointed out that northward shifts of southern species are a globally observed phenomenon corresponding to the warming trend over recent decades. Although direct influences of global warming are not certain, the northward shift of southern plankton species associated with northern intrusion of warm water has been reported both in the eastern North Pacific (Mackas *et al.* 2004) and eastern North Atlantic (Beaugrand *et al.* 2002a).

Various statistical and multivariate analysis methods have been used for time series decomposition not only of zooplankton but also phytoplankton communities. However, selection of the methods and application of them in an appropriate manner is crucial to extract a 'pattern' of temporal and spatial variation. Beaugrand *et al.* (2003b) reviewed and discussed a series of multivariate methods, including a range of Principal Component Analyses (PCA), non-metric multidimensional scaling (MDS), cluster analysis, and spectral analysis (Table 6.4). These methods were used to effectively extract

information from the data collected by the CPR survey in the North Atlantic, which are a highly extensive, both temporally and spatially, 50-year zooplankton data set. Biodiversity of the zooplankton community can be used as an indicator of variation in physical and climatic conditions. Using a similar approach based on species richness and the Shannon-Weaver Index, Hooff and Peterson (2006) detected a relationship between copepod biodiversity and transport of coastal subarctic waters into

Table 6.4 Types of multivariate analysis performed on CPR data.

Multivariate techniques	Ecological goal	Authors
Standardized PCA	Identification of species assemblages. Examination of the relations between species. Geographical locations of species associations.	Colebrook (1964, 1984)
Centred PCA	Determination of seasonal and diel patterns of months to hour diversity of calanoid copepods. Quantification of two scales of variability of diversity of calanoids at a mesoscale resolution in the North Atlantic. Examination of the spatial variation of the diversity of calanoids at diel and seasonal scales.	See Table 6.2 in Beaugrand *et al.* (2003b)
Seriation	Examination of the relations between species based on their annual fluctuation in abundance.	Colebrook (1964) Colebrook and Robinson (1964) Colebrook (1969)
Cluster analysis, single linkage agglomerative (nearest-neighbour) clustering method	Grouping of species or taxa.	Lindley (1987) Lindley and Williams (1994)
Cluster analysis, hierarchical agglomerative flexible clustering technique (Lance and Williams 1967)	Clustering of pixels or geographical areas to identify regions with similar year-to-year or annual patterns in the abundance of species.	Planque and Ibñez (1997) Beaugrand *et al.* (2000a)
Cluster analysis, complete linkages agglomerative clustering	Partition of the North Atlantic Ocean based on the diel and seasonal patterns of diversity of calanoid copepods.	Beaugrand *et al.* (2000b)
Indicator-value method (Dufrêne and Legendre 1997)	Determination of species associations based on the relative abundance and presence of species in distinct areas in the North Atlantic.	Beaugrand *et al.* (2000b)
Non-metric MDSg	Ordination of species or taxa based on the similarity of their spatial distribution.	Lindley (1987) Lindley and Williams (1994)
Mantel correlogram	Study of relationships between the size of spatial structures and their temporal variability.	Planque and Ibañez (1997)
Generalized additive models	Spatial and temporal modelling of the abundance of species.	Beare and McKenzie (1999a, 1999b)
Three-mode PCA	Analysis of biological tables structured in space and time. Evaluation and quantification of the interactions between biology, space, and time.	Beaugrand *et al.* (2000)

Source: From Beaugrand *et al.* (2003b).

the northern California Current, which was driven by a basin-scale climatic forcing.

At higher trophic levels, studies have shown that long-term variations of fish abundance are species-specific, and the mechanisms of variation have been investigated for the respective target species in a number of ecosystems. It is thought that life history strategy is a key determinant of productive success and survival of the year class when an environmental perturbation occurs. With the aim of providing a conceptual framework for fisheries management, King and McFarlane (2003) classified fish species based on their life history strategies, Periodic, Equilibrium, Opportunistic, Salmonic and Intermediate, and examined how long-term variation patterns could differ among these groups.

6.6.2 Detecting phenology

The fourth IPCC report mentioned changes in seasonality, phenological change, also as a result of the impact of the warming trend on global ecosystems. Community-level or species-level analysis of zooplankton populations enables phenology to be detected at lower trophic levels of marine ecosystems. These changes, induced by climatic and physical forcing, affect productivity of higher trophic levels through match-mismatch mechanisms, bottom-up, or top-down controls.

The Cumulative Sum (CuSum) technique is a simple method developed to visualize the extent and duration of change of time series variables by cumulating the value at each data point consecutively in a temporal order. When the temporal resolution of data is high enough (e.g. monthly or more), CuSum has been applied to the time series abundance/biomass of target species to detect shifts of peak abundance/biomass and reproductive timing. Greve et al. (2001) studied the phenological shift of appendicularians in the North Sea in the late twentieth century using the CuSum technique, by setting the interval between the 15 and 85% percentiles of the annual cumulative abundance as the productive season of each year (Fig. 6.11).

Interannual variation in seasonal developmental stage composition of target species is useful in understanding lower trophic level phenology.

Calculating the timing at which the CV copepodid stage reached 50% of the total abundance and assuming this to be an indicator of peak reproductive timing of *Neocalanus* species, Mackas et al. (1998) detected a decadal-scale shift of the peak in the North Pacific which was closely related to the water temperature anomaly associated with the PDO.

Remote sensing techniques are relatively new and an extensive data set only exists since the late 1990s. However, this tool is extremely useful for better understanding of detailed spatio-temporal variations of marine ecosystems, such as phenology, due to its high-observation frequency. Yamada et al. (2006) developed a method to estimate timing and duration of the spring bloom during recent decades based on chlorophyll *a* variation from satellite ocean color, ship and buoy data obtained for the period 1998–2003 in the Japan Sea. These authors found a decadal-scale phenology of phytoplankton productivity: an early and short productive season during the 1980s, as inferred from seasonal ship observation-based study. This method will be applicable for other regions as remote sensing data accumulate.

6.6.3 Stable isotope analysis

Compared to zooplankton, information on primary production other than total chlorophyll *a* data are scarce for the past decades, and community-level and functional-level analyses of phytoplankton were limited in most of the GLOBEC regions. This makes it difficult to clarify the response of phytoplankton to long-term environmental variation. However, chemical properties of secondary and higher trophic level organisms sometimes tell us cumulative information about phytoplankton they have fed on: its availability, physiological conditions etc. The nitrogen stable isotope ratio ($\delta^{15}N$) indicates the trophic level of organisms, and carbon stable isotope ratio ($\delta^{13}C$) of higher trophic level organisms mirrors the condition of primary producers. Therefore, $\delta^{15}N$ and $\delta^{13}C$ of target species of secondary producers and the higher trophic levels can be a proxy of primary productivity and its response to environmental variation in the past.

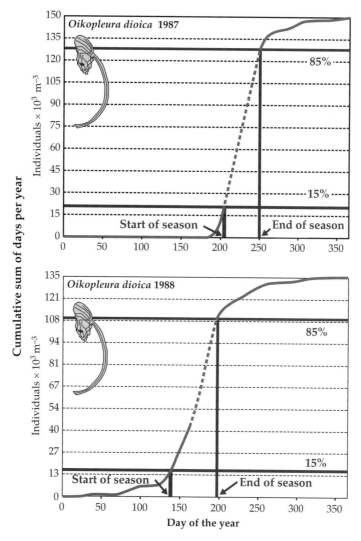

Figure 6.11 CuSum methods of detecting phenology; phenological determination on the season of *Oikopleura dioica* in 1987 (cold) and 1988 (warm). (From Greve *et al.* 2001.)

This method is useful for detecting changes in a food web structure and mechanisms of bottom-up control.

Observing a 50 year decreasing trend of $\delta^{13}C$ in the baleen of bowhead whales, even after considering the degree of $\delta^{13}C$ decline due to anthropogenic fossil fuel consumption Schell (2000) suggested a decline of primary productivity in the Bering Sea during the latter half of the twentieth century. In the California Current system, long-term variation of $\delta^{15}N$ in several key zooplankton species has been examined together with physical and climate indices (Rau *et al.* 2003). Although no trend was observed in $\delta^{15}N$ of those species, *El Niño*-related $\delta^{15}N$ enrichment was conspicuous, which was considered to be a result of (1) reduced nutrient supply due to weak upwelling, or (2) increasing advection of $\delta^{15}N$-enriched nitrate from southern water. Stable Isotope ratios were also applied to fish scale analysis for the past 2 millennia (Finney *et al.* 2002).

6.6.4 Making various time series comparable

Regional information on long-term changes at lower and higher trophic levels has accumulated during the last decades of the twentieth century. The next phase of GLOBEC retrospective programmes is integration and synthesis of regional ecosystem responses to common, large-scale climatic forcing. Having recognized synchrony in the long-term variation in commercial fish species (Kawasaki *et al.* 1991), scientists next attempted basin- to global-scale comparison of existing zooplankton time series to understand regionally specific mechanisms of ecosystem change. There were, however, a number of impediments for comparison of these time series (Perry *et al.* 2004), which were collected at various sampling frequency with a wide variety of sampling gear and at different target depths. Many time series were based on a single-season (mostly during the high-productivity period) or seasonal observation, and sometimes temporarily intermittent. Sampling gear and methods were sometimes changed during a single time series. Systematic solutions are required to tackle these impediments to inter-time-series comparison.

One of the most extensive zooplankton time series, the Station Papa time series in the Gulf of Alaska was collected with three different sampling gears, the NORPAC net and SCOR net, from 1956 to 1981, and the Bongo net after 1997. Mckinnell and Mackas (2003) recalibrated these three nets and found that previous biomass estimation based on the early calibration method overestimated the SCOR net catch. According to the new criteria, the average time series biomass reduced especially in the early period, resulting in changes in the long-term trend in the zooplankton time series.

A simple solution for comparison of regional time series with different characteristics is to standardize annual biomass data and environmental variables, for example, water temperature, for each time series. Mackas *et al.* (2004) applied log-scale anomalies for comparison of interannual variation of three independent zooplankton time series covering 850 km distance along the continental margin of the eastern North Pacific. They found spatial coherence in low-frequency variation of the zooplankton community in the three regions with a marked transition coinciding with the North Pacific Regime Shift (1988–91, 1998/9) and/or the ENSO event (Fig. 6.12). Standardization methods other than the log-scale anomalies have been applied for other regional comparison studies depending on characteristics of the time series.

6.6.5 Time series analysis: red-noise or shift, linear or non-linear

Although long-term changes in marine ecosystems and the influence of climatic forcing on these changes are now widely recognized, determination of the 'type' of change (oscillation, shift and/or trend) has been an area of debate in retrospective studies.

By composite analysis of 100 physical and biological time series, Hare and Mantua (2000) suggested that environmental and ecological regime shifts occurred in the mid-1970s and the late 1980s in North Pacific. Rudnick and Davis (2003) challenged the definition of the regime shift based on the composite of time series by demonstrating that composite analysis of random, independent red-noise time series could generate such a step-like change. However, non-linear, step-like changes were indeed demonstrated to be a characteristic of biological time series by Hsieh *et al.* (2005). They tested non-linearity of a series of physical and biological time series in the California Current system by comparing the out-of-sample forecast skill of a linear and equivalent non-linear models, and concluded that, while all of the climatic and physical time series showed a linear red-noise pattern, biological time series responding to such linear changes in physical and climatic forcing almost exclusively varied in a non-linear manner. This result suggests the capacity for dynamic changes in marine ecosystems responding to the low-frequency fluctuation of physical environments. To statistically test the timing of the 'shift', Rodionov and Overland (2005) developed an improved regime shift detection method based on the sequential t-test analysis (STARS) (Fig. 6.13), and an application tool is available at (http://www.beringclimate.noaa.gov/regimes/).

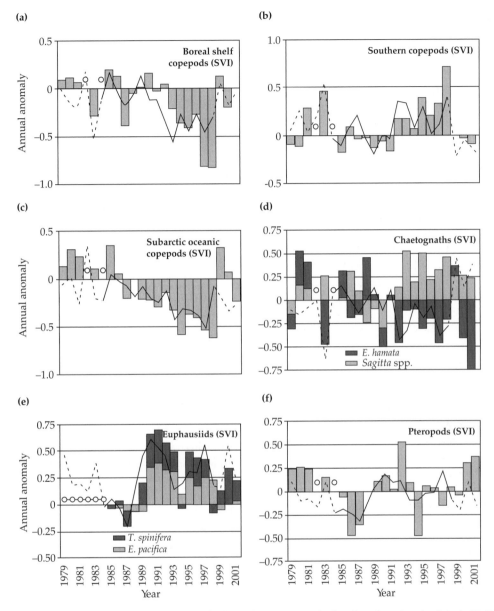

Figure 6.12 Zooplankton anomaly time series, 1979–2001, for the southern Vancouver Island continental margin region (latitude 48–49°N). Species groups for averaging and/or comparison are (a) 'boreal shelf' copepods, (b) 'southern' copepods, (c) 'subarctic oceanic' copepods, (d) chaetognaths (*Sagitta* spp. vs. *E. hamata*), (e) euphausiids (*Euphausia pacifica* and *T. spinifera*), and (f) thecosomatous pteropods (*L. helicina* and *Clio pyramidata*). Bar graphs are the annual zooplankton anomalies, averaged over the entire southern Vancouver Island region (anomalies in shelf and offshore regions are highly correlated). See Mackas *et al.* (2001, 2004) for calculation and year averaging methods. SVI anomalies include samples from all seasons, however about two-thirds of data are from spring and summer (April–September). Circles show years with no anomaly estimates due to low sample numbers or gear bias. Lines show regression fits of the anomalies to ocean climate indices; solid lines are the 'predicted' anomalies for a 'learning set' of years (1985–98) used to estimate the regressions; dashed lines show 'predictions' for the remaining years (1979–84 and 1999–2001). Note the strong inverse correlation of the 'southern' versus the 'boreal shelf' and 'subarctic' copepod groups, and the rapid change in sign of the anomalies 1998/9.(Reprinted from Mackas *et al.* 2004 with kind permission from Elsevier).

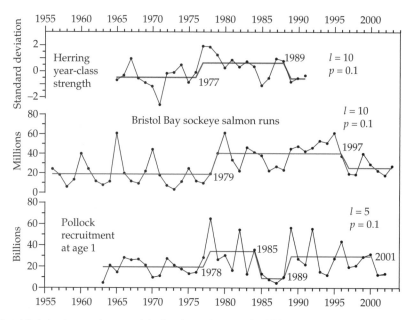

Figure 6.13 Regime shifts in herring year-class strength (*top*), sockeye salmon runs (*middle*) and pollock recruitment (*bottom*) in the Bering Sea. (From Rodionov and Overland 2004. Application of a sequential regime shift detection method to the Bering Sea ecosystem. ICES Journal of Marine Science, **62**, 328–32. By permission of Oxford University Press).

6.7 Future directions

The need for programmes such as GLOBEC and comparable future programmes is becoming ever more compelling on the international stage as the public becomes evermore aware of the effects of global change on marine ecosystems. Increasing occurrences of red tides, the collapse of major fisheries, loss of coral reef communities, global warming effects on ecosystems, and ocean acidification, are just some of the emerging problem areas (e.g. Dickey and Bidigare (2005): table 6.2). Thus, there is strong justification for increases in financial resources that can be directed to fundamental and applied oceanographic research, ocean observing systems, and predictive oceanographic models that bear on these issues. It will be important for marine ecologists and oceanographers to utilize research funding most effectively to solve these problems and to assist in management decisions.

A wide variety of technological advances have contributed to improved understanding of the structure and functioning of the global ecosystem as it is affected by and in turn affects the physics and chemistry of the ocean. GLOBEC research,

spanning the past decades has been influential in this regard (e.g. Dickey 1993). This chapter has highlighted many relevant technological and methodological advances either as a direct result of the programme, or stimulated by it. Clearly, the groundwork has been laid for future breakthroughs that will benefit succeeding research. GLOBEC has also contributed significantly to the general science of life in the ocean and how it may change in the future as the physical and chemical environments evolve in response to natural and anthropogenic forcing. We conclude this chapter with some conceptual ideas concerning future directions for sampling and observing ocean systems, experimentation, as well as the necessary linkage with models.

At the outset of GLOBEC, it was recognized that a very large number of interdisciplinary variables needed to be sampled on space- and time scales covering over 10 orders of magnitude (e.g. Dickey 1988, 1990). Several new *in situ* sampling platform technologies emerged and/or became more available during GLOBEC. These include autonomous sampling fixed-depth and profiling moorings (e.g. Dickey *et al.* 2006), profiling floats (e.g. Davis *et al.*

2001; Argo Science Team 2001; Bishop *et al.* 2002; Perry and Rudnick 2003), autonomous underwater vehicles (Griffiths 2003; Perry and Rudnick 2003), and gliders (e.g. Davis *et al.* 2003; Perry and Rudnick 2003). Many of these platforms include near-real-time data telemetry systems. Looking towards the future, miniaturized, low-power biological, optical, chemical, and acoustic 'chip-based' sensors will be developed and will be suitable for interfacing to these and other novel platforms (e.g. Dickey and Bidigare 2005). Already, there is great promise as indicated by microelectromechanical systems (MEMS; Tokar and Dickey 2000) and nanotechnologies (e.g. Bishop *et al.* 2001), which are being developed at a rapid pace for a host of applications. While there will always be a need for ship platforms for certain observations, oceanographers will need to be especially vigilant in following these developments and will need to form partnerships to facilitate the widespread availability and application in field programmes of these new sensors and telemetry methods.

While GLOBEC developed much new technology that is reviewed in the chapter it is interesting to reflect that many of the key advances in understanding came from conventional approaches to sampling and experimentation. 'Working with the animals' through traditional net sampling and shipboard and laboratory experimentation has been a fundamental foundation of the programme. The advances based on these approaches are well represented in the studies reviewed.

Particularly, promising future directions involve the application of molecular and biochemical techniques to complex species assemblages and their interactions within marine ecosystems (e.g. DeLong *et al.* 1999). A specific example is provided in Box 6.1 demonstrating the potential of the application of techniques from the rapidly advancing field of molecular biology.

The shipboard, laboratory and *in situ* process studies have provided fundamental insight and data needed to formulate and parameterize essential processes controlling abundance of zooplankton and ichthyoplankton species targeted in GLOBEC programmes. Coupled physical-biological modelling is still at an early stage. Nevertheless it shows the potential to extract from the complexity

of the ecosystem and population dynamics processes, simplifying formulations and approaches. These will allow evaluation of effects of environmental forcing on species abundance and distribution and on processes determining ichthyoplankton survival, with implications for use as a tool in ecosystem approaches to management. The challenge for the future is to build on this foundation of methodological approaches and integrative tools. A key issue is to establish appropriate observing systems and gather knowledge of population dynamics and ecosystem production processes that will provide us with useful predictions of change in the coastal and open ocean ecosystems.

The utilization of satellites and aircraft for biological applications has been well developed for phytoplankton and primary productivity. Recent advances in hyperspectral optical sensing of the ocean for both *in situ* and satellite platforms bode well for identifying at least groups if not species of phytoplankton (e.g. Dickey 2004; Dickey *et al.* 2006). However, major development is needed in order to capitalize on these platforms, which can in principal provide large-scale upper-ocean sensing, for studies of higher trophic level organisms and their distributions. Satellite-based data telemetry (i.e. for near-real-time data transmission) and positioning information will continue to advance both in quality and quantity. Tracking of organisms from aircraft and satellites can be especially powerful as has been demonstrated (see Section 6.3.4.). Sound transmission in the sea remains one of our most important *in situ* sensing methods for zooplankton and higher trophic level organisms (Foote *et al.* 2000; Chu and Wiebe 2003; Wiebe and Benfield 2003). In analogy to hyperspectral optics, broadband, multi-frequency acoustics has perhaps one of the greatest potentials and can in principle allow studies of trophic interactions, especially if deployed in conjunction with video and holographic methodologies.

There is growing consensus that observationalists and modellers need to coordinate their efforts and the studies reviewed in this chapter illustrate good progress in this regard (e.g. Robinson and Lermusiaux 2002). For example, development of sampling strategies, adaptive sampling, and more traditional inter-comparisons of data and model results are key to future breakthroughs in understanding as well as to prediction (e.g. Dickey 2003). Predictions of the state of global ecosystems on short as well as very long time scales are clearly needed for a host of societally relevant issues involving the stewardship of the world ocean resources and human health. Advances in computational capabilities will offer modellers opportunities to make high temporal and spatial simulations of complex ecosystems and the physical and chemical environment. Education of the next generation of oceanographers would be well served by not only interdisciplinary training, but also exposure of students to both theoretical and observational research modes, regardless of thesis emphasis.

Finally, it is important to recognize that international cooperation and coordination has been a hallmark of GLOBEC. Globalization of ocean sciences through expanded efforts to share remote sensing and *in situ* data and models as well as predictions is especially important for the future of interdisciplinary oceanography and its applications for societal benefit.

CHAPTER 7

Dynamics of marine ecosystems: ecological processes

**Coleen L. Moloney, Astrid Jarre, Shingo Kimura,
David L. Mackas, Olivier Maury, Eugene J. Murphy,
William T. Peterson, Jeffrey A. Runge, and Kazuaki Tadokoro**

7.1 Introduction

The ecological processes that underpin marine ecosystem dynamics are influenced by physical and chemical driving variables and by the internal properties of the ecosystem. These ecological processes include primary production as a bottom-up process, trophodynamic interactions within the food webs, and the influences of top predators, including humans, as top-down processes. The responses of marine ecosystems to global change have been of particular interest over the past few decades. Global change is considered here to include both rapid and slow changes in the variables driving ecosystem dynamics, resulting at least partly from human activities (GLOBEC 1999). These include the gradual impacts of global warming as well as changes resulting *inter alia* from human population growth in coastal areas, increased pollution, eutrophication, extensive fishing, and the effects of invasive species. GLOBEC science programmes used a 'target-species' approach (Gifford *et al.*, Chapter 4, this volume), which focuses on studies of the physical and biological factors influencing key species. Although these were primarily zooplankton or fish species, there has been a rich diversity of studies involving plants and animals across all trophic levels.

In the period before GLOBEC, the intense focus of the Joint Global Ocean Flux Study (JGOFS) and its precursors provided a solid foundation for understanding the flows of carbon and other elements in oceanic food webs (Fasham 2003), particularly for lower trophic levels. The distribution and dynamics of dissolved organic carbon in the oceans was described, along with its roles in fuelling bacterial metabolism and heterotrophic microbial processes (Ducklow 2003). These studies highlighted the complexity of ecological interactions within microbial systems, which recycle production and provide alternative routes for energy flow to higher trophic levels, although the key roles of microzooplankton in these trophic pathways remained to be fully elucidated (GLOBEC 1997). There were major advances in understanding the factors determining large-scale primary production, assisted by improvements in satellite remote sensing and modelling techniques. Underlying these technological advances was a significant contribution from the JGOFS programme, where a series of ground-breaking field experiments demonstrated that primary production in high nutrient, low chlorophyll regions of the world's oceans is often limited by iron availability (de Baar and Boyd 2000) and that this micronutrient can be limiting in other regions as well (Behrenfeld and Kolber 1999). Satellite-derived data on oceanic chlorophyll concentrations provided comprehensive global-scale views of the general dynamics of planktonic production systems (Longhurst 1998; Carr *et al.* 2006). Analyses of these data alongside global-scale coupled ocean-biogeochemical modelling studies have resulted in a greatly improved understanding of the major controls on primary production (Friedlingstein *et al.* 2006). These studies have also provided a valuable basis for analysing the general pattern of energy flow from primary producers to zooplankton and higher trophic level organisms.

Zooplankton play linking roles in marine food webs both as conduits of primary production from phytoplankton to vertebrate predators, and in the vertical transport of primary production to deep water ecosystems and sediments through a combination of surface feeding, diel vertical migration (DVM), production of faeces at depth and sinking of carcasses. In comparison with lower trophic level ecosystem operation and dynamics, little was known at the start of GLOBEC about overall food web dynamics at intermediate (zooplankton) and higher trophic levels. For unexploited species such as larger plankton, nekton and marine predator species, studies were dependent on spatially restricted sampling (e.g. based on nets) or on data from other sources (e.g. commercial fisheries). These limitations still apply to some extent, but targeted studies under GLOBEC have generated significant improvements in our understanding of the large-scale distribution of a number of zooplankton species. New sampling and analysis techniques (Harris *et al.*, Chapter 6, this volume), such as the optical plankton counter, the continuous underway fish egg sampler, and video recognition hardware and software, have greatly enhanced the spatial and temporal coverage of zooplankton and ichthyoplankton samples. Long-term sampling by ships of opportunity (as part of the Continuous Plankton Recorder Survey) has produced valuable perspectives of how zooplankton community structures have changed over decades and large geographical areas (Barnard *et al.* 2004; Beaugrand *et al.* 2007; Heath and Lough 2007). Syntheses of historical data sets are being used for large-scale characterization of distribution and abundance of many zooplankton species (e.g. NMFS-COPEPOD database, http://www.st.nmfs.noaa.gov/plankton/; Atkinson *et al.* 2004), contributing to understanding food web structure. There remain, however, few comprehensive large-scale data sets for many key zooplankton species including small zooplankton (<1 mm), carnivorous zooplankton, and many gelatinous zooplankton (Lynam *et al.* 2004; Hay 2006; Box 7.1). Fisheries data have provided large-scale views of the distributions of exploited species, but for most species of fish that are not exploited, as well as many groups of cephalopods, marine mammals and seabirds, there has been little observation-based integration across all levels in the ecosystems, although some modelling studies attempted this, as described below. Broad characterization of marine food web structures remains, therefore, limited. In contrast, basic understanding of physical-biological interactions influencing food web operation has advanced greatly during the last decade, based mainly on relatively local and regional analyses of species distributions and food web structures.

This chapter aims to summarize advances in understanding ecological processes as a result of the research carried out in connection with GLOBEC. It is not possible to separate 'GLOBEC' results from the more general ecological paradigms within which the studies were conducted and to which they contributed. However, some attempt is made to focus on results emerging from the GLOBEC regional programmes, which covered several major ecosystems representative of polar and subpolar seas, shelf waters including coastal upwelling regions (see Box 7.2), and oligotrophic oceanic areas (GLOBEC 1997). Global synthesis also requires that mid and high trophic level studies (the focus of GLOBEC) be integrated with those of low trophic levels to understand the dynamics of major marine ecosystems. The chapter is divided into three main sections: Section 7.2 summarizes current understanding of ecological processes affecting food webs, particularly for GLOBEC study regions. Section 7.3 examines the current understanding of ecosystem dynamics in relation to global change, and Section 7.4 highlights some of the advances that were made in understanding ecological processes during the period of GLOBEC's influence and some new issues that have emerged as topics for further study.

7.2 Ecological processes and food webs

Marine ecosystem structures vary on different space- and time scales and according to different factors, the roles of which vary according to the ways in which the ecosystem is viewed. The GLOBEC Science Plan (GLOBEC 1997) used

Box 7.1 Roles of gelatinous zooplankton

Our understanding of the biology and ecology of jellyfish (pelagic cnidarians and ctenophores) and of the role they play in the ecosystem lags far behind that of other zooplankton groups, perhaps because of the limited, disjointed, and geographically scattered research that has been conducted. Although our knowledge is still limited (see below), these organisms received increased coordinated research attention during the GLOBEC era, assisted in part by the considerable amount of global press coverage they have received.

Jellyfish have a suite of life history, biochemical, and physiological attributes that distinguish them from the better-studied crustacean zooplankton, and which lend them an edge in a changing pelagic ecosystem (Richardson *et al*. 2009) submitted). Although markedly seasonal in abundance, ctenophores are hermaphrodites that can reproduce within days of hatching (Martindale 1987); cnidarians are always present in one life history stage or another. Jellyfish are ~96% water and carbon constitutes ~17% of dry weight (Arai 1997), which means that (all rate processes being equal) almost 19 times as much jellyfish as equivalent finfish can be supported on a given food base. Jellyfish are extremely tolerant of low oxygen water (Purcell *et al*. 2001a); their polyps can encyst for prolonged periods when environmental conditions are unfavourable (Bouillon *et al*. 2004) and they can regenerate from fragments if damaged (e.g. by trawl nets) (Arai 1997).

Jellyfish are normal components of healthy marine ecosystems, and peaks in jellyfish abundance usually reflect the seasonal life cycles of individual species (Nicholas and Frid 1999), as well as physical processes of advection that concentrate individuals in enclosed embayments or at oceanic discontinuities (Graham *et al*. 2001). There is often a strong underlying climatic signal to population outbreaks (Goy *et al*. 1989; Gibbons and Richardson 2009), which in the North Sea have been linked to the NAO and associated changes in Atlantic inflow (Lynam *et al*. 2004, 2005a; Attrill *et al*. 2007). In the Pacific

Ocean, fluctuations in some species have been linked to ENSO (Dawson *et al*. 2001; Raskoff 2001) and decadal scale (Anderson and Piatt 1999) events. In the Bering Sea, Brodeur *et al*. (2008a) have linked interannual fluctuations in jellyfish population size to changes in sea ice cover, which, it has been suggested, has impacts on the timing of the spring bloom and the amount of food available for jellyfish and their polyps (Box 7.1, Figs. 1 and 2).

Increasingly severe jellyfish blooms have been reported in a variety of largely temperate coastal marine systems, including those of North America (Graham 2001; Mills 2001; Link and Ford 2006), eastern Europe (Shiganova 1998; Purcell *et al*. 2001b; Daskalov *et al*. 2007; see Box 7.1, Fig. 1), Africa (Lynam *et al*. 2006), and the Far East (Ishii and Tanaka 2001; Kang and Park 2003; Uye and Ueta 2004; Cheng *et al*. 2005; Uye and Shimauchi 2005; Purcell *et al*. 2007). These outbreaks have generally occurred in ecosystems that have become disturbed in some way, such that 'normal' population controls no longer function effectively.

Jellyfish are carnivores that can feed on a diversity of prey, ranging in size from unicellular protists to other jellyfish, but it is now known that they can also include benthic items in their diet (Flynn and Gibbons 2007; Pitt *et al*. 2008). They can be voracious, if selective, predators (Purcell and

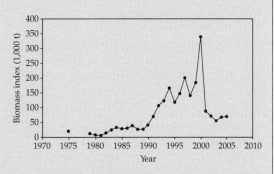

Box 7.1, Figure 1 Changes in jellyfish biomass in the Bering Sea, from standardized trawl surveys. (Adapted from Brodeur *et al*. 2008a with kind permission from Elsevier.)

continues

Box 7.1 *continued*

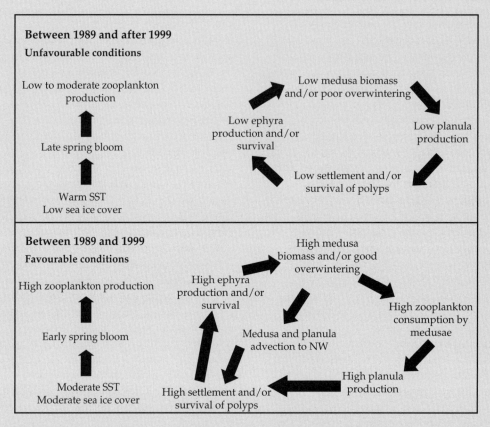

Box 7.1, Figure 2 Conceptual model of linkages between the biomass of medusae and their biophysical environment during warm (*upper*) and cool periods (*lower*) in the Bering Sea. (Adapted from Brodeur *et al.* 2008a with kind permission from Elsevier.)

Mills 1988) (no studies have yet demonstrated feeding satiation under natural food densities), simply shrinking in size when starved (Hamner and Jenssen 1974). Seasonal changes in jellyfish abundance can thus result in top-down changes to the structure and functioning of local ecosystems (e.g. Feigenbaum and Kelly 1984; Pitt *et al.* 2007).

Because they can consume the eggs and larvae of organisms at the same or higher trophic positions in pelagic food webs (e.g. Purcell 1985), jellyfish outbreaks are also implicated in the recruitment failure of some fisheries (Möller 1984; Lynam *et al.* 2005b), a situation that can be exacerbated by the substantial dietary overlap that can exist between jellyfish and the adults of these same species (e.g. Purcell and Sturdevant 2001; Brodeur *et al.* 2008b).

Whereas large jellyfish have relatively few common predators, small individuals and benthic polyps are subject to predation (Arai 2005), although our understanding of this is extremely poor. Top-down control of jellyfish is not unlikely in natural systems, as evidenced perhaps by the very recent nature of the increased incidence of jellyfish blooms.

Following the overexploitation of fish in some systems, jellyfish have been observed to increase in abundance (Lynam *et al.* 2005b, 2006). This may

continues

Box 7.1 *continued*

be in response to reduced predation pressure exerted by pelagic finfish on jellyfish and by demersal finfish on polyps. It is undoubtedly assisted by a release of niche space that these organisms then occupy (Bakun and Weeks 2006). Persistent top-down control of fish populations by jellyfish is not inconceivable through a predator-pit model (Bakun 2006), especially if other changes that favour jellyfish are also being noted in the ecosystem.

Eutrophication is becoming increasingly evident in coastal ecosystems around the world (e.g. Diaz and Rosenburg 2008). The altered Si:N ratios observed in such systems favour long food chains based on small particles, which in turn favour jellyfish (Parsons and Lalli 2002), which can tolerate, and indeed continue to reproduce in, low oxygen water (Grove and Breitburg 2005) and which are able to continue feeding in the often highly turbid conditions that result (Eiane *et al.* 1999). But eutrophication is not the only factor that can tip the balance within an ecosystem towards jellyfish. Other factors that have been invoked include climate change (Attrill *et al.* 2007; Richardson and Gibbons 2008), a proliferation of hard substrata for polyp attachment (Lo *et al.* 2008), as well as alien introductions (Graham and

Bayha 2007) and deliberate seeding and mariculture operations.

Much of our knowledge about the dynamics and role of jellyfish within ecosystems is still based largely on hypothetical supposition (e.g. Parsons and Lalli 2002) or model outputs using scant empirical data (Pauly *et al.* 2009). This needs to be redressed by an increased investment in research, focusing on diet, which is traditionally obtained as a snapshot (e.g. Suchman *et al.* 2008), but which needs to be extended to include a greater temporal resolution. There is a need to more fully understand the predators of jellyfish, although this is being improved by new online databases (Pauly *et al.* 2009). The hidden stage in the life cycle of cnidarians (the polyp) has to be studied and a start needs to be made to moving some studies into the laboratory, even though this will be frustrating until such time as culture techniques have been perfected. There is an urgent need to establish long-term monitoring programmes for jellyfish, similar to the long-term monitoring in operation for crustacean zooplankton (Hays *et al.* 2005), and models need to capture the complexity of jellyfish life history and ecology (Boero *et al.* 2008).

Mark J. Gibbons

differences among regional ecosystems to try and advance ecosystem understanding. This section will examine similarities and differences among some key ecosystems from different geographic regions, based largely on the GLOBEC study regions. It will also contrast the various ways in which marine food webs are represented, and how these representations can influence understanding of ecosystem dynamics.

7.2.1 Key ecological processes in regional ecosystems

A fundamental aspect of a marine ecosystem is where it is located in the world's oceans. The science of GLOBEC was carried out in ecosystems categorized

under six regional programmes located in (1) the Southern Ocean, (2) the North Atlantic, (3) the North Pacific, (4) sub-Arctic Seas, (5) shelf seas, and (6) oceanic waters (Fig. 7.1). Most GLOBEC studies were located in polar and temperate regions; tropical regions were less intensively studied. High-latitude ecosystems exhibit strong seasonality of plankton productivity and biomass, and of upper trophic level reproduction, migration, somatic growth, and/or diet. This influenced the processes that were selected and highlighted in GLOBEC studies. There are differences among the GLOBEC study regions in terms of environmental forcing, the scales of temporal variability, community structures, productivity, vertical structure, biomasses, and fisheries. The challenge here is to link different ecosystem structures to salient

Box 7.2 Spatial and temporal influences on production cycles in the Oregon upwelling region (north-east Pacific GLOBEC)

Seasonal cycle of production

The annual cycle of plankton production begins with a spring bloom that can come at any time between February and April (Box 7.2, Fig. 1), depending entirely upon the frequency of occurrence of winter storms, cloud cover, and day length, all of which are related to the sign and strength of the Pacific Decadal Oscillation (PDO). The PDO is an oscillation in the east-west pattern of sea surface temperatures in the North Pacific-when the PDO is in negative

phase, the Gulf of Alaska, California Current, and eastern equatorial Pacific are cooler than normal whereas the central and western portions of the North Pacific are warmer than normal, and vice versa. These spatial patterns are maintained by large-scale winds-negative phase is driven by winds from the north and west over the north-east Pacific during winter, positive phase by south-west winds over the north-east Pacific in winter.

The end of winter is marked by the advent of the 'spring transition', the time of the year when winter ends and coastal upwelling begins. This event usually occurs in April and results in moderate phytoplankton blooms, 5 µg Chl a L^{-1}; the largest phytoplankton blooms are not seen until July and August, (order 10–30 µg Chl a L^{-1}) (Box 7.2, Fig. 1). The upwelling season usually ends in September and the water column begins to de-stratify. An autumn bloom is observed in October of most years. The length of the upwelling season seems to be related to the PDO as well in that the season is short during positive phase and long during negative phase.

The timing of the initiation of the production cycle in spring is critical because many resident higher trophic level species in the northern California Current have timed their reproduction cycles in anticipation of the adults and/or young finding a high biomass of prey items upon which to feed. Other species are resident elsewhere, but make annual feeding migrations to the northern California Current timed so they arrive by late spring. Any disruptions or delays in the initiation of the seasonal production cycles can have disastrous consequences on food chain productivity, for example as seen during the summer of 2005. Copepod biomass fell to very low levels, and the species were dominated by offshore subtropical species (Mackas *et al.* 2006). Moreover, due to the collapse of the food chain, massive deaths of seabirds were seen coast-wide (Sydeman *et al.* 2006) and mortality rates of chinook salmon, which entered the ocean that summer, were

Box 7.2, Figure 1 Climatology of sea surface temperature, nitrate, chlorophyll, and copepod biomass at a station 10 km offshore of Newport, Oregon, based on biweekly samples collected from 1997–2006. Note that a 'spring bloom' can occur at anytime between February and May, but that the largest blooms are in July, during the upwelling season.

continues

Box 7.2 *continued*

extraordinarily high. In fact, returns of adult salmon to their natal streams in 2007 and 2008 were so low that the salmon fisheries along the California and Oregon coast were closed during the summer of 2008 (http://www.nwfsc.noaa.gov/).

Timing of the production cycle

Many fish in the Oregon upwelling zone spawn in winter months, for two reasons. First, winter downwelling ensures that fish larvae are transported northwards and to shelf waters where food concentrations are higher than in offshore waters. Second, by the time upwelling starts in spring, larvae have reached the juvenile stage, a time in their life when they need high concentrations of food for growth. Thus they anticipate the climatological start of the production cycle. Examples for the northern California Current include osmeriids (smelts) and sand eels (both of which are important prey for young salmonids), most of the shelf flatfish (soles and flounders), slope rockfish, sablefish, and coho, spring chinook, chum, and sockeye salmon. The salmon are a somewhat different case in that they spawn in freshwater, but regardless, they migrate to the ocean from mid-April through mid-May in anticipation of finding a bountiful harvest of young smelts and sand eels as well as krill.

Pacific hake, the species that supports the fishery with the highest biomass in the California Current, perform a feeding migration to Oregon/Washington from southern California, usually arriving off Oregon by June. The rate and extent of their poleward migration is believed to be controlled by euphausiid biomass-if biomass is low and late in developing, as in warm years with weak upwelling, the migration continues to move north into Canadian waters. Similarly, Pacific sardine perform a feeding migration to the waters off Oregon and Washington in late spring in order to take advantage of the high biomass of lipid-rich zooplankton.

Most local seabirds lay eggs in April/May anticipating availability of large stocks of euphausiids and/or small pelagic fish upon which the adults feed, which in turn are fed to the chicks through regurgitation. Important prey items include the euphausiid *Thysanoessa spinifera*, and juvenile smelts, sand eels, and rockfish. Migratory birds (shearwaters and albatrosses) appear off Oregon in April/May, again in anticipation of the results of early and high production.

Gray whales begin their poleward migration in March, from Mexico, arriving off Oregon by May where they expect to find high numbers of their preferred prey, mysids. Humpback whales also appear in the northern California Current in late spring, hoping to find an abundant biomass of their favourite prey, euphausiids, and herring.

Spatial variations in production

Inner-mid-shelf waters. The upwelling process brings nutrient-rich water to the sea surface very near to the coast, within a zone ranging from the beach out to 10–15 km from shore. Massive phytoplankton blooms composed of large chain-forming diatoms occur here. This is where hungry herbivores such as copepods prosper. Offshore of that zone, in water depths > ~75 m, upwelled waters do not reach the sea surface because the water column is stratified, thus the source of phytoplankton to these mid-shelf waters is offshore advection in the Ekman layer. Zooplankton species that dominate the nearshore zone include copepods (*Pseudocalanus mimus*, *Centropages abdominalis* and *Acartia hudsonica*), mysids, cladocerans (*Evadne* spp. and *Podon* spp.), and the larvae of benthic invertebrates. Phytoplankton are predominantly large, chain-forming diatoms and, because of their extraordinarily high growth rates as well as large cell size (i.e. many species form chains that can often reach lengths of 1 mm), only 10–20% of the production is consumed per day, with the bulk of the biomass sinking to the benthos where it fuels a rich benthic community.

continues

Box 7.2 *continued*

Adult *Euphausia pacifica* can be abundant in mid-shelf waters because they are transported there by the upwelling process in the same manner as described by Peterson *et al.* (1979) for *Calanus marshallae* (i.e. by staying in the deep waters that move across the shelf from the shelf break to the nearshore zone). Spawning by *E. pacifica* is very intense in mid-shelf waters during July and August following massive phytoplankton blooms associated with upwelling relaxation events, with egg abundances on the order of 200–1000 eggs m^{-3} (Feinberg and Peterson 2003). During late summer, the zooplankton 'biomass' can be dominated by large jellyfish (*Chrysaora fuscens*) for several months. These scyphomedusae are zooplanktivores, feeding on copepods and euphausiid eggs (Suchman *et al.* 2008).

Outer shelf and shelf break. Zooplankton biomass in outer shelf/shelf break waters is comprised of copepods (*C. marshallae*, *P. mimus*, *Acartia longiremis* and *Metridia pacifica*) and euphausiids (*E. pacifica* and *T. spinifera*), with euphausiids being the dominant contributor to total biomass. *Oithona similis*, in contrast, is abundant everywhere, but given their small size they contribute only a small percentage of mass to the total copepod biomass. Phytoplankton size structure is chiefly cells <10 μm and overall phytoplankton biomass is about 10% of the mass observed in inshore waters. Reduced biomass compared with the inner shelf waters is likely due to higher grazing pressure observed in outer shelf waters.

The shelf break region is the zone of the euphausiid. One species, *E. pacifica*, is a keystone species, being the chief prey of Pacific hake (the largest biomass of any fish in the northern California Current at 1–2 million tons), most of the rockfish (*Sebastes* spp.) as well as coho and chinook salmon. Euphausiid biomass at the shelf break often exceeds copepod biomass by 5–10 fold.

Oceanic waters. Oceanic waters adjacent to and outside of the coastal upwelling region are dominated by salps (*Salpa fusiformis*) and doliolids (*Dolioletta gegenbaurii*). Important copepods in offshore waters include the subtropical species, *Eucalanus californicus*, *Calanus pacificus*, *Mesocalanus tenuicornis*, *Paracalanus parvus*, *Ctenocalanus vanus*, and several *Clausocalanus* spp. During January through May, two subarctic species can dominate, *N. plumchrus* and *N. cristatus*. The production cycle in offshore waters is the 'classical' cycle with a spring bloom followed by a long period of oligotrophy with zooplankton production fuelled by the microbial loop (through microflagellates and ciliates). The dominant nekton in these waters are myctophids, squids, and sauries which are in turn fed upon by the albacore tuna which cruise these waters during summer.

Benthic ecosystems

Although very little is known about ecology of the benthos off Oregon, we can deduce from fisheries information and from first principles the following:

The inner-mid-shelf region, where phytoplankton production is extraordinarily high, is the home of amphipods, clams, mysids, Dungeness crabs, several flatfish species and juvenile fall chinook salmon, resident seabirds (murres), and gray whales. It is likely that at least 80% of the primary production is exported to the benthos, fuelling a benthic food web that results in massive numbers of mysids (the chief food of gray whales which forage only in the inner shelf region within 5 km of the shore) and amphipods and bivalves (the chief food of Dungeness crabs which are also most abundant within 5 km of shore). The Dungeness crab fishery is in many years the most lucrative of any fishery in the Pacific Northwest, producing approximately 10 000 t of catch in a good year. The mid-outer shelf region is also characterized by a rich benthic community and this is the home of the semi-pelagic pink shrimp (*Pandalus jordanii*) fishery, a stable fishery that recently received an award for 'excellence in management'.

William T. Peterson

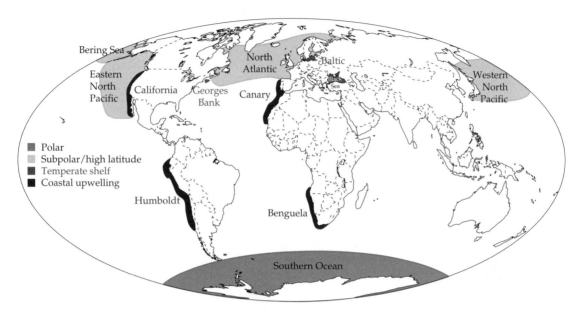

Figure 7.1 Regional ecosystems mentioned in the text where some GLOBEC-related studies were carried out.

processes identified largely (but not exclusively) from studies of life histories of target species.

7.2.1.1 Polar ecosystems

Polar ecosystems experience marked seasonal extremes, particularly in terms of light. In the Southern Ocean the basic structures of marine food webs are relatively well established, varying between summer and winter and among the different neritic and oceanic environments (Moloney and Ryan 1995; Fig. 7.2). Animals at all trophic levels have different life history, feeding, and behavioural adaptations to these seasonal changes and different trophic pathways operate according to the changing architecture of the habitat, the strong seasonality of the environment and of primary productivity, and the movements and migrations of the organisms. At present, areas of the Southern Ocean are being impacted strongly by climate change, affecting all components of the ecosystem. To understand and ultimately predict the consequences of these changes, which are acting at the same time as selective fishing, requires mechanistic understanding of physical-biological interactions and the ability to quantify ecological processes over a range of scales.

Antarctic krill *Euphausia superba* was the zooplankton target species of GLOBEC studies in the Southern Ocean (Gifford *et al.*, Chapter 4, this volume), focusing on physical and biological factors influencing growth, reproduction, recruitment, and survival (Hofmann *et al.* 2004b), especially during winter. Antarctic krill are the dominant macrograzers over large parts of the Southern Ocean and a key component of the food web supporting seabirds, seals, and whales. Their life history involves complex interactions of life stages with bathymetry, advection, and a strongly seasonal physical environment (Hofmann and Hüsrevoglu 2003; Smetacek and Nicol 2005). In summer, adult krill spawn in current systems that transport their larvae onto inshore shelf regions. The larvae and adults overwinter in association with sea ice, relying on the associated under-ice communities for food, and the adults have a flexible diet that allows them to feed opportunistically (e.g. Schmidt *et al.* 2006; Clarke and Tyler 2008). In high northern latitudes, key zooplankton species (e.g. amphipods) also use the under-ice habitat for feeding, but other dominant zooplankton use diapause as a strategy to survive winter. A key difference in the south is that different life stages of a single, long-lived species (Antarctic

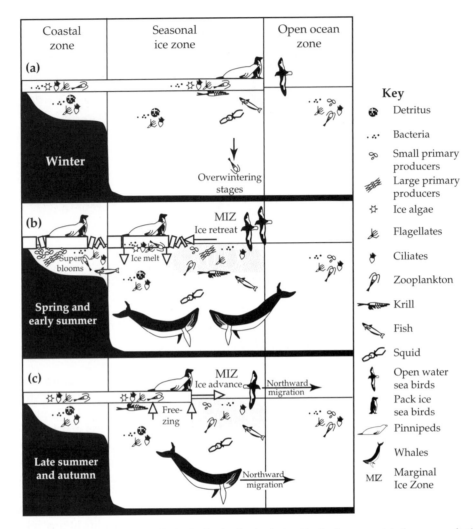

Figure 7.2 Schematic of generalized Antarctic marine food webs, showing the dominant food web components in three seasons for three environmental zones. Horizontal and vertical distances are not drawn to scale. Icons represent major feeding or taxonomic groups and are not all-inclusive. (Reprinted from Moloney and Ryan 1995. Copyright Elsevier 1995).

krill) integrate the strategies adopted by many species in the north. The extent, timing, movement, and duration of sea ice are important elements determining Antarctic krill recruitment. The timing and distribution of krill larvae influence connections between habitats needed by different life stages, ultimately affecting population size (Atkinson et al. 2004; Ducklow et al. 2007; Murphy et al. 2007). Interannual variability in the population sizes of Antarctic krill in the Scotia Sea is believed to cause changes in food web operation (Murphy et al. 2007). During periods when populations are large, most of

the energy to top predators is channelled through krill (Fig. 7.3), but when krill populations are small, alternative pathways via copepods and other zooplankton become important.

Interannual variability in food webs is strongly linked to variations in the extent and duration of sea ice in both southern (e.g. Ducklow et al. 2007) and northern (e.g. Hunt et al. 2002) polar ecosystems. However, whereas alternative energy pathways are important in ecosystem dynamics of the Scotia Sea, at similar latitudes in the northern Hemisphere alternating trophic controls are

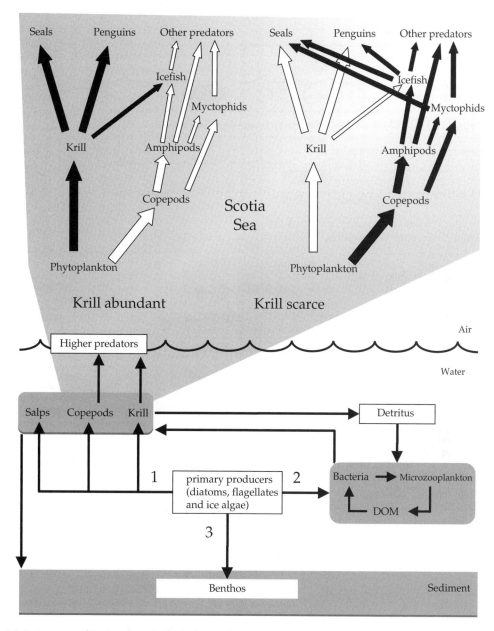

Figure 7.3 Basic structure of Southern Ocean food webs showing three main pathways for primary production (adapted from Clarke *et al.* 2007). The relative importance of different pathways varies with area and season. Pathway 1 has two alternative structures in the Scotia Sea, depending on the abundance of krill, with the major pathways shown in black and the alternative minor pathways in white. (From Murphy *et al.* 2007).

believed to affect food web functioning. Variations in the pelagic food web of the south-eastern Bering Sea have been related to temperature effects on zooplankton growth, influenced by the timing and extent of ice retreat (Hunt *et al.* 2002; Fig. 7.4). During periods of no ice or early retreat, phytoplankton

blooms are delayed and occur in warm water, which favours rapid zooplankton growth and enhanced larval fish survival. During periods of late ice retreat, phytoplankton blooms occur at or near the ice-edge in cold water, where zooplankton growth is slow and survival of larval fish is reduced. On

(a) Physical forcing

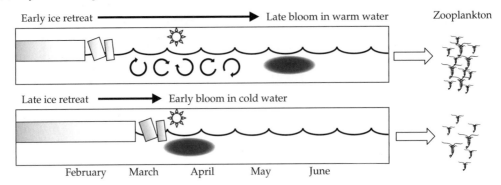

(b) Oscillating control hypothesis

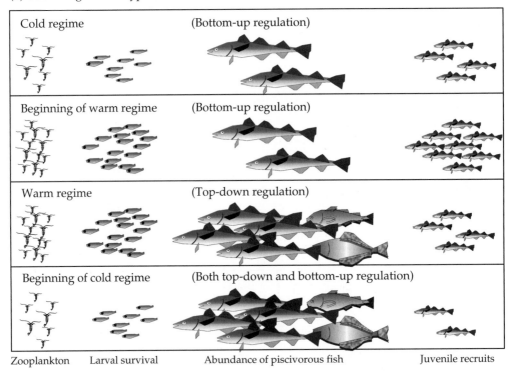

Figure 7.4 Hypothesized food web dynamics in the south-east Bering Sea (after Hunt *et al.* 2002 with kind permission from Elsevier). (a) Physical forcing influences the timing of ice retreat. When ice retreats early (in late winter) the spring bloom is limited by light until solar heating stratifies the water column. Zooplankton grow rapidly in the warm conditions. When ice retreats late there is sufficient light and the spring bloom develops early in cold water, slowing zooplankton growth. (b) The oscillating control hypothesis links zooplankton dynamics to biological controls in the food web. With reduced zooplankton, larval/juvenile pollock survival is limited by food availability. When zooplankton increase, so do the larval/juvenile pollock, which have low levels of predation from large piscivorous fish. When the juvenile fish grow into adults they exert predatory/cannibalistic control on juvenile fish, even in the presence of abundant zooplankton food. When zooplankton are again reduced, decreased juvenile growth results in reduced adult populations.

decadal scales, Hunt *et al.* (2002) suggest that these bottom-up effects on zooplankton and ichthyoplankton propagate through the food web, and that cannibalism (among walleye pollock *Theragra chalcogramma*), predation (of piscivores on planktivores), and competition (among piscivorous fish, seabirds, and pinnipeds) all interact to produce food web structures that are characteristic not only of periods of late or early ice retreat, but also transitions between these environmental states (Fig. 7.4).

7.2.1.2 Subpolar high-latitude ecosystems

The general life cycle patterns of planktonic copepods, particularly the genera *Calanus* and *Neocalanus*, have been the focus of northern Hemisphere GLOBEC programmes. Zooplankton in high-latitude ecosystems have two main strategies for coping with food shortages during the dark, cold winters: vertical migration associated with diapause and switching of diets and trophic levels. In *Neocalanus* and in some polar *Calanus* species, internal body stores (including lipids) are used to produce eggs without the need for a contemporaneous external food supply.

In the subarctic northeast Pacific the plankton community is dominated by nanophytoplankton and microzooplankton, the latter having high feeding rates on the nanophytoplankton, leading to top-down grazing control during the season of high primary productivity (Miller *et al.* 1991). Four target species dominate mesozooplankton biomass in the region: three species of *Neocalanus* (*N. flemingeri*, *N. cristatus*, and *N. plumchrus*) and *Eucalanus bungii*. The four species spend part of the year growing and feeding in surface waters, and part of the year at depth, with season and depth partitioning among them. The three *Neocalanus* species have a 1-year life cycle. They grow in the upper layers, then migrate into the mid-deep layer at stage C5 (*N. cristatus* and *N. plumchrus*) or female stage C6 (*N. flemingeri*) for diapause and spawning. They leave diapause in late winter, the C5s moult to adults which produce eggs, and the nauplii subsequently migrate into the surface waters (Miller *et al.* 1984). The timing of this ontogenetic vertical migration is different among the three species. Copepodite stages of *N. cristatus* mainly appear in the surface layer in late winter-spring (Miller *et al.* 1984), whereas copepodites of *N. plumchrus* and *N. flemingeri* mainly appear in

spring-summer (Miller *et al.* 1984; Miller and Clemons 1988). In contrast, *E. bungii* have a 1–2-year life cycle, but spawn in the surface layers rather than at depth (Miller *et al.* 1984). The major mesozooplankton species thus use different strategies to exploit the spring-bloom productivity.

In the oceanic ecosystems of the Gulf of Alaska, the peaks and troughs of the annual cycles of primary productivity and phytoplankton biomass are dampened compared to the subpolar Atlantic, the Oyashio or the Gulf of Alaska shelf regions. The winter minima are not as low (because there is a year-round halocline that limits deep winter mixing) and the spring maxima not as high (probably because of iron limitation) (Whitney *et al.* 2005). At the shelf margins in winter, high productivity is supported by macronutrients that are carried coastward by downwelling, mixing with coastal water that is more stratified (therefore having less light limitation in early spring) and contains more iron and silicate (because of land run-off) (Childers *et al.* 2005).

In the Oyashio Current region significant phytoplankton blooms occur during the boreal spring (April to May) (e.g. Kasai *et al.* 1997). The same three species of *Neocalanus* as were found off the Gulf of Alaska (*N. flemingeri*, *N. cristatus*, and *N. plumchrus*) dominate mesozooplankton in these waters, with a similar life cycle for *N. cristatus* and *N. plumchrus* (Kobari and Ikeda 1999, 2001a,b; Tsuda *et al.* 1999). However, the pattern for *N. flemingeri* is different in the Oyashio compared to the Gulf of Alaska (Fig. 7.5), with the species mainly appearing in the surface layers in late winter-spring (rather than spring-summer; Kobari and Ikeda 2001b). In addition, some *N. flemingeri* populations have a 2-year life cycle in these waters (Tsuda *et al.* 1999). These differences possibly reflect a bet-hedging strategy that extends the spawning season for *N. flemingeri*, and they indicate remarkable flexibility in patterns of growth and reproduction among the species and according to the environments. These nuances are difficult to incorporate into food web and ecosystem models, which have to sacrifice species-level details for model simplicity, consequently reducing their predictive ability. One way of addressing these shortcomings is to ensure that models are closely linked to field data, so deviations from the 'norm' can be detected and the models adjusted for new processes; the need for field studies will be ongoing.

In the North Atlantic, primary productivity increases in the boreal spring (April), when the mixed layer depth becomes shallower than the critical depth, and peaks in summer (July) (Parsons and Lalli 1988). The seasonal variation in chlorophyll *a* is synchronous with primary productivity (but peak chlorophyll *a* concentrations are approximately one-tenth of those in Oyashio waters). Species of *Calanus* are prominent in the region and were important target species. *C. finmarchicus* dominates the zooplankton assemblage of the subarctic North Atlantic and *C. glacialis* and *C. hyperboreus* become abundant a bit further poleward. *C. helgolandicus*, their warm-temperate congener (Fleminger and Hulsemann 1977), dominates in the North Sea and eastern boundary current region off Europe and North Africa. Populations of *C. finmarchicus* are characterized by one to several generations per year, with an overwintering, dormant phase typi-

cally in the last pre-adult copepodite stage C5 (Conover 1988). Individuals grow during spring-summer in the surface layer and stage C5s migrate into the subsurface layer in summer and autumn for dormancy (Aksnes and Blindheim 1996). Dormant C5s typically reside in deep water and have reduced metabolism and arrested or slowed development. Summer feeding and lipid storage determines their survival through winter, affects the depth at which they overwinter (Heath *et al.* 2004) and sustains metabolism during moulting and development of gonads (Rey-Rassat *et al.* 2002a). In mid to late winter, typically, C5s leave dormancy, moult into adults, and mate. Females reproduce and offspring grow in response to food levels and temperature (e.g. Campbell *et al.* 2001b; Stenevik *et al.* 2007). There is considerable variability in the timing of dormancy and the numbers of generations across regions (e.g. Planque *et al.* 1997; Hind *et al.* 2000; Johnson *et al.* 2008). For example, in the coastal Gulf of Maine *C. finmarchicus* emerges from dormancy in late December (Durbin *et al.* 1997)

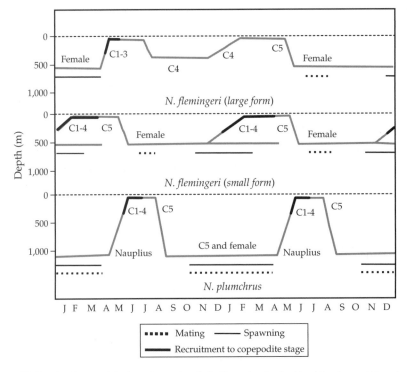

Figure 7.5 Illustrative life histories of copepods in the western subarctic Pacific, contrasting the biennial and annual life cycles of two forms of *Neocalanus flemingeri* and *N. plumchrus*. (Modified from Tsuda *et al.* 1999 with kind permission from Springer Science and Business Media.)

and produces two to three generations over the course of a season. In contrast, in the adjacent slope water, exit from dormancy is 2–3 months later and only one generation is produced.

7.2.1.3 Shelf ecosystems

In productive shelf seas, seasonal cycles of high production can be linked to physical processes that resupply nutrients into the upper water column, thereby prolonging the productive season. Two of the more important and regular processes are upwelling (both coastal upwelling (Box 7.2) and open-ocean upwelling), which resupplies nutrients on a daily-weekly basis, and tidal mixing, which can mix the water column and replenish nutrients on a daily-fortnightly basis. Since such regions sustain high zooplankton and fish abundance and are the focal point of many of the world's industrialized fisheries, it should not be a surprise that many GLOBEC studies focused on these regions, with target fish species being small pelagic fish from upwelling regions and mostly cod and salmon in northern regions.

Georges Bank, off the north-east coast of the United States, is a shallow region of the temperate North Atlantic, with bottom depths of 30–100 m. Tidal mixing supplies nutrients on Georges Bank from the open ocean (Horne *et al.* 1996), with other sources of nutrients being from upwelled water from the deep ocean and the Gulf of Maine and transport from the Scotian Shelf. Increased stratification of the surface layer of the outer bank occurs in spring-summer. A tidal front between the bank water and the open ocean is reinforced by stratification and maintained by tidal mixing. The tidal front plays an important role in controlling nutrient supply to the region, and on-bank transport is balanced by water (and plankton) moving off the Bank (Steele *et al.* 2007). Chlorophyll *a* concentrations peak in late winter-early spring with diatoms being the dominant phytoplankton group (Townsend and Thomas 2002). The dominant mesozooplankton species is *C. finmarchicus*, with biomass peaks contributing more than 90% of total mesozooplankton biomass during April-June (Davis 1987). The dynamics of the shallow-water ecosystem over Georges Bank are strongly influenced by physics, as is highlighted by Steele *et al.* (2007; Fig. 7.6). There are also strong links between the pelagic and benthic biota. Feeding interactions transfer material upwards to benthivorous fish and detrital showers transfer material downwards to suspension feeders. The benthos is also responsible for rapid nutrient recycling, which fuels primary production in the surface waters. To fully understand the responses of this ecosystem to change requires integration of physical and biological processes across a range of scales from the water surface to the seabed (Steele *et al.* 2007).

In eastern boundary upwelling ecosystems, sardine and anchovy have been target species for GLOBEC studies. Understanding their large variations in recruitment has been linked strongly to physical processes, extending Bakun's triad hypothesis (1996) (deYoung *et al.* Chapter 5, this volume) to include transport between spawning and nursery grounds (e.g. Parada *et al.* 2003; Miller *et al.* 2006). Seasonal influences are reflected in the timing of various biological processes throughout the food web. For example, in the Oregon upwelling zone many fish spawn during winter, for two reasons. Winter downwelling ensures that their larvae are transported to shelf waters where food concentrations are high, and larvae will have reached juvenile stage in summer when they need high concentrations of food for growth. Thus they anticipate the climatological start of the production cycle. Traditional seasonal (i.e. winter-summer) influences in upwelling regions can be overshadowed both by intraannual upwelling variability (Roy *et al.* 2001) and interannual variability. Carr and Kearns (2003) showed that large-scale circulation was the main factor determining environmental variability in eastern boundary systems, affected at least in part by El Niño (Humboldt and Californian systems) or Benguela Niño (northern Benguela) events. In the Humboldt region, the species composition of mesozooplankton is influenced by ENSO variability. Large copepods and *Euphausia mucronata* appear during La Niña periods (Hidalgo and Escribano 2001) and small copepods such as *Paracalanus parvus* and *Oithona* spp. increase during El Niño periods (Escribano *et al.* 2004). In the southern Benguela region, there is evidence of long-term (decadal) changes in the composition of the zooplankton community, manifested also as changes in size composition, abundance, and overall biomass (Verheye *et al.* 1998). The mechanisms for these changes remain speculative in the absence of

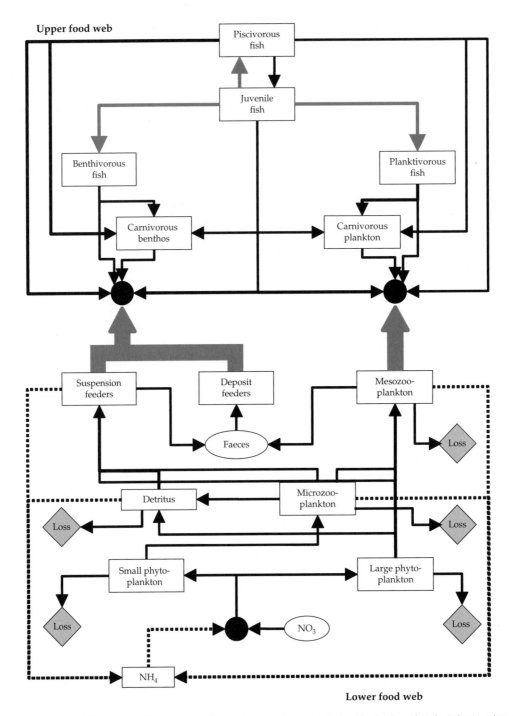

Figure 7.6 Georges Bank food web showing components (boxes) of lower and upper trophic level food webs, with circles indicating where the two join and diamonds indicating physical losses. Connections among components include fluxes (solid black lines), growth of fish (solid grey lines), and nutrient recycling (dotted lines). Note that arrows in the lower food web show the flows of nutrients up to plankton and benthos, and in the upper food web they show the consumption by fish down to plankton and benthos. (Modified from Steele *et al*. 2007 with kind permission from Elsevier.)

sufficient long-term data sets to test hypotheses, although the trends correspond with changes in environmental variables and other biota, including small pelagic fish (Howard *et al.* 2007).

Changes in relative abundance between anchovy and sardine species have been ascribed to biophysical influences but are also strongly affected by fishing. Trophic differences between the two species have been used to explain different ecosystem states (van der Lingen *et al.* 2006a), with food chains dominated by small phytoplankton favouring sardine and large phytoplankton favouring anchovy (Fig. 7.7). For the southern Benguela, the water column stabilizes during warm periods when upwelling events are weak and less frequent. The phytoplankton community is dominated by small cells, leading to a zooplankton community dominated by small copepods

that favour filter-feeding sardine (van der Lingen *et al.* 2006a). In cool periods when upwelling is strong, phytoplankton biomass is dominated by large, chain-forming diatoms and zooplankton are large and ideal for the biting feeding behaviour of anchovy.

The juxtaposition of different scales of environmental variability along with fishing pressure has made it difficult to understand the mechanisms of ecosystem change in upwelling regions. Comparisons of mass-balance trophic models of the northern and southern parts of both the Benguela and Humboldt ecosystems indicated that intensive fishing could cause changes in ecosystem structure (Moloney *et al.* 2005; Coll *et al.* 2006b), weakening the pelagic links in the food web, enhancing sedimentation, favouring gelatinous zooplankton, and driving feeding connections deeper, towards a more benthic mode

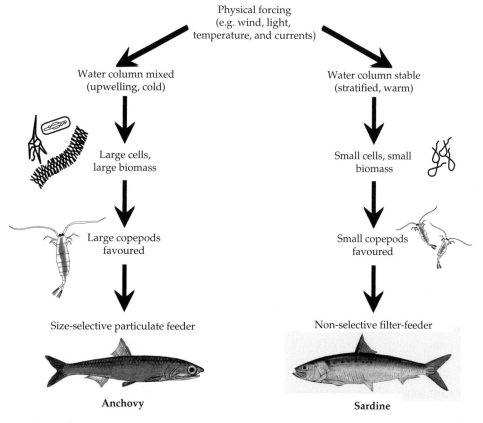

Figure 7.7 Hypothesized food web dynamics in the southern Benguela coastal upwelling ecosystem (Van der Lingen *et al.* 2006a © NISC (pty) Ltd, reproduced with permission). Physical forcing influences the size composition of the phytoplankton community, which influences zooplankton sizes. Large zooplankton favour anchovy, whereas small zooplankton favour sardine.

(Coll *et al.* 2006b). However, environmental change is likely to be an overwhelming factor, with fishing effects playing important but supplementary roles because of the large magnitude and variability of environmental signals, particularly as they affect pelagic fish (Shannon *et al.* 2009).

7.2.1.4 *Tropical and subtropical oceanic ecosystems*
Tropical ecosystems were not a focus of early GLOBEC studies, but they assumed increased importance through studies of top predators. Oceanic areas tend to have low productivity but the ecosystems nonetheless support viable food webs with important fish, cephalopod, turtle, seabird, and marine mammal top predator species. This is partly due to spatial enhancements of productivity, often associated with mesoscale features. The ecological importance of these features, a key focus of GLOBEC, has become increasingly evident as new remote sensing tools have allowed detection and tracking of both the physical features themselves and the top predators that use them for spawning and feeding.

7.2.2 Feeding interactions and marine food webs

Feeding interactions are fundamental ecological processes. They determine the structure of marine food webs through the dynamics of population interactions, the structure (or pattern) of trophic connections among species, and the transfer of energy and cycling of nutrients (e.g. carbon, nitrogen, and phosphorous). Present-day hypotheses about trophic interconnections and how they affect marine ecosystem structure and function derive from much earlier concepts and observations. A key early recognition (well illustrated in Hardy 1924) was that most zooplankton and fish feed on multiple prey species and are eaten by multiple predator species. Although true for individual life stages, diversity of linkage is especially strong for species that undergo large ontogenetic changes in size and morphology. A follow-on concept is that flowcharts of trophic connections, pool sizes (biomass), or rates/paths of energy transfer are all best viewed as interconnecting 'food webs' rather than as one or a few linear 'food chains'

(Elton 1927; Pomeroy 1974). The nodes in food webs can represent life history stages, species, groups of taxa, functional groups, feeding guilds, size classes, or trophic levels. The flows can represent energy, mass, nutrients, or particles. The nodes and flows are often determined by the data available for their analysis. Measurements of pool sizes for lower trophic levels are often available only in highly aggregated units: chlorophyll concentration as a proxy for primary producer biomass; wet weight, dry weight, or displacement volume as a proxy for mesozooplankton biomass, itself a mix of herbivores and first/second-stage carnivores. In contrast, measurements of higher trophic levels are often made at the species level in units of wet mass or numbers, but are all too soon aggregated to guilds for non-commercial fish species.

Constructing food webs and investigating their dynamics are integrating activities, usually (but not always) involving models, with some degree of quantification being required to understand ecosystem operation. This section describes different approaches for constructing food webs, using GLOBEC studies as examples. The advantages and disadvantages of each approach will be discussed and some lessons learned are highlighted.

7.2.2.1 *Species-based approaches*
The use of target species was a core aspect of the GLOBEC approach (Gifford *et al.*, Chapter 4, this volume). Although detailed studies of individual species cannot be extended readily to the ecosystem level, which includes all species, the value of the target-species approach for food web studies has been in recognizing the importance of population-level processes for ecosystem dynamics. A number of key population concepts were developed and refined during the years of GLOBEC research, and it is difficult (if not impossible) to try and predict the consequences of change in ecosystems without taking these population processes into account. For example, Bakun and Cury (1999) hypothesized that individuals of a less abundant species could be adversely impacted by an abundant species if they schooled together, with the less abundant species being forced to adapt to different modes of swimming and foraging. They suggest that the 'school

trap' could be an important population-regulating mechanism for small pelagic species like sardine and anchovy that alternate in abundance.

Another example involving feeding interactions is ontogenetic reversal in the roles of prey and predator in food webs. Adults of a forage species can eat the larvae/juveniles of their main predator, thereby escaping predatory control of their own adult populations by strongly suppressing recruitment of their predator (the 'depensation' effect; Walters and Kitchell 2001). In contrast, when the predatory species is sufficiently abundant it crops down the adults of the forage species, thereby releasing its larvae and juveniles from predator pressure (the 'cultivation' effect; Walters and Kitchell 2001). Cultivation/depensation has been used to explain non-recovery of the north-west Atlantic cod population (Rose 2004; Bundy and Fanning 2005) and maintenance of specific ecosystem regimes (Möllmann et al. 2005).

Also at the population level, the BOFFFF (big, old, fat, fecund, female fish) hypothesis (Marteinsdottir and Steinarsson 1998; Berkeley et al. 2004) emphasized the disproportionate contributions of large fish to successful recruitment (Jarre et al. 2000), producing more eggs per unit mass over a longer period than small fish. These eggs often have enhanced survival probabilities because of good energy reserves (Morita et al. 1999) and/or a more suitable timing of spawning (e.g. Wieland et al. 2000). Similar arguments apply for recruitment in short-lived species when parents are in poor or good condition, and density-dependent effects have been proposed to explain recruitment patterns in sardine populations in the north-west Pacific (Kawasaki and Omori 1995; Kim et al. 2006) and south-east Atlantic (van der Lingen et al. 2006c), and in haddock in the north-west Atlantic (Friedland et al. 2008). The basis for these hypotheses is that fast-growing larvae are able to grow rapidly out of the 'predator pit' (Bakun 2006).

The species-based approach adopted by GLOBEC considerably advanced the theoretical-observational understanding of population dynamics of target species. To some extent it is difficult to generalize the detailed processes involved, because characteristic life history and behavioural strategies of individual species determine the outcomes of these species-species interactions. However, there are a finite number of life history

traits that influence ecological dynamics, and future work comparing ecosystems and species could help identify any general patterns. One significant outcome of GLOBEC has been the realization of the need to expand on studies of krill in the Antarctic as well as in the north Atlantic and north Pacific. Thus, krill were important elements of several GLOBEC regional programmes and a great deal has been learned about these target species; this work requires synthesis (e.g. Hofmann et al. 2008).

7.2.2.2 Mass-balance approaches

In the 1980s, dedicated working groups under the auspices of the Scientific Committee on Oceanic Research (SCOR) consolidated theory around ecosystem networks and trophic flow budgets (Ulanowicz and Platt 1985; Ulanowicz 1986), and software became available for their construction and analysis (Polovina 1984; Wulff et al. 1989; Christensen and Pauly 1992). These approaches are now used widely (e.g. Jarre et al. 1998; Brander et al., Chapter 3, this volume) in the fisheries-ecosystem scientific community (Christensen and McClean 2004; Link et al. 2005). During the GLOBEC era, the equilibrium assumptions in mass-balance models were partially relaxed (Walters et al. 2000) to allow biomass pools to evolve in response to varying external losses (i.e. fishing harvest), and mass-balance approaches to dynamic ecosystem studies have been applied in a variety of GLOBEC study regions (e.g. Shannon et al. 2008a). The mass-balance models of food webs tend to use functional groups to represent lower trophic levels and top predators, feeding guilds for non-commercial fish species and individual species or life stages for commercial fish species (Fig. 7.8). The advantage of the mass-balance approach is that it constrains the number of possible solutions to food web flows, because of assumptions about conservation of mass. One disadvantage is that it is not able to capture the diversity of processes that occur at different temporal and spatial scales (e.g. those at lower trophic levels) in a model with a time resolution of 1 year. Thus it is difficult to use the approach for simulations, but it can be useful for synthesis and for integrating into large temporal and spatial scales. Such integrative studies in turn have fuelled the discussions about trophic controls in food webs (see below).

To integrate across scales for the Georges Bank ecosystem, Steele *et al.* (2007) constructed two linked food web models. They used a mass-balance, carbon-budget approach to summarize planktonic and benthic processes, and a species-based, dietary approach for carbon flows and interactions in the fish community. The role of physical exchange in the ecosystem was explicit in their food web diagram (Fig. 7.6). The two approaches for representing the food web involved different assumptions and tradeoffs (Steele *et al.* 2008; Table 7.1). The species-centric approach emphasized the effects of physics on recruitment, essentially ignoring ecological processes such as feeding and predation; the timescale was short term. The trophic approach

simplified the environment in favour of apportioning energy among taxa; it was therefore integrative with a long timescale.

7.2.2.3 Size-based approaches

Size-based approaches for pelagic food webs have a strong theoretical foundation (e.g. Peters 1983; Platt 1985; Brown *et al.* 2004), supplemented by empirical descriptions of biomass spectra (Bianchi *et al.* 2000) and stable isotope measurements (Fry and Quiñones 1994; Jennings *et al.* 2002). They can be applied from an individual/species level to the community/ecosystem level (Shin and Cury 2004; Enquist *et al.* 2003). Individual body size increases with trophic level (Fig. 7.9) because primary

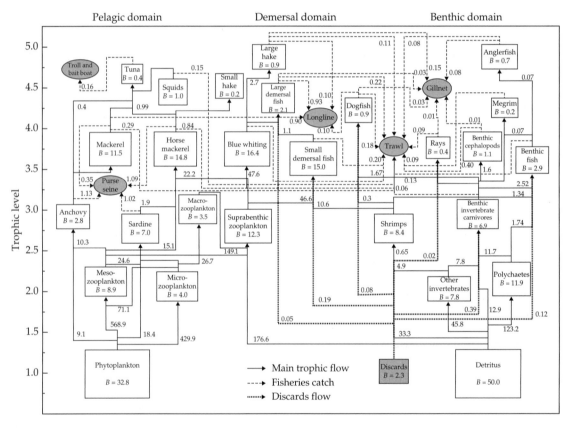

Figure 7.8 Mass-balance model of the food web in the Cantabrian Sea showing main trophic interactions (Reprinted from Sánchez and Olaso 2004 with kind permission from Elsevier). The boxes (size roughly proportional to biomass) are arranged on the vertical axis by trophic level and along the horizontal axis on an approximate pelagic to benthic scale. Main flows are expressed as t km⁻² year⁻¹ and the biomass of each trophic group (B) in t km⁻². Minor flows, respiration, catch, and all back flows to detritus are omitted. Note the increased levels of aggregation at low trophic levels and disaggregation into two size classes for commercial hake species.

Table 7.1 Comparison of a species-centric versus trophic-centric approach for modelling food webs (Steele *et al.* 2008).

	Species-centric approach	Trophic-centric approach
Assumptions regarding critical physics	Direct action on planktonic individuals such as copepods or fish larvae	Indirect effects through processes of nutrient supply for new production
Assumptions regarding critical ecology	Population dynamics at critical life stages such as diapause	Energetics of trophic groups and fluxes between groups
Complexity in the…	…physics: detailed dynamics for regions with complex topography	…trophodynamics: food webs contain many groups from microbes to fish
Simplicity in the…	…trophodynamics: spatial demography of one species or one life stage	…physics: linear steady states average over time and space

producers are mostly microscopic photosynthetic algae and cyanobacteria, and herbivores are mostly protist microzooplankton or small metazoan mesozooplankton. In general, large organisms eat smaller ones, and prey are often ingested alive and more or less intact. Prey selectivity is mostly based on size, overlap of spatial distribution, and behaviourally mediated encounter rates, and not necessarily specific taxonomic or chemical preference. Detrital particles (e.g. faeces, discards, and carcasses) larger than a few millimetres sink rapidly, and therefore tend to be exported to benthic or mesopelagic environments rather than remaining in the epipelagic consumer/decomposer food webs. The extent of this sinking loss varies with organism size, and therefore also with trophic level. Both observations and models usually place the boundary between producers of sinking 'export' detritus versus suspended 'recycled' detritus at the microzooplankton-mesozooplankton transition (see papers in Fasham 1984).

In marine food webs size ranges span approximately six orders of magnitude in linear dimension and approximately 20 in volume/weight; on a logarithmic scale the mesozooplankton occupy the middle of this range. Size-based analyses provide a useful basis for reducing some of the complexity of real food webs, and are valuable for examining the general operation of oceanic food webs and the

impacts of fishing (e.g. Cury *et al.* 2003; Jennings and Blanchard 2004; Shin and Cury 2004). They are also useful for examining key species interactions in a general food web framework (Cury *et al.* 2003; deYoung *et al.* 2004a). However, it is clear that for many systems simple views based on fixed size ratios of predators and their prey do not always apply. Top predators occupy the highest trophic levels during a substantial fraction of their lives, but can experience dramatic ontogenetic size changes that result in them spanning several trophic levels. This is especially the case for large predatory fish for which adult length can be more than three orders of magnitude greater than the length of their eggs and larvae. The complications introduced by changes in feeding interactions and hence trophic position during ontogenetic development can be accommodated in more complex size-based frameworks (Shin and Cury 2004), which can provide extremely powerful generalizations (Fig. 7.9), but they require simplifications and species-level assumptions that restrict their application.

7.2.2.4 Moving towards integrating the approaches
Networks representing energy or material flows in ecosystems need to have defined nodes, but because these nodes cannot represent the full range of possibilities represented by species, body size, and functional groups simultaneously, any emergent

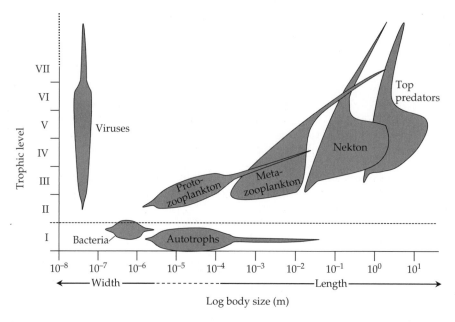

Figure 7.9 Schematic showing the relationship between trophic level and body size. The widths of the objects along the trophic level axis indicate abundance of organisms (not drawn to scale). (Based on Sieburth *et al.* 1978; Moloney *et al.* 1991).

properties will be restricted, with a limited perspective. Species-based approaches help in identifying key species in ecosystems, in understanding adaptive differences and similarities among species in the context of ecosystem change, and in linking species diversity to ecosystem attributes. However, there are too many species for this to be a feasible basis for modelling food webs, and interspecific competition and predation can only be handled in a very coarse or incomplete fashion. In a trophic context, species-based approaches can be problematic because aquatic animals change their diets dramatically as they develop and grow. Size-based approaches can solve this dilemma but are unable to address issues of biodiversity, cannot cope easily with characteristics of organisms that are not linked to body size, and cannot represent different growth patterns of different species. Functional group approaches aid in integrating and understanding energy and material flows but cannot capture the richness of potential responses made possible by differences among species, including linkages among functional groups when species change their functions with body size.

Can we make useful progress under these different, contrasting levels of simplification for understanding food web dynamics? Yes, but we need to be mindful of the limitations of each approach, to have clear objectives, and to define the spatial and temporal scope of the analyses. No single approach can capture food web dynamics adequately, and studies under GLOBEC have illustrated that there is something to be learned from each approach. Future work needs to extend the use of existing methods to more ecosystems and find new ways of integrating the key aspects of each.

7.3 Marine ecosystem dynamics in relation to global change

The GLOBEC programme aimed to improve understanding of ecosystem dynamics in the context of global change. This section focuses on four ecological topics to which GLOBEC studies make significant contributions: timing of events, variability of feeding, natural and fishing mortality, and spatial structure, movements, and migrations. Many of these studies

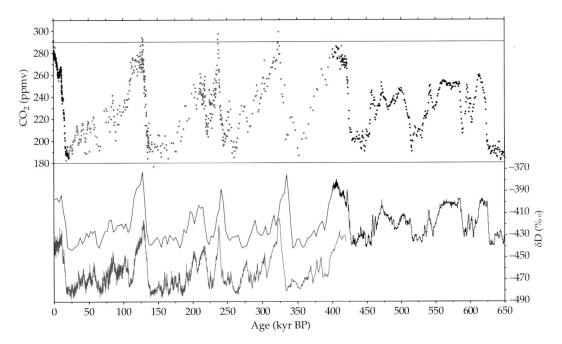

Plate 1 A composite CO_2 record over six-and-a-half Ice Age cycles, back to 650,000 years before present. The record results from the combination of CO_2 data from three Antarctic ice cores: Dome C (black), Vostok (blue), and Taylor Dome (light green). High resolution deuterium record (δD), a proxy for temperature, covaries with CO_2 concentrations, thus linking the climate and atmospheric chemical composition. (From Siegenthaler *et al.* 2005.) (See page 2).

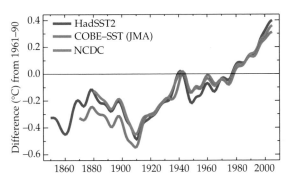

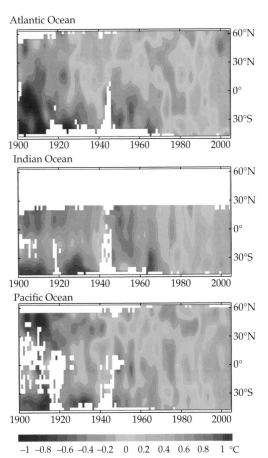

Plate 2 (Left panel) Smoothed annual global SST anomalies, relative to 1961–90 (°C), from the UK Meteorological Office sea surface temperature data set (HadSST2) (Rayner *et al.* 2006), the USA National Climatic Data Centre (NCDC) (Smith *et al.* 2005; red line) and the Japan Meterological Agency COBE-SST dataset (Ishii *et al.* 2005; green line). (Right panel) Sea Surface Temperature (°C) annual anomalies across each ocean by latitude from 1900 to 2005, relative to the 1961–90 from HadSST2 (Rayner *et al.* 2006). (From Trenberth *et al.* 2007.) (See page 3).

Early ice retreat ⟹ Late bloom   Small copepods   Pelagic pathway
                    warm water   abundant

Late ice retreat ⟹ Early bloom   Small copepods
                   Cold water   Scarce

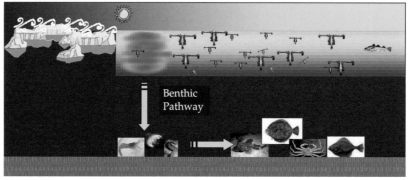

February   April   June

Plate 3 The relationship between the timing of the ice retreat and the spring bloom. When ice retreat is early, the bloom occurs later, the water is warm, a large copepod biomass develops, and a pelagic food web is favoured (top panel). When the ice retreat is late, the bloom occurs earlier, the water is cold, much of the production goes into a benthic food web and the copepod biomass is small. (Modified from Hunt *et al.*, 2002 with kind permission from Elsevier; F. Mueter, University of Alaska, Fairbanks, personal communication.) (See page 30).

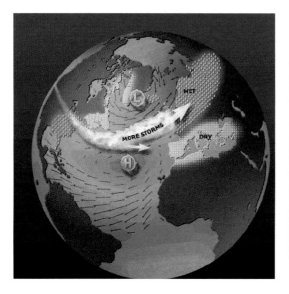
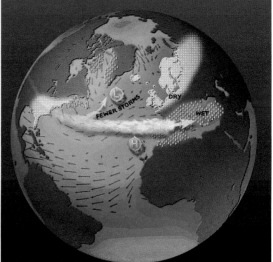

Plate 4 The atmospheric conditions during the positive (*left*) and negative (*right*) phases of the NAO. (Taken from http://www.ldeo.columbia.edu/res/pi/NAO/.) (See page 14).

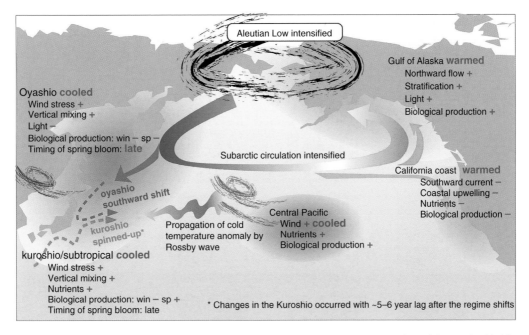

Plate 5 The atmospheric and oceanic conditions as well as the phytoplankton response during the positive phase of the PDO. (Modified from Chavez *et al.* 2003 with permission from AAAS.) (See page 16).

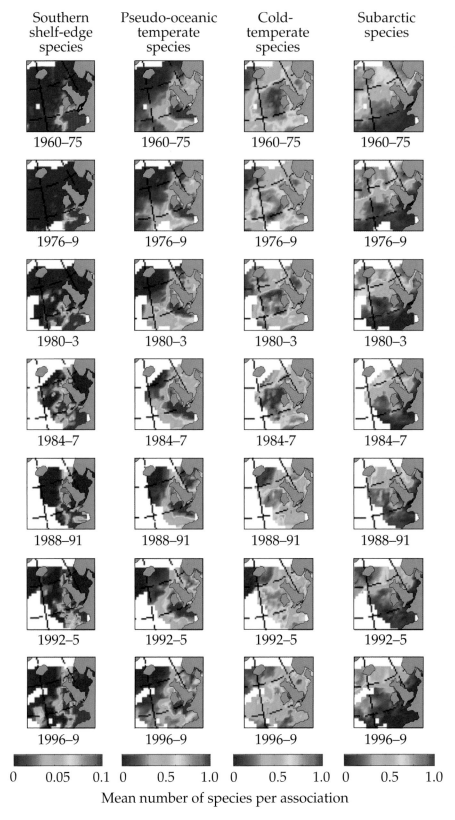

Plate 6 The mean number of southern shelf edge, pseudo-oceanic temperate, cold temperature, and subarctic species in the north-east Atlantic for different time periods from the 1960s to the late 1990s. (From Beaugrand *et al.* 2002c.) (See page 21).

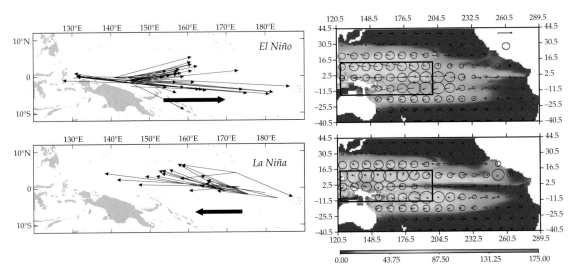

Plate 7 ENSO and movements of skipjack tuna in the Pacific Ocean. *Left*: Observed movements of tagged skipjack during *El Niño* and *La Niña* phases. (Redrawn from Lehodey *et al.* 1997.) Tagging data were compiled from records of a large-scale tagging programme carried out by the Secretariat of the Pacific Community. *Right*: Predicted distribution of biomass of skipjack (age cohort 9 months) in November 1997 (*El Niño* phase) and November 1998 (*La Niña* phase) in the Pacific Ocean. Circles and arrows represent random (diffusion) and directed (advection) movements of population density correspondingly and averaged by 10 degree squares. (Redrawn from Lehodey *et al.* 2008 with kind permission from Elsevier.) (See page 27).

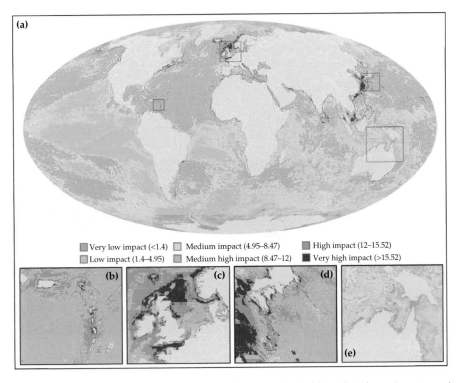

Plate 8 Global map from Halpern *et al.* 2008 reprinted with permission from AAAS showing (a) cumulative human impact on marine ecosystems. Insets show highly impacted regions including (b) the Eastern Caribbean, (c) the North Sea, (d) Japanese waters, (e) one of the least impacted regions, in northern Australia, and (e) the Torres Strait. The methodology uses expert judgement, standardization, and weighting to combine the 17 anthropogenic drivers. Terms describing degree of impact correspond to the 'per cent degraded' scheme of Lotze *et al.* (2006). (See page 47).

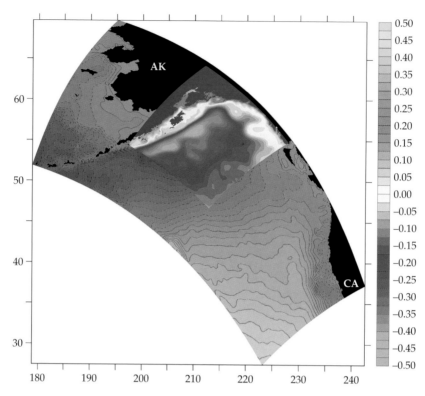

Plate 9 The grid for the ROMS model structure for the north-west Pacific showing the high-resolution sub-grid set inside a coarser resolution large-scale domain. The colour scala on the right shows the anomaly of sea surface height in metres. (Courtesy of A. Herman). (See page 106).

Plate 10 Modelled particle tracks (white dots) in the Benguela current system transported over a 6-week period; the land mass is in red and the depth contours are the remaining shaded areas. (Courtesy of Christian Mullon.) (See page 103).

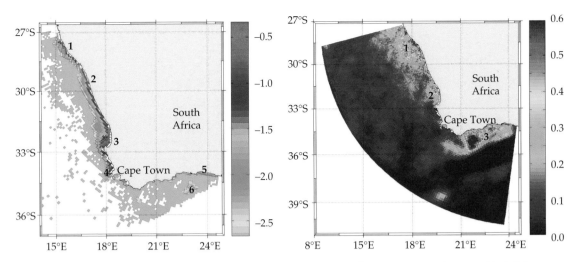

Plate 11 On the *left* is the enrichment intensity obtained through simulated particle upwelling-regions 1–4 are the most enriched off the west coast. On the *right* is the map of the simulated pattern of retention. Values correspond to the proportion of particles retained averaged over the period 1992–9 and depth. Only the areas 1 and 2 of retention match the regions of enrichment of the left-hand panel. Although recruitment for both anchovy and sardine is considered to occur predominantly off the west coast; the high retention predicted for region 3 may not result in observed successful recruitment. (From Lett *et al.*, 2006.) (See page 103).

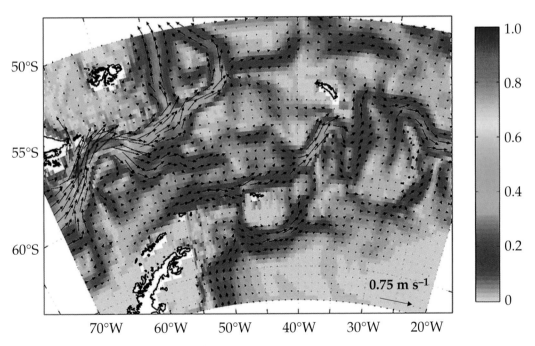

Plate 12 Six-month average of the depth weighted mean velocity fields for the upper 182 m of the circulation model OCCAM in the Scotia Sea for October 1999–March 2000. The colours show the magnitude (m/s), arrows (not shown at every point for clarity) have been capped at 0.75 ms⁻¹. (Reprinted from Murphy *et al.* 2004 with kind permission from Elsevier.) (See page 114).

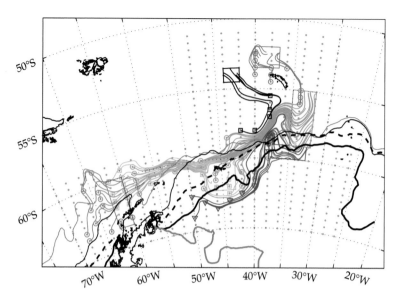

Plate 13 Lagrangian tracks of particles released at the beginning of October 1999 that passed through the high krill biomass areas during January 2000. Based on releases in a 0.5° latitude by 2° longitude grid on the depth-weighted mean velocity field over the upper 182 m. Different colours are related to different regions of high biomass and the associated particles tracks. The September to January ice-edge is shown-September 1999 with a thin grey line; October 1999 with a solid black line; November 1999 with a broken black line; December 1999 with a thick black line; and January 2000 with a thick grey line. (Reprinted from Murphy *et al.* 2004 with kind permission from Elsevier.) (See page 114).

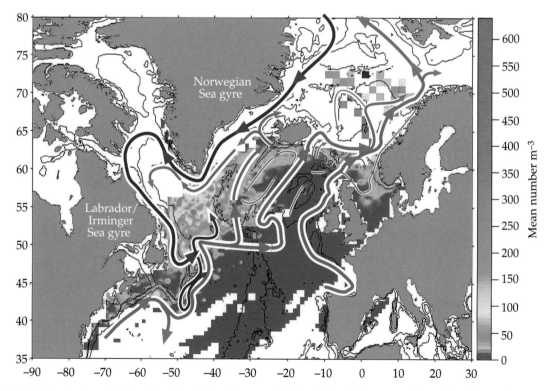

Plate 14 The abundance of *Calanus finmarchicus* in the surface waters of the North Atlantic. The arrows and lines indicate the major features of the sub-polar gyre. The thickness of the line approximates the strength of the current and the colour indicates whether it is a locally warm (red) or cold (blue) current system. The peak concentrations of *Calanus* occur in the eastern and western cores of the gyres. (From deYoung *et al.* 2004 reprinted with permission from AAAS.) (See page 125).

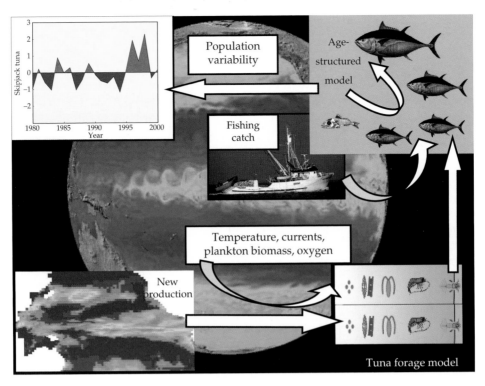

Plate 15 The lower left image shows a snapshot of the simulated time-varying distribution of NO_3-based primary production (referred to as 'new production') integrated over the euphotic zone in the Pacific, from a coupled physical-biogeochemical model (Chai *et al.* 2003 with kind permission of Springer Science and Business Media). These data together with simulated temperature and velocity data drive the tuna population model (insert *top right*). The basin-scale population of skipjack is represented by a (horizontally) spatially resolved age-structured cohort model, which tracks the numbers of individuals-at-age over space and time (From Lehodey 2001 with kind permission from Elsevier). The relative spatial distribution of recruitment, and age-specific migration rates are the only two aspects of the tuna model that are coupled to the underlying physical-biogeochemical system. Coupling is achieved through the dispersal pattern of eggs and larvae, and dynamic linkage to an intermediate model of the prey of tuna (referred to as 'forage', lower right panel). Coupling to the tuna model is achieved by spatial scaling of the overall mortality rate of forage biomass. (See page 127).

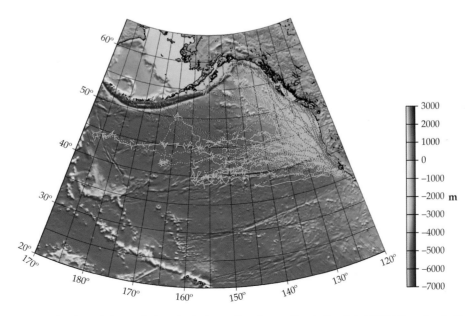

Plate 16 Distribution of northern elephant seals determined using satellite telemetry. (From Le Boeuf *et al.* 2000.) (Reproduced with permission of Ecological Society of America, from Foraging ecology of northern elephant seals, Le Boeuf, B. J., Croker, D. E., Costa, D. P. *et al.* Ecological Monographs, **70**, 2000; permission conveyed through copyright clearance centre, Inc.). (See page 150).

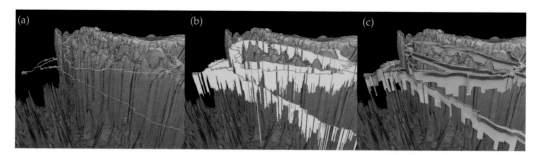

Plate 17 Track of southern elephant seals in the western Antarctic Peninsula obtained using the SMRU CTD-SRDL 9,000 tag. (a) Surface track. (b) Shows the surface track and with diving behaviour. (c)The temperature and salinity profiles obtained from animal dives. (From Costa, Goebel, and McDonald, unpublished data.) (See page 150).

Plate 18 Differences in the movement patterns of southern elephant seals (yellow), crabeater seals (red), and Weddell seals (green) along the Antarctic Peninsula. The tracks cover the same time period during 2007. (From Costa, Goebel and McDonald, unpublished data.) (See page 151).

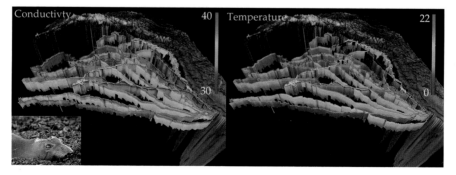

Plate 19 Left: conductivity and right: temperature profiles obtained from seven female elephant seals migrating across the North Pacific Ocean. The different coloured lines refer to the tracks of individual seals and the 'curtain' effect shows the depth over which the CTD data were obtained. The coloured bars are the scale for conductivity (mS/cm) and temperature (°C). Inset lower left: female elephant seal with CTD tag on her head. (From Costa, unpublished data.) (See page 152).

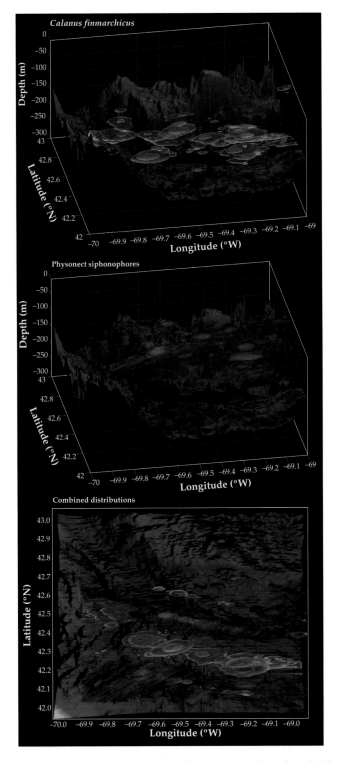

Plate 20 Spatial distributions of diapausing *Calanus finmarchicus* (top panel) and physonect siphonophores (middle panel) in Wilkinson Basin during December 1999 as measured with a towed Video Plankton Recorder (VPR). Observations of each taxon were converted to abundances and interpolated in three-dimension using GLOBEC EasyKrig 3.0 software. Isosurfaces in this visualization correspond to the highest densities. For *C. finmarchicus* these densities were 100–300 individuals per m³ and for siphonophores (1–4 colonies m⁻³). Combined distributions in plan view are illustrated in the bottom panel, which demonstrates that the patches of *C. finmarchicus* are generally absent from regions of high siphonophore densities (Christian Briseno). (See page 136).

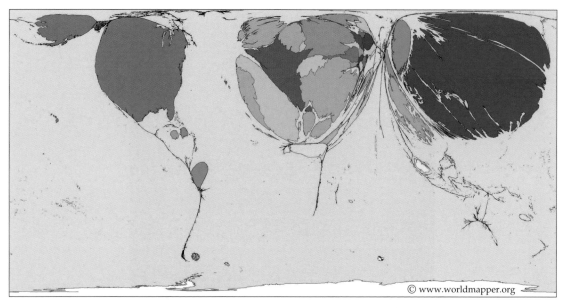

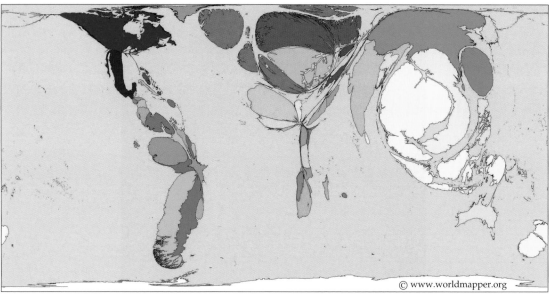

Plate 21 Map of the world illustrating net fish imports (top panel) and net fish exports per country. Territory size shows the proportion of worldwide net exports (imports) of fish (in US$) that come from that territory. Net exports (imports) are calculated as exports (imports) minus imports (exports). When imports (exports) are larger than exports (imports) the territory is not shown. (*Source*: www.worldmapper.org © 2006 SASI Group (University of Sheffield) and Mark Newman). (See page 224).

Projections of Surface Temperatures

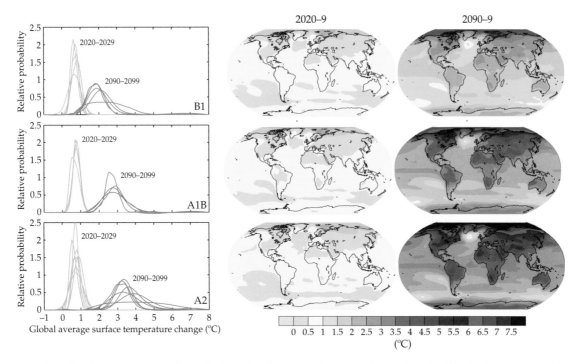

Plate 22 Projected surface temperature changes for the early and late twenty-first century relative to the period 1980–99. The central and right panels show the atmosphere-ocean coupled general circulation model (AOGCM) multi-model average projections for B1 (*top*), A1B (*middle*) and A2 (*bottom*) IPCC Special Report on Emissions Scenarios (SRES) scenarios averaged over the decades 2020–9 (*centre*) and 2090–9 (*right*). IPCC has assumed four storylines that encompass a marked portion of the underlying uncertainties in the main driving forces (demographic change, economic development, and technological change). Four qualitative storylines describe four sets of scenarios called families: A1 (world market), A2 (provincial enterprise), B1 (global sustainability), and B2 (local stewardship). The A1 family was divided into three groups characterizing alternative developments of energy technologies: A1FI (fossil fuel intensive), A1B (balanced), and A1T (predominantly non-fossil fuel). The scenario used historical data for atmospheric CO_2 between 1860 and 2000, and then assumed atmospheric CO_2 concentrations will reach 850 ppm (A2), 720 ppm (A1B), 500 ppm (B1) in the year 2100, respectively. The left panels show the corresponding uncertainties as the relative probabilities of estimated global average warming from several different AOGCM and Earth System Model of Intermediate Complexity studies for the same periods. (From IPCC 2007a.) (See page 288).

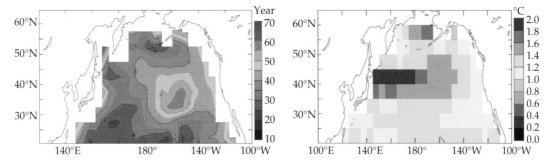

Plate 23 *Left*: Estimated year (+2000) when the net warming exceeds the magnitude of natural variability. This year for each location was defined by dividing twice the standard deviation of the observed sea surface temperature (SST) by the modelled temperature trend. This threshold is generally reached for the North Pacific before the middle of the twenty-first century. *Right*: Projected winter SST change for 2040–9 minus 1980–99. Changes are in the range of 1–2°C. (From Overland and Wang 2007a.) (See page 290).

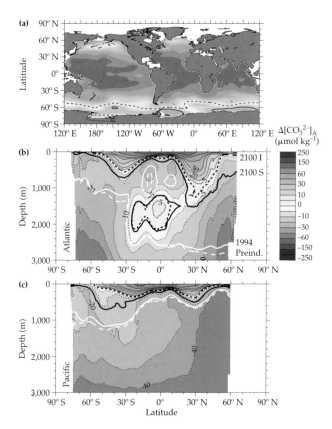

Plate 24 The aragonite saturation state in the year 2100 as indicated by $\Delta[CO_3^{2-}]_A$. The $\Delta[CO_3^{2-}]_A$ is the *in situ* $[CO_3^{2-}]$ minus that for aragonite-equilibrated sea water at the same salinity, temperature, and pressure. Shown are the OCMIP-2 median concentrations in the year 2100 under scenario IS92a: (a) surface map; (b) Atlantic; and (c) Pacific zonal averages. Thick lines indicate the aragonite saturation horizon in 1765 (pre-industrial; white dashed line), 1994 (white solid line) and 2100 (black solid line for S650; black dashed line for IS92a). Positive $\Delta[CO_3^{2-}]_A$ indicates supersaturation; negative $\Delta[CO_3^{2-}]_A$ indicates undersaturation. (After Orr *et al.* 2005 Reprinted by permission of Macmillan Publishers Ltd: Nature copyright 2005.) (See page 295).

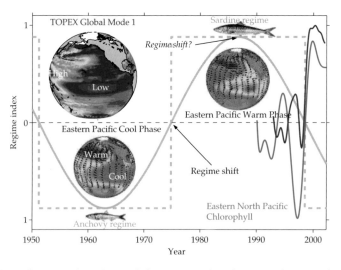

Plate 25 Hypothetical oscillation of a regime index with a period of 50 years. From the early 1950s to about 1975, the Pacific was cooler than average, and anchovies dominated. From about 1975 to the late 1990s, the Pacific was warmer, and sardines dominated. The spatial patterns of SST and atmospheric circulation anomalies are shown for each regime (Mantua *et al.* 1997). Some indices suggest that the shifts are rapid (dashed), whereas others suggest a more gradual shift (solid). The first empirical orthogonal function (EOF) of global TOPEX sea surface height (SSH) is shown above the cool, anchovy regime. The coefficient is shown in blue together with surface chlorophyll anomalies (green, mg m^{-3}) for the eastern margin of the California Current system from 1989 to 2001 (Pearcy 2005), also low-pass filtered. (From Chavez *et al.* 2003.) (See page 305).

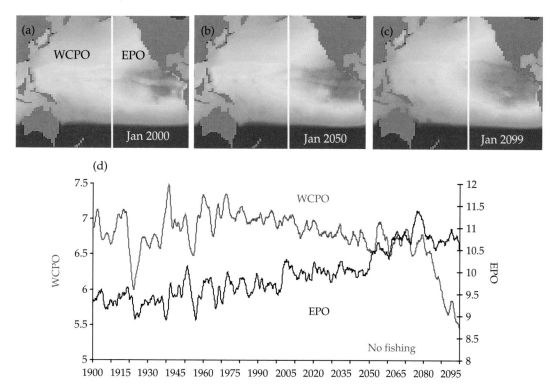

Plate 26 Predicted spatial distribution of bigeye tuna biomass in the Pacific Ocean (a) current, (b) in 2050 and (c) in 2099 under a global warming scenario, and (d) comparison of predicted biomass (million tons) of bigeye tuna in the western central (WCPO) and eastern (EPO) Pacific Ocean under a climate change scenario (without fishing). (From Lehodey *et al.* 2007.) (See page 311).

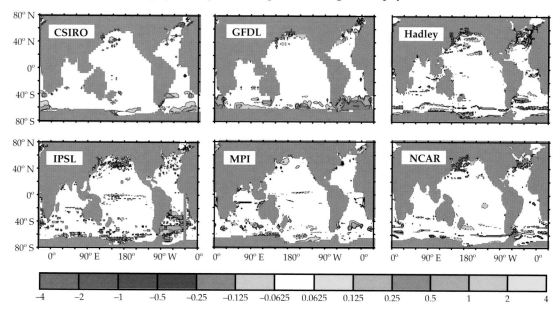

Plate 27 Simulations of the impact of global warming on the chlorophyll concentration in mg m⁻³. Chlorophyll is calculated for both the control and the warming simulations using the empirical model. The figure shows the difference between the warming simulation and the control simulation averaged over the period 2040 to 2060 (except for Max Planck Institut (MPI), which is for the period 2040 to 2049). Areas in white are those for which the chlorophyll change is smaller than ±0.0625 mg m⁻³. (From Sarmiento *et al.* 2004.) (See page 315).

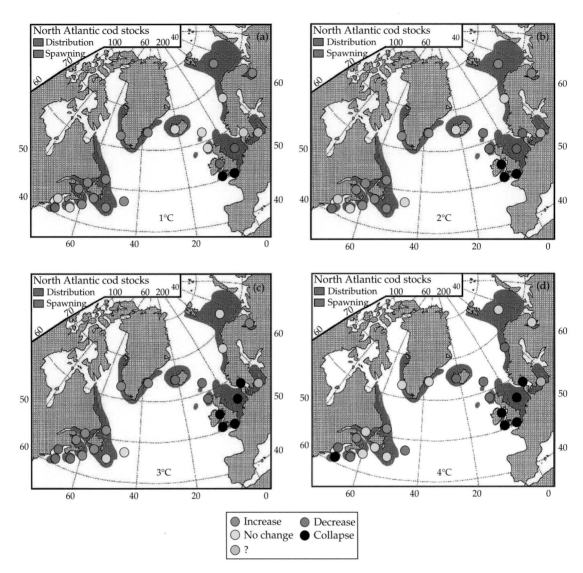

Plate 28 The expected changes in the abundance of the cod stocks with a temperature increase of (a) 1°C, (b) 2°C, (c) 3°C, and (d) 4°C temperature increase. Note the changes are relative to the previous change, that is, 4°C represents the changes from those at 3°C. (From Drinkwater 2005.) (See page 307).

were conducted at the species-population level and many of the ramifications at the ecosystem level are still to be explored.

7.3.1 The importance of timing

Trophic transfer rates are often seasonally modulated, and overall efficiency will depend on whether or not seasonal windows align between taxa and trophic levels. With its focus on target species, GLOBEC programmes and related research provided considerable evidence of strong climate-driven variation in zooplankton seasonal timing (Mackas and Tsuda 1999; Greve *et al.* 2001; Edwards and Richardson 2004; Valdés *et al.* 2006), and of regionally dominant zooplankton life history strategy and timing match-mismatch with other trophic levels (Bertram *et al.* 2001; Hunt *et al.* 2002; Beaugrand *et al.* 2003a; Edwards and Richardson 2004).

The life cycles of many zooplankton species are closely tied to the main period of phytoplankton blooms (see Box 7.2), and variation in timing and magnitude of new production can significantly affect zooplankton development (e.g. Tadokoro *et al.* 2005), determining their abundance and distribution. The general timing of phytoplankton blooms in turn affects the timing of exit of copepods from diapause and their migration from depth (Speirs *et al.* 2006), and changes in this timing can affect survival and development of zooplankton populations and food web structure. GLOBEC programmes have contributed to recent discoveries of shifts in the seasonal timing of life cycle events and associated changes in climate-forced environmental variables. The copepod *N. plumchrus*, which by itself can constitute about half of the mesozooplankton biomass in the upper layer of the north-east Pacific Ocean, grows and develops through its life stages during a brief, 1–2-month period in spring and early summer before descending to deep water to overwinter. This period of maximum biomass in the upper ocean has shifted markedly from late June in the early 1970s to early to mid-May in the late 1990s (Mackas *et al.* 1998). The shift is strongly and almost linearly associated with an increase in upper ocean temperature (Fig. 7.10), consistent with climate-change scenarios presented by the Intergovernmental Panel on Climate Change

(Mackas *et al.* 2007). The processes linking the timing of the life cycle with upper ocean temperature likely involve not only direct, positive effects of increasing temperature on copepod development rate, but also secondary effects of a warmer ocean on food supply and predators, affecting seasonal rates of survivorship (Mackas *et al.* 1998).

Similarly, in the North Atlantic, GLOBEC researchers improved understanding of the control in *Calanus* species of timing of the dormancy period, which has been a major challenge for modelling copepod population dynamics (e.g. Runge *et al.* 2005). Johnson *et al.* (2008) reviewed the various hypotheses put forward to explain what causes a developing copepodite stage to enter into and then emerge from a dormant state and concluded that only one, the lipid accumulation window hypothesis (Irigoien 2004), was consistent with observations of variable timing of entry and exit from dormancy in the north-west Atlantic. The fundamental premise is that a developing individual will only initiate hormonal/physiological processes to prepare for dormancy if its lipid stores are above a certain threshold level assuring sufficient energy and mass to sustain overwintering, moulting, and early gonad development. If this lipid level is not attained, the individual develops to the adult stage. Under the hypothesis, the duration of dormancy is also variable, dependent on ambient temperature and the level of lipid stores. As temperature and food supply are the primary exogenous factors controlling lipid accumulation, this is a potentially important mechanism by which climate change may influence abundance of the regional *Calanus* populations. This conceptualization of *Calanus* life history has been formulated into a life cycle model with species-specific parameterization of responses of growth, development, and reproduction to environmental variables (Johnson *et al.* 2008). Embedding the consequences of these population-level processes into ecosystem models remains a challenge for future studies, although some of the implications for the food web have been documented from a variety of observational studies (see below).

The influence of climate change on zooplankton life histories impacts not only zooplankton population dynamics but also leads to trophic mismatches, shifts in food web structure, and other ecosystem-level

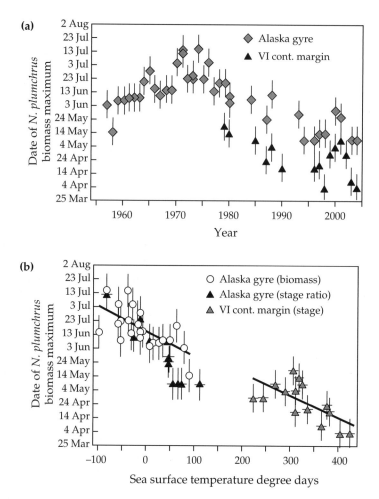

Figure 7.10 Seasonal timing of the upper ocean biomass maximum of *N. plumchrus* in the Alaska Gyre (ocean station P and outer line P) and along the Vancouver Island (VI) continental margin. (a) Time series of the estimated date of the biomass maximum in each year. Although timing is 3–5 weeks earlier along the Vancouver Island continental margin, interannual fluctuations are strongly correlated with those in the Alaska Gyre. (b) Timing of the biomass maximum plotted against cumulative warmth of the upper ocean during March-May (as degree days relative to a 6°C baseline). Warm water during the growing season is strongly associated with early timing. (From Mackas *et al.* 1998, 2007).

changes (Edwards and Richardson 2004). For example, reproduction and nestling growth of the planktivorous seabird, Cassin's auklet *Ptychoramphus aleuticus*, is reduced during warm springs off the coast of British Columbia, related to a mismatch in seasonal timing of the life cycle of *N. cristatus* upon which they prey (Bertram *et al.* 2001; Hedd *et al.* 2002). In the North Sea, decreases in cod recruitment have been explained by a shift in dominance in zooplankton communities from *C. finmarchicus*, whose naupliar stages are prey for larval cod, to *C. helgolandicus*, which reproduce

later in the season and consequently are unavailable to cod larvae (Beaugrand *et al.* 2003a). Studies of larval fish trophodynamics highlight the need for good observational data for modelling the early life cycle and feeding interactions of fish (Gallego *et al.* 2007; Heath 2007). Integrated seasonal analyses are required to identify the mechanisms involved. For example, in a combined study of pink salmon and herring in Prince William Sound, Cooney *et al.* (2001) showed that juvenile herring were affected by winter food availability, whereas juvenile pink salmon population sizes were linked to availability of alternative prey for their fish and bird predators. The two species also exploit different components of the lower trophic level plankton system with differences in timing of peak requirements for prey.

Spatial mismatches and distribution shifts also influence predator-prey interactions, with knock-on effects for ecosystem dynamics. The dynamics of cod feeding on northern shrimp populations off West Greenland can only be understood if spatial overlaps of predator and prey are considered along with the dynamics of another predator, Greenland halibut (Wieland *et al.* 2007). Basin-scale changes in the abundance and distribution of Antarctic krill probably influenced recent shifts in population dynamics of Antarctic seals and marine mammals (Atkinson *et al.* 2004; Ducklow *et al.* 2007). Correspondingly, salps, notably *Salpa thompsoni*, the other dominant macrograzer in low productivity zones of the Southern Ocean, appear to be on the increase (Atkinson *et al.* 2004), with implications for the control of the carbon cycle in this region (Le Fèvre *et al.* 1998). Off southern Africa, an eastward shift in sardine *Sardinops sagax* populations has influenced breeding populations of Cape gannets *Morus capensis*, with breeding colonies on the west coast decreasing and those on the east coast increasing in response to availability of sardine prey (Pichegru *et al.* 2007). For studies in this region, Grémillet *et al.* (2008) show how spatial distributions of key environmental variables could not be used to infer ecosystem dynamics, because the spatial patterns result from complex ecological processes occurring in different areas and at different times (as detailed in various contributions in Shannon *et al.* 2006b).

7.3.2 Feeding variability

Zooplankton feeding interactions are variable within and between seasons, regions, and ecosystems. Studies have shown how flexibility in feeding by many species provides strategies for coping with seasonal, interannual, and decadal environmental variability. The main focus of many studies of the role of zooplankton in oceanic food webs has been on the larger-sized meso- and macro-zooplankton

because of their importance as prey for fish and large pelagic predators, and their dominance in temperate and polar seas.

Within zooplankton communities, the full range of feeding strategies is displayed: herbivory, omnivory, carnivory, and cannibalism. Copepods have been shown to feed on a wide range of groups including ciliates, diatoms, dinoflagellates, and cyanobacteria (Schmidt *et al.* 2003; Stibor *et al.* 2004; Peters *et al.* 2006; Saiz *et al.* 2007). Euphausiids also have variable diets (Schmidt *et al.* 2006) and gelatinous species, such as Cnidaria, Ctenophora, and salps, have carnivorous and phytoplanktivorous diets (Nelson *et al.* 2000; Phleger *et al.* 2000). This flexibility within communities is also demonstrated within many species and there are indications that a mixed diet is a fundamental requirement for zooplankton. Varied diets provide the appropriate nutritional balance for growth and development of zooplankton, and can be especially important for reproductive development and successful spawning (Vargas *et al.* 2006; Andersen *et al.* 2007; Bell *et al.* 2007). Food quality and quantity are important for growth and development of individual zooplankton species (Runge *et al.* 2005), fish larvae (Buckley and Durbin 2006), adult fish (Corten 2001; Rose and O'Driscoll 2002), and top predators (Grémillet *et al.* 2008). At the ecosystem level, variability in the diets can strongly affect the stoichiometric basis of nutrient exchanges involved in ecological interactions, having a major influence on food web structure over a wide range of time scales (Elser and Hessen 2005).

Complex interactions in oceanic food webs can generate significant predation impacts within zooplankton communities. These effects can be negative, such that smaller species are suppressed, or positive by reducing predators or competitors in the microplankton (Stibor *et al.* 2004; Saiz *et al.* 2007). The complexity of feeding interactions in zooplankton communities has been highlighted through a range of stable isotope analyses (Schmidt *et al.* 2004; Bode *et al.* 2007), which emphasized that, although simple, fixed trophic level views can be useful for general analyses, species-based analyses require detailed information. Such detailed studies have improved our understanding of relationships between feeding rate and food availability (Gentleman *et al.* 2003; Cook *et al.* 2007), but many of these studies have been based on either single species food cultures or on mixed, general, local assemblage water samples. Prey preferences can change as community composition changes (Sparre 1991), having a major impact on both predator and prey dynamics. Determining these impacts requires analyses of how relative prey preference and availability interact to affect realized feeding rate on any given component of a prey assemblage (Gentleman *et al.* 2003).

As with zooplankton, analyses of the feeding interactions of key fish species and other predators have highlighted their complex and dynamic nature (Miller and Brodeur 2007). Microbial groups have been shown to be important in the diets of larval fish, for example, in the Irish Sea (de Figueiredo *et al.* 2005, 2007). There have been major advances in understanding the complexity of feeding interactions during ontogenetic development of many fish species (Tadokoro *et al.* 2005; Sara and Sara 2007; Latour *et al.* 2008). Kaeriyama *et al.* (2004) suggested that high seas Pacific salmon *Onchorhyncus* spp. switch diets as their prey resources change in response to climate change. In the Kuroshio region of the North Pacific, the myctophid fish *Myctophum nitidulum* feeds mainly on euphausiids, copepods, and amphipods (Watanabe and Kawaguchi 2003a, b), but at times the larvae of the Japanese sardine *Sardinops melanostictus* can be a common prey item, affecting the dynamics of regional populations of sardine. Such interactions can determine survival of larval fish, illustrating the potential importance of ichthyoplankton-zooplankton-fish interactions. Large zooplankton can negatively affect young fish through predation (Agostini *et al.* 2007) and jellyfish can be significant predators of herring larvae (Lynam *et al.* 2005b). Pre-recruit fish can be significant grazers on lower trophic level microbial and small zooplankton species (Heath 2007). Complex feedback effects are possible in oceanic food webs, influenced by zooplankton community structure and detailed interactions within plankton systems (Kaeriyama *et al.* 2004; Möllmann *et al.* 2004, 2005; van der Lingen *et al.* 2006a; Dickmann *et al.* 2007).

There are also whole groups of organisms whose importance in food webs is only beginning to be recognized. For example, mysids can be important in regional food webs. They have wide-ranging diets, including microbial groups, and they also feed in benthic systems, showing marked diurnal vertical migration and therefore linking benthic and pelagic ecosystems (Jumars 2007). The role of gelatinous zooplankton received increasing attention during the GLOBEC era, and this zooplankton group has been implicated strongly as indicators of major ecosystem changes (Box 7.1). Small planktonic copepods (<1 mm length) are substantially under-sampled in most conventional zooplankton studies because they readily pass through 200 μm mesh nets (Gallienne and Robins 2001). The cyclopoid genus *Oithona*, one of these small taxa, occurs throughout the world's oceans and can be extremely abundant in oceanic systems (Gallienne and Robins 2001; Castellani *et al.* 2007). These small zooplankton can be important grazers of microbes and phytoplankton, and provide prey for larger zooplankton including larval fish (Turner 2004; Castellani *et al.* 2005).

7.3.3 Predation and fishing mortality

Mortality is a driving force in life history responses, and GLOBEC programmes have provided new approaches and insights into mortality patterns among life stages of key zooplankton and fish species. The significance of high and likely density-dependent mortality on eggs and early naupliar stages of zooplankton caused by predation by older life stages as well as other invertebrate predators has been shown for *C. finmarchicus* in the North Atlantic (Ohman *et al.* 2004, 2008). Cannibalistic predation has also been shown for cultures of *C. helgolandicus* in the laboratory (Bonnet *et al.* 2005), although the importance of this source of mortality in the field is uncertain. The *Calanus* mortality pattern contrasts with the stage-dependent mortality of *Pseudocalanus* species, which, unlike *Calanus* species, carry and protect their eggs until hatching (Ohman *et al.* 2002). Life cycle models can incorporate these observations in combination with simulations of temperature and primary production fields. For example, physical circulation models have been linked to lower trophic level (nutrient-phytoplankton-zooplankton) models (e.g. Helaouët and Beaugrand 2007; Ji *et al.* 2008) and used to understand transport of life stages (e.g. Speirs *et al.* 2006). These models provide a tool for predicting shifts in the distribution and abundance of *C. finmarchicus* and *C. helgolandicus* as the temperature of the upper ocean warms, although they are unable to incorporate the effects of predation mortality on population dynamics.

Notwithstanding their limitations in an ecosystem context, detailed analyses of the life cycles of key zooplankton species have been shown to be crucial for understanding their population dynamics (Heath *et al.* 2004, 2008b; Perry *et al.* 2004; Speirs *et al.* 2005, 2006). Variability in predation mortality and unresolved spatial covariability of predator and prey, however, remain significant knowledge gaps, especially in the context of global change. Detailed analyses and models have been developed for some key species, for example *C. finmarchicus*, but there is little general understanding of the seasonal life cycle and trophodynamics of many other important species. Many analyses of food web structure and dynamics are based on a homogeneous view, but spatial and temporal variability of zooplankton and microbial system interactions, as well as interactions with large predators, are significant determinants of overall ecosystem dynamics. Potential impacts of such spatial interactions horizontally or vertically (Brentnall *et al.* 2003; Morozov *et al.* 2007) are likely to be important, but have yet to be determined fully in studies of oceanic food webs (Melian and Bascompte 2004).

The strong emphasis on understanding the role of climate change in marine ecosystems led naturally into an initial bottom-up focus for GLOBEC, but as the programme matured studies that extended throughout the food web received increasing emphasis. There has been increasing recognition that the impacts of fishing need to be considered in marine ecosystems, and the role of top predators determined in ecosystem dynamics (Maury and Lehodey 2005). Top predators are subject to both bottom-up (resource-driven) and top-down (consumer-driven) controls during their early life stages, but for most of their adult life they are almost entirely liberated from predation pressure, except from fisheries. Fish typically become too large to be preyed upon when they are bigger than the largest size that the largest predator can eat, ~60–70 cm for a typical tropical pelagic fish community (Maury *et al.* 2007). Fisheries, however, have typically targeted the biggest fish throughout their history (Jackson *et al.* 2001). The removal by fishing of large predators from ecosystems has consequences for population and probably food web dynamics (Frank *et al.* 2007).

To ensure access to a regular food supply, marine top predators often rely on a variety of different prey species. In regions where prey diversity is low, such as upwelling systems, top predators can experience food limitation because of the temporary disappearance of their main prey species, which cannot be replaced. This is especially true for bird colonies, which are often attached to a particular geographic location (Frederiksen *et al.* 2006; Sydeman *et al.* 2006; Pichegru *et al.* 2007). To maximize their chance to feed on a variety of different potential prey effectively, predators need to be able to find and use the opportunities that present themselves. Fish top predators display strong trophic opportunism (e.g. Ménard *et al.* 2006) and are characterized by omnivory. They generally select their prey on a size and spatio-temporal co-occurrence basis, eating a large range of prey sizes across a broad spatial range, with potentially extensive movements and migrations at various scales (e.g. Dingle 1996) to find the highest prey concentrations in the ocean (Block *et al.* 2003; Weimerskirch *et al.* 2004; Fonteneau *et al.* 2008).

By occupying the highest positions in food webs, top predators exert pressure on their prey either directly through prey mortality or indirectly through behavioural modifications (the so-called risk effect; e.g. Heithaus *et al.* 2008). For example, Wirsing *et al.* (2008) found that bottlenose dolphins *Tursiops* sp., harbour seals *Phoca vitulina*, and dugongs *Dugong dugon* all modified their behaviour to minimize risky situations, avoiding productive feeding areas or desisting from risky feeding behaviour. They suggest that non-lethal effects of predation can have important impacts on energy usage, an opportunity cost that needs to be factored into energy budgets.

More generally, top predators act as couplers between slow (low productivity) and fast (high productivity) food chains, leading to structural asymmetry of food webs, which appears to be a generic feature of ecosystem organization (Rooney *et al.* 2006). They also link different mid-trophic level prey species potentially belonging to different functional taxonomic groups (myctophids, euphausiids, mesopelagic squids, small pelagics, etc.) and different prey sizes, often ranging over more than two orders of magnitude for a given predator, hence linking several trophic levels together. Having wide distribution ranges and exhibiting extensive migrations, they also link different regional ecosystems (Young *et al.* 2001), and their physiological capabilities allow them to couple different bathymetric ecosystems as well, because some top predator species, such as toothed whales, are able to feed both on epipelagic and mesopelagic prey (Ménard *et al.* 2007; Potier *et al.* 2007).

By coupling together many components of the ecosystem, top predators can multiply the 'weak links' in trophic networks (McCann *et al.* 1998). In that sense, they could act as stabilizing factors, regulating the variability of lower levels and increasing the stability of the whole ecosystem (McCann *et al.* 2005; Sala 2006). In their absence, prey populations are hypothesized to fluctuate with greater amplitudes because of bottom-up factors. Examples of large predatory fish removal, such as occurred during the Canadian cod collapse, show that their elimination can cause marked cascading effects in the pelagic food web, potentially causing major ecosystem changes (e.g. Frank *et al.* 2005; Scheffer *et al.* 2005).

Organisms package and store energy derived mainly from solar inputs into parcels that can be used by other organisms. Trophic interactions both transfer and dissipate energy within food webs (Lindeman 1942). Energy dissipation associated with biological processes (exothermic chemical reactions of metabolism) can be distinguished from losses of material through mortality or egestion. Losses are reincorporated in decomposer food webs, and much of this material sinks. Top predators help to maintain energy packets in the water column; without top predators, large proportions of biomass of mid-trophic level organisms, not preyed upon, can be lost from pelagic to benthic ecosystems, potentially shifting the trophic structure of the whole ecosystem from a typical phytoplankton-mesozooplankton-predator food web to an alternative bacteria-microzooplankton-

jellyfish food web (Daskalov *et al.* 2007; Oguz and Gilbert 2007).

Fishing is one of the causative agents of global change, but responses of marine ecosystems to fishing cannot be considered in isolation from other factors. There can be synergistic effects when different forcing factors act together, and fish stocks have collapsed when intensive fishing occurred at the same time as unfavourable environmental conditions (e.g. Peru in early 1970s, Newfoundland in early 1990s, and Namibia). There has been a wide international effort to develop ecosystem approaches to fisheries management (Barange *et al.*, Chapter 9, this volume), with part of the research effort concerning the impacts of fisheries on ecosystems (e.g. Hollingworth 2000), the development of ecosystem indicators (e.g. Daan *et al.* 2005), and attempts to understand the single and combined effects of fishing and other forcing factors on ecosystems. This paradigm, which emerged at the beginning of the GLOBEC era and is now generally accepted, has shifted perspectives of anthropogenic top-down impacts to a large scale, encouraging focus on ecosystem-level understanding of food web interactions as well as interactions with social systems (Perry *et al.*, Chapter 8, this volume).

7.3.4 Spatial structure, movements, and migrations

Horizontal and vertical movements of animals can link disparate ecosystems. Vertical movements also influence the functional role of zooplankton in transporting material from the surface to the deep ocean and link pelagic with benthic ecosystems. Spatial connections determine the response of oceanic food webs to variability and change. Spatial structure generates segregation of different age groups, which may be an important factor in the avoidance of cannibalism during development of some fish species (e.g. Yamamura 2004). The structure of food webs also varies across the spatial range of many exploited fish populations, influencing the response of the species to climate-related changes in prey availability. For example, Heath and Lough (2007) showed that larval cod in the North Atlantic at the northern edge of their range feed mainly on developmental stages of the copepod *C. finmarchicus*, while those at the southern edge feed on *Paracalanus* and *Pseudocalanus* species. In contrast, juvenile cod feed on a wide range of taxa with euphausiids as the main target prey. The implication is that climate-related shifts in plankton distribution can drive spatial shifts in cod population dynamics. Understanding the factors determining the distribution of key zooplankton (Helaouët and Beaugrand 2007) and forage fish (Grémillet *et al*. 2008) species has highlighted the importance of considering spatial structure in oceanic food webs. Advection processes can supplement food webs in which key species of commercially exploited fish develop (Yamamura 2004). Analyses of *C. finmarchicus* in the North Sea have shown the importance of understanding the factors determining the advective dispersal of copepods into shelf habitats in the east. In the Southern Ocean, interdisciplinary analyses have demonstrated that dispersal of Antarctic krill maintains many of the large colonies of predators during summer breeding seasons (Murphy *et al*. 2004a, b; 2007).

Spatial effects are fundamental to the operation of food webs but the biological and physical processes vary with scale (Skogen 2005; Murphy *et al*. 2007). Many species at tertiary trophic levels show marked seasonal shifts in distribution and migration-related spatial changes in diet. For example, spatial movement and probably changes in diet of small pelagic fish has occurred off southern Africa (Roux and Shannon 2004; van der Lingen *et al*. 2006a). In the Southern Ocean, fur seals shift from a diet centred on Antarctic krill during summer to a diet of mainly fish and squid while dispersing over large areas during winter (Murphy *et al*. 2007). In the western North Pacific, stable isotope analyses of minke whale feeding during migration indicate a dietary shift from krill to fish (Mitani *et al*. 2006). Understanding the physical interactions determining the spatial operation of oceanic food webs is crucial (Murphy *et al*. 2007). Munk (2007) showed that larval growth of North Sea cod tends to be enhanced in areas of fronts where prey are also concentrated. Predators can also switch their prey sizes when feeding at fronts because small prey can be present in greater than normal densities, making them energetically attractive to feed on (Vlietstra *et al*. 2005).

More generally, we are just beginning to understand that spatial as well as temporal mismatches in feeding interactions within food webs can be crucial in determining the annual dynamics of key species and overall food web structure (Morozov *et al*. 2007; Nottestad *et al*. 2007; Grémillet *et al*. 2008). A mismatch in space not only influences diets and feeding, but can have other direct physiological effects, for example if temperature is suboptimal, or if organisms are eaten by predators they would not normally encounter. When organisms are transported or migrate to waters where they cannot survive, these migrations have been termed 'death migrations' (Kimura *et al*. 1994). Migrations usually are behavioural adaptations to enhance the survival of a species; death migrations lead to a mismatch in space and can have a great impact on recruitment and population dynamics.

Experimental studies have shown that the feeding efficiency of fish larvae varies in response to differences in turbulence in the ocean, even with the same prey density, and that efficient feeding requires a certain level of turbulent flow. These types of studies originally began as investigations of turbulence in the surface layer (Rothschild and Osborne 1988), caused by winds and tides, and how they relate to biomass fluctuations of zooplankton and phytoplankton (Mackenzie and Leggett 1991). Several studies have investigated the role of turbulent mixing in production and growth of organisms (e.g. Petersen *et al*. 1998; Peters and Marrasé 2000), including its effects on feeding conditions for fish larvae (Bailey and Macklin 1994; Dower *et al*. 1997). The 'optimum environmental window' was proposed by Cury and Roy (1989) to describe the dome-shaped relationship between fish reproduction and the strength of upwelling, with maximum reproduction at medium upwelling strength. This model can be applied also to the relationships between growth/survival of fish larvae and juveniles and the strength of turbulent flow (MacKenzie *et al*. 1994). Experiments using larvae of bluefin (Kato *et al*. 2008) and yellowfin (Kimura *et al*. 2004) tunas, both of which are important fishery species, revealed a dome-shaped survival curve between different levels of turbulent flow and survival rates. Concurrence of peak feeding rate and survival rate, and the strong relationship between them, indicates that success or failure in early feeding is an important factor that determines survival rates. The results suggest that appropriate levels of turbulence are as important as early feeding conditions, and highlight the importance of sub-mesoscale processes in ecosystem dynamics.

In the sea, every organism passively experiences the influence of currents at every stage of its life, and species are dispersed by currents during a short period relative to the time required for speciation, without rapid expansion of their habitats. Species have thus tended to maintain ancient distribution patterns, including spawning locations, habitats, and feeding areas. As a consequence, we cannot necessarily explain stock variations of marine resources simply by events (such as overfishing or pollution) that happen in a specific time and space. They are results of physiological-ecological responses to environmental change combined with mechanisms of variation inherent in every fish species. This tendency is strong among migratory pelagic fish that spawn, mature or feed in the open sea, although it has also been explored for shelf species. Mullon *et al.* (2002) used an evolutionary, individual-based model to explore the concept of environmental constraints driving observed spatial and temporal spawning patterns for anchovy *Engraulis encrasicolus* in the southern Benguela. They found that constraints linked to a threshold temperature for larval development and the avoidance of offshore currents gave results that best matched the current observed mean-spawning pattern.

7.4 Advances in understanding marine ecosystems

There have been major advances in understanding ecological processes during the period of GLOBEC. Two important issues where substantial progress has been made have been the role and importance of trophic controls and of alternative trophic pathways (including regime shifts).

7.4.1 Trophic controls

Trophodynamic and food web analyses have had a major influence on the developing debate about the impacts of climate change on ocean ecosystems and potential controls in food webs. JGOFS and related biogeochemical research programmes showed that upper trophic levels account for only a tiny fraction of the total energy and mass flux in pelagic food webs. Nevertheless, the historic economic-sociologic importance of fish and fisheries, combined with greater data availability, has focused most marine food web analyses on plankton-fish, benthos-fish, and fish-fish interactions. There are multiple possible pathways of effect in food webs, with bottom-up effects related to changes in physical conditions and primary production, and top-down effects related to changes in higher trophic level species, often associated with harvesting (Hunt and McKinnell 2006; Loeng and Drinkwater 2007). For zooplankton-fish interactions, GLOBEC results have provided examples of both bottom-up and top-down controls in food webs, and it is increasingly recognized that such controls are likely to be context-dependent, with different controls operating in different states of an ecosystem and at different times (Cury *et al.* 2003).

In general, bottom-up processes tend to dominate but are modulated by diffuse top-down effects that occasionally become dominant (Box 7.3). Zooplankton populations typically show strong and rapid positive response to physical environmental factors that enhance phytoplankton productivity (Richardson and Schoeman 2004; Mackas *et al.* 2006). In many regions, positive spatial and temporal correlations extend upwards from phytoplankton through zooplankton to fish production and/or survival rate (Batchelder *et al.* 2002a; Ware and Thomson 2005; Frederiksen *et al.* 2006). Consumption budgets based on predator diet and abundance often indicate that, even for fish that are planktivorous as adults, total ration estimates are often quite small compared to regional secondary production (e.g. Mackas and Tsuda 1999). Zooplankton are important prey for early life stages of many fish species, but total biomass-density and dietary demand of larval/juvenile fish are small in comparison with both zooplankton and adult fish. Zooplankton availability can therefore control the early (numeric) mortality of fish populations, without much direct negative feedback from the larval fish predators. This is an extreme degree of bottom-up control referred to as 'donor control', in which bottom-up influences are not matched by commensurate top-down effects (Strong 1992).

Box 7.3 Trophic controls in food webs: inferring processes from patterns

To predict ecosystem responses to global change requires some understanding of the ways in which predator-prey relationships influence the overall structure and dynamics of a food web. These trophic controls can manifest as observable patterns, which can be explained by one or more of the three dominant hypotheses for mechanisms of trophic control.

The 'bottom-up' hypothesis states that productivity at all levels of the system is approximately proportional to total primary production; biomass/abundance of adjoining trophic levels covary positively (Box 7.3, Fig. 1). Global-scale spatial

comparisons (Iverson 1990) broadly support this hypothesis-oligotrophic regions of the ocean have low fish yields per unit area, and low biomass and productivity at all intervening trophic levels. Productive regions such as eastern boundary upwelling systems have high biomass and productivity of zooplankton, forage fish, and large predators.

Conversely, there is evidence from some food webs that historic absence (or removal) of upper trophic level predators is accompanied by differences in the amount and composition of lower trophic levels (Estes and Palmisano 1974; Carpenter *et al.* 1985). The 'top-down' control hypothesis

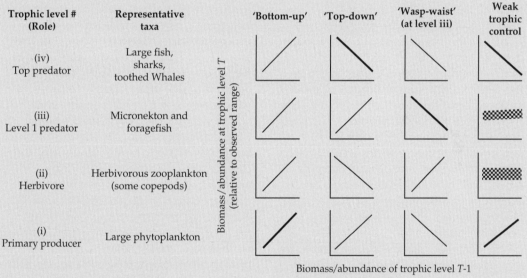

Box 7.3, Figure 1 Schematic responses of a 'four trophic level' food web, under differing scenarios for trophic control. Left text column is the nominal trophic level, second column lists examples of pelagic taxa often assigned to these levels. The four columns of graphs show responses of the four trophic levels, with the *y*-axis an index of population abundance or biomass within each level. Heavy lines indicate trophic level(s) that are responding directly to external perturbation, light (diffuse) lines those that are linked to the perturbation primarily by strong (weak) trophic control. The *x*-axis measures strength and direction of forcing, but units/interpretation differ among studies and authors (common choices are biomass/abundance at the next lower level, biomass/abundance at a hypothesized control level, or time sequence after a disturbance). Figure is a composite. (Based mostly on Cury *et al.* 2001; McQueen *et al.* 1986 © 2008 NRC Canada or its licensors. Reproduced with permission.)

continues

Box 7.3 *continued*

states that absence or severe depletion at a high trophic level greatly reduces predation mortality on the trophic level below it, allowing biomass/ abundance to increase, which in turn leads to increased predation on the next trophic level. In principle, the sequence can continue in a downward 'trophic cascade' (Pace *et al*. 1999) in which positive and negative responses alternate (Box 7.3, Fig. 1).

In some systems food web dynamics seem to be regulated by changes in key species at intermediate trophic levels. The 'wasp-waist' control hypothesis (Rice 1995; Bakun 1996; Cury *et al*. 2000) states that top-down control will operate at trophic levels below the wasp waist and bottom-up control will operate on their dependent predators so correlation is positive above the wasp waist and alternate below the wasp waist (Box 7.3, Fig. 7.1).

Identifying dominant trophic controls can be done in various ways. The most conclusive tests would be controlled experimental manipulations (predator addition or removal for top-down; fertilization or diversion of nutrient inputs for bottom-up), but in aquatic systems these are difficult to carry out except in small and confined bodies of water such as ponds, small lakes, or mesocosms, or in some benthic food webs in which all trophic levels share limited mobility. Although severe overfishing, cultural eutrophication, and outbreaks by introduced exotic species can arguably be viewed as experimental 'treatments' in large, open systems, there is seldom a nearby 'untreated' control.

When controlled experiment is not feasible, the alternative is comparison of patterns of abundance across mean trophic levels among locations, or among different time periods at single locations. Comparisons can be purely correlative, but are greatly strengthened if they also incorporate quantitative budgets of diet, growth, and mortality, and (for time series comparisons) consideration of frequency spectrum and lag-lead relations.

In practice, cross-trophic level abundance comparisons often yield less clear and conclusive evidence of trophic control (especially top-down control) than implied by the idealized schematic in

Box 7.3, Fig. 1. In these cases, the results are often indistinguishable from what might be expected under weak trophic control (Box 7.3, Fig. 7.1), where response at each trophic level is independent of the other levels. There are several reasons for this. First, predator (prey) taxa can and often do interact with prey (predators) that span two or more lower (higher) trophic levels. Wide trophic breadth within species, and high species diversity within trophic level, will spread and dilute downward predation pressure, converting a strong trophic cascade to a weak and diffuse 'trophic trickle' (Strong 1992).

Second, an assumption implicit in Box 7.3, Fig. 1, is that population densities at all trophic levels are more or less at current 'equilibrium' targets. Differing response time scales can prevent synchronous equilibration of all trophic levels, and can prolong or induce cycling between alternating intervals of resource limitation and underutilization. Brief time series (i.e. those shorter than a few generations for the slowest responding taxa) might observe only a portion of a cycle, and conclude that a transient mode of regulation is the norm (Power 1992).

Third, analyses may omit 'hidden' taxa with intermediate trophic roles. For example, in many trophic classifications, marine mesozooplankton continue to be characterized as dominant herbivores, and 5–20 µm phytoplankton as dominant primary producers, despite recent evidence that much smaller nano- and picoplankton are the dominant photosynthesizers in many ocean regions, and heterotrophic protists the dominant herbivores and the dominant prey for particle-feeding mesozooplankton.

Finally, top-down pressure may alter lower trophic level species composition more strongly than lower trophic level biomass if fitness advantage shifts from attractive prey species that are efficient utilizers of resources to unattractive prey species that have low food value and/or active chemical defenses (examples reviewed in Strong 1992). Shifting-community composition is a very significant outcome of such interactions, but requires more complex assessment than altered biomass pools.

continues

Box 7.3 *continued*

Despite these difficulties and limitations, cross-food web comparisons prior to the GLOBEC era produced several important and robust generalizations, well summarized by a quartet of papers in 1992 (Hunter and Price 1992; Menge 1992; Power 1992; Strong 1992). First, bottom-up influences appear to be universal or nearly so. Second, clear-cut trophic cascades are quite rare (for reasons discussed above), and are more likely to be found:

- in aquatic than terrestrial food webs
- in systems/trophic levels with low species and prey diversity

- in small/closed/isolated/freshwater rather than large/open/spatially connected/marine systems (Strong 1992; Shurin *et al.* 2002)
- in benthic rather than pelagic food webs (Shurin *et al.* 2002)
- in systems in which prey defence mechanisms (chemical and physical) are absent or ineffective

Even in aquatic freshwater systems, bottom-up appears to be the dominant control over much of the pelagic food web, with top-down effects only evident as inverse correlations between the piscivorous and planktivorous fish at the top of the food web (McQueen *et al.* 1986).

David L. Mackas

Evidence of intermittent or persistent top-down control of zooplankton by fish includes three main observations. First, there has been quantification of episodic, very intense predation mortality in years when small pelagic fish stocks are at peak levels (Flinkman *et al.* 1998; Verheye *et al.* 1998; Tadokoro *et al.* 2005). Second, out-of-phase biennial oscillations have been observed for pink salmon and copepods (Shiomoto *et al.* 1997; note however that the interpretation of this oscillation reverses if the 'hidden' micronekton trophic step is as important to pink salmon annual success as it is to total pink salmon diet). Third, changes in species composition have been found within the zooplankton community of the North Sea when herring had been depleted by heavy fishing (Reid *et al.* 2000). Changes in food web structure and biomass have also been found in systems where the historically dominant top predators have been substantially reduced by fishing (Frank *et al.* 2005, 2006). Downward trophic controls need not originate with fish or fisheries. Some of the most extreme predator alterations of pelagic food webs are from gelatinous zooplankton (ctenophores and medusae) preying on crustacean zooplankton, fish larvae, and/or other gelatinous zooplankton. The Black Sea is perhaps the most dramatic documented example (see Box 7.4).

For the North Sea, Heath (2005) studied fish food webs for the period 1973–2000, reconciling estimates of secondary production with the estimated consumption demands of various fish feeding guilds. His results suggested fishing had caused decreases in fish populations such that demands for secondary production decreased over time. Effects of climate manifested mainly in the pelagic food web where there were correlations among trends in production for omnivorous zooplankton, planktivorous fish and pelagic piscivorous fish, indicating bottom-up forcing. In contrast, demersal fish production had been impacted by fisheries and there had been no compensatory increase in production of non-exploited species as commercial species decreased. It appears that the benthos responded to released top-down predatory control by increased production, resulting also in increased landings in invertebrate fisheries (Heath 2005).

An important insight into trophic control that emerged during the GLOBEC era was based on hypotheses of wasp-waist control in pelagic food webs (Rice 1995; Cury *et al.* 2000, 2001; Bakun 2006). The shared structural characteristic is that an intermediate trophic level is heavily dominated by very few species relative to the species diversity above and below in the food web. The important dynamic behaviour is that the abundance/biomass of the wasp-waist species fluctuates at decadal time scales by one to several orders of magnitude, and that the identity of

Box 7.4 Effects of introduced species

Understanding the effects of introduced species on ecosystems was one of the original tasks identified within GLOBEC but which subsequently received limited attention. This was probably because most documented examples of introductions are in coastal and estuarine environments (Grosholz 2002; Occhipinti-Ambrogi and Savini 2003), associated with marine benthic habitats or affect air-breathing marine predators on their breeding islands. There is ample evidence of seabird extinctions as a result of introduced terrestrial, mammalian predators, but no evidence of 'truly' marine species extinctions occurring as a result of invasions of non-indigenous marine species. However, there is evidence of widespread impacts (Gurevitch and Padilla 2004), and it is likely that these impacts will increase rather than decrease in the future (Bax *et al.* 2003). Three examples in pelagic ecosystems are described.

Range expansions are associated with environmental change, and can cause introductions of non-indigenous species to new areas. Although these introductions can have anthropogenic drivers (climate change) as their ultimate cause, they can arguably be regarded as 'natural' introductions

because the proximate factors are not human-related. For example, the marine cladoceran *Penilia avirostris* has increased in occurrence and numbers in the North Sea since 1999, corresponding with increased sea surface temperatures (Johns *et al.* 2005), although artificial introductions cannot be ruled out. The species was also first recorded in the southern Gulf of St. Lawrence in 2000, likely introduced through ballast water. In contrast, there are a number of direct, human-linked vectors for introducing non-indigenous species to marine environments (e.g. ballast water and aquaculture activities) and these can cause inadvertent, large-scale, ecosystem disruptions.

In the Black Sea there was an explosive increase in the 1980s of an introduced ctenophore, *Mnemiopsis leidyi*, followed/accompanied by collapse of crustacean zooplankton and planktivorous fish populations. The large-scale ecosystem effects accompanying this invasion have been interpreted as resulting from a number of factors. In the early 1990s, phytoplankton productivity in the Black Sea was much higher than in the late 1970s, probably in response to eutrophication during the 1980s. Small pelagic fish populations decreased, primarily as a result of overfishing, and it appears that *M. leidyi* proliferated in response to

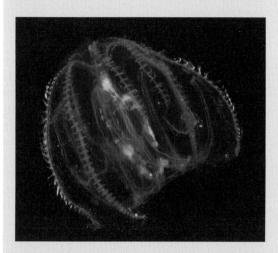

Box 7.4, Figure 1 A specimen of *Mnemiopsis* from the Baltic Sea, a little smaller than 20 mm in length. (Photo: Nicklas Wijkmark in Hanson (2006).)

Box 7.4, Figure 2 Gross signs of Pilchard *herpesvirus* infection. Pilchards that died *in extremis* with mouth open and opercular covers adducted, New South Wales, 1995. (Photograph by Richard Whittington (Whittington *et al.* 2008.)

continues

Box 7.4 *continued*

vacation of the pelagic ecological niche and increased plankton production. In the late 1990s, the by then overwhelmingly dominant *Mnemiopsis* populations were in turn driven down by another introduced ctenophore, *Beroe ovata* (Shiganova *et al.* 2001; Kideys 2002). *Mnemiopsis* has since been found in the eastern Mediterranean (Shiganova and Panov 2003), the southern North Sea, and the Baltic Sea (Hansson 2006; Box 7.4, Fig. 1), with unknown consequences to date.

Aquaculture operations provide vectors for introduction of non-indigenous species, including disease-causing organisms. Off western Australia, fishfeed for tuna mariculture is believed to be the most likely route of introduction of the *Pilchard herpesvirus*, which infected sardine *S. sagax neopilchardus* in 1995 and 1998–9 (Box 7.4, Fig. 2). During these events there was up to 75% mortality of sardine populations (Ward *et al.* 2001), one of the largest recorded such impacts, which also affected piscivorous predators such as Little Penguins *Eudyptula minor* (Dann *et al.* 2000). The effect was massive but acute, as there do not appear to have been long-term ecological effects. However, the pathogen is now endemic to the Australian sardine population (Whittington *et al.* 2008).

Coleen L. Moloney

the dominant species can switch back and forth among two or three candidate taxa (sardine versus anchovy in mid- and low-latitude boundary current systems; but in other places clupeoids versus gelatinous predators, euphausiids versus planktonic tunicates, herring versus sprat, flatfish versus crabs, etc.). The histories and internal workings of several wasp-waist pelagic systems have been documented, and it is believed that many wasp-waist taxa are sensitive to direct forcing by physical climate (temperature and transport/retention anomalies) and to indirect effects of climate on food availability (both productivity and degree of spatial aggregation). Strong (2002) argued that low species diversity within a predator trophic level is an important and perhaps essential enabler for a downward trophic cascade. Examples of top-down control of marine zooplankton (coming mostly from systems in which the quantitatively dominant planktivores are either a small pelagic fish or a gelatinous predator) support this view. An important additional factor is that life history strategies and trophic connections of the wasp-waist species (schooling, reliance on small and hydrodynamically defined spawning and nursery zones, large reproductive potential, and ontogenetic reversals of prey-predator roles) introduce positive feedbacks that act to maintain food webs in one or other state between major disturbances.

Leading on from the development of the wasp-waist hypothesis is the generic, important recognition that useful and robust trophic control hypotheses are situation-dependent and incorporate combinations of top-down and bottom-up pathways. For example, Frank *et al.* (2006) argue for a latitudinal gradient of control mechanisms, with top-down control dominating in high-latitude regions with low diversity of top-predators, and bottom-up control becoming increasingly dominant at lower latitudes. Worm and Myers (2003) report a latitudinal gradient in the top-down control of shrimp by cod. Hunt *et al.* (2002) document time-dependent 'oscillating control' of walleye pollock recruitment in the Bering Sea (Fig. 7.4). Heath (2005) proposed that bottom-up effects in pelagic fish communities can also generate top-down effects into benthic systems.

7.4.2 Alternative trophic pathways and regime shifts

There can be a number of different configurations for ecosystem structure, and alternative trophic pathways can occur at different times and in different places. These trophic pathways vary seasonally (e.g. in the Southern Ocean; Murphy *et al.* 2007), are episodic (e.g. related to El Niño events) or persist for prolonged periods (regime shifts; Shannon *et al.* (2008b). Forcing factors and mechanisms also vary, with natural environmental forcing occurring in both the short and the long term, and fishing either causing or exacerbating changes (e.g. Coll *et al.* 2006b). Alternative pathways of energy flow in food webs are crucial for determining responses of species, species-pairs, and wider food webs. Such alternative pathways can buffer overall system structure, but future change can result in apparently weak pathways of energy flow becoming important. These alternative pathways in food webs contribute to overall properties of ecological network stability and can determine their resilience to fluctuation and change (Odum 1969; Neutel *et al.* 2002).

In examining the current operation of oceanic food webs, it is important to remember that these systems have been disturbed through extensive long-term harvesting across a range of trophic levels. Effects of such disturbances are likely to be significant in determining their present-day operation and dynamics. It is also important to note that fluctuations are a fundamental aspect of natural systems, determining both the success of individuals of particular species and overall food web structure. Natural changes in the timing and pattern of fluctuations in physical processes will generate fluctuations in food web operation that are difficult to interpret with short time series. Some of the major insights have come from analyses of extended time series, examining how natural variability affects the dynamics of key organisms and ecosystems. For Georges Bank, analyses of time series of Atlantic cod and haddock in relation to zooplankton prey have been used to examine impacts of changing zooplankton biomass on larval fish. These have shown a strong correlation between zooplankton biomass and larval fish growth rates (Buckley and Durbin 2006). It is important to progress from correlations to mechanistic understanding of underlying factors and processes (Cooney *et al.* 2001; Beaugrand *et al.* 2007; Murphy *et al.* 2007), and climate-related fluctuations can cause changes in trophic interactions (Hunt and McKinnell 2006). For example, stable isotope analyses suggested that the diet of Pacific salmon shifted in relation to climate-induced changes from squid and gelatinous species to a wider range of zooplankton species. Analyses of nekton species in the northern California Current revealed not only that many of the species have a diverse diet, but also that most variation occurred

because of fluctuations in the occurrence of a few groups such as euphausiids (Miller and Brodeur 2007). Organisms respond to fluctuations in prey availability by switching feeding interactions within and between trophic levels. Such impacts can generate changes in diet of major predators between years, which can also involve fluctuations in prey size and result in a shift in the major pathways of energy flow through food webs (Brodeur *et al.* 2007; Durant *et al.* 2007; Moss and Beauchamp 2007; Murphy *et al.* 2007). Switching in diet can minimize changes in species compositions in ecosystems, but it also involves changing patterns of energy flow that can have long-term consequences for overall food web structure. Non-linear responses of feeding interactions (and hence of food webs) to climate-related change can potentially generate rapid, unexpected regime changes in ocean ecosystems. The current understanding of the term 'regime shift' is 'a relatively abrupt change in marine system functioning that persists on a decadal timescale, at large spatial scales, and observed in multiple aspects of the physical and biological system' (Shannon *et al.* 2008b). This definition stands in contrast to the concept of species alternation, which describes an alternation in species dominance but which does not upset the overall flows or functioning of an ecosystem. Regime shifts also should be distinguished from interannual-scale changes, such as *El Niño* events, the effects of which typically last only 1 to 5 years. The concept of regime shifts or 'abrupt discontinuities' was first formally referred to in the fisheries context in the mid-1970s and has since been the topic of much debate and discussion (see Box 7.5).

Global synchrony has been suggested for several upwelling systems characterized by alternating periods of anchovy and sardine dominance (Schwartzlose *et al.* 1999). This implies global climatic forcing of some sort, impacting small pelagic fish or their main prey species. Although the detailed mechanisms are poorly understood, these have been linked to ocean-atmospheric forcing (for an overview, see Schwing 2008). Because small pelagic fish are important forage species in pelagic systems, and also can control their own prey when abundant, these environmental effects can be propagated up and down pelagic food webs (Cury *et al.* 2000; Cury and Shannon 2004). Impacts of regime

Box 7.5 Regime shifts: physical-biological interactions under climatic and anthropogenic pressures

Ecosystem regime shifts are documented for many areas (Box 7.5, Fig. 1), and investigated particularly well in the (1) North Pacific, (2) North Atlantic, (3) Baltic Sea, (4) Black Sea, and the (5) Benguela. Shannon *et al.* (2008b) provide a summary of ecological aspects. The mechanisms underlying regime shifts are diverse and still not well understood. This makes it impossible to predict reliably how climate variability, eutrophication, fishing, or a combination of these drivers manifest as ecosystem change. Environmental control of regime shifts is hypothesized to operate in two ways: through continuous environmental change and/or through episodic events. An example of continuous change is a prolonged period of warming, permitting expansion of a species' spawning range and thus increased population sizes due to favourable habitat conditions, for example, cod off West Greenland (Jensen 1939). Alternatively, the environment can operate directly on fish recruitment (Cury and Roy 1989). Climatic

regime shifts have been associated with changes in winds, temperature, rainfall, storm intensity, sea levels, and ice volume (affecting salinity, sea temperature, mixing, and water circulation) (Harris and Steele 2004). Environmental factors that are considered to play roles in sustaining, rather than initiating, regime shifts include changes in circulation, temperature, upwelling intensity (winds), availability of plankton (related to turbulence), and extent of suitable habitat for spawning or recruitment.

Decadal-scale ecosystem regime shifts are well documented from the North Pacific (e.g. Hare and Mantua 2000; Overland *et al.* 2008; Tian *et al.* 2008) and the North Sea, a semi-enclosed shallow shelf sea in the eastern North Atlantic. Changes in the North Atlantic Oscillation (NAO) cause changes in wind intensity and direction, which in turn act on currents and change the magnitude of the oceanic inflow into the North Sea. This impacts not only on the sea temperature but also on the

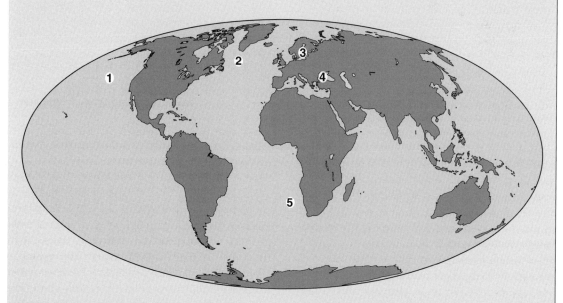

Box 7.5, Figure 1 Some regions mentioned in the text for which ecosystem regime shifts have been documented.

continues

Box 7.5 *continued*

stratification of the water column. Changes in the 1980s triggered changes in the phytoplankton community and the diversity of calanoid copepod species, and were also related to fish, for example, recruitment of cod at age 1 and flatfish (Beaugrand 2004).

In addition to such decadal-scale changes in the forcing, a short series of good year classes can result from an episodic, favourable environmental event. These can switch an ecosystem into a different state. Off South Africa, two favourable events of wind-induced upwelling in the summer of 1999/2000, are thought to have been the reason for the sudden upsurge in abundance of small pelagic fish in the early 2000s (Roy *et al.* 2001), leading to a period of coexistence of anchovy and sardine both at high levels (Howard *et al.* 2007).

The eutrophication of the enclosed, brackish Baltic Sea triggered a discontinuous regime shift (with hysteresis effect) in the 1950s, when an oligotrophic sea with cod as top predator was affected by increasing eutrophication in combination with reduced inflow of oxygenated, saline water from the North Sea, causing conditions of widespread hypoxia in deep waters and frequent algal blooms (Österblom *et al.* 2007). High temperature and nutrients, together with low salinity and oxygen towards the end of the 1980s, led the transition from a system dominated by *Pseudocalanus* and cod to one dominated by *Acartia* and sprat (Möllmann

et al. 2009). This transition, too, is hypothesized to be discontinuous, because improving salinity and oxygen conditions, as well as decreasing temperature and nutrients, have not led to a re-establishment of the *Pseudocalanus*-cod dominance. The ongoing heavy fishing pressure on cod is thought to play a role in this respect.

Similarly, in the western North Atlantic, the warming of subarctic areas of the North Atlantic (e.g. the Icelandic Shelf and the Irminger Current off West Greenland) from the mid-1990s did not result in cod re-establishing itself. Similar to observations on the Newfoundland shelf and in the Baltic, the ecosystem off West Greenland has not simply switched back to its previous state of high cod biomass, suggesting an alteration of top-down control mechanisms, possibly due to the pressures of fishing and hunting (Wieland *et al.* 2007; Hovgård and Wieland 2008).

The combination of long-term change and an episodic event has been documented to initiate the regime shift from an anchovy-dominated Black Sea to one dominated by the ctenophore *Mnemiopsis leidyi* (Oguz *et al.* 2008). In a general context of eutrophication and high fishing pressure on anchovy, a series of milder winters induced higher temperature and weaker mixing, which in turn led to over-enrichment and induced the gelatinous zooplankton outburst.

Astrid Jarre and Lynne J. Shannon

shifts are not restricted to fish and their prey but extend across the whole ecosystem to top predators such as seabirds, seals, and predatory fish. Seabirds, many of which are specialist feeders, are strongly affected by regime shifts. Diets of several seabird species have been found to reflect shifts in small pelagic fish prey, and there are many examples where seabird colony size, the number of chicks successfully fledged per pair, and the proportion of adult birds breeding in a year are significantly related to small pelagic fish availability.

Environmental changes can initiate regime shifts by changing the community composition of phytoplankton and/or zooplankton. These changes can

initiate or sustain shifts in planktivores that, in turn, propagate up the food web to predatory species (e.g. Shannon *et al.* 2004c; van der Lingen *et al.* 2006b; Taylor *et al.* 2009). Climatic shifts cause changes in zooplankton biomass within 1–2 years (Hooff and Peterson 2006), affecting fish, seabird, and mammal predator populations on different scales, with responses lagging several years. For example, changes in zooplankton availability has been found to affect groundfish recruitment on fairly short time scales via changing the enrichment and concentration processes (changes observed within 1–2 years), whereas responses are largely indirect and lagged for up to 9 years in spawner stock biomass.

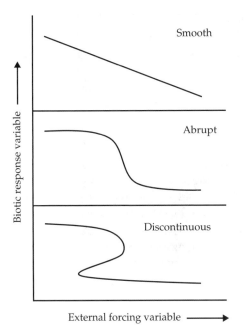

Figure 7.11 Three types of regime shift (described by Collie *et al.* (2004) with kind permission from Elsevier), showing the transition in ecosystem state in response to some forcing variable.

In addition, climatic shifts can have direct impacts on groundfish recruitment within 0–7 years because of environmentally induced changes in larval survival and 0–10 years for spawners through effects on adult growth and spawning (Lees *et al.* 2006). To sustain different regimes, complex ecological interactions come into play and are evident in changes that occur throughout the food web (Box 7.5).

Regime shifts have been classified into three categories (Collie *et al.* 2004): smooth transitions, abrupt shifts, and discontinuous shifts (Fig. 7.11). Smooth and abrupt shifts are often reversible whereas discontinuous regime shifts are not immediately reversible and pose serious challenges for fisheries resource management. In turn, fisheries themselves can induce or at least contribute to regime shifts.

7.5 Conclusions and future directions

GLOBEC studies have helped provide a global context for detailed, regional projects, facilitating cross-system comparisons, and firmly entrenching the importance of long time series of ecological data in understanding global change. Studies of ecological processes during the GLOBEC era have resulted in major advances in our understanding of how feeding interactions of particular species vary with ontogeny, season, and across different regions. They have also shown that the species present in different regions are important in determining ecosystem function and dynamics; biodiversity is important. The scope of the problem of predicting change in marine ecosystems is recognized, and there has been progress in understanding the complexity of physical-biological systems. The importance of alternative pathways of energy flow has been recognized, affecting ecosystem responses to change by influencing nutrient cycling and carbon flows. Gaining an understanding of the controls generating temporal variability in food web structure is clearly important. It is also clear that spatial links are crucial, including both horizontal and vertical interactions between epipelagic, mesopelagic, bathypelagic, and benthic systems.

The development of models of plankton systems has shown that feeding interactions affect overall ecosystem dynamics. Simple models based on a small number of components can do reasonably well at representing the general seasonal dynamics of phytoplankton and zooplankton, but the inclusion of several phytoplankton and zooplankton groups and their complex feeding interactions can change the dynamics of the model system. The form and parameterization of feeding interactions can affect levels of primary and secondary production, influencing nutrient cycles and hence biogeochemical dynamics. GLOBEC research has shown that variations among different marine ecosystems can often be attributed to species-specific adaptations of key organisms. These nuances are difficult to incorporate into generic food web and ecosystem models, which sacrifice species-level details for model simplicity. There is a need to improve the ability to extract key aspects of population dynamics (e.g. flexibility in temperature adaptation, spatio-temporal overlaps of predator and prey) for understanding food webs and ecosystems.

Contrary to earlier beliefs, oceanic ecosystems are inherently complex, which is fundamental to determining their responses to change. Simplified views of ecosystem operation are required for projections of future climate change. However, simplification needs to be achieved while taking account of relevant complexity, and new approaches are required. Mechanistic analyses (especially management-related) based on reductionist approaches are needed to address some questions, but global-scale analyses are also required, identifying generic structures for 'typical' systems but recognizing departures from the norm. Comparative analyses based on different ocean regions can be helpful. Similarities among different food webs have been highlighted, both in terms of taxa and functional groups, illustrating the value of the comparative approach for understanding ecological processes while acknowledging system-specific characteristics. Important differences can help indicate ecological processes that are susceptible to change and can cause disruptions to ecosystem dynamics. At present, global comparisons among ocean regions focus mainly on physical, chemical, or lower trophic level properties, which do not necessarily apply to higher trophic levels. It is necessary to understand how change might affect species compositions, and the life history strategies that might cope best with different aspects of change. These intriguing questions are being addressed through a variety of studies, and we look to the future to see further progress made in this arena.

The paradigm of climate forcing of ecosystem structure and the response of key species to climate change, an important legacy of GLOBEC, will continue to be key to future research. The consequences of global change for plant and animal communities can be direct and/or indirect. Indirect effects are likely to be complex, with many possible responses and different degrees of response to different combinations of factors. Living organisms can adapt and evolve, and some aspects of ecosystem responses to change are essentially unpredictable for this reason. Perturbations to an ecosystem can propagate upwards and/or downwards and be amplified and/or dampened, affecting living organisms and feeding back to biogeochemical cycles. To make progress in understanding the consequences of global change for marine ecosystems, it is necessary to adopt a perspective that synthesizes food web dynamics of all trophic levels and on all space- and time scales simultaneously. This developing focus on understanding the dynamics of end-to-end food webs, which span the full range of organism sizes and functional groups, requires the combined efforts of researchers across a range of disciplines. Integrating the different temporal and spatial scales within and among ecosystems remains a challenge as we develop the legacy of GLOBEC science in this area.

Simplified views of ecosystems can be enriched by detailed analyses of particular interactions or processes. Large-scale integrated analyses of climate impacts on ecosystems are feasible and will involve detailed knowledge of the life histories of key species combined in a multi-scale, physical-biological arena that takes into account behaviour and population dynamics. The switching of ecosystems into different states is difficult to accommodate in current ecosystem models, because it can involve abrupt changes, discontinuities at thresholds, and major changes in ecosystem parameters. New modelling approaches are needed to accommodate such switches, and developing research and synthesis strategies that capture the important complexity without attempting to include all the detail remains a major challenge for the future.

Acknowledgements

We thank John Field, Roger Harris, Jim Churchill, and two anonymous referees for helpful comments. Financial and logistical support was provided by the SEAChange Programme of the National Research Foundation and Department of Environmental Affairs (CLM), the South African Research Chair Initiative (AJ), and Fisheries and Oceans Canada (DLM).

PART III

The human dimensions of changes in marine ecosystems

The first two sections of this book are concerned with the natural science of marine ecosystems: physics, chemistry, and biology, and their interactions. We cannot manage marine ecosystems, but we need to manage how humans interact with ecosystems. Therefore, we need to understand how humans behave as an integral component of marine ecosystems, so that we can manage human activities to both secure the goods and services that we obtain from marine ecosystems and protect them for the benefit of future generations. Chapter 8 explores the reciprocal interactions between natural marine ecosystems and human communities within the context of global change, by incorporating social sciences to under-

stand these interactions. One of the key elements is interactivity; marine ecosystem changes impact on marine-dependent human communities and humans affect marine ecosystems by harvesting or over-harvesting important components of ecosystems. Thus, governance is highlighted as an important element of human responses to ecosystem changes. Chapter 9 takes this forward, reviewing how management needs have changed in the last decades and what forms of management are likely to be needed in future. This leads to a discussion of the convergence of developments in both ecosystem science and ecosystem management and their marriage in an ecosystem approach to fisheries management.

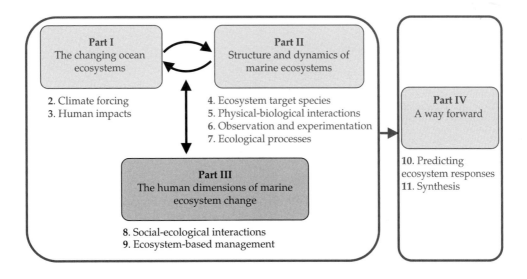

CHAPTER 8

Interactions between changes in marine ecosystems and human communities

R. Ian Perry, Rosemary E. Ommer, Edward H. Allison, Marie-Caroline Badjeck, Manuel Barange, Lawrence Hamilton, Astrid Jarre, Renato A. Quiñones, and U. Rashid Sumaila

8.1 Human-marine ecosystem interactions: a social-ecological perspective

Humans are integral components of marine systems, not 'exogenous' to an otherwise 'natural' ecosystem. Such systems have marine (including physical-biological) and human (including cultural, management, economic, and socio-political) components which are highly interconnected and interactive. Changes in marine ecosystems have impacts on, and consequences for, the human communities that depend on these systems, and vice versa. How these human communities respond to changes in their environment can have reciprocal impacts on marine ecosystems. However, 'natural' marine ecosystems are usually studied separately from their human components, and by different scientific disciplines with different scientific traditions (e.g. natural scientists, economists, social scientists, and humanists). Understanding the important issues and collaborating with other disciplines is essential for correctly interpreting the causes and dealing with the consequences of global changes in these coupled marine social and ecological systems, and for developing practical and successful methods for managing marine resources using ecosystem-based approaches.

The humanities and social sciences have much to contribute to understanding the social-ecological interactions of humans with the ocean. The Global Ocean Ecosystem Dynamics (GLOBEC) programme, although rooted in the natural sciences, developed a recognition that humans and the marine environment *interact*, and paid attention to the reciprocity between people and the marine environments on which they depend and in which coastal human communities are embedded. Humans are among the significant drivers of changes in marine ecosystems and as such can alter, guide, or even withdraw their ocean-changing activities through governance mechanisms of various kinds ranging from fisheries management to laws on pollution or overfishing in the world's oceans. In addition, they can adapt their behaviour when the drivers of marine ecosystem change are environmental. The bidirectionality of the drivers of ocean ecosystem change, particularly with respect to impacts and responses between marine environments and human communities, is now increasingly being recognized (e.g. Ommer *et al.* 2007; Bolster 2008). While the overall goal of GLOBEC was to advance understanding of the structure and functioning of marine ecosystems and their responses to physical forcing, it was also understood that humans are integral components of marine ecosystems. Therefore, GLOBEC recognized two subsystems: the human, and the non-human (sometimes referred to as 'natural' although this term tends to downplay the fact that humans are part of the natural world).

One focus of GLOBEC has been on small-scale fisheries, fishing communities and their related social-ecological systems, a term that has been used by Berkes and Folke (1998) to describe general systems that involve social and ecosystem interactions, particularly in the contexts of their resilience and vulnerability to large-scale changes. We have applied this perspective to highlight the reciprocity of interactions between natural marine ecosystems and human communities within the context of global change. The definition of social-ecological systems has since been expanded to include a multi-scale culturally embedded pattern of resource use, in which humans have organized themselves into particular social structures (settlement pattern, occupational pluralism, sustainable resource management, and associated socio-cultural norms and rules; see also the Resilience Alliance web site: http://www.resalliance.org). This chapter focuses on the reciprocity inherent in social-ecological thinking. Other chapters in this volume consider related but separate issues: the chapter by Brander *et al.* (Chapter 3, this volume) considers the specific impacts of fishing on fish populations in the marine environment, and the chapter by Barange *et al.* (Chapter 9, this volume) concentrates on specific aspects of fisheries and marine resources management.

Worldwide, fisheries and marine resources provide a significant source of animal protein to about 1 billion people (Fig. 8.1). Marine-derived protein was historically particularly important to the poor, as it is one of the cheapest and most accessible sources of protein available: fish were, for example, a mainstay of slave populations, with cheap codfish driving trade patterns between eastern Canada and the rest of the world in the nineteenth century (see, e.g, Ommer 1990, p.165). In addition, it has been proposed that the essential fatty acids from aquatic ecosystems are critical components for human health (DeWailly and Knap 2006; Ommer *et al.* 2007) and have been significant contributors to human evolution (e.g. Arts *et al.* 2001). Marine resources also provide important direct

and indirect sources of employment and financial income to human communities, whether for domestic consumption or for export. As potential sources of significant wealth, depending on their management and wealth distribution, both industrial and small-scale fisheries can contribute globally to economic growth and to poverty reduction. The FAO (2007) has estimated that 41 million people worldwide worked as fishers and fish farmers in 2004 (Table 8.1). The estimated global consumption of fish products in 2005 was 107.2 million tons, providing an average per capita food fish supply of 16.6 kg (FAO 2007; Table 8.1). About two-thirds of this is provided by wild-capture fisheries of various kinds and scales (World Health Organization, http://www.who.int/nutrition/topics/3_food-consumption, last accessed 19 November 2008). Projected annual growth rates to 2020 of fish for human consumption range from 0.2% in developed countries to 2% in developing countries, with a world average of 1.5% (Delgado *et al.* 2003). Since the mid-1990s, developing countries have led the production of fish from wild-capture fisheries, accounting for over 70% of total production for food (Delgado *et al.* 2003). The increasing importance of industrial-type fisheries (often foreign-based and owned, e.g. Alder and Sumaila 2004) for developing countries has meant that they too can become large exporters of fish (Fig. 8.2).

Small-scale local fisheries, on which this chapter concentrates, are culturally significant (as well as essential) to many communities around the world (see, e.g, McGoodwin 1990; Turner *et al.* 2000; Ommer *et al.* 2007). The economic importance of an artisanal fishery, while less statistically obvious in the formal economy, can be critical to community survival. For example, the Newfoundland and Labrador (Canadian) small-scale 'inshore' fishery came to be recognized as economically crucial only when the north-west Atlantic groundfish fisheries collapsed (see Box 8.1), with over 40 000 jobs lost. While the industrial ('offshore') fleets fishing in eastern Canada survived by fishing elsewhere, the place-based inshore fisheries of the province came to a standstill, with the resulting costs of food substitution and unemployment being felt in major economic and social disruptions. Small-scale fisheries in

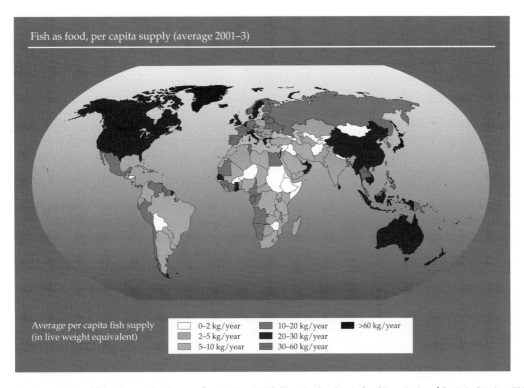

Figure 8.1 Contribution of fish to human diets (average from 2001 to 2003). (From Food and Agricultural Organization of the United Nations 2007.)

Table 8.1 Total (in million tons live weight equivalent) and per capita (kg/year) food fish supply by continent, and numbers (in thousands) of fishers plus fish farmers, in 2003.

	Total fish food supply	Per capita fish food supply	Numbers of fishers plus fish farmers
World	104.1	16.5	41 293
Africa	7.0	8.2	2870
North and Central America	9.4	18.6	841
South America	3.1	8.7	689
China	33.1	25.8	13 163
Asia (excluding China)	36.3	14.3	23 026
Europe	14.5	19.9	653
Oceania	0.8	23.5	50

Source: From FAO (2007).

Mauritania, north-west Africa, represent only 10% of national fish production but provide 80% of the jobs (FAO 2006), thus alleviating poverty. Fish can also be among the few products available for rural economies with which to earn cash when other food products are mostly consumed within the household (FAO 2005). It has been estimated that about 50% of all food fish comes from small-scale fisheries, most of which is used for human consumption (FAO 2005).

The focus of this chapter, on small-scale fisheries where the reciprocity of humans and marine

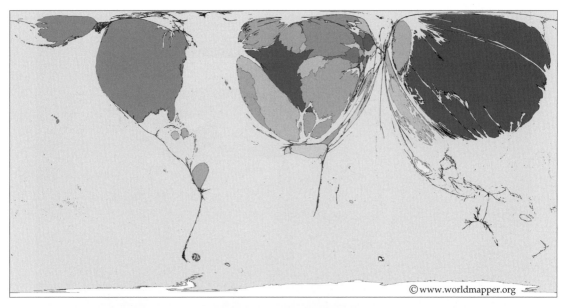

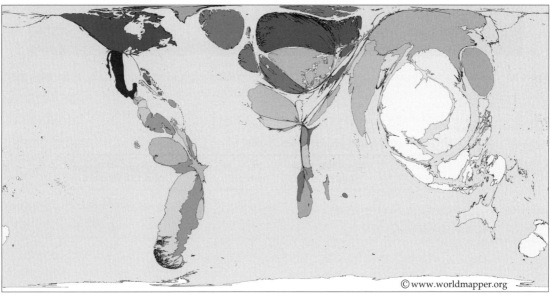

Figure 8.2 Map of the world illustrating net fish imports (top panel) and net fish exports (bottom panel). Territory size shows the proportion of worldwide net exports (imports) of fish (in US$) that come from that territory. Net exports (imports) are calculated as exports (imports) minus imports (exports). When imports (exports) are larger than exports (imports) the territory is not shown. (*Source*: www.worldmapper.org © 2006 SASI Group (University of Sheffield) and Mark Newman). (See Plate 21).

ecosystems can more easily be investigated and understood, is important, even though the definition of small-scale or artisanal fisheries is debated and can vary widely among countries. Traditional, artisanal, or small-scale fisheries can include (Mathew 2003): fisheries for sedentary molluscs in

coastal waters, migratory tuna in high-seas regions, subsistence fisheries in the South Pacific, export fisheries in Chile, and vessels ranging from canoes in West Africa to 30 t ships in Peru. In this chapter, we consider fishers who originate from a specific caste, tribe and/or community, use spe-

Box 8.1 The collapse of groundfish in Newfoundland

The Newfoundland and Labrador waters of the north-west Atlantic host a large coldwater ecosystem with strong seasonal and interannual environmental variability. Human population density is relatively low and dwindling in most fishing-dependent areas. Major groundfish stocks collapsed in the early 1990s, after which moratoria were imposed in many areas. Some stocks have declined further since the moratoria and none are recovering quickly (as of 2006). Some remain under moratoria except for by-catch allowances; some have small to moderate directed fisheries. The Committee on the Status of Endangered Wildlife in Canada (COSEWIC) considers one population of Atlantic cod (*Gadus morhua*) to be endangered, and other populations of cod plus cusk (*Brosme brosme*), northern wolf fish (*Anarchichas denticulatus*), and spotted wolf fish (*A. minor*) to be threatened.

Since these moratoria, the dominant commercial fish species have changed from demersal finfish to shellfish (snow crab, *Chionoecetes opilio*, and shrimp, *Pandalus borealis*), small pelagics and seals (Box 8.1, Fig. 1). Production is now primarily for sale (and for export) rather than for subsistence or local and national sale, which makes the industry vulnerable to fluctuations in exchange rates, international markets, and trade. In employment terms, the dominant sector is the small boat sector (vessels less than 20 m in length).

Since the early 1990s, the fish processing labour force has decreased by ~50%, while overall seasonality in this sector has increased. Fish processing takes place in industrial environments (fish plants) primarily onshore, and such processing plants are often the only major employer in many local communities. Although the value of fish and shellfish landings in Newfoundland has increased since the early 1990s, the species mix and distribution of this wealth is now very different.

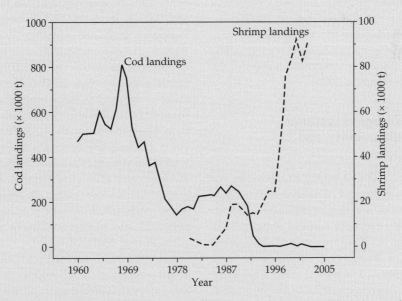

Box 8.1, Figure 1 Landings (× 1000 t) of Atlantic cod (solid line) and northern shrimp (dashed line; × 1000 t) from Newfoundland, north-west Atlantic.

continues

Box 8.1 *continued*

The small boat fisheries, some large boat fisheries, and most fish plants are challenged by economic and resource availability issues. Harvesters have developed a variety of fishing and non-fishing livelihood strategies to deal with variations in their social, economic, political, and natural environments. Strategies include:

- Prey switching.
- Migrations to more distant fishing grounds.
- Concentrating fishing income in a single household.
- Multiple household and employment insurance incomes from fishery and other sectors (local work, migrant work, make work projects).
- Trust agreements with processors and leasing of quota (a threat to the fleet separation policy that prohibits processors from owning quotas in inshore and nearshore sectors).

- Increased hunting of terrestrial mammals such as moose and caribou for subsistence.

Similar strategies exist among processing workers although they have fewer options. By 2003, the number of fish harvesters had almost recovered to early 1990s levels, although with different species being processed and with increases in occupational health issues (Dolan *et al.* 2005). Although the landed value of seafood in Newfoundland in 2003 was greater than that in 1992 (with invertebrate species largely replacing vertebrates), the social distribution of this wealth was quite different, underlining the scale mismatch between household, community, and provincial economic statistics and impacts.

**Matt Windle, Barbara Neis,
Rosemary Ommer**

cific types of boats or gears, and fish locally and sometimes, but not always, using 'traditional' methods. We recognize the importance of this kind of fishing in terms of community sustainability, poverty reduction, and food security for coastal peoples (Mathew 2003).

8.2 Social-ecological systems: resilience, vulnerability, and adaptive capacity

Communities of fish and fishing communities are nested systems composed of natural marine environments, humans, and human institutions. Bakun and Broad (2002) proposed a model of two nested systems to deal with natural marine ecosystems and their human interactions. They used the term *fish habitat system* (*fhs*) to refer to what is typically considered by natural scientists to be the marine ecosystem, that is the characteristic climatology, physico-biochemical habitat, and the fish resources of any particular region (Table 8.2). They defined the *Marine Resource System* (*MRS*) as being the

entire interacting non-human and human (eco)system, that is including the *fhs* plus the fisheries, associated economic activities, the management frameworks and institutions, and the socio-political contexts in which the management and economic activities operate. This is a useful way to think about marine social-ecological systems, but we need to be able also to deal with the interactivity between the *fhs* and the *MRS* (Perry and Ommer 2003), because the two parts are interconnected, with the non-human subsystem (those parts of the marine social-ecological system without direct human involvement), and the human subsystem (those parts which involve human culture, social networks, and institutions) being interdependent (Ommer *et al.* 2007).

GLOBEC was concerned primarily with marine ecosystem dynamics. In regards to the combined impacts of climate changes and the pressures that global fishing efforts are putting on these marine ecosystems, and the dependence of human communities on their fisheries, the specific questions that we asked of small-scale fisheries were:

Table 8.2 Definition of the broad and interlinked marine resource system (MRS).

fish habitat system (fhs)	• fish resource	Marine Resource System (MRS)
	• Regional marine ecosystem	
	• Characteristic climatology and physico-chemical habitat	
	• The fishery (or fisheries) on the stock	
	• Associated economic activities (including other uses of the regional marine ecosystem)	
	• Relevant management framework and institutions	
	• Socio-political context in which the management and economic activities operate	

Source: Bakun and Broad (2002).

- What characteristics of social-ecological systems contribute to high or low vulnerability (or resilience) to disturbances?
- What can we learn from marine social-ecological systems regarding the sustainability of their interactive subsystems?

Historically, resilience, vulnerability, and adaptive capacity have been characteristic of human small-scale resource-based communities (Pollnac 1980; Acheson 1981; Ommer *et al.* 2007). Biologically, *resilience* was recognized in a seminal paper by Holling (1973) who defined it as the ability of a natural system to 'absorb and accommodate future events in whatever unexpected form they may take'. Ecosystem resilience has come to be defined by the amount of information the system can absorb without change to its structure, function, identity, and feedbacks; its capacity for self-organization; and capacity for learning and adaptation. This thinking has been applied to social systems using the concept of 'panarchy' (Gunderson and Holling 2001). Recently, resilience, vulnerability and adaptive capacity, taken together, have become important in discussions of human communities and global changes, in particular through the work of the Intergovernmental Panel on Climate Change (IPCC) and the Resilience Alliance (http://www.resalliance.org). These concepts apply to interactive social-ecological systems (Adger *et al.* 2005), with resilience being defined as the capacity of linked social-ecological systems to absorb distur-

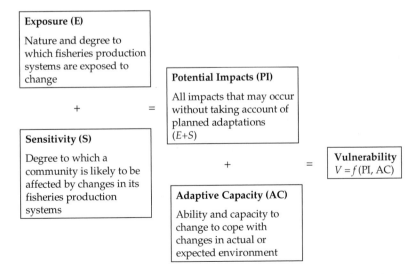

Figure 8.3 Definitions of exposure, sensitivity, impacts, and adaptive capacity and how they contribute to defining vulnerability. (Redrawn after Allison *et al.* 2005.)

bances and still retain essential structures, processes, and feedbacks. *Adaptive capacity* is understood as the capacity of a system to cope with environmental contingencies, that is to be able to maintain or improve its condition when exposed to changes, so as to be able to promote life, reproduction and, in the context of human systems, retain viable social and economic activities (Galopin 2006; Fabricius *et al.* 2007). In the IPCC context, adaptive capacity is 'the ability of a system to adjust to climate change (including climate variability and extremes) to moderate potential damages, to take advantage of opportunities, or to cope with the consequences' (IPCC 2007b, p. 21). *Vulnerability* is at the other end of the same scale as resilience. In the IPCC framework, vulnerability has been defined as the combination of the potential impact (sensitivity plus exposure) and the adaptive capacity of a system (Fig. 8.3). It is a function of the character, magnitude, and rate of change and variation to which a system is exposed, its sensitivity, and its adaptive capacity (IPCC 2007b).

8.3 Communities of fish, and fishing communities: issues of scale and value

A detailed discussion of the linkages between adaptive capacity, resilience and vulnerability, and how they contribute to sustainability, which has the merit of combining the social and ecological perspectives on these matters, is to be found in Berkes *et al.* (2003). Here we apply this thinking to small coastal fishing-dependent human communities, and the fish communities on which they depend. We do not consider the large urban areas which operate major-scale industrial fisheries, although we recognize that fishing communities are arrayed along a spectrum from small to large and do not fall simply into two totally discrete categories. Indeed, there are many different types of fishing communities, ranging from the small fishery-based coastal communities of the developed and less-developed world through large fishing centres that are still fishery-dependent, to urban areas where the fishery is merely one part of a diversified economy. Globally, and theoretically, one size cannot fit all, and much work remains to

be done to understand the complexities involved at each point along this spectrum. The work of GLOBEC is a step in this direction, concentrating on the small resource-dependent communities of the developed and less-developed world. We provide examples where appropriate.

Within the context of marine social-ecological systems comprised of two interrelated subsystems, communities of fish and human fishing communities share some basic characteristics, but they also have significant differences. Communities of fish are characterized by their biophysical environments (water properties, bottom and surface habitats, currents, hydrographic conditions); the species which form their assemblages (differing among demersal and pelagic habitats as well as soft and hard bottom, slow and fast currents, etc.); and their functional properties (age and sex structure, life history and reproductive strategies, predator-prey relationships). Significant characteristics of human fishing communities include their networks of individuals, families and communities, their particular technologies (built environments) and social institutions, their culture and the larger socio-economic, political, and cultural systems of which they are a part. There are issues of power and agency (i.e. who can do what, whose voices have influence), language and knowledge systems involved, including traditional ecological knowledge and scientific knowledge which, in contrast to non-human systems, are able to be transferred non-genetically.

Defining and identifying fisheries-dependent human communities is not straightforward (e.g. see papers in Symes 2000). Brookfield *et al.* (2005, p. 57) defined a fishing-dependent community as 'a population in a specific territorial location which relies upon the fishing industry for its continued economic, social and cultural survival'. The United States Magnuson-Stevens Fishery Conservation and Management Act mandated a formal definition of a fishing-dependent community as 'a community which is substantially dependent on or substantially engaged in the harvest or processing of fishery resources to meet social and economic needs, and includes fishing vessel owners, operators, and crew and US fish processors that are based in such a community' (Jacob *et al.* 2001, p. 16). These are implicitly or explicitly place-based, but there are no distinctions of scale involved in these definitions. Their relationship to the sea is implicit, but gives limited attention to the interconnected dynamics of the natural and social worlds. It is better to call them 'coastal communities' in recognition of the fact that people both live and meet their daily needs in an interconnected social-ecological system, even though their actual fishing activities can take place over a relatively large area, quite distant from the community itself.

In defining human fishing communities for the purposes of analysis and model building, consideration has been given to a range of characteristics, which may include ethnicity, culture ('a culture of fishing'), and *modus operandi* (crew members may be kin, or may come from different places), for example. Fish-dependent communities may be thought of as only those that are involved in traditional fishing, or may include other fisheries and their linkages such as recreational fishing, industrial fishing, and the associated activities of bait and tackle shops, hotels, restaurants (e.g. tourist-oriented activities); commercial fish processors, vessel supply industries, etc.; or aquaculture, fish farming and fish ranching industries. Communities of interest or occupation may also be considered: the crews of fishing vessels who may come from a variety of (land-based) places but still form a fishing community of a kind (*vide* Acheson 1981), but note that their characteristics are likely to differ somewhat from those of place-based fishing communities.

GLOBEC work on the human dimensions of marine ecosystem changes concentrated on one 'type' of coastal fishing community, the small-scale fish-dependent community, which is place- and kin-based, culturally rooted in fishing as its *raison d'être*, and operates relatively small boats, although these may be able to travel quite far out to sea and may employ modern technologies. Such communities are often thought of as 'traditional', although they may use modern technologies, and they exist the world over. We examine some features of the type and its variants here. First, we consider the interactions within and among the natural and human subsystems of marine social-ecological systems, which can be examined by taking a nested-

scale approach (Ommer *et al.* 2007; Fig. 8.4). In the natural subsystem, climate conditions and circulation affect the physical characteristics of the regional and local ocean. Atmospheric conditions influence water properties (such as temperature and salinity) and the mixing and circulation patterns which supply nutrients to the productive upper ocean layers. This physical enrichment of nutrients controls the biological production of plankton, which form the base of the marine food web. The production of fish then depends upon the production of plankton, and the distribution of fish mainly depends upon the spatial patterns of both plankton production and the physical circulation. The chapter by Drinkwater *et al.* (Chapter 2, this volume) provides much more detail on climate impacts on the natural subsystems of marine ecosystems.

In the human fishing subsystem, the impacts of global and national markets, capital and labour, and legal agreements normally flow through successively smaller spatial and lower organizational scales from national policies and practices to region, community, fleet sector, processing plant, household, and to individual vessels and fishers: a top-down process. As described by Charles (2001), socio-economic development and the legislative and marine management regimes that affect these communities are largely defined at 'regional' scales. Demographic and cultural characteristics are important, with the 'household' comprising extended kin relationships. Post-harvest activities, and other supplementary occupations support the community fisheries and help to determine the level of dependency. Individual 'fishers' are characterized by the species they choose to target and the fishing methods (including vessels and gears) they employ. It is the fishers, through their selection of fishing vessels, fishing gears, and target species (in the human subsystem) and the production and distribution of fish (in the natural subsystem) that interact most directly (Fig. 8.4). More diffuse interactions between subsystems occur at other levels, ranging from local impacts of point-source effluent release, to much larger-scale impacts such as the 22,000 km^2 anoxic 'dead zone' due to run-off in the Gulf of Mexico (e.g. Justić *et al.* 2005), or the long-distance transport of contaminants that accumulate in long-lived species and biomagnify in marine food webs.

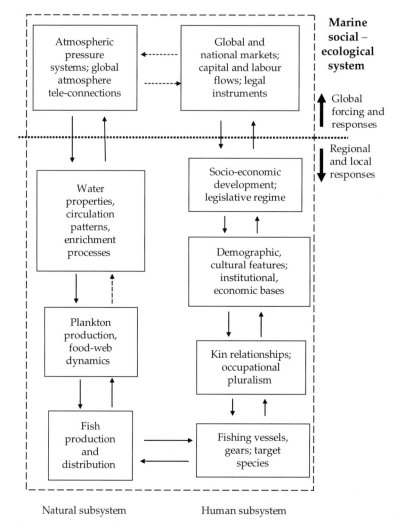

Figure 8.4 Schematic illustrating characteristics and processes within the natural and human subsystems of marine social-ecological systems, and how they are interconnected. Predominant connections between the natural (non-human) subsystem and the human subsystem occur at large scales (regional to global) and at the local scales (local to regional) at which fish production and distributions interact with fishing. Solid arrows represent stronger interactions; dashed arrows represent weaker effects.

Issues of scale, their differences between natural and human subsystems, the down-scaling and up-scaling effects of events at one scale to others, are significant problems for cross-disciplinary research in marine social-ecological systems. For example, what are the spatial, temporal, and organizational scales at which marine and human populations interact? What are the scales of environmental and socio-economic change compared with the scales on which humans and human institutions have the ability to change and adapt? Recent work on coastal social-ecological systems (Ommer *et al.* 2007) points to the top-down nature of management in the human subsystem, and the distance that can occur between upper-level management and legal components of the system and the people at the 'bottom' of the hierarchy and the marine ecosystem in which they operate. This scale mismatch can create significant disconnects between policy, management, and the healthy functioning of the natural marine subsystem. Thus, disturbances in the social-ecological system at local or large scales may not be recognized as problematic by people viewing the system from different scales until it is too late. Two striking twentieth-century examples are the collapses of Northwest Atlantic groundfish fisheries in the late 1980s and early 1990s (Box 8.1) and Northeast Atlantic herring fisheries in the mid-twentieth century (Box 8.2), in both cases due to combined pressures from environmental change and intensive fishing. Much analysis has been devoted to these crises, with their complex interacting aspects of ocean environment (e.g. Ástthórsson *et al.* 1983; Vilhjálmsson 1997a; Drinkwater 2002), fisheries management (e.g. Finlayson 1994; Harris 1998; Haedrich and Hamilton 2000), and societal impacts (e.g. Jentoft

1993; Palmer and Sinclair 1997; Hamilton *et al.* 2000, 2003, 2004b,c, 2006; Hamilton and Butler 2001; Hamilton 2007; Ommer *et al.* 2007). In each case, pre-crisis symptoms of resource decline seem obvious in retrospect, but at the time did not translate into management actions. The Atlantic herring and cod stories stand as a warning about scale mismatches in social-ecological systems and the connectivity between social and ecological systems.

Given these warnings, how can problems be studied in which the scales of the drivers are national or global but the impacts are local? Perry and Ommer (2003) note how humans interact with their environments at many different scales,

---

Box 8.2 Iceland's herring adventure

The twentieth-century saga of Iceland's herring adventure well illustrates the complex interactions that can occur between marine ecosystems and fishing societies. Norwegian spring-spawning herring, the largest stock of the Atlanto-Scandian herring complex (*Clupea harengus*), was known and fished for centuries on its spawning areas along the Norwegian coast. In the mid-nineteenth century, Norwegian fishermen discovered that the same herring were abundant on feeding grounds north and east of Iceland in summer. Norwegians, soon followed by Icelanders, began fishing these Icelandic waters too.

Herring fishing built up with warming North Atlantic conditions during the early part of the twentieth century. For Iceland, this helped set into motion a remarkable climb from poverty to affluence. The international fishery in Icelandic waters took between 10 000 and 25 000 t/year during the first decades of the twentieth century. Initially, Icelandic vessels accounted for only a small fraction of the Icelandic-waters catch, but after 1915 they became dominant. Total catches continued their uneven increase, reaching peaks above 200 000 t several times in the 1930s and 1940s. These good herring seasons contributed to

Iceland's achievement of economic and then political independence in the 1940s (Kristfinnsson 2001).

During these years strong markets, improving technology and increasing effort led to rising success in exploiting this stock throughout the north-east Atlantic. Total catches fluctuated around a general upward trend, exceeding 1 million tons per year during the 1950s. Expanding markets together with technological innovations, sonar to locate herring schools, and power-block assisted purse seines of nylon mesh to catch them, propelled a mid-1960s spike that reached almost 2 million tons.

Collapse followed quickly after this 'killer spike,' as catches fell below 100 000 t in 1969 and 10 000 in 1973. In retrospect, it was clear that the golden years had been times of unsustainable overfishing. Estimated spawning biomass of the spring-spawning stock declined from 14 million tons in 1950 to less than half a million in 1972. The 1960s spike in catches occurred at a time when biomass had already dropped by 74% in just 16 years. Rising catches combined with falling population to produce an abrupt jump in fishing mortality, effectively killing off the

continues

Box 8.2 *continued*

resource. Only a coastal remnant of the stock survived around Norway.

Overfishing drove the steady decline of herring after 1950. Climate change, however, played a role in the terminal decade of the 1960s. From 1920 until 1965, during the herring fishery's high years, relatively warm conditions prevailed over the northern North Atlantic. Cold, low-salinity Arctic surface water, which formed a boundary for the herring feeding area, generally stayed north of Iceland. In 1965, sudden change occurred, and this front shifted to the south-east. Northwesterly winds associated with a prolonged negative state of the North Atlantic Oscillation drove unusual volumes of polar surface water and ice through Fram Strait into the Greenland and Iceland Seas (Dickson *et al.* 1988). The cold, stratified water reduced phytoplankton production, and hence the zooplankton on which herring needed to feed (Ástthórsson *et al.* 1983). The north Iceland feeding grounds became a virtual desert (Vilhjálmsson

1997a). The herring thus lost their main feeding area while under intense fishing pressure, removals exceeding a million tons per year. This combination of overfishing and environmental change led to a total collapse lasting more than 2 decades. On land, the herring towns faced a crisis as well (Hamilton *et al.* 2004c; Hamilton 2007).

Box 8.2, Figure 1 tracks some key dimensions of this story, above and below the water. The upper-most time series, salinity at 0–200 m along the Siglunes oceanographic section north of Iceland (5-year means, based on data from Iceland's Marine Research Institute), marks the massive pulse of cold, relatively fresh Arctic surface water in the 1960s that brought ecological changes which, coming atop overfishing, effectively ended the herring adventure. Herring catches themselves show two peaks: a ragged but prolonged lower one in the 1930s and 1940s, marking a labour-intensive phase of the fishery and the heyday of north Iceland's

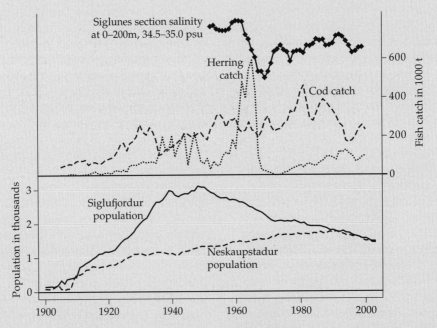

Box 8.2, Figure 1 Salinity off North Iceland, Iceland region herring and cod catches, and the population of two Icelandic fishing towns over the course of the twentieth century. (From Hamilton (2007) with kind permission from Elsevier; see Hamilton *et al.* (2004c, 2006) for details and discussion.)

continues

Box 8.2 *continued*

'Herring Capital', Siglufjörður; and a higher, sharper peak in the early 1960s marking a more technological phase as shrinking stocks were pursued farther offshore out of east Iceland towns such as Neskaupstaður (data from ICES www.ices.dk). The human populations of Siglufjörður and Neskaupstaður are graphed in the bottom of Box 8.2, Figure 1 (data from Statistics Iceland). Siglufjörður peaked during the 'golden years' of the herring fishery in the 1930s and 1940s. As herring jobs declined, outmigration steadily reduced the human population as well. Neskaupstaður participated less in that labour-intensive phase but gained

more from the 1960s terminal boom, and did not decline right afterwards because (unlike Siglufjörður) it was also a cod fishing port. As Icelandic cod stocks and catches weakened in the 1980s and early 1990s, however, Neskaupstaður's population declined too. The population changes in both towns hint at broad shifts in social conditions, and drove searches for economic alternatives. The Siglufjörður story is told with more detail in Hamilton *et al.* (2004c). Hamilton *et al.* (2006) look also at how the herring collapse affected towns in east Iceland and Norway.

Lawrence Hamilton

---

including on organizational (individuals, families, neighbourhoods, communities), spatial (cities, regions, provinces, nations, and supranational organizations) and temporal (minutes, hours, seasons) dimensions. Significantly, the choice of scale affects the identification of patterns. Theoretical propositions derived at one level on a scale may (or may not) be generalized to another level (smaller or larger; higher or lower). Perry and Ommer (2003) derived four scale lessons: (1) methods need to be found to manage fisheries that reach from the global to the local; (2) appropriate scales of natural science analysis must be used when building management policies; (3) both social and natural science analyses need to be aware of the 'shifting baseline' problem; and (4) caution needs to be taken over the use of predictive models, in particular when dealing with human societies, because there are usually more variables involved (such as those concerning motivation) than can be readily quantified and included in the models. Pauly (1995) coined the term 'shifting baseline' to refer to situations in which the expectation of some quantity, such as fish abundance, can differ significantly depending on the time period experienced by the observer.

Issues of scale lead to issues of interactivity within social-ecological marine systems. The processes driving changes in the natural and human subsystems can be different between these subsystems and

among their various nested scales, but they can also interact in important and complicated ways. For example, the drivers of changes in natural marine subsystems include:

- climate variability and trends (biophysical changes);
- internal ecosystem dynamics, such as predator-prey relationships and disease;
- fishing;
- habitat degradation;
- contaminants; and
- introductions of exotic species and new diseases.

Regime shifts (e.g. deYoung *et al.* 2008) represent change in the nature of climate variability; global warming represents persistent directional change. All these drivers of change can interact, as for example between climate change and fishing, and between fishing and habitat degradation, which can alter the resilience of the natural subsystem to external drivers of change. In fishing-dependent human communities, the drivers of change include:

- resource changes;
- environmental changes;
- demographic changes;
- economic and market changes;
- technological innovations;
- law and property relations;

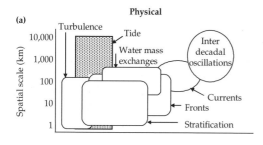

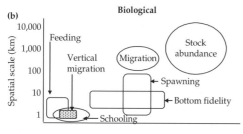

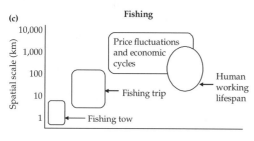

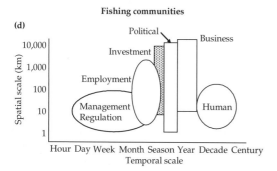

Figure 8.5 Space/timescale diagram of characteristic processes from the natural sciences: (a) physical; (b) biological; and from the social sciences: (c) fishing; (d) fishing communities. Processes and activities at overlapping spatial and temporal scales are likely to interact. (From Perry and Ommer 2003.)

- policy changes;
- gender and ethnic relationships; and
- shifting societal values.

As with the natural subsystem, these changes can also interact, as for example the interactions between environmental changes and changes in policy towards environmental issues. In general, change in marine social-ecological systems has spatial, temporal, and organizational dimensions (Fig. 8.5). Change can also be bidirectional, such that marine ecosystem changes may affect fishing-dependent human communities, but how these human communities react can also amplify or mitigate the marine ecosystem changes.

As stated in the Millennium Ecosystem Assessment (2003), decision making concerning ecosystems and their services can be challenging because different disciplines, philosophical views, ideologies, and schools of thought conceive of the value of ecosystems differently. At the same time, ecosystem valuation is an important tool for decision makers when they evaluate trade-offs between alternative ecosystem management regimes. The way human communities interact with changes in marine ecosystems depends a great deal on how these communities assign values to different ecosystem states. To understand how marine ecosystem changes may affect human societies (and vice versa) and what to do about these changes, we need to understand how people value ecosystems. Valuation of marine ecosystems captures the way that people address the trade-offs of different ecosystems and their states.

One method of valuation of marine ecosystems is solely economic; it is usually the way that we value commodities derived from the ecosystem that are sold in the market and those that are not (Sumaila 2008). A key challenge is how to value benefits from marine ecosystems in a broad and comprehensive manner, and in a way that captures their long-term cultural and conservation benefits (Sumaila 2004; Sumaila and Walters 2005; Berman and Sumaila 2006). The economic theory of valuation is based on what people as economic actors want: it is limited to their economic preferences (Brown 1984; Arrow *et al.* 1993). Preferences are seen to be expressed through the economic choices and trade-offs they make given the resource and time constraints that they face. However, more than economic preferences need to be captured if we are to understand human behaviour and how it can successfully be regulated. A comprehensive compilation of all kinds of values is needed (not only market values) before we can talk about total value (Goulder and Kennedy 1997). The current practice of valuation largely concentrates on economics and thinks of marine ecosystems only in terms of those marine ecosystem 'services' (a utilitarian perspective) from which commodities can be derived which are then sold in the market (but see also National Research Council 2005). Under this view, it has been noted that changes in species composition from groundfish to invertebrates in the north-west Atlantic Ocean now brings in more dollars to Newfoundland than before the groundfish collapse (Schrank 2005). This is a partial way of looking at ecosystem values because ecological, cultural, and social values have not been considered (e.g. Sala and Knowlton 2006). Economic valuations, whether market or non-market, in fact are likely to miss important socio-cultural values related to fish and fishing. The 'self-actualization' component of fishing can mean that fishers are so attached to their occupation that they refuse to leave the fishery even though their incomes and livelihoods are declining (McGoodwin 2001; Pollnac and Poggie 2008).

8.4 Fisheries, food and economic security: vulnerability and response

Fisheries used to be the cheapest source of rich protein in the world, and as such could reduce malnutrition and starvation among the world's vulnerable populations, but in the modern world fisheries have become increasingly vulnerable to such things as environmental degradation, climate change, and economic shocks (e.g. Vilhjálmsson *et al.* 2005; Hannesson *et al.* 2006). Changes in marine ecosystems have affected fishing activities and therefore the food, cultural, health, and economic security of human small-scale fishing communities (Ommer *et al.* 2007). Such changes in marine ecosystems, their drivers, pathways, and outcomes, are described in detail in other chapters in this book (e.g. Drinkwater *et al.*, Chapter 2, this volume). Briefly, small fishing communities can be affected (as are workers and firms in industrial fisheries) by marine ecosystem changes that arise from changes in fish distributions, changes in the behaviour of fish which

may change their vulnerability to fishing gear, changes in the growth and productivity of fish, changes in fish recruitment, and changes in the species composition of marine communities (that may arise through a number of processes such as climate change, intensive fishing, etc.). The act of fishing itself can change how marine populations, communities, and ecosystems are able to respond and adapt to global changes (Perry *et al.* 2010a; Planque *et al.* 2010; Brander *et al.*, Chapter 3, this volume).

Human fishing-dependent communities have developed a variety of coping and adaptive strategies to overcome changes and crises in marine ecosystems. The detailed characteristics of these strategies vary (Hamilton and Butler 2001; Hamilton *et al.* 2003, 2004a, 2006; Brashares *et al.* 2004; Brookfield *et al.* 2005; Dolan *et al.* 2005; Ford *et al.* 2006; McGoodwin 2007; Ommer *et al.* 2007): three summary case studies are provided in Box 8.1 (Newfoundland), Box 8.2 (Iceland herring), and Box 8.3 (Chilean jack mackerel). Such studies illustrate the general statement that resource-dependent communities seek to increase their adaptive capacity through flexibility in the human subsystem that enables responses to significant changes in the natural subsystem. Perry *et al.* (in revision) suggest that there are short- and longer-term strategies employed by small-scale fishing communities. Shorter-term coping strategies include:

- *Intensification* by fishing harder or with new technologies, and targeting new fishing grounds previously considered too distant or difficult. (e.g. Neis and Kean (2003); Essington *et al.* 2006).
- *Diversification* of fishing to other areas, other times (seasons), and possibly to other habitats and/or marine ecosystems (e.g. pelagic to benthic or reef systems; inshore to offshore). It may also include diversifying to other species, especially if the vessel size and gear types are similar requiring little new capitalization. Intensification and diversification might set in motion the 'fishing down the food web' process described by Pauly *et al.* (1998).
- *Migration*, whether to 'follow the fish' or to seek new livelihoods elsewhere (e.g. Hamilton *et al.* 2004a-c). The propensity of individual fishers and families to use this strategy will largely depend on the fish species normally targeted and whether

they are naturally migratory, for example on seasonal time scales, whether the species are altering their distributions because of changes in, for example, sea temperatures, and whether the spatial organization of fisheries management permits movements of fishers to other areas. It will also depend on whether fishing still seems a desirable occupation, and on the physical, legal, and economic access to better fish supplies elsewhere. It depends too on the range of alternative options available. Outmigration towards non-fishing alternatives in urban areas has been a common and rapid response, especially for young adults and families, in some recent North Atlantic fisheries crises (e.g. Boxes 8.1 and 8.2), but is less possible in remote fishing communities.

- *'Riding out the storm'* ('hibernation') through increased reliance on subsistence self-employment, employment in other sectors, and/or transfer and support payments from governments. This may be a common short-term strategy among fishing peoples (and other natural resource-dependent communities) because of their experiences with naturally variable fish populations and the need for self-reliance. In developed countries such as Norway and Canada, the institutional social safety net and public services and programmes introduced in response to fisheries crises can provide important forms of support for fishing families while they wait for these crises to pass. Where these support mechanisms are not available or are seen as insufficient, this strategy may include short-term political actions such as political lobbying and street demonstrations.

At longer time scales, adaptive strategies include:

- *Diversification* to other fisheries and industries. These may include moving from the local small boat and inshore fisheries to the large highly mobile and industrial offshore vessels, or shifting into other maritime activities such as aquaculture or (fishing) tourism. In both Norway and Newfoundland, the oil and gas industry has provided an important alternative source of employment for some displaced workers from the fishery.

- *Networking*, that is building social relations and partnerships, is a strategy which can provide greater cooperation and sharing of ideas, and technological and product development. These often involve interactions among fish processing plants and firms, and can include enhanced links with university academics.

- *Education* and skills upgrading is a powerful adaptive strategy for individuals, families, and communities. This strategy could be as simple as parents encouraging their children to attend college instead of seeking work in a fishery. Often, it involves the networking between firms and universities described above. It is a particularly important strategy to reduce dependency on fisheries (e.g. Corbett 2007).

- *Political action* is also a longer-term strategy for introducing new policies and changing the human subsystem to adapt to and fit better with changes in the natural subsystem.

- *Termination*, such as mass outmigration, depopulation, and loss of communities, is the ultimate long-term strategy. It is not a 'coping' strategy since it represents a fundamental restructuring of the human subsystem, and recognition of a severe crisis from which recovery (to the previous state) may not be possible. Short of termination, many small fishing communities have experienced a loss and aging of their populations, as young adults move away.

These strategies are all broadly part of what have been called 'sustainable livelihoods' (Allison and Ellis 2001; Hamilton 2003) in which rural communities cope with continually changing natural resources. Allison and Ellis (2001, p. 379) defined a livelihood as 'comprising the assets (natural, physical, human, financial, and social), the activities, and the access to these (mediated by institutions and social relations) that together determine the living gained by the individual or household' (Table 8.3). Over the short term (Perry *et al.*, in revision), these coping strategies can lead some fisheries and their dependent human communities to become more diverse and variable. Over the long term, however, short-term strategies such as fishing harder and diversifying fishing targets may not be sustainable. Intensive fishing on previously lightly or un-fished species, often about which very little is known, can put these species at risk of collapse (e.g. Devine *et al.* 2006). Increased ecological damage can result from the use of destructive fishing practices-for example dynamite on coral reefs (e.g. Roberts *et al.* 2002). The expenses required to change gears or vessels to catch alternative species in different environments may increase the debt load of fishers and processing facilities, and create serious occupational health and safety issues. Migration of humans to 'follow the fish' can lead to over-exploitation of resources in the new locations (e.g. Bremner and Perez 2002). In addition, the diversity of economic activities may be decreased at national scales by the collapse of fishing (and fishing-dependent) sectors of the economy and the consequent need to import fish to maintain the domestic supply (e.g. Perry and Sumaila 2007). State subsidies and transfer payments can be considered as mechanisms to increase the initial adaptive capacity of human subsystems during marine environmental crises. In the longer term, however, they may inhibit the willingness and capacity to adapt to changing environments that at some later point may be essential to the subsystem. Among the outcomes of marine ecosystem crises interacting with policy changes can be consolidation of fisheries into fewer, larger enterprises (which has major consequences for small communities), globalization (with a consequent reduction in the variety of novel approaches to problems of marine ecosystem change), and standardized organizational systems and approaches to fisheries problems and management.

Box 8.3 Social adaptive responses: the case of the jack mackerel fishery off central-southern Chile

The eastern South Pacific (ESP) is, in terms of exploitable fish biomass, the most productive region of the global ocean. This region is directly affected by the *El Niño* Southern Oscillation (ENSO) phenomenon, which is the strongest natural interannual climate fluctuation known. Therefore, the ESP provides a 'natural experimental setting' to study the performance of different fishery management strategies as well as human community responses to collapses or drastic changes in fisheries resource abundance or distribution in ecosystems with strong environmental variability and high rates of exploitation.

The fishery for the jack mackerel (*Trachurus symmetricus murphyi*) is one of the most important resources of the ESP, especially off Chile. This pelagic species is distributed from the equator to southern Chile and to New Zealand and Tasmania. It exhibits strong seasonal migrations (Arcos *et al.* 2001), moving to offshore oceanic waters to spawn in early spring (mainly in oceanic waters off central Chile from 32 to 40°S and beyond 90°W).

During the summer, the schools migrate to coastal waters for feeding (Quiñones *et al.* 1997).

The majority of this fishery is conducted by the Chilean purse-seine fleet and the jack mackerel is used mostly for fishmeal production. The fishery reached its maximum in 1995 with catches of 4.4 million tons. Since 2001, a foreign fleet has been operating again in the high seas areas adjacent to the Chilean exclusive economic zone. From 1978 to 1991, the jack mackerel was also exploited by an international mid-water trawl fishery operating mainly between 20 and 50°S, from the outer limit of the Chilean Exclusive Economic Zone (EEZ) to 105°W. This fleet was catching up to 1 million tons (Grechina 1998).

In January-February 1997 a major change in the size structure of jack mackerel was observed in the catches off central-southern Chile. Juveniles (<26 cm) became the dominant fraction in the fishing grounds (Arcos *et al.* 2001). In March 1997, a first fishing ban to protect the stock was imposed. In parallel, the strongest *El Niño* event of the twentieth century started in March-April 1997.

continues

Box 8.3 *continued*

Annual catches declined from 4.4 million tons in 1995 to about 1.2 million tons in 1999. It is important to note that, just after the *El Niño* event, a *La Niña* event took place in 1997/8. Accordingly, it was extremely difficult to assess the real situation of the stock. Major uncertainty arose regarding whether some of the signs of stock deterioration were produced by overfishing or by major changes in stock distribution caused by strong environmental variability.

In 2001, the central government proposed an annual quota of 800,000 t which triggered major social conflict in central southern Chile (Quiñones *et al.* 2003). The different scientific views as to the causes of the size structure changes were also the factors that increased social tensions. Scientists not only disagreed upon the cause of the crisis (environmental change and/or overexploitation) but also regarding the amount of biomass left in the ocean. Today it is clear that since the mid-1990s there was a significant deterioration of the parental structure (older than 4 years) due to overfishing and also the stock was affected by low recruitment in comparison to those observed in 1986/7 (SUBPESCA 2007). It has also been postulated that *El Niño* may have changed the distribution patterns of jack mackerel in the ESP (Arcos *et al.* 2001).

A key characteristic of the jack mackerel crisis was not that the fishermen were unable to find fish. The problem was the drastic decrease of adults and the dominance of small fish in the catches. During the first 2 months of 1997 the fleet had already caught about 500,000 t of jack mackerel. In other words, the crisis was produced by an important reduction of catches (imposed by conservation regulations) but there was no collapse of the fishery.

To confront this crisis, the Bio-Bio Regional Government (central-southern Chile) created an Ad-hoc Conflict Committee on which all sectors (public and private) were represented. This committee realized that negotiations were not advancing due mainly to the lack of scientific agreement regarding the status of the stock. They created a Scientific Commission with scientists (stock assessment, fisheries oceanography) from different institutions which, after a 1-week-long intensive meeting, reached a consensus regarding quotas for 2002. Subsequently, the Ad-hoc Conflict Committee agreed on a quota of 1.2 million tons for 2002 which was then approved by the National Fisheries Council. In addition the Committee proposed changes to executive and legislative powers to improve fisheries management and crisis mitigation which were critical to confront the deteriorating economic and social situation. The consensus reached in 2001 by the Ad-hoc Conflict Committee marks a turning point in Chilean fisheries management. The jack mackerel crisis generated such economic and social impacts that all stakeholders had no choice but to adapt to a new scenario where the catches were only one-half to one-third of values typical during the early 1990s. This reduction in catches generated major impacts in the fishing sector of central southern Chile in employment, income, industrial structure, investment, fleet characteristics, social conditions and fisheries management (Quiñones *et al.* 2003).

The crisis lasted 6 years (1997 to 2002) and resulted in a loss of 2,900 and 10,700 direct and indirect jobs, respectively, in central southern Chile (Quiñones *et al.* 2003). The fleet was reduced from 182 vessels in 1997 to 64 vessels in 2001 and the number of trips decreased from 7,600 in 1998 to 4,000 in 2001 (Quiñones *et al.* 2003). Presently, the annual quotas/catches in Chile are about 1.5 million tons.

In order to deal with the drastic reduction in the jack mackerel quotas, each component of the social system had to show adaptive responses. Box 8.3, Table 1 presents a summary of these social responses (Quiñones *et al.* 2003). Since 2003, Chile has been confronting a new fishery crisis regarding the common hake (*Merluccius gayi*). However the adaptive social responses developed during the jack mackerel crisis have been extremely useful in addressing this new fishery crisis.

Renato A. Quiñones

Box 8.3, Table 1 Main adaptive responses of the Chilean government, Chilean Parliament, and stakeholders as a result of the jack mackerel (*Trachurus symmetricus murphyi*) crisis in central-southern Chile. A detailed analysis of each of these social responses can be found in Quiñones *et al.* (2003).

Sector or stakeholder	Adaptative social responses during the jack mackerel crisis
Parliament	• Generation of a new legal framework including the incorporation of a new regime of property rights (maximum allowable catch per ship owner) for the industry and a new access right for managing artisanal resources. Passage of laws 19 713 and 19 849 • Creation of the Fund for Fisheries Administration. This fund has the flexibility to allocate resources to confront different aspects of fisheries crises mitigation (e.g. employment programmes, training for reconversion, etc.) • Inclusion of fixed percentages of quota allocation between artisanal and industrial fisherman for shared fisheries (Law No. 18 949)
Executive government	• Promotion of the approval of the Laws 19 713 and 19 849 in the parliament • Generation of major social mitigation programmes • Increment in the capacity of government control on industrial fishing activities through externalization (privatization) of some of the functions of the National Fisheries Service (SERNAPESCA)
Industry	• Re-engineering of productive processes and finances • Company amalgamations • Reduction of the fleet • Closing some plants • Increasing the production of human consumption products. Products of higher economic value • Fleet contract changes and externalizing services • Strengthening of the dependence between industrial and artisanal fishermen due to the need of raw materials for producing fishmeal
Artisanal fishermen[1]	• Increment in fishing effort on sardine and anchovy to compensate for the decrease in jack mackerel landings. The employment in the artisanal fleet increased from 600 to 2,900 man per trip equivalent units from 1997 to 2001 (i.e. 380%) • Strengthening of the dependence between industrial and artisanal fishermen due to the need for raw materials for producing fishmeal • Expansion (amnesty) of the Government's registry of artisanal fishing permits • Increment in total number of registered artisanal fishermen and, consequently, diminution of potential allocation of per capita quotas in other resources (e.g. common hake) • Increment in the media exposure of the artisanal demands and socio/economic importance of this subsector
Crew members of the jack mackerel fleet	• Transfer of crew members from the industrial fleet to the artisanal fleet • Transfer of crew members to merchant fleet
Others	• Changes in the relations between and within guilds (Artisanal federations and unions, Plant unions, fleet unions, industrial associations)

[1] According to Chilean Fishing Law, an artisanal fisherman is an owner, skipper, or crew member of an artisanal boat. An artisanal boat is one having a maximum length of 18 m and a maximum hold capacity of 50 t. It is important to note that the artisanal fishermen caught less than 1% of the jack mackerel at the time of the crisis. However, they participated actively in the whole conflict due to the implications emerging from the legislative discussions and also because sardine and anchovy are often caught together with jack mackerel. Sardine and anchovy are captured mainly by the artisanal purse-seine fleet.

8.5 Governance

Governance is fundamental to fisheries, and determines the manner in which power and influence are exercised over their management. Governance is more than management; it is, according to Kooiman and Bavinck (2005, p. 17) 'the whole of public as well as private interactions taken to solve societal problems and create social opportunities. It includes the formulation and application of principles guiding those interactions and care for institutions that enable them.' Fisheries governance includes the

Table 8.3 Sustainable livelihood components and their elements. Redrawn after Allison and Ellis (2001).

Livelihood platform	Access modified by	In the context of	Resulting in	Composed of	With effects on
	Social relations: Gender Class Age Ethnicity	Trends: Population Mig Technological change Relative prices Macro policies National economic trends World economic trends		Natural resource-based activities: Fishing Cultivation of food Cultivation (non-food) Livestock Non-farm natural resources	Livelihood security: Income level Income stability[n] Seasonality Degree of risk
Assets: Natural capital Physical capital Human capital Financial capital Social capital	Institutions: Rules and customs Land and sea tenure Markets		Livelihood strategies		
	Organisations: Associations NGOs Local administration State agencies	Shocks: Storms Recruitment failures Fish distribution changes Diseases Civil war		Non-natural resource based: Rural trade Other services Rural manufacturing Remittances Other transfers	Environmental sustainability: Soil and land quality Water Fish stocks Forests Biodiversity

legal, social, cultural, economic, and political arrangements used to manage fisheries, and has international, national, and local dimensions (FAO 2001). As noted by FAO (2001), the focus on and need for governance of fisheries has increased over the past decades with the increasingly global extent of fisheries and collapses of high-profile fish stocks. Governing fisheries involves consideration of the many factors affecting fish and fishing, including:

- *Environmental*: global climate change, capacities of the ocean ecosystem to produce fish, and human impacts on fish production such as by fishing and contaminants.
- *Social*: issues of cultural survival, poverty reduction, gender equity, rights of movements and migration, sustainable livelihoods, world sustainable development goals, and the FAO code of conduct for responsible fisheries.
- *Legal*: the implementation of international agreements to modify fisheries practices (e.g. FAO code of conduct for responsible fisheries) and others such as the United Nations Convention of the Law of the Sea; trade regulations including bans on trade of endangered species, health and other standards for handling and quality of seafood products, etc.

That said, within the 2000s, there have been increasing global attempts towards governance through 'soft' global voluntary codes of conduct, market incentives, and partnerships between fishers and governments.

This new governance approach faces significant problems, not least those of cross-scale interactions, and will require sustained international political commitment and complex sensitivities if it is to succeed (Allison 2001; Oosterveer 2008).

Jentoft (2007) sees fisheries governance as a relationship between two systems: a 'governing system', and a 'system-to-be-governed'. The governing system is the human social system, comprising its institutions and mechanisms, and the system-to-be-governed has both natural and social elements: the natural ecosystem and the human system of users and stakeholders. Characteristics of the system-to-be-governed include diversity, complexity, dynamics, and vulnerability. In order to be effective, the governing system must assume similar traits, except for vulnerability. Instead, the governing system must be robust, but there are limits to how diverse, complex, and dynamic the governing system can become before itself becoming ungovernable. Institutions affect whether or not fisheries can continue to provide economic, food security and livelihood benefits for fishers, but they can suffer severely from (often unrecognized) mismatches in scales between natural systems, fisheries systems, management, and policy systems (Ommer *et al.* 2007). Fisheries are complex and dynamic and fisheries management must take account of a range of information, as well as changing policy priorities, but many institutions face severe financial constraints and there is a great need for capacity development and improved governance, particularly in developing countries.

In the face of global change, in particular in the context of the main GLOBEC focus on global environmental change, what are the characteristics and challenges of an appropriate governance system for small-scale marine social-ecological systems? First, management must recognize that all fisheries, including the small-scale, are dynamic and complex, with difficult issues of cross-scale (spatial, temporal, and organizational) interactions (Box 8.4 provides an example of concurrent natural and human system changes in South Africa with respect to fisheries). Objectives therefore need to be multiple and set rather broadly, and then be translated into specific conditions for local management; they must also seek to maximize socio-economic and cultural benefits while sustaining the productivity of the marine ecosystem. Governance systems also must struggle with uncertainty and poor information concerning the current state of the natural or human systems and the effectiveness of the regulations and regulatory processes that are in place. The emergence of the ecosystem-based approach to management (EAM; see also Barange *et al.*, Chapter 9, this volume) includes a wide suite of objectives that address the cumulative impacts of fisheries regulations across sectors. Such a framework has to be scale-nested, consisting of objective guidance from top (government), medium-scale (regional or EAM) planning and bottom (sector: small-scale firms and communities) levels of strategy and creative planning. This hierarchy ensures that objectives defined at the highest (legislation and policy) level are transparently linked, not only to implementation efforts at the lowest (operational) level, but also to suggestions for future strategy and planning that are generated at this lowest level. Governance is not merely a one way (top-down) process, but a productive interchange between creative strategizing at all levels: this is what is meant by 'panarchy' (Gunderson and Holling 2003).

The design of management systems that must operate with imperfect knowledge is a challenge facing fisheries around the world, especially in the context of climate variability and global changes. When assessing how fishing communities are affected by, and respond to, climatic stresses, the role of governance is essential. Box 8.3 provides a discussion of the jack mackerel (*Trachurus symmetri-*

cus murphyi) crisis in Chile, in which scientific uncertainty due to the combined effects of overexploitation and *El Niño* provoked a significant management challenge. Miller (2000) described the situation between Canada and the United States in the fishery and management of Pacific salmon (*Oncorhynchus* spp.), and how this agreement is made more difficult by the uncertainty in salmon return migration routes caused by increasing sea temperatures. She concluded that adapting governance regimes is difficult for resources exploited by multiple competing users who each have incomplete information on which to base their decisions. In Peru (see Box 8.5), open access fisheries facilitate migration as an important coping strategy, but changes in property rights regimes (restriction of access as a result of the development of aquaculture) may limit such a response and increase the vulnerability of fishers. Coping more effectively with climate variability and change requires governance systems and policies that foster flexibility (Kennedy 1990; Glantz and Feingold 1992; Lenton 2002). Knapp (2000) and Troadec (2000) argue that flexibility is a crucial tool for responses to climate change, but the ability of fishers and fishing communities to be flexible is unequally distributed. Small-scale fishers have less capacity for following fish migrations or of migrating themselves to new fishing grounds if local stocks fail than do industrial fleets (Troadec 2000; World Bank 2000), and they may also have less capital with which to purchase new gear (Knapp *et al.* 1998; Broad *et al.* 1999) unless provided with government subsidies.

Ommer *et al.* (2007) have argued that interactive governance could be achieved (at least in Canada) for small-scale fisheries by following these strategies:

- Paying attention to habitats, species interactions, and ecology.
- Investing in recovery.
- Developing the social and economic structures needed for effective co-management, that is top-down and bottom-up systems working together to provide joint management.
- Strategic identification, development, monitoring, and ongoing adjustment of spatially and temporally appropriate management initiatives.

Box 8.4 Fisheries governance under variability and change in natural and social systems: an example from the new South Africa

South African society has undergone huge changes in connection with the adoption of democracy in 1994. The new policy and regulatory framework for fisheries management was laid down in the Marine Living Resources Act which initiated a revision of the fisheries management system (Kleinschmidt *et al.* 2003). A new system for allocation of fishing rights was implemented in 2001, with medium-term fishing rights granted for the period 2002–5; long-term fishing rights currently are being implemented. During the transition phase in the late 1990s-early 2000s, transformation of all South African fishing sectors from a few white-owned businesses to a mix of large and small businesses with substantial representation of historically disadvantaged people was emphasized. This process was tedious and difficult for all stakeholders, and is regarded as a mixed success (e.g. Isaacs (2006); Raakjaer Nielsen and Hara 2006). At the interface between human and natural systems, however, it is important to note that the transformation of the fishing industries has created a situation characterized by high uncertainties for their stakeholders, which affected many coastal communities (e.g. Hauck and Sowman 2003).

These huge societal changes were paralleled by large changes in the southern Benguela Current, the natural system on which the most important South African fisheries rely. This system extends along the west and south coasts of South Africa and is a very productive region that has been supporting important fisheries for a long time (e.g. Griffiths *et al.* 2004). It exhibits large variability on interannual and multi-annual time scales which affects plankton production as well as recruitment in its pelagic and demersal resources. In addition, long-term (decadal-scale) change has been documented which affects the entire ecosystem (van der Lingen *et al.* 2006b). Transitions between periods of relative stability ('regimes') have occurred in the second half of the 1950s and in mid-1990s (Howard *et al.* 2007), and the available data indicate additional changes in the

early 1970s and mid-1980s. Global climate change is expected to have an additional effect both on resources and ecosystems (Clark 2006).

The revision of the fisheries system in the late 1990s, therefore, was paralleled by an ecosystem-scale change in the southern Benguela. The beginning saw pelagic fish stocks (notably sardine *Sardinops sagax* and anchovy *Engraulis encrasicolus*) increasing since the mid-1990s to reach record highs in the early 2000s when the medium-term rights were being allocated. However, concurrent with the increase in abundance of pelagic fishes was a displacement of their distribution from the central west coast (the traditional fishing grounds) towards the south coast, and the geographical distribution of catches changed accordingly (Fairweather *et al.* 2006). By the time the long-term rights application process was in full swing, resource abundance was decreasing again, and its displacement persists until the present. West coast communities therefore had to deal with the uncertainty of the rights allocation process as well as the negative effect of apparent long-term change in the ecosystem. This latter change affected not only fisheries but also the tourism industry, for example through changes in the migrations of seabirds following their food. In many of these communities, alternative sources of income are in short supply. Communities on the south coast, on the other hand, feel that they have not profited sufficiently from the increase of resources off their front door.

In this climate of widespread frustration among stakeholders, Marine and Coastal Management (the government branch tasked with research and management of South Africa's fisheries) has been working to revise its fisheries management systems, not only aiming to increase co-management arrangements (Kleinschmidt *et al.* 2003) but also to move towards an ecosystem approach to fisheries. Important issues with respect to ecological well-being, social and economic well-being, as well as the ability to achieve, were raised in a sequence of

continues

Box 8.4 *continued*

workshops for several fisheries. These issues were ranked by the risk that their persistence would pose to the long-term sustainability of the fishery, and recommendations on a way forward were developed (Nel *et al.* 2007). As a contribution towards the transparency of the management process in this complex setting, the use of decision support tools,

such as knowledge-based ('expert') systems are being explored (Paterson *et al.* 2007; Jarre *et al.* 2008). Their value lies not only in structuring complex problems, but also, notably, in increasing communication among stakeholders and the public at large.

Astrid Jarre

- Managing for multiple generations of people and fish with a goal of promoting the health of people, communities, and environments.

Bavinck *et al.* (2005) proposed the transparent, accountable, comprehensive, inclusive, representative, informed and empowered (TACIRIE) procedural principles to facilitate the process of achieving effective governance (Table 8.4). The main goal is to ensure the involvement of all stakeholders in the development of marine resources governance. Co-management has been proposed as one approach to provide for increased local control of fisheries, and therefore increased adaptability of human communities to how global changes are manifest locally. Various types of co-management arrangements have been developed to share the responsibility and authority between government and fishers to manage a fishery (e.g. Pomeroy and Berkes 1997; Symes 1998; Ommer *et al.* 2007). Pomeroy *et al.* (2001) developed a suite of conditions that they found affected the success of fisheries co-management projects in Asia (Table 8.5). They concluded that these conditions were derived from experiences in Asia and may not be directly transferable to other parts of the world with different cultural norms. To this caveat must also be added that they were developed during environmental and socio-economic conditions prevailing during the 1990s, again stressing the dynamic and complex nature of these governance challenges. That said, the suite of conditions developed by Pomeroy *et al.* (2001) for Asia can be fruitfully compared with those of Ommer *et al.* (2007) which were developed for Canadian coastal communities.

8.6 Climate change and an uncertain future

Global climate change is upon us (IPCC 2007b), and we are now faced with dealing with issues of adaptation as well as impact. Potential impacts of climate change for the ocean and marine ecosystems are reviewed by Barange and Perry (2009) and discussed further by Ito *et al.* (Chapter 10, this volume). Increasing temperatures, changes in salinity, and change in patterns of circulation and nutrient replenishment predict changes in the abundance, distribution, and productivity of marine ecosystems, including fish which are of livelihood-importance to humans. Each species, however, will be affected differently, with some benefiting in some areas from these changes and other species being disfavoured. The type of response is likely to vary with the timescale, such that abundances and distributions are expected to change on scales from a few years to a decade, with ecosystem productivity changes occurring on multi-decadal scales (Barange and Perry 2009). Such changes will have impacts on the amount of catch, the catch composition (Hannesson *et al.* 2006), and even the collapse of populations (Roessig *et al.* 2004), with these being different for different species and different countries. Sissener and Bjørndal (2005) provide an example of the management problems created by changes in the migration patterns of Norwegian herring, and K.A. Miller (2007) provides an example of spatial alternation of tuna catch locations due to environmental changes (Fig. 8.6) and the consequent problems for international governance of high-seas fisheries.

Box 8.5 *El Niño* and the scallop fishery in Peru

The total harvest of the Peruvian artisanal nearshore fishery is relatively low compared to the pelagic harvest, the latter mainly destined to fishmeal and fish oil. However, the livelihoods of nearly 40,000 artisanal fisherfolk, more than 40% of the total workforce involved in the fisheries sector, are dependent on this diverse fishery (IMARPE 2008). These fisherfolk operate in a highly fluctuating environment, particularly modulated by *El Niño* events which affect the distribution and abundance of fisheries resources. One resource highly suscepti- ble to increases in sea surface temperatures (SST) triggered by *El Niño* events is the Peruvian bay scallop (*Agropecten purpuratus*). Diving followed by sea ranching is the main method of harvesting this highly valuable commercial species. The dive fleet is growing quickly in Peru, representing only 5.1% of the artisanal fleet in 1995 compared with 13.7% in 2005 (IMARPE 2008); scallops in 2005 represented 41% of total aquaculture exports (PROMPEX 2006).

During the 1982/3 *El Niño*, the warm water conditions in Independence Bay (Pisco) in the south of Peru enhanced growth rates and recruitment for these scallops, as well as the bay's carrying capacity (Wolff and Mendo 2000), resulting in the highest ever recorded scallop harvest (Box 8.5, Figure 1). These environmental changes took place within the context of changing international markets. Before the 1980s the international demand for Peruvian scallops was limited but in 1983 Canada implemented a scallop moratorium in the north- east Atlantic resulting in a high demand for Peruvian products in North American markets. The combination of a dramatic increase of the scallop population due to high SST and growing interna- tional demand resulted in the first 'scallop boom', with both intensified individual fishing effort and more fisherfolk, taking regulatory agencies by surprise (Badjeck *et al.* 2009). The scallop 'boom' resulted in an economic bonanza not only for divers, but also other fisherfolk using other gears, small-scale processors (mainly women, Box 8.5, Figure 8.2), and local businesses (Badjeck 2008).

The boom in Pisco also created the incentive to develop national legislation regulating the harvest and culture of this now valuable resource. After the 1982/3 *El Niño*, the success of the fishery encouraged divers in the south to try other places, some of them settling in the northern part of the country (Sechura Bay) where scallop harvests and diving had not been practised traditionally. Due to the warmer local climate, scallop production proliferated here and consequently, in the early 1990s Sechura started developing, luring more and more migrant fisherfolk for seasonal fishing and settlement (Box 8.5, Figure 2).

Whereas in Pisco *El Niño* is synonymous with thriving environmental conditions for scallop resources, in Sechura it is characterized by a combination of increased SST, heavy rainfall and increased river discharge resulting in the collapse of the scallop stock in 1997 (Box 8.5, Figure 1). Indeed, according to Wolff *et al.* (2007) and Vadas (2007) there is a significant correlation between scallop catches, spawning stock size, and river discharge in Sechura Bay, the latter increasing mortality probably due to decreased salinity and/or increased sedimentation rates. The collapse of the stock in Sechura in 1997, combined with flooding, drove diving fisherfolk to change target species, occupation and/or location, many migrating to Pisco.

In Pisco, proactive measures to regulate entry into the fishery in the face of increased migration and fishing effort were implemented. However, the measures proved insufficient: allocation of resources from the national to the local level to enforce rules were inefficient, and coordination and collaboration with local actors was limited. Adding to this bureaucratic crisis, a management plan was never developed leading to the failure of enforcement (Badjeck *et al.* 2009). Additional pressure was put on the stock with fisherfolk rushing to extract scallops instead of waiting for the cohort biomass to increase (see Box 8.5, Figure 1), showcasing how adaptive responses can have adverse conse- quences on marine ecosystems.

continues

Box 8.5 *continued*

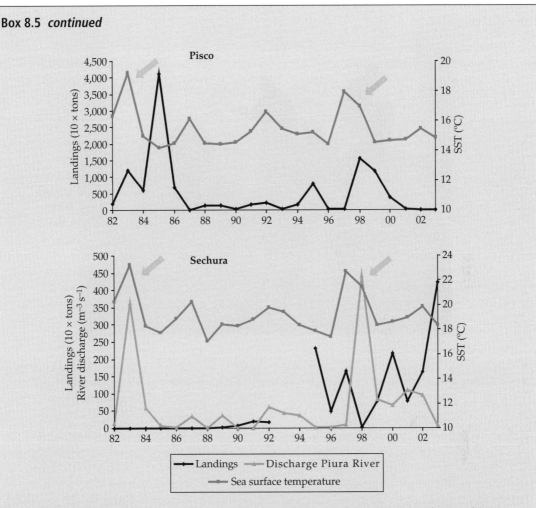

Box 8.5, Figure 1 Annual catch, sea surface temperature (SST) changes in Pisco and Sechura between 1982 and 2003, and river discharge in Sechura (lower panel). Arrows show the arrival of the strongest *El Niño* events (1982/3 and 1997/8) of the represented time period. The effect is a lagged boom in Pisco (upper panel) and immediate collapse of the fishery in Sechura in 1998 (lower panel). The lagged boom can be explained by the fact that fishing effort reached its peak in 1984/5 once the population cohort of 1982/3 reached its marketable size. In 1997/8, the scallop stock increased but small juvenile scallops of low market value were extracted leading to lower landings. (Badjeck *et al.* 2009 with kind permission of Springer Science and Business Media.)

Since 1999/2000 Sechura is again experiencing a rapid development of the scallop fishery due to optimal growth conditions and increases in fishing effort and migration. Additionally, good international and national markets and lower landings in the south of the country favoured this expansion of the fishery. However, the lack of management plans which take into account climate variability, the lack of regulations limiting access to the fishery, and measures to mitigate the impact of restricted access to resources due to the development of sea ranching are jeopardizing the sustainability and resilience of this important fishery.

The story of the scallop fishery in Peru illustrates how individual fisherfolk (mal)-adaptation

continues

Box 8.5 *continued*

Box 8.5, Figure 2 These scallop divers in Sechura Bay, Peru are members of a fisherfolk association led by a migrant diver from Casma (800 km south of Sechura) who uses his skills and knowledge to implement diving and sea ranching techniques around the bay. (Photo © M-C Badjeck.)

is occurring and will occur spontaneously in the face of environmental and market changes. The 1997/8 scallop boom in Pisco also revealed that formal institutional responses are slow and are not strongly shaped by past experience, forgoing knowledge building and the importance of 'institutional memory' in adaptation processes. Additionally, the dynamics of the fishery are shown to be modulated by multiple drivers: exogenous factors such as climate variability, and international market forces, but also endogenous drivers such as the open access nature of the fishery (permitting migration and easy entry) at the regional level. With *El Niño* being a recurrent phenomenon in Peru, and expected to increase in frequency due to global climate change, better governance through adaptive management strategies focusing on migration, property rights, and limiting extraction are imperative.

**Marie-Caroline Badjeck, Flora Vadas,
and Matthias Wolff**

Such changes are likely to have significant impacts on small-scale fishing communities, whose people depend to a considerable extent on fish for their food and income. Changes in catches will affect the price of fish, which will alter the value of economic benefits from marine ecosystems. Similarly, while the cost of fishing will be affected differently for different fish stocks and in different countries, fuel costs will likely continue to be variable and somewhat unpredictable (e.g. due to economic conditions, natural events such as hurricanes, etc.) as climate change affects the economy and oil industry in a variety of ways. Requests for government support are likely to increase globally, not only in the

Table 8.4 The transparent, accountable, comprehensive, inclusive, representative, informed, and empowered (TACIRIE) procedural principles to facilitate the process of achieving effective governance of marine social-ecological systems.

Transparent	Everyone sees how decisions are made and who makes them.
Accountable	Decision makers are answerable to those they represent.
Comprehensive	All interest groups are consulted from the outset in defining the nature of the problem.
Inclusive	All those with a legitimate interest are involved.
Representative	Decision makers are representative of all interest groups.
Informed	All interest groups understand the objectives of the participatory process and have the relevant information.
Empowered	All interest groups are capable of participating in decision making.

Source: From Bavinck *et al*. (2005).

Table 8.5 Nested suite of conditions found by Pomeroy *et al.* (2001) to affect the success of fisheries co-management projects in Asia.

- Supra-community level
  - Enabling policies and legislation
  - External agents of change
- Community level
  - Appropriate scale and defined boundaries
  - Clearly defined membership
  - Group homogeneity
  - Participation by those affected
  - Leadership
  - Empowerment, capacity building, and social preparation
  - Community organizations
  - Long-term local government support
  - Property rights over the resource
  - Adequate financial resources
  - Sense of partnership/ownership in the co-management process
  - Accountability
  - Conflict management mechanism
  - Clear objectives
  - Management rules are enforced
- Household and individual level
  - Individual incentives

industrial sector but also in the small boat sector, and in small remote resource-dependent communities. Similarly, the expected redistribution of fish populations will likely create new 'fish rich' and 'fish poor' countries thereby generating new patterns of fish trade, some of which may benefit small-scale fisheries of various countries, although it will be matters of ownership and access (derived from governance structures) that will have primary effects here. In those small-scale fisheries that have a significant subsistence component, an already dif-

ficult situation is likely to be exacerbated. Some studies suggest that there may be a few general policy rules available, while recognizing that optimal response policies may be strongly dependent upon the particulars of individual fisheries (Arnason 2006).

Many small-scale fishing communities, particularly in developing countries, are considered to be economically poor, although 'poverty' is a relative term which can have different interpretations depending on the viewer and context. Fishing is often seen as a means to

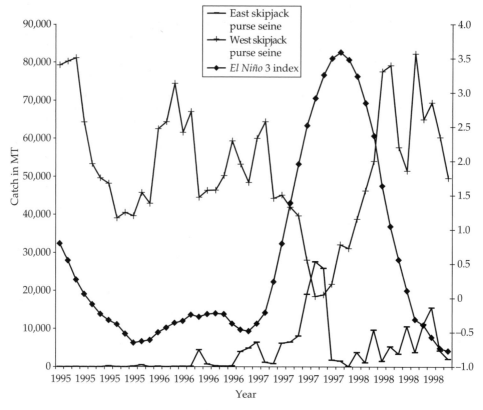

Figure 8.6 Alternation of catches of skipjack tuna (*Katsuwonus pelamis*) in the West Central Tropical Pacific in conjunction with changes in environmental conditions (*El Niño* index, right axis). (From K.A. Miller 2007 with kind permission from Elsevier)

help alleviate this poverty (e.g. Béné 2003). In the context of understanding the effects of the marine environment on poverty alleviation, Allison *et al.* (2009), using the idea that vulnerability is a function of exposure of a community to hazards, the sensitivity of that community to hazards, and the capacity of that community to adapt to changes, explored the potential impact of climate change on the sustainability of fisheries that are important to people in developed and developing countries. They built an index of vulnerability that used air temperature to assess exposure; the number of fishers and their economic and nutritional dependence on fish to assess sensitivity; and measures of overall health, education, governance, and the size of the economy for the country being considered to assess adaptive capacity (Fig. 8.7). Note that these statistics are at the national level, and therefore do not distinguish among the scales of the various fishing communities. They subjected these assessments to models of future climate change using air temperature projections to 2050 from the IPCC (2001) assessments, and concluded that Asia, the Amazon and western Sahara regions would be most at risk from increasing temperature changes. The largest numbers of poor fishers occur in south and south-east Asia, which is also where the majority of fishers (greater than 80%) live. The low per capita GDP of most African countries suggests that Africa has a large number of fishers living in poverty, but the most *sensitive* countries appear to be China, Indonesia, India, Vietnam, Mauritania, and Peru. Iceland was the most sensitive country in the developed world. Combining these assessments into a composite index of vulnerability, they concluded that Africa was the region that is most vulnerable to the impacts of climate change on its fishing communities (Table 8.6).

Fisheries sensitivity

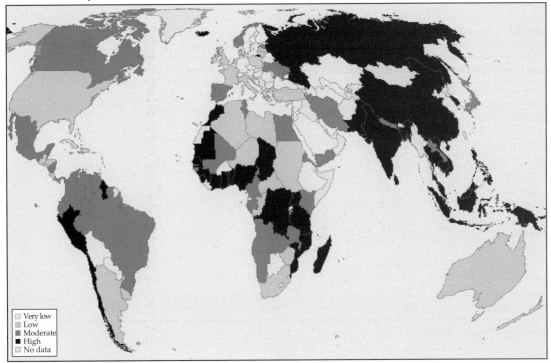

Proportion of fishers

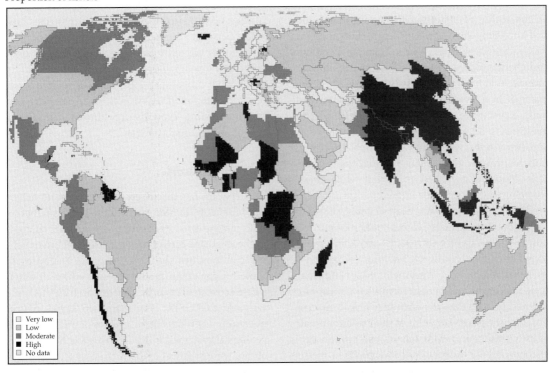

Figure 8.7 Composite index of sensitivity of national economies to fisheries sector changes (*top*). Proportion of fishers in the population (*bottom*). (From Allison *et al.* 2009.)

Table 8.6 Countries with the highest indices of vulnerability based on their exposure to hazards, the sensitivity to these hazards, their capacities to adapt to changes, and their exposures to rising temperature changes predicted by two models from the IPCC (2001) assessment.

Rapid development, high emissions scenario (IPCC A1F1)		Local development, lower emissions scenario (IPCC B2)	
Angola	81.97	Mauritania	83.10
Mauritania	81.18	Angola	82.15
Niger	79.24	Zimbabwe	79.32
Congo, Democratic Republic	78.82	Niger	78.95
Mali	78.01	Congo, Democratic Republic	76.03
Sierra Leone	77.09	Mali	75.92
Burkina Faso	76.01	Mozambique	75.13
Burundi	74.96	Russian Federation	74.33
Mozambique	74.86	Sierra Leone	73.61
Zimbabwe	74.55	Senegal	73.31
Senegal	73.70	Botswana	72.96
Guinea-Bissau	72.97	Zambia	71.78
Côte d'Ivoire	71.18	Burundi	71.68
Sudan	70.68	Burkina Faso	71.57
Russian Federation	70.57	Peru	70.98

Source: From Allison et al. (2005).

The resilience of fisheries systems also depends on the ability of institutions to adopt information and knowledge in order to organize their necessary transformations in the face of climate variability and change (Badjeck *et al.* 2009). Lebel *et al.* (2006) propose that the ability to detect significant system changes in a timely manner, and the capacity to build knowledge about ecological processes into institutions, should improve the fit between rules and ecosystems, and allow societies to take measures to prevent ecosystems from crossing thresholds. Allison *et al.* (2009) argue that while models are being developed to be more accurate and better predict climate anomalies, and accuracy allows the reduction of uncertainty, interpretation of information is still an issue and forecast providers should consider explicitly how the fisheries sector and all its stakeholders (i.e. government, large- and small-scale sectors) would benefit from such information. The ability to do this successfully will require recognition of the issues inherent in their different scales of information and operation. Greater emphasis is also needed for downscaling the results of models to management scales (e.g. from global models to regional and local levels). Furthermore, models should not only include links between projected climate change, environmental responses, and marine ecosystems and fish stocks, but also take into account human community responses. Techniques and approaches developed for responding to and managing human interactions with large-scale environmental changes such as regime shifts (e.g. deYoung *et al.* 2008) may point the way and provide testing grounds for human community responses to global changes.

8.7 Conclusions

This chapter has discussed how small-scale fisheries dependent human communities are interactive with marine ecosystems. It has shown how changes in these marine ecosystems can impact human fishing communities, and also how the responses to these human communities can exacerbate or ameliorate ecosystem changes. The research on this topic within the GLOBEC programme asked two central questions: (1) What characteristics of marine social-ecological systems contribute to high or low vulnerability (or resilience) to disturbances? and (2) What can be learnt from marine social-ecological systems regarding the sustainability of their interacting non-human ('natural') and human subsystems?

In answer to the first question, this chapter has shown that human responses depend on the extent of the marine ecosystem change, the exposure of the human community to these changes, the sensitivity of the community to changes in fishing-related activities, and the extent to which the community can cope with or adapt to these changes. Options that human communities have for responding to marine ecosystem changes include intensifying their exploitation of local and distant ecosystems (thereby possibly 'exporting' local problems to larger scales) to withdrawal from fishing and, in extreme situations, abandonment of the entire fishing community. Scale is an important consideration in these issues, both in terms of the intensity and duration of the ecosystem changes, the spatial extent of the ecosystem change and human responses, and the organization (e.g. above the human community) and governance levels that are affected and become involved. In fact, governance is a particularly important element of human responses to ecosystem changes. It must have the broad-based participation and the flexibility to respond to both foreseen and unforeseen ecosystem changes, changes that may be concurrent with additional alterations taking place in purely the human subsystem (e.g. globalization of trade, economic crises, etc.). Value is an additional consideration, including both the monetary and non-monetary (especially cultural) values that are placed on ecosystems by humans or that are provided to humans in the form of ecosystem goods and services.

In answer to the second question, it is clear that studies of the impacts of global change on marine ecosystems must consider humans as both significant drivers of change (e.g. Brander *et al.*, this volume) and recipients of these changes (e.g. Ommer *et al.* 2008, 2009). While the focus of this chapter has been on small-scale fisheries and issues of their human dimensions of marine ecosystem changes, we expect the conclusions of what supports the sustainability of marine social-ecological systems to be more widely applicable. Recognition of the interactive nature of these systems is key, and therefore it is imperative that we bridge the gap between the social and natural sciences on the issues of marine ecosystem changes. Such studies will include natural and social scientists working together with resource users, managers, and the larger resource community to improve stewardship of their marine ecosystems. In a changing and uncertain world, we need to learn the lessons of resilience and vulnerability, and apply these globally if we are to manage the world's marine ecosystems responsibly in the future.

Acknowledgements

We are pleased to acknowledge the review comments of Keith Brander, Sophie desClers, and Kevin Cochrane on this manuscript, as well as the editorial advice of Roger Harris and Cisco Werner. Financial support of GLOBEC to Working Group 4 on the human dimensions of global change and of the Social Sciences and Humanities Research Council and the Natural Sciences and Engineering Research Council of Canada to the Coasts Under Stress project is also greatly appreciated.

CHAPTER 9

Marine resources management in the face of change: from ecosystem science to ecosystem-based management

**Manuel Barange, Robert O'Boyle, Kevern L. Cochrane,
Michael J. Fogarty, Astrid Jarre, Laurence T. Kell, Fritz W. Köster,
Jacquelynne R. King, Carryn L. de Moor, Keith Reid,
Mike Sinclair, and Akihiko Yatsu**

9.1 How have resource management needs changed during the life of GLOBEC?

As it is well accepted in scientific, policy, and public arenas, despite some successes global fisheries management has generally failed to achieve the biological, ecological, and socio-economic objectives of marine resource exploitation. The underlying problems leading to this failure have been reflected in the evolving nature of fisheries management and research.

The overexploitation of marine resources in the northern Hemisphere started after the Second World War. The main management measures in vogue at the time were largely technical, such as mesh size, minimum fish size limits, closed areas, and seasons. The impact of these measures was limited, overwhelmed by an increasing fleet size (both local and distant) and unresolved access to resources. From the mid-1950s to the 1990s, fisheries research and management developed along the idea that resources were deterministically predictable and thus could be optimized (Degnbol 2001). The basic unit of management was the 'stock', whose dynamics and fishing impacts could be understood and managed by averaging life history parameters and stock abundances over the total stock area. The main non-explained source of variation was assumed to be stock recruitment, which was to be handled to optimize long-term yield. Research thus focused on estimating life history parameters of individual stocks, including recruitment variability, on the basis of data sampling schemes and estimation models.

The fisheries management landscape was fundamentally changed with the implementation of the 1982 United Nations Convention on the Law of the Sea (UNCLOS) to govern the use of the oceans and seas. UNCLOS provided a new legal basis for marine resources conservation, management, and research. It recognized the right of coastal states to take full control over fisheries in their Exclusive Economic Zones (EEZ), but it also placed new responsibilities on these states over their requirements for scientific advice, including the determination of allowable catches of living resources and protection against overexploitation.

In the early 1990s, the precautionary approach and the need to include considerations of the effects of fishing on the marine ecosystem at large into fisheries management entered the scene. This was underpinned by two new policy developments. The first was Agenda 21, a programme of action for sustainable development at global, national, and local levels, adopted at the UN

Conference on Environment and Development (UNCED) in 1992. Chapter 17 of Agenda 21 deals with the protection of the oceans, all seas, coastal areas and their living resources, and proposes the adoption of the precautionary approach. The second, the 1995 Food and Agriculture Organization Code of Conduct for Responsible Fisheries (FAO 1995b), explicitly called for conservation of aquatic ecosystems and protection of living aquatic resources and their environments and coastal areas, as well as due respect for the ecosystem and biodiversity. It also emphasizes the need to take uncertainties into account, through the application of the precautionary approach, and calls for research on the effects on fishery resources of climatic, environmental, and socio-economic factors. Two other binding agreements of relevance to GLOBEC were also concluded at UNCED: the Convention on Climate Change and the Convention on Biological Diversity (CBD). The CBD, in particular, focuses on the conservation and sustainable use of all components of biodiversity, bringing a new set of concepts to the fisheries research and management dialogue.

The above policy developments influenced the main fisheries research discourse, which increasingly focused on risk avoidance in relation to stock conservation, internalizing uncertainty in an approach that has been labelled 'stochastic predictability' (Degnbol 2001). The basic principles of the advice did not change-to predict outcomes of management measures over large scales with the 'fish stock' as the basic unit-but the concept of uncertainty was more formally introduced, not just in model parameters and sampling schemes but also in terms of the impacts of management measures. In addition, management moved from an emphasis on maximizing sustainable production to also incorporating conservation and risk management.

This was the policy environment when the GLOBEC programme was established by the Scientific Committee on Oceanic Research (SCOR) and the Inter-governmental Oceanographic Commission (IOC) of UNESCO in late 1991. Rather than following the dominant fisheries research discourse, GLOBEC was primarily concerned with describing and understanding the role of natural variability in

the functioning of marine ecosystems, as an essential premise to effective management of global marine living resources (GLOBEC 1997, 1999). The main GLOBEC research focus was on marine zooplankton and their primary predators (fish larvae and juveniles), under the premise that zooplankton was the link connecting physical variability in marine ecosystems to fish recruitment. The emphasis on understanding recruitment success was the most direct link between fisheries management advice and the ecosystem science encapsulated in GLOBEC.

GLOBEC became a core project of the International Geosphere-Biosphere Programme in 1995, and its goal broadened to understand the structure and functioning of global ocean ecosystems, and their response to physical forcing, *so that a capability can be developed to forecast the responses of the marine ecosystem to global change* (GLOBEC 1999). While the major objectives of GLOBEC remained the same, its research focus, particularly at the regional scale, enlarged and progressively incorporated fishery resources. For example (see Ashby 2004), the ICES-GLOBEC Cod and Climate Change programme (CCC) was established to understand and predict variability in cod recruitment, both in the short (annual forecasts) and long term (climate effects). The Small Pelagic Fish and Climate Change (SPACC) programme was established to forecast how changes in ocean climate was altering the productivity of small pelagic fish populations (Checkley *et al.* 2009), and the GLOBEC Climate Impacts on Oceanic Top Predators (CLIOTOP) was created to conduct a large-scale worldwide comparative effort to identify the impacts of both climate and fishing on open ocean ecosystems and their top predators (Maury and Lehodey 2005).

The paths of GLOBEC and the fisheries research and management communities came together in 2001 with the signature of the Reykjavik Declaration on Responsible Fisheries in the Marine Ecosystem. The Reykjavik Declaration was a turning point in that it introduced more ecosystem considerations into fisheries management, recognizing that fisheries impact marine ecosystems and that ecosystems impact the status and productivity of fishery resources. The Declaration made strong points in relation to the research needs for fisheries management, recognizing the importance to advance the scientific basis for incorporating ecosystem considerations, and pledging, *inter alia*,

- to advance the scientific basis for developing and implementing management strategies that incorporate ecosystem considerations and which will ensure sustainable yields while conserving stocks and containing the integrity of ecosystems and habitats on which they depend;
- identify and describe the structure, components and functioning of relevant marine ecosystems, diet composition and food webs, species interactions and predator-prey relationships, the role of habitat, and the biological, physical, and oceanographic factors affecting ecosystem sustainability and resilience; and
- build or enhance systematic monitoring of natural variability and its relationship to ecosystem productivity.

The Reykjavik Declaration provided a foundation for the development of an evolving new paradigm, the Ecosystem Approach to Management (EAM), which was further enforced at the 2002 World Summit on Sustainable Development (WSSD). The new paradigm required that research advice be broadened. This included, among others, adding new layers of complexity to models, enabling ecosystem effects to be predicted with stochasticity included. Recognition of the difficulty in measuring all relevant ecosystem processes in detail, also led to the development of indicators of ecosystem pressure and state as an alternative to pretending to understand the entire ecosystem. At this point GLOBEC's research, which had largely focused on understanding ecosystem dynamics as driven by climate and physics, became increasingly significant to the new fisheries management paradigm, up to then which had particularly focused on the footprint of fisheries only.

How this significance was translated from research to management, however, is far from simple. In the next section, we will illustrate how scientific developments in GLOBEC have responded to this evolutionary convergence and are contributing to the scientific development of the ecosystem approach to fisheries.

9.2 How has ecosystem science been used to identify and address resource management needs?

As mentioned above, traditional fisheries management developed along the line that predictions of stock dynamics can be based on early observations of year class strength and on explanatory models in which biological parameters are averaged over a large area and generally considered constant over some time period (Beverton and Holt 1957). This approach recognizes that there are significant limits to understanding (and predicting) ecological processes and focuses on developing management procedures (MPs) robust to uncertainty (Walters and Collie 1988; Ludwig *et al.* 1993). In recent times, however, some practitioners have recognized that process uncertainty can and should be reduced (Ulltang 1996; King *et al.* 2001), arguing that lack of information about environmental conditions or lack of use of relationships between stock parameters and environmental conditions is a major limiting factor for fish stock predictions, and thus successful fisheries advice (Ulltang 2003). While it is true that fisheries management has increasingly incorporated some ecological information into management decision making (Hilborn 2003; see Payne *et al.* 2008), in general, environmental and ecological information is not applied in management, even after the increased recognition of the unidirectional nature of climate change. In this section, we discuss a number of examples from GLOBEC and GLOBEC-like research initiatives where key ecosystem processes have provided substantial support for expanding management systems to include ecosystem issues.

9.2.1 Predicting recruitment success

Understanding the relationship between parental stock size, recruitment, and environmental forcing is fundamental to fisheries management, but quantifying this relationship is not simple. Although a number of environment-recruitment relationships have been published, Myers (1998) found that, given further years' data, nearly all of these correlations did not hold, except for populations at the limit of a species' geographical range. The examples below explore recruitment estimation and its use in management.

The stock size of Baltic Sea sprat (*Sprattus sprattus*) declined during the 1970s, remained at a stable low level throughout the 1980s and then increased in the first half of the 1990s to an historical high. This growth was due to a reduction in predation pressure by cod, but also to increasing reproductive success at relatively low fishing mortality (Köster *et al.* 2003b). Current stock assessments employ a long-term geometric mean of recruitment for making short- and medium-term predictions of catch and spawner biomass. These predictions could be improved with the use of environmental information. It has been shown that recruitment predicted by environmental variables (i.e. winter water temperature, North Atlantic Oscillation index and Baltic Sea ice coverage) yields smaller and less variable deviations from observed recruitment than recruitment estimated with standard methodology (specifically, MacKenzie and Köster 2004). MacKenzie and Köster (2004) confirmed that temperature in the Baltic intermediate water layer during spawning time (spring) explains a significant proportion of the variability in sprat recruitment. This observation is consistent with Myers (1998), who reported that in general there is a good relationship between water temperature and recruitment for data sets of a species corresponding to temperature range extremes. Furthermore, this approach linked recruitment success to large-scale climatic indices, such as the North Atlantic Oscillation (NAO) and ice coverage in the Baltic Sea, potentially giving several months of forecast lead-time for sprat stock assessment compared to the present procedure.

MacKenzie and Köster (2004) also concluded that standard stock and catch predictions would have performed better if environment-recruitment relationships had been used. Projections based on a temperature-recruitment relationship rather than on the traditional geometric mean in recruitment suggest that fishing at precautionary approach fishing mortality reference points (F_{PA}) under unfavourable hydrographic conditions has a 20% chance to drive the stock below precautionary approach biomass reference points (B_{PA}) within 10 years (Fig. 9.1), even though the stock is presently at an historical high (MacKenzie and Koster 2004). A better

understanding of the impact of environment on recruitment could lead to more precise estimates of recruitment, with particular gains when making medium term resource productivity projections.

Predicting recruitment is particularly important for short-lived fish species, where the size of the stock depends on the incoming annual recruitment. In these cases, environmental indices that provide short-term predictions of recruitment have the potential to improve average yields without an associated increase in risk of collapse. Cochrane and Starfield (1992), for example, estimated that the average annual catch of South African anchovy (*Engraulis encrasicolus*) could theoretically increase by almost 50% if a very precise recruitment prediction could be made at the start of the fishing season. Such potential benefits, however, depend on resolving uncertainties to avoid 'overfitting' data, adequately selecting the explanatory or proxy variables and including all associated errors. In a subsequent simulation study of the South African anchovy, for example,

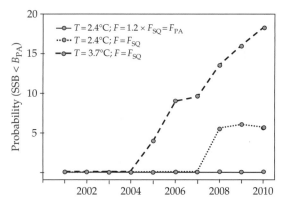

Figure 9.1 Probability that simulated sprat spawner biomass falls below the 'precautionary-approach biomass' (B_{PA}) under different temperature and exploitation scenarios. The simulations assume that recruitment depends on both temperature (T) and spawner biomass (B). Key: solid line, temperature = 3.7°C (mean temperature during 1973–99) and status quo exploitation during 1998–2000 (F_{SQ}); dotted line, temperature = 2.4°C (temperature during 1973–99, mean = −1, SD = 3.7°–1.3°C = 2.4°C) and status quo exploitation during 1998–2000; dashed line, temperature = 2.4°C and 1.2 × status quo exploitation corresponding to F_{PA}. Abbreviations: PA = precautionary approach; SQ = status quo; F = fishing mortality rate. (Reproduced with permission of Ecological Society of America from Fish production and climate: Sprat in the Baltic Sea, Mackenzie, B.R. and Köster, F.W., Ecology, 85, 2004. Permission conveyed through Copyright Clearance Center, Inc.).

De Oliveira and Butterworth (2005) concluded that an environmental index needed to explain at least 50% or more of the total variation of recruitment before the MP starts showing benefits. The species' actual recruitment is measured a few months after the start of the fishing season, which in addition to severe constraints in year-on-year variation in total allowable catches, explains why prediction uncertainties limit the application of early recruitment estimates.

The Bay of Biscay anchovy, *Engraulis encrasicolus* (Box 9.1), on the other hand, lacks direct annual measurements of incoming recruitment. Therefore, substantial effort has been invested in estimating recruitment using environmental variables. From historical data, Borja *et al.* (1996, 1998) concluded that recruitment can be favoured by the occurrence of spring wind-driven upwelling along the French and Spanish coasts. Allain *et al.* (2001, 2004) proposed an environmental model that combines upwelling intensity and stratification breakdown events, which explained about 75% of the interannual variability between 1986 and 1997. In 1999, the International Council for the Exploration of the Sea (ICES) assessment working group used Borja's upwelling index to predict a failure of recruitment at age 1 for 2000. However, subsequent observations indicated that this prediction was incorrect, and so that the reduction of the initial TAC adopted by managers was unnecessary. This experience caused intensive debates among the management bodies and scientific advisers, not only about the use of environmental indices to predict recruitment but also about the nature of the advice required for this short-lived species. Generalizing their conclusions, we can summarize that recruitment predictions should only be used in management if they are supported by adequate process understanding, and followed by an adequate framework to define criteria against which prediction success or failure would be judged. In recent years, the Bay of Biscay anchovy has collapsed and the fishery closed in 2005 after several recruitment failures. Recent research has identified a number of recurrent and occasional spawning sites for anchovy, which have recently changed due to changes in environmental conditions (Bellier *et al.* 2007). Anchovy now increasingly use offshore waters for recruitment,

apparently to avoid predation (Irigoien *et al.* 2007). Because of these uncertainties in process understanding, De Oliveira *et al.* (2005) concluded that precautionary approaches may better ensure successful management than the application of environmental estimators of recruitment success (see Box 9.1).

Research on fish recruitment continues to be a crucial part of GLOBEC (see Ashby 2004), and while environmental estimates of recruitment are not generally used in management, applications of this research continue to develop. For example, monthly egg production census surveys and pre-recruit surveys are used as predictors of Japanese sardine, *Sardinops melanostictus*, Spawning Stock Biomass (SSB) and recruitment (R), respectively (Nishida *et al.* 2006). From these surveys, the extension of the sardine spawning area is calculated as a measure of SSB by integrating the areas in which early developmental stage eggs are present (Zenitani and Yamada 2000), while a temperature-weighted pre-recruit index corresponds well to the VPA-derived recruitment estimate.

9.2.2 Incorporating habitat change

It is well understood that environmental considerations affect the survival success of fish populations by changing habitat conditions. However, these are not explicitly included in most fisheries assessment and management practices. In the central Baltic Sea, the biomass of Atlantic cod (*Gadus morhua*) declined throughout the 1980s to the lowest level on record in 1992, and has still not recovered. While causal mechanisms are still debated, there is agreement that the decline is driven by a combination of increasing fishing pressure and recruitment failure, the latter caused by inadequate habitat for eggs and larvae. Atlantic cod eggs require seawater of a sufficiently high salinity (>11 psu; Westin and Nissling 1991) and oxygen content (>2 ml/l; Wieland *et al.* 1994) to get fertilized and survive. Plikshs *et al.* (1993) introduced the concept of Reproductive Volume (RV) as a measure of the habitat volume with adequate conditions for Baltic cod egg survival. RV is mainly determined by the number and magnitude of the seawater inflows from the North Sea into the Baltic

Box 9.1 The benefits and pitfalls of incorporating environmental estimates of fish recruitment in management: the case of the Bay of Biscay anchovy

The European anchovy, *Engraulis encrasicolus*, occurs from the Mediterranean Sea and the south coast of Spain to the North Sea and the western Baltic Sea in the north (Uriarte *et al.* 1996; Beare *et al.* 2004). The anchovy in the Bay of Biscay is isolated from populations to the north and to the south, and thus considered a single stock for assessment and management purposes (ICES 2007c). Anchovy catches were high from the mid-1950s to the early 1970s, and progressively diminished up to the 1980s, followed by a similar decline in the Spanish Purse seine fishery. Fleet capacity grew again in the late 1980s with the addition of French pelagic trawlers, as did catches. In this century, catches have declined steadily to historical minima until the recent fishery closure period started in 2005 (ICES 2007c).

The bulk of the population and catches are from 1- to 2-year-old fish, demonstrating the strong dependence of the population on incoming recruitments. However, at the end of the calendar year, when Total Allowable Catch (TAC) limits are set and the fishery commences, direct estimates of recruitment are not available. Therefore, short-term management advice is based on assumptions rather than assessments of the strength of recruitment. In the late 1990s, the possibility of forecasting anchovy recruitment on the basis of environmental indices was explored, thus potentially enhancing the TAC at a time when year-class strength was still unknown. In autumn 1999, ICES decided to use an upwelling index in a quantitative manner to predict a failure of recruitment for 2000. This measure was taken after Borja *et al.* (1998) established a statistical relationship between wind-driven upwelling and recruitment. However, the predictions failed and the reduction of the TAC to 16 000 t in 2000 turned out to be unnecessary. The TAC was raised to 33 000 t on 1 July, after direct estimates of SSB from surveys became available. This experience caused intensive debates among the management bodies and the scientific advisers, not only about the use of environmental indices to

predict recruitment but also about the nature of the advice required for this short-living species. Eventually, it was decided to abandon the use of the environmental indices for this species given the limited predictive power of the model used. It has since been recognized that a semi-quantitative use of this, and an alternative model that uses upwelling and stratification breakdown events (Allain *et al.* 2001, 2004), could perform as well as a purely quantitative one (ICES 2006).

De Oliveira *et al.* (2005) investigated under what circumstances incorporating environment-based recruitment indices would lead to management improvements in the Bay of Biscay anchovy, compared to using the average of past recruitments or a precautionary low-recruitment estimator. The analysis considered a range of correlation values between recruitment and the environmental index used in the prediction, as well as both direct and indirect ways of using recruitment predictions in TAC formulation. Because of uncertainties in process understanding, it was concluded that precautionary approaches may better ensure successful management than consideration of uncertain (moderate to weak in terms of the relationship to recruitment) environmental effects.

Assessment scientists of this stock now consider that any future attempts to use environmental estimates of recruitment require prior definition of criteria for which success or failure would be judged against, a process that should apply to other stocks and species. In the meantime, plans are in place to revise the current management regime so as to take into account recruitment fluctuations. This may be achieved by developing a decision rule directly using the information from the existing time series of spring surveys or by taking into consideration the results of a recently established autumn juvenile survey programme. In the meantime, environmental estimators of recruitment continue to be developed and improved (ICES 2007c).

Manuel Barange

Sea in any given year (Hinrichsen *et al.* 2002a). Since the mid-1970s, there have been only sporadic major inflows into the Baltic, very few reaching eastern spawning regions (Fig. 9.2). Sporadic inflows do not necessarily result in good recruitment though. For example, in 1993–94, egg survival was enhanced by a major water inflow (MacKenzie *et al.* 2000), but the long-term salinity

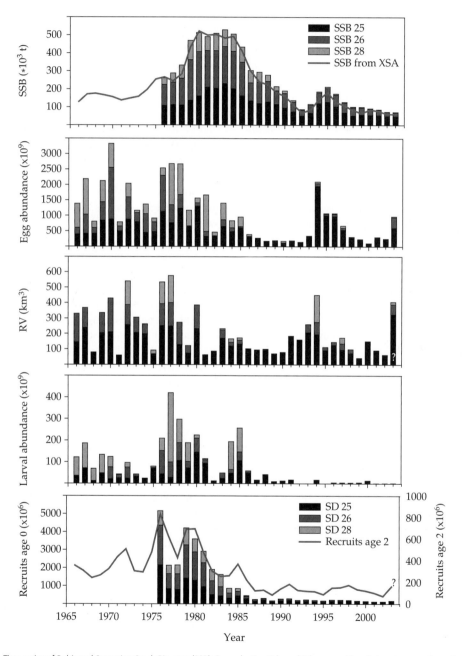

Figure 9.2 Time series of Baltic cod Spawning Stock Biomass (SSB), Reproductive Volume (RV), egg and larval abundance, and recruitment at age 0 and age 2, in subdivisions 25 (East Baltic), 26, and 28 (West Baltic). (Modified from Köster *et al.* Baltic cod recruitment and the impact of climate variablility on key processes. ICES Journal of Marine Science 2005, **62**, 1408–25. By permission of Oxford University Press).

reduction prior to this event affected the main food for larval cod, the copepod *Pseudocalanus* sp. (Hinrichsen *et al.* 2002b), resulting in low cod recruitment and continued stock depletion (Köster *et al.* 2005). Climate predictions from global circulation models suggest that the frequency of inflows will not increase in the foreseeable future, thus questioning the sustainability of this resource on the basis of habitat considerations alone (MacKenzie *et al.* 2007a).

Since 1999, the International Baltic Sea Fisheries Commission (IBSFC) has adopted both a long-term management strategy and recovery plan. Despite these measures, the Atlantic cod stock remains in a critical status. In addition to ignoring scientific advice when setting quota and overshooting approved TAC (ICES 2007a), lack of suitable spawning habitat in eastern spawning areas hinders recovery (ICES 2007d). Long-term simulations indicate that management alone may not facilitate an increase in stock size to precautionary approach levels (Köster *et al.* 2008) without adequate spawning habitats, illustrating the need to develop management systems that are responsive to habitat changes, by incorporating up-to-date environmental considerations and by modifying management goals, for example by adjusting biological reference points.

The California sardine (*Sardinops sagax caerulea*) is another exploited resource whose productivity is directly influenced by habitat. Contrary to the management of Baltic cod, this is considered in its management system. Historically, the California sardine was known to exhibit large fluctuations in abundance, with cooling periods (including *La Niña* events) believed to negatively influence production (Lluch-Belda *et al.* 1989; Baumgartner *et al.* 1992) by reducing the sardine's reproductive habitat. Sardine assessment employs environmentally-dependent surplus production models, which take sea surface temperature and habitat area into account. Such models use mean 'three-season' SST (sea surface temperature) records at Scripps Pier in San Diego, California, as a proxy for the environmental conditions that affect recruitment (Fig. 9.3). The subsequent spawner-recruit model (Jacobson and MacCall 1995) pro-

vides a substantially better fit to the data than conventional surplus production models (Jacobson *et al.* 2005). Ecological support for this approach is based on the observation that Maximum Sustained Yield (MSY) and stock biomass at MSY (B_{msy}) are dependent upon habitat area (Jacobson *et al.* 2005). If the core habitat for California sardine is defined by mean SST, a larger area can support a larger and more productive stock (Jacobson *et al.* 2005). Monitoring this area helps anticipate periods of high and low productivity, and thus permits adjustments to management reference points (Jacobson *et al.* 2005). Taking the relationship between environmental conditions and stock production one step further, SST has also been explicitly included in the harvest control rule for California sardine, both to maintain consistent catches and to protect the stock, which is foraged by fish, bird, and marine mammals. The rule decreases harvest rates with decreasing average three-season SST (ICES 2007e).

The role of habitat in promoting or constraining fish productivity also has important consequences in light of climate change. Sarmiento *et al.* (2004) estimated that climate change would reduce sea ice biomes in both hemispheres by up to 40%. Polovina *et al.* (2008b) reported that the permanent stratified subtropical gyre habitats worldwide have grown 15% since 1998. Adaptations in suitable habitat linked to climate change have been estimated for freshwater fish (Mohseni *et al.* 2003), sockeye salmon (*Oncorhynchus nerka*; Welch *et al.* 1998) and Atlantic cod (Drinkwater 2005), among others. These results suggest that climate change may be an important consideration in defining suitable habitat. However, not all species have the same dependency on suitable habitat. Comparing the dynamics of anchovy and sardine stocks in the Benguela, Kuroshio, California and Humboldt Currents, Barange *et al.* (2009b) suggested that anchovies have a larger dependency on available habitat than sardine populations. Such observations show that habitat health, in the context of climate change, brings a broader ecosystem perspective to traditional fisheries management.

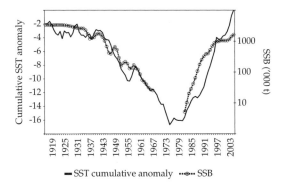

Figure 9.3 California sardine Spawning Stock Biomass (log scale) and Scripps Pier sea surface temperature anomalies 1916–2005. (From ICES 2007e).

9.2.3 Considering multi-species interactions in single-species management

The examples used so far illustrate how the environment influences single-species dynamics and should thus be considered in their management. In the context of single species management, multi-species models can be very useful to update estimates of natural mortality, capturing environmentally-driven changes in species interactions. Attempts have been made to incorporate cannibalism in the stock assessment for cod populations in the north-east (ICES 2007b) and Baltic (Sparholt 1994), predation by arrowtooth flounder (*Atheresthes stomias*), Pacific halibut (*Hippoglossus stenolepis*) and Steller sea lion (*Eumetopias jubatus*) in the stock assessment of Gulf of Alaska walleye pollock (*Theragra chalcogramma*; Hollowed *et al.* 2000) and predation by Pacific cod (*Gadus macrocephalus*) and northern fur seals (*Callorhinus ursinus*) in the eastern Bering Sea walleye pollock assessment (Livingston and Methot 1998), to name a few. The consideration of multi-species interactions in single-species management appears to be limited to relatively simple boreal ecosystems though. In more complex systems, where many different prey resources exist and where predator-prey interactions are weak (Link 2002), this approach appears to be more difficult to implement (Plagányi and Butterworth 2005).

Another example of multi-species management in a single-species context comes from the Japanese small pelagic fishery. This fishery lands between 2 and 6 million tons of small pelagic fishes annually, with quasi-decadal alternations in dominant species (sardine, anchovy, *Engraulis japonicus*, chub mackerel, *Scomber japonicus*). These species are important prey of minke whale (*Balaenoptera auctorostrata*) and major predatory fishes (Yatsu *et al.* 2003). Katsukawa and Matsuda (2003) proposed a MP to prohibit the catch of the most depleted species in favour of the alternative, replacement species, which would result in a lower risk of stock depletion and higher long-term yield compared to traditional single-species management.

9.2.4 Recognizing ecosystem productivity cycles

Population productivity fluctuations are often associated with particular climate-ocean regimes that affect the dynamics of entire ecosystems (Kawasaki and Omori 1986; Mantua *et al.* 1997; King 2005) and even ocean basins (Lluch-Belda *et al.* 1989; Chavez *et al.* 2003). It has been proposed that these fluctuations may reflect multiple ecosystem stable states or 'regimes' (Knowlton 2004). A shift in regime can be climate-driven (e.g. North Atlantic, Beaugrand *et al.* 2008; North Pacific, Mantua *et al.* 1997), originating from removals of specific species (e.g. Scotian Shelf; Frank *et al.* 2005) or a combination of both (e.g. eastern Baltic Sea; Köster *et al.* 2003b). The likelihood of a climate-driven regime shift seems to increase when humans reduce ecosystem resilience by removing key functional or age groups (Folke *et al.* 2004; Anderson *et al.* 2008).

To date, very few attempts have directly incorporated the dynamics of regimes into the management of fish populations, notwithstanding its advisability. The particular complication posed by productivity regimes is that these can affect different species in opposite directions, and thus isolated single-species management measures could have unexpected outcomes on associated species. There is diversity of opinion as to what management strategy is best to account for regime shifts.

Walters and Parma (1996) considered that a constant rate single-species harvest strategy performs better, while Spencer (1997) and Peterman *et al.* (2000) concluded that a regime-specific harvest rate strategy is optimum. A middle way was proposed by MacCall (2002), who suggested that, for short-lived species, a regime-specific harvest rate was optimal for maintaining high yield and low variation in spawning stock biomass, while for a long-lived species (lifespan greater than 30 years), a constant harvest rate was more appropriate. However, MacCall noted that to avoid overfishing during low-productivity regimes, a constant harvest rate strategy would have to be well below traditional reference points (i.e. ~10% of the exploitable biomass) to achieve its objectives. Polovina (2005) considered that a regime-specific harvest rate strategy would increase yield in high-productivity regimes, but warned that it may result in fishery closures during low-productivity regimes.

In general, regime-specific management requires that regimes be recognized and, ideally, their emergence predicted. This is not a straightforward mat-ter (deYoung *et al.* 2008; Ito *et al.*, Chapter 10, this volume). Recently, King and McFarlane (2006) proposed a framework (Fig. 9.4; Box 9.2) that requires that regime-shifts be detected but not necessarily predicted. The management response time needs only to mirror the biota response time to changes in productivity, which is lagged by the age of maturity or recruitment. King and McFarlane (2006) observed that variable harvest rates that reflected the relative levels of stock productivity produced the best balance between benefits (high yield) and trade-offs (fishery closures) and allowed for rebuilding of the spawning stock when productivity improved.

De Oliveira (2006) investigated harvesting strategies for the South African fishery for sardine (*Sardinops sagax*) and anchovy. These species have been suggested to roughly alternate in dominance in many ecosystems (Schwartzlose *et al.* 1999), although this has been contested (Fréon *et al.* 2003). In South Africa, the fisheries for both species interact, thus creating a potential positive feedback loop to climate-driven shifts. De Oliveira (2006) evaluated whether Management Procedures (MP,

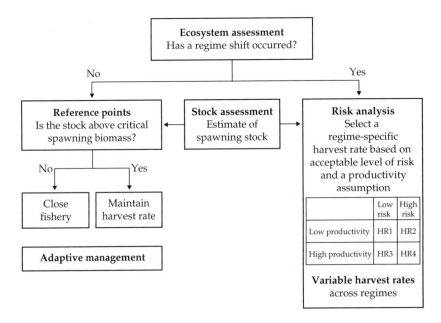

Figure 9.4 Conceptual decision-rule framework to include the application of regime-specific harvest rates (HR) in fisheries management. (From King and McFarlane 2006 with permission from Wiley-Blackwell.)

Box 9.2 Regime-specific harvest rates

Several studies have investigated the use of different harvest rates as management tools for fish populations subjected to climate-driven regime shifts. When benefits (yield) and trade-offs (fishery closures and conservation concerns) are explicitly accounted for, population simulations with regime-like shifts in productivity have illustrated that varying harvest rates by productivity regime is an effective tool for contending with decadal-scale climate or environmental

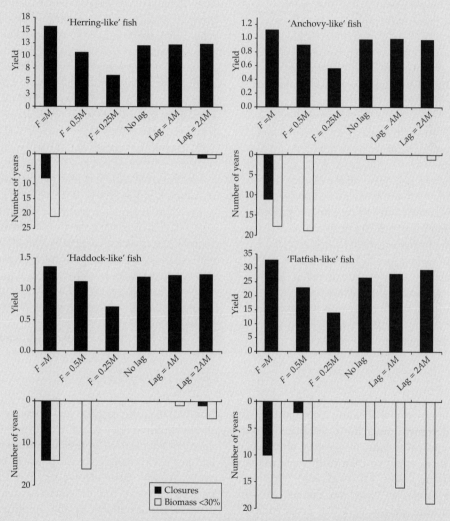

Box 9.2, Figure 1. Top panels: Total yield (thousands of tons for 'herring-like' and 'flatfish-like'; millions of tons for 'anchovy-like' and 'haddock-like' examples) based on four population simulations incorporating regime impacts on recruitment and varying levels of exploitation: constant F that does not vary by productivity regime ($F = M$; $F = 0.5M$; $F = 0.25M$) and regime-specific F that does vary by productivity regime with a change in F coincident with the regime shift (no lag) or lagged by age of recruitment (AM) or twice age of recruitment (2AM). Bottom panels: The number of years in which the fisheries for the four populations were closed (biomass <25% of maximum biomass) or in which there was a conservation concern (<30% of maximum biomass) for the same exploitation scenarios in the top panels (From ICES 2007e.)

continues

Box 9.2 *continued*

variability (King and McFarlane 2006; ICES 2007c). These simulations applied regime impacts on recruitment for short-lived species that were 'anchovy-like' (maximum age = 5; age of maturity = 1; M = 1.2) and 'herring-like' (maximum age = 10; age of maturity = 3; M = 0.4) and for long-lived species that were 'haddock-like' (maximum age = 15; age of maturity = 3; M = 0.2) and 'flatfish-like' (maximum age = 50; age of maturity = 5; M = 0.1). The populations were exploited at levels of fishing mortality (F) that are typically employed in fisheries management (i.e. $F = M$; $F = 0.5M$; $F = 0.25M$). If spawning stock biomass fell below 25% of the maximum biomass, the simulated fisheries were closed until the populations increased above this threshold. A 60-year period was simulated beginning with a 20-year regime of good recruitment, followed by 20 years of poor recruitment and ending with an average recruitment regime of 20 years.

Overall, the management strategies that employed regime-specific harvest rates that reflect relative levels of productivity (i.e. good recruitment, poor recruitment, and average recruitment) outperformed those that employed constant fishing mortalities by providing a balance between benefits (high yield) and trade-offs (fishery closures or years with a conservation concern, i.e., biomass <30% of maximum biomass). Regime-specific fishing mortality management strategies allowed for rebuilding of spawning stocks from periods of low productivity to periods of improved productivity. For both short-lived and long-lived species, dramatic impacts of regime shifts on productivity means that constant harvest rates that are typically employed ($F = M$) cannot be maintained. It is possible to use constant harvest rates across productivity regimes, but these harvest rates must be low (e.g. F = to $0.25M$ to ensure no fishery closures or conservation concerns) and are partnered with reduced yield (between 42 and 50% lower). The adjustment of harvest rates do not need to coincide with regime shifts in productivity, but can be lagged by the age of recruitment to the fishery (Box 9.2, Fig. 1).

These results require changes in regimes to be detected, with possibly a short lag. However, for short-lived species with very early ages of maturity (e.g. anchovy), the difficulty to detect regime shifts and to adjust harvest rates within 1 or 2 years suggest that constant low harvest rates might logistically be the most appropriate management strategy.

Jacquelynne R. King

sensu the International Whaling Commission; see Box 9.4), designed under the assumption of out-of-phase cyclical trends in species abundance could be more effective than traditional modelling approaches. The MP evaluated long-term harvesting strategies based on the net present value of future profits and the proportion of future years that experience zero or negative profit. 'Regimes' were modelled as out-of-phase cycles for the two species, with a range of amplitudes (from zero to halving/doubling of recruitment) and periods (30 and 50 years). The author concluded that, without the operational interaction between the sardine and anchovy fisheries, there would be little value in incorporating management adaptations to regime shifts, even if one had perfect knowledge of the phase in the regime cycle at any time. However, the interaction between species makes adaptive management necessary, particularly as the amplitude of the cycle increases (Fig. 9.5). In comparing regime-shift estimators, De Oliveira (2006) concluded that, in the case of the South African anchovy, a 6-year running mean of the proxy estimator outperforms precise knowledge of the actual position in the cycle.

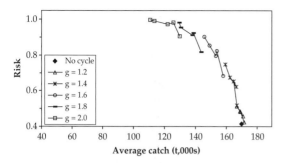

Figure 9.5 Performance of the operational management procedure, in terms of 20-year average annual catch and risk of stock collapse, for different cycle amplitude g, ranging from $g = 1$ (no cycle) to $g = 2$ (a doubling or a halving of the functions used to generate recruitment at the peak or through the cycle, respectively). Points within each curve correspond to cycle duration, from 20 to 60 years, in steps of 10 years, with 60 being closest to the bottom-right corner for all curves. The result illustrates the deterioration of the management procedure as the amplitude of the regime cycle increases and the period decreases. (From De Oliveira 2006.)

An additional problem caused by regime shifts is the mismatch between the longevity of fishing fleets and that of productivity cycles (Yatsu 1995). Fleets expand following increases in fish biomass, usually some time after the start of a new high-productivity regime. Since the duration of the regime is generally shorter than the lifespan of the new fishing fleet, this can lead to fleet overcapacity at the start of a subsequent low-productivity regime (Fréon *et al.* 2008).

9.2.5 Climate change and the management of transboundary fishery resources

Many of the world's marine fisheries are shared between the fleets of two or more nations, requiring exploitation agreements and, occasionally, common management. In most cases, agreements are based on records of past access to the common resource, assuming that the resource distribution and thus availability to the different players remains reasonably constant. However, the warming of the oceans is changing the distribution and migratory patterns of fish species worldwide (Perry *et al.* 2005; Dulvy *et al.* 2008; Barange and Perry 2009). Such changes have the potential to affect fisheries management, requiring adaptations to unforeseen scenarios.

The Norwegian spring-spawning herring, *Clupea harengus*, has been one of the largest fish stocks in the North Atlantic, and an important source of food for centuries. Historical records show dramatic fluctuations in harvests of this stock over the past 500 years (Alheit and Hagen 1997). Large-scale migrations have been observed, which appear to be related to stock size as well as the environment. After the stock crashed around 1970, it changed its migratory pattern and became confined to the Norwegian EEZ. As it recovered, combined with a habitat increase due to ocean warming, it progressively migrated into open areas and other countries' EEZs. Hannesson (2006) investigated whether it would be advantageous for Norway to keep the stock for itself by fishing it down to a level low enough to prevent it from migrating out of the Norwegian zone. He concluded that a cooperative solution would make all parties better off, as long as Norway (the main exploiter) obtained a large share of the total catch quota which was not determined only by the zonal attachment of the stock. Future ocean warming may make Norway less 'attached' to the stock, strengthening the bargaining position of other parties and making a competitive harvesting solution more attractive to them.

Another example of the consequences of climate-driven change on transboundary stocks involves the Pacific sardine, *Sardinops sagax caerulea*. The Pacific sardine stock is shared between Mexico, the United States, and Canada, but there is currently no agreed management policy between the three countries. Instead, the United States assumes that 87% of the stock resides in its waters and allocates quotas accordingly. Canada sets annual catch limits based on a 10% fraction of the US harvest guideline, while Mexico does not limit total harvest (ICES 2007e). The best management approaches of a single nation will not insure long-term stability of the sardine resource and fishery in the absence of international agreements, because the latitudinal distribution of sardine is dependent on decadal climate regimes (McFarlane and Beamish 2001; Rodriguez-Sanchez *et al.* 2002). Simulation results, in fact, suggest that a non-cooperative strategy would reduce the size of the stock by 65% within 30 years (Ishimura *et al.* 2008), illustrating how this issue may become increasingly conflictive in the future.

The extent to which management solutions require predictive capacity to respond to climate changes is worth exploring. Miller and Munro (2004) differentiate between 'prediction', the fore-telling of an event with some level of detail and precision, and 'anticipation', the expectation of the possibility of an event at some unspecified time in the future. Some of the research highlighted in this book shows how investments in data collection and modelling can yield reliable predictions, but also that complex dynamics or unobservable components of the system may make accurate predictions virtually impossible. Yet, one can anticipate the possibility of significant change. Recent advances in scientific understanding of climate regime shifts, climate-change impacts, and their impacts upon fisheries have created new opportunities for managers to anticipate such shifts and take appropriate precautionary measures, including pre-agreements that outline actions to be taken under a variety of contingencies (Hilborn *et al.* 2001; deYoung *et al.* 2008). Clearly, *anticipation* of possible changes in the condition or distribution of the stock would be a necessary condition for workable pre-agreements.

9.2.6 Incorporating environmental information into single-species management

The above results illustrate two important points that justify incorporating ecosystem science in fisheries management: good fisheries management alone may not be able to recover a depleted stock under unfavourable environmental conditions, and poor management can prevent the recovery of a stock even if environmental conditions are favourable. While it has long been accepted that fisheries advice is provided within the context of a varying environment, the need to take into account not only environmental variability, but also trends and shifts in the environment in management advice, has been increasingly recognized. A recent ICES/GLOBEC workshop identified concepts for improving existing fisheries management strategies and advice by incorporating environmental information (WKEFA; ICES 2007e). The conclu-

sions and recommendations of this initiative are particularly relevant to improving traditional single-species management.

WKEFA (WorKshop on the integration of Environmental information into Fisheries management strategies and Advice) concluded that, regarding short-term changes, these can be brought into management advice, even if the complexity of the drivers is unknown. For example, changes in growth and maturation could be brought directly into methods for estimating spawning stocks 1 or 2 years ahead, and for estimating catch where catch quotas are required. However, errors in the information need to be included and indicator testing via simulation is required before they are formally applied, including developing implementation frameworks that are informative and robust to errors. Medium-term changes cannot be predicted in the same way as short-term effects. Where explicit relationships exist between stock and the environment, the mean of stochastic projections should be modified, such as average temperature dependence, species interactions, and food availability for different exploited stocks. Where no explicit relationships exist or there is no basis for predicting environmental drivers into the future, advice should be based on scenario testing, as part of evaluations of management plans (e.g. see ICES 2007b). As a corollary, in the light of climate change, rather than assuming that the mean of a given parameter derived from the past will apply in future, recent trends should be estimated and used instead. This calls for the development of a number of tools that evaluate estimates of current values and current trends in the presence of noise in both measurement and environment.

WKEFA made specific recommendations with regard to issues such as habitat, growth and maturation, recruitment, and regime shifts. Environmental effects on habitat need to be investigated to ensure that they do not result in catchability biases in surveys or commercial catch rates used for tuning assessments. However, the development of bias corrections requires investigation of causal links in addition to the evaluation of the effects. If available, measures of reproductive habitat should be integrated into the evaluation of reference points, and construction of stock-recruitment relationships, to be used in medium- to long-term projections. The value of using such information in short-term predictions will, however, depend on the understanding and predictability of processes affecting reproductive success.

Both growth and maturation are sensitive to environmental change in the short and medium term. These parameters affect management through estimation of SSB and catch. WKEFA recommended that methods be developed to evaluate how variable weights and maturation should be projected forward in the short-term. Retrospective analysis should be used to monitor the performance of the chosen methods. For improving medium- to long-term simulations, coupled growth and maturation models need to be developed, considering changes in ambient abiotic and biotic environment.

Knowledge and prediction of recruitment is obviously one of the most critical aspects for setting short-term catch quotas, particularly for short-lived species, medium-term management plan evaluation, and for the determination of targets and biological reference points. Regarding short-term estimation, if direct estimation is not possible, the use of environmental indicators could be considered, but they should be evaluated and incorporated in short-term projections only if performance is shown to improve prediction. This evaluation should consider the error structure, relative information content, and functional relationship of any indicator in selecting its use. If recruitment estimates are required in the absence of any direct or indirect estimation method, the use of geometric mean can be replaced by time series and autoregressive models of recruitment to provide values. Regarding medium- to long-term simulations, these should include evaluations under different environmental regimes, considering environmentally dependent, stock-recruit relationships, and parameter uncertainties. The consideration should be primarily for robustness to different plausible possibilities, rather than expecting to optimize management under all conceivable options.

Regarding both naturally occurring and fishery-induced regime shifts, regime-specific fishing mortality management strategies can be used as a tool for contending with decadal-scale climate variability, outperforming constant fishing mortality management strategies by providing a balance between benefits (high yield) and trade-offs (fishery closures). However, it is unlikely that a single management strategy will be optimal under different regimes. Short-lived species with low age of recruitment and high exploitation rates would require very rapid detection, but their management may already be more adapted to conditions of regime shifts. In contrast, long-lived species exploited at a low rate and with older age of entry to the fishery allow for slower management response. In these circumstances, it is necessary to re-evaluate previously defined biomass or fishing reference points, considering not only the current carrying capacity but also the potential for further stock depletion, and the ability of the stock to recover should the habitat return to previously observed conditions.

9.3 From ecosystem science to ecosystem-based management

In the previous section, it has been shown that the performance and reliability of fish-assessment models could improve by incorporating more ecosystem and environmental considerations in the models. Some recommendations have also been provided on how to incorporate such information in single-species stock assessment as they become available (see also Payne *et al.* 2008). The historical separation between fish stock assessment experts and experts within various fields of basic marine sciences needs to be bridged if we are to better develop and incorporate environmental variability and change into resource management. Parallel developments also needed include a re-balance between the use of both short-term forecasts (annual) and medium- to long-term trends (decadal) in management, and take better account of the synergistic effects between harvesting pressure and environmental variability and change (Frank *et al.* 2005; Anderson *et al.* 2008; Perry *et al.* 2010a; Planque *et al.* 2010). Such developments would considerably improve our ability to account for patterns and cycles in environmental and ecosystem variability in fisheries management.

While the above developments were limited to the fisheries science and management arenas, a broader debate is underway regarding the economic, social, and ecological value of all the goods and services provided by marine ecosystems, including fisheries (World Resource Institute 2005). In this debate, fisheries management is contextualized by the increasing recognition of the functional value of ecosystems (beyond extraction) and a number of conflictive issues, *inter alia*,

- the importance of interactions among fishery resources and between fisheries resources and the ecosystem in which they exist (see Brander *et al.*, Chapter 3, this volume);
- the wide range of societal objectives for and values of fisheries resources and marine ecosystems within the context of sustainable development (see Perry *et al.*, Chapter 8, this volume); and
- the perceived poor performance of current management approaches as witnessed by the poor state of many of the world's fisheries and supporting ecosystems.

This debate acknowledges that maintaining and protecting ecosystem function may provide a much larger contribution to society than destructive fisheries (Costanza *et al.* 1997). Influenced by this debate, most advisory and management bodies are increasingly considering fisheries as part of an integrated management strategy of human exploitation and use of the marine environment, in a move broadly labelled the ecosystem approach. In its application to fisheries (Garcia *et al.* 2003), the ecosystem approach would be able to address many of the shortcomings described in Section 9.2, and will therefore provide a framework on which to base recommendations on how to improve the links between ecosystem science and management.

9.3.1 Developing a new management framework

The concept of the 'Ecosystem Approach' was first coined in the early 1980s, but found formal acceptance at the Earth Summit in Rio in 1992 where it became an underpinning concept of the Convention on Biological Diversity as 'a strategy for the integrated management of land, water and living resources that promotes conservation and sustainable use in an equitable way'. The Ecosystem Approach to Fisheries (EAF) has the specific purpose to plan, develop, and manage fisheries in a manner that addresses the multiplicity of societal needs and desires, without jeopardizing options for future generations to benefit from a full range of goods and services provided by marine ecosystems (FAO 2003). It strives to balance diverse societal objectives by taking account of the knowledge and uncertainties regarding biotic, abiotic, and human components of ecosystems and their interactions, in an integrated approach (Fig. 9.6).

The Ecosystem Approach to Fisheries is an evolutionary and incremental extension of traditional fisheries management, based on precautionary principles, ecological understanding, and including the management of cumulative and long-term impacts (Rosenberg *et al.* 2008). Although GLOBEC was not directly linked to the development of the EAF, the scientific requirements of an EAF cannot be achieved without the consideration of GLOBEC research. This is because the EAF demands that research advice be broadened, requiring more comprehensive ecological understanding or at least a broader consideration for ecosystem issues. Garcia and Cochrane (2005) argued that the main difference between conventional and EAF management approaches is that while both are aimed at resource conservation, in practice the former has focused on conserving livelihoods and employment, using resource conservation as a weak constraint.

The implementation of the EAF framework requires a nested set of processes at regional, national, sectoral, and local (e.g. fishery) levels, where interconnected policies (the high-level conceptual goals and constraints), strategies (the translation of conceptual goals into operational objectives), and plans (the specific measures and enforcement mechanisms agreed among stakeholders) can be developed (Fig. 9.7). While the main conceptual steps may be similar for all levels, the

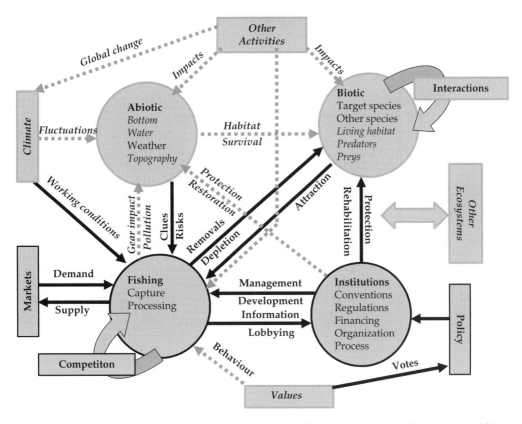

Figure 9.6 Ecosystem components and interactions addressed by EAF. Elements in black and emboldened specify the conventional fishery management approach. Elements in grey and italics represent elements to add for EAF. (Garcia and Cochrane, Ecosystem approach to fisheries: A review of implementation guidelines. ICES Journal of Marine Science, 2005, **62(3)**, 311–8. By permission of Oxford University Press.)

focus, scope, means, and approaches may be different, requiring top-down guidance as well as an institutional environment within which the lower-level processes can develop. The high level of uncertainty inherent in an EAF also requires a capacity for the system to adapt as information improves.

The role of ecosystem science in contributing to the decision making is fundamental under the EAF. Additional ecosystem science is required to extend the scope of assessments, better understand ecosystem functioning, tackle issues of uncertainty and risk, improve forecasting capacity, apprehend natural variability, and navigate between neutral scientific advice and advocacy (Garcia and Cochrane 2005). However, the increase in the scope and sophistication of the science in proportion to ecosystem complexity and escalating uncertainty is likely to be constrained by human and financial resources. Addressing inter-sectoral issues in a multidisciplinary manner, dealing with data-poor situations, broadening the scope of information used to include fishers' knowledge, assessing their relevance and reliability and resolving apparently conflicting signals, to name a few, are significant challenges that require credible yet pragmatic approaches. It is in this context that the role of indicators and reference values become fundamental as a necessary, but not a sufficient, condition for EAF implementation (Garcia and Cochrane 2005).

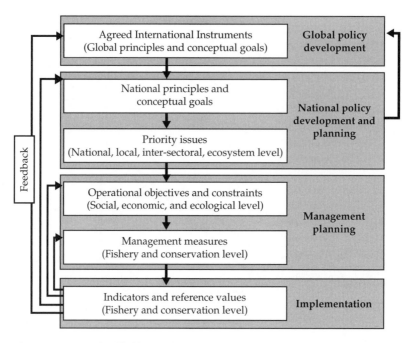

Figure 9.7 The EAF implementation process. (Modified from Food and Agricultural Organization of the United Nations 2007).

9.3.2 The establishment of objectives for an ecosystem approach to fisheries

In this section, we explore in more detail the objective setting process in the implementation of the EAF, including the determination of which ecosystem components are both most important to ecosystem functioning and most at risk to human impacts. This then guides development of the indicators and reference points which would be linked to management actions required to mitigate the risks. The need to incorporate risk emanated from the precautionary approach (FAO 1995a), with its requirement for explicit accounting for uncertainty and the use of reference points as an indicator of risk and performance. The assessment and management of risk to all components of the marine ecosystem has thus become a significant requisite towards an EAF. A comprehensive risk-assessment framework will allow the involvement of stakeholders in the assessment of impacts of fisheries and other human activities, as well as mitigation measures in an objective way. Again, this approach is an extension of con-

ventional management, particularly fisheries which are primarily based on single-species assumptions, so that a broader range of issues, including ecological, social, and economical aspects can be assessed in an integrated way.

What are the steps required to identify operational objectives for an EAF? It is evident that there are many ecosystem components and many risks that must be prioritized. A promising approach which provides an overall framework is the three-levelled hierarchical risk-assessment approach used by Braccini *et al.* (2007) and Smith *et al.* (2007a), in which qualitative, semi-quantitative, and quantitative assessments are used, depending upon data availability and quality. In these, the first step is to develop an understanding of the key ecosystem components and processes that may be at risk to human impacts, perhaps using input expert judgement and stakeholder engagement (Fletcher 2005). The threats posed by the moderate to high-risk impacts are then further analysed using semi-quantitative empirically based analyses and/or ecosystem modelling to evaluate which ecosystem components are most at risk and by how much. This allows prioritization of work on indicators/reference points and associated management actions to mitigate the impacts of the high-risk human impacts, thus excluding ecosystem components at low-risk from resource-intensive assessments.

Provided below are two experiences of this objective-setting process. The first, from South Africa, took a qualitative approach, engaging stakeholders, scientists, and managers in the identification of priorities for an EAF, strongly influenced by ecosystem research through the GLOBEC-affiliated Benguela Ecology Programme and BENEFIT programmes (see Hampton *et al.* 2009). The second, from the Southern Ocean, took a quantitative approach, through the development of ecosystem models, to inform priorities for management. Each case has responded to the situation that they were faced with and both were successful in identifying the key issues faced in their regions, emphasizing that progress towards an EAF can be made under a wide spectrum of conditions and circumstances.

9.3.2.1 An example from South African coastal fisheries

During the past decade, it has been increasingly emphasized that the identification and prioritization of issues facing an EAF and thus the setting of management objectives is a societal process, in which all stakeholders need to be engaged. Ecological Risk Assessment (ERA), as carried out in Australia (Fletcher 2005) and South Africa (Nel *et al.* 2007) is an example of a holistic and participatory approach which involves structured stakeholder workshops, and which results in a list of issues associated to specific fisheries, and an evaluation of the risk associated with each issue. In South Africa, ERA workshops have been carried out for the small pelagic, demersal, west coast rock lobster, large pelagic and squid fishery sectors (Nel *et al.* 2007). Issues considered were grouped into categories: ecological well-being (where most of the ecosystem science is considered), human well-being, governance, and external (such as influences of climate change and global economy).

To address each of the issues, indicators were identified that could be used to guide an EAF in South Africa. Decision-support tools, such as the fuzzy-logic knowledge-based system developed by

Paterson *et al.* (2007) were found to be helpful for monitoring implementation, although have yet to be incorporated in the formal decision-making process. Shannon *et al.* (2006a) considered ways in which ecosystem science could contribute to addressing the high priority issues of an EAF in South African pelagic, demersal, and rock lobster fisheries.

9.3.2.2 An example from the Southern Ocean Krill Fishery

When the Commission for Conservation of Antarctic Marine Living Resources (CCAMLR; Box 9.3) agreed to an annual catch limit for the krill (*Euphausia superba*) fishery in the Antarctic (FAO Area 48), it recognized the potential ecosystem risk that could arise should a substantial part of the catch be taken from a small area. To address this risk, the catch limit was severely restricted until a suitable approach could be found to avoid negative localized impacts (CCAMLR 2001). The process of subdividing the area into small-scale management units (SSMUs) and the subsequent process of evaluating catch allocation options provide an example of a qualitative process to implement conservation objectives in an ecosystem-based fisheries management regime. The boundaries of the SSMUs were defined based on maps of krill distribution and predator foraging areas, historical fishery distribution as well as bathymetry. These maps then informed consensus agreement on a parsimonious set of SSMUs (CCAMLR 2002) by both fishing and non-fishing nations, leading to proposed catch allocation options (Hewitt *et al.* 2004).

However, it was soon realized that developing the ecological understanding that would allow assessing which of these allocation options would most likely achieve the objectives of the Commission would be financially and logistically impossible to achieve. Therefore, a suite of ecosystem function models were developed to explore the trade-offs between different catch options, and their sensitivity to structural and numerical ecosystem uncertainties (Constable 2005; Plagányi and Butterworth 2005; Watters *et al.* 2005). In particular, the models can be used to consider the sensitivity to the degree of transport of krill between SSMUs via the Antarctic Circumpolar Current. This approach allows the

---

Box 9.3 Ecosystem Approach to Management in the Southern Ocean: The Convention for Conservation of Antarctic Marine Living Resources

An ecosystem approach has been adopted for management in the Southern Ocean under the Convention for Conservation of Antarctic Marine Living Resources, which came into effect in 1982 (Constable *et al.* 2000). High level objectives for management of fishery resources have been established as:

"Any harvesting and associated activities in the area to which this convention applies shall be conducted in accordance with the provision of this Convention and with the following principles of conservation:

1. Prevention of decrease in the size of any harvested population to levels below which ensure its stable recruitment. For this purpose, its size shall not be allowed to fall below close to that which ensures the greatest net annual increment.
2. Maintenance of the ecological relationships between harvested dependent and related populations of Antarctic marine resources and the restoration of depleted populations to the levels defined in [sub]paragraph (1) above.
3. Prevention of changes or minimization of the risk of change in the marine ecosystem which are not potentially reversible over 2 or 3 decades, taking into account the state of available knowledge of the direct and indirect impact of harvesting, the effect of introduction of alien species, the effects of

associated activities on the marine ecosystem, and of the effects of environmental changes, with the aim of making possible the sustained conservation of Antarctic marine living resources."

Harvesting of Antarctic krill has been accorded particular attention in an ecosystem context. Krill occupy the nexus of Southern Ocean food webs and their role as prey for higher level predators has been recognized in the specification of operational objectives for management of this species, these being:

- To keep krill biomass at a level higher than would be the case for single-species harvesting considerations and, in so doing, to ensure sufficient escapement of krill to meet the reasonable requirements of predators.
- Given that krill dynamics have a stochastic component to focus on the lowest biomass that might occur over a future period, rather than the average biomass at the end of the period, as might be the case in a single-species context.
- To ensure that any reduction of food to predators which may arise out of krill harvesting is not such that land-breeding predators with restricted foraging ranges are disproportionately affected compared with predators in pelagic habitats.

Keith Reid

---

provision of risk-based advice on the catch allocation options that take account of uncertainty in the absence of full ecosystem understanding. Similar to the ERA (Smith *et al.* 2007a), this approach clearly recognized the importance of including uncertainty as an integral part of the assessment of risk, rather than just acknowledging it. The models used reflect a long history of modelling ecological interactions in the Southern Ocean directed towards the sustainable management of its resources (see Hill *et al.* (2006) for review).

Regarding the identification of threats, the Southern Ocean lacks the range and scale of direct human influences in coastal and land-use inputs. However, given the relationship between krill population dynamics and physical changes associated with climate change, the greatest 'threat' to the region may come from a fisheries management approach that does not account for the uncertainty associated with future climate change. Within the Convention that established CCAMLR, ecosystem level conservation objectives embody the requirement to 'avoid changes that are not reversible in two-three decades' (CCAMLR 2005). While the precise operational definition of this has yet to be fully specified, the essence is that a large (catastrophic) change in an ecosystem that produces an irreversible change, or at least a change that has a strong likelihood of not being reversible, should be avoided. It should be noted that while these high-level conservation objectives provide a basis for defining species-based objectives, the application of value to different species, in order to evaluate trade-offs among different strategies, remains controversial.

9.3.3 The management of risk

Once the multiple objectives of an EAF manage-ment plan have been identified, it is then necessary to develop a suite of strategies (combination of indicators, reference points, and management actions) that can achieve these objectives. In the last 2 decades, considerable progress has been made on how this can be achieved. It is now generally accepted that a management system needs to be considered as just that-the interaction of a number of elements (ecosystem, stock, monitoring, decision making, and implementation), each with its uncer-tainties, configured to robustly meet a defined set of objectives (McAllister *et al.* 1999). As discussed above, the nature of scientific advice in support of fisheries management has changed from the provi-sion of a single strategy (or a few strategies) based on optimizing yield to the provision of a suite of strategies, with associated levels of uncertainty or risk to the stock. Several developments have been influential in directing this change: (1) the precau-tionary approach (FAO 1995a), with its requirement for explicit accounting for uncertainty and the use of reference points as an indicator of risk and per-formance; (2) adaptive assessment and manage-ment (Hilborn and Walters 1992), which involves the identification of major uncertainties, or alterna-tive hypotheses and models, about stock response to various policy options; (3) risk assessment (Francis and Shotton 1997) which accounts for uncertainty and assesses probability of the possible outcomes of various management decisions; and (4) the development of MPs (Butterworth *et al.* 1997) which use operating models to project future

catches and stock status to assess the performance and selection of a set of decision rules (Cochrane *et al.* 1998). Projections are conducted for plausible ranges of assumptions underlying the operating models. MPs also define what data are to be col-lected and used during the proposed time frame for implementation, and how these data are to be analysed (Butterworth *et al.* 1997).

Logistical and financial limits to the development of large-scale experiments on ecosystems and associ-ated management strategies have led towards the use of Management Strategy Evaluation (MSE) using computer simulation. MSE, pioneered by the Scientific Committee of the International Whaling Commission (IWC; Hammond and Donovan, in press), is now being used in fisheries manage-ment, particularly in South Africa and Australia (De Oliveira *et al.* 2008). The benefit of MSE is that the performance of alternative stock and ecosystem esti-mation methods, data collection regimes, reference points, and management options can be evaluated via simulation tools before being implemented. In particular, alternative mitigation measures can be evaluated with respect to the trade-offs among objec-tives. This is of particular importance in the imple-mentation of an EAF since stakeholders need to agree on aims and find acceptable ways to implement management measures. Importantly, MSE is explic-itly precautionary through its requirement for simu-lation trials to demonstrate robust performance across a range of uncertainties about ecosystem and resource status and dynamics.

9.3.3.1 MSE

As implied above, MSE requires *a priori* speci-fication of clear and prioritized management objectives each with quantifiable performance measures. These performance measures are used to evaluate the performance of alternative man-agement strategies against the range of objectives, while taking into account uncertainty. Sources of uncertainty include structural and parameter uncertainties in the models, error in data and observations, decision making, and implementa-tion uncertainty (Francis and Shotton 1997). The alternative management strategies are evaluated with Monte Carlo simulations that test a reason-able array of operating models of the system (i.e. dynamic models of the resource and fishery), each of which reflect known (or likely) uncertainties. The performance results from these simulations form the basis of the provision of scientific advice for managers.

It is worthwhile to ask whether or not MSE is superior to previous approaches. Schnute *et al.* (2007) regard MSE to be a dominant issue in the future of fisheries stock assessment and management, since the technique requires evaluating all aspects of a management system-the data to be collected, methods of analysis, management actions that follow from the analyses, and the uncertain consequences of these actions (implementation error). Most importantly, MSE allows identification of those management strategies that are most robust to uncertainties in the scientific and management advice framework and have the highest likelihood of leading to achievement of the management objectives. O'Boyle (2009) considers MSE as a prerequisite to the implementation of an EAF, highlighting the fact that an EAF by itself cannot resolve issues (e.g. poor implementation due to fishing fleet overcapacity) that MSE can identify and provide the consequences of not addressing.

Owing to the intensive nature of the approach, MSE has to date been limited to relatively simple stock assessments. However, recently, frameworks for MSE are being developed (Kell *et al.* 2007) that can be used to evaluate the consequences of environmental change on management advice (Kell *et al.* 2004; Fromentin and Kell 2007; Kell and Fromentin 2007) as well as to incorporate multi-species effects, some of the aspects that GLOBEC research has identified as key ecological issues for management. It is likely that the approach will increasingly be used to incorporate environmental considerations and evaluate the trade-offs among multiple objectives as various parts of the world move towards an EAF. For example, Fulton *et al.* (2004a,b; 2005) have applied MSE to evaluate the performance of indicators in an Australian fishery, using a relatively complex deterministic model to describe ecosystem dynamics. Those authors then used a sampling model to generate data with realistic measurement uncertainty for a given sampling design to produce the data required to calculate these indicators. Simulated data were collected for different levels of fishing, and for fishing combined with other activities. The performance of the indicators derived from the data was then assessed in terms of the indicators' capacities to track properties of interest. In Europe, the approach is being extended to incorporate stochastic processes (Kell *et al.* 2007).

Smith *et al.* (1999) list several single-species examples for which MSE is currently being employed in the provision of scientific advice for fisheries management in Australia, such as southern bluefin tuna (*Thunnus maccoyii*), southern shark, eastern gemfish (*Rexea solandri*), orange roughy (*Hoplostethus atlanticus*), eastern tuna, and billfish. The MP approach, which is analogous with MSE (Plagányi *et al.* 2007; Rademeyer *et al.* 2007), has been used extensively in the management of South African fisheries (Cochrane *et al.* 1998; Plagányi *et al.* 2007). Both MSE and MP are frameworks for developing a management strategy that meets management objectives which is robust to a wide range of uncertainties. Other fisheries for which single-species MPs have been employed include southern African hake (*Merluccius* sp.; Punt 1992), South African anchovy (Butterworth and Bergh 1993), Icelandic cod (*Gadus morhua*; Baldursson *et al.* 1996), and eastern gemfish in Australia (Smith *et al.* 1996; Punt and Smith 1999).

MSE has not yet been used in Europe to develop formal management procedures as elsewhere where pre-defined data are input into a simulation-tested algorithm to provide a value for a TAC or effort control measure. Although harvest control rules (e.g. for North Sea herring, *Clupea harengus*, Irish Sea, West Scotland, Kattegat and North Sea cod, and northern hake, *Merluccius merluccius*) have been agreed to, these have not been tested using MSE (Punt 2006). However, implicit management procedures have been assessed using MSE (Kell *et al.* 2005a,b, 2006), in particular to evaluate the effect of restricting interannual variability in catch limits to bring stability to the industry.

9.3.3.2 MSE and multi-species fisheries management
Multi-species fisheries management can involve conflicting objectives, and MSE could be used to evaluate which of the possible management strategies

provides a compromise between these conflicting objectives. The focus of multi-species assessments is typically to estimate the impact of exploiting a predator or prey on the future exploitation of that species' prey or predator, that is, a tropho-dynamic or biological interactions. Alternatively, incidental catch of non-target species is another reason for applying a multi-species assessment approach. This is illustrated by the joint management of South African sardine and anchovy (De Oliveira and Butterworth 2004; Box 9.4). In this example, juvenile sardine by-catch in the South African anchovy fishery, which occurs because the two species school together as juveniles (Crawford *et al.* 1980), is explicitly accounted for.

There are only a few cases in which MSE has been applied to multi-species assessment and management. The food web model MULTSPEC for fish and marine mammals in the Barents Sea (Bogstad *et al.* 1997; Tjelmeland and Bogstad 1998) is an operating model in a MSE for single-species management strategies of cod, herring, and minke whale fisheries (Schweder *et al.* 1998). An MSE has been applied to the southern and eastern scalefish and shark fishery which is comprised of numerous separate fisheries operating in south-eastern Australia (Smith *et al.* 2007a). The Atlantis model (Fulton *et al.* 2005) provides the underlying dynamics, which includes linked biophysical, economic, and management models. The

---

Box 9.4 Management of South African sardine and anchovy using an Operational Management Procedure

The South African sardine (*Sardinops sagax)* and anchovy (*Engraulis encrasicolus*) fisheries are managed jointly using an Operational Management Procedure (OMP). Butterworth (2007) defines a management procedure to be "a formula to provide, say, a TAC recommendation, where the inputs to the formula (essentially resource monitoring data) have been pre-specified . . . This formula has been tested by simulation to confirm that it can be expected to . . . [achieve] appropriate trade-offs amongst the mutually conflicting objectives of maximizing catches, minimizing interannual catch variability . . . and minimizing the risk of substantial depletion of the population." The joint management of these species is necessary due to the unavoidable sardine by-catch occurring with anchovy. This means that anticipated average-directed sardine and anchovy catches cannot be simultaneously maximized. A 'trade-off' between these two catches (the latter with an associated juvenile sardine by-catch) is a key factor of the South African pelagic OMP (Box 9.4, Fig. 1).

The first joint OMP for the South African pelagic fishery was implemented in 1994 and revised in

1999, 2002, and 2004. Between 1994 and 1999, a sardine-only OMP was implemented in 1997 and a 'hybrid' OMP was used in 1998 while the joint OMP for 1999 was still under development. Over the last decade, the TACs set by the responsible Minister for the South African pelagic fishery have exactly matched output from the OMP (Plagányi *et al.* 2007).

The OMP is used to set a Total Allowable Catch (TAC) for directed sardine and anchovy fishing and a Total Allowable Bycatch (TAB) for sardine by-catch. The directed sardine TAC is set at the start of the fishing season (January) based on the sardine adult biomass estimated from the preceding November acoustic survey (De Oliveira and Butterworth 2004). An initial anchovy TAC is set at the same time, also based on the anchovy adult biomass estimated from the preceding November acoustic survey and assuming forthcoming recruitment will be average, though this calculation is reduced by 15% to allow for the fact that the TAC cannot be lowered in midseason if recruitment is found to be poor, because the initial TAC may already have been taken by that time (De Oliveira and

continues

Box 9.4 *continued*

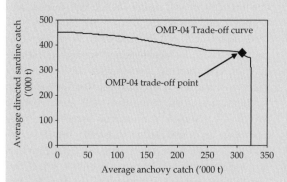

Box 9.4, Figure 1. The trade-off curve generated when testing 'OMP-04', the OMP used to provide TAC recommendations for the South African sardine and anchovy fisheries from 2004 to 2007. This curve is obtained by adjusting OMP control parameters such that pre-specified risk criteria for sardine and anchovy are satisfied. The trade-off point indicated on the curve represents that chosen for OMP-04 (Cunningham and Butterworth 2005).

Butterworth 2004). As the bulk of the coming year's anchovy catch consists of incoming recruits, the initial TAC is revised as soon as an estimate of recruitment becomes available from the mid-year recruit survey. The computation of the revised or final TAC replaces the average recruitment assumed earlier with the actual estimate from the recruit survey. An 'additional sub-season' extending from September until the end of the year was introduced into the anchovy fishery in 1999, with the goal of targeting anchovy that no longer shoal with juvenile sardine. Targeting such 'clean' anchovy allows a greater anchovy catch to be made without increasing the risk to the sardine resource through an associated by-catch of juvenile sardine. This sub-season TAC is set using the same data and at the same time as the revised TAC for the normal season, after the mid-year recruit survey.

The initial sardine TAB is based on a fixed tonnage for by-catch (mainly adults) linked to a small fishery for round herring (*Etrumeus whiteheadii*), and a component proportional to the initial anchovy TAC, using a conservative estimate for the ratio of juvenile sardine to anchovy catch for the coming season. This ratio is increased if the sardine biomass from the preceding November acoustic survey is large. This is to avoid a premature closure of the anchovy fishery due to the sardine TAB being filled in years of high juvenile sardine abundance, prior to the anchovy TAC and sardine TAB being revised. As for the anchovy TAC, the sardine TAB is revised after the recruit survey using an updated estimate of the ratio of sardine to anchovy juveniles. A TAB for the 'additional sub-season' from September onwards is proportional to the additional sub-season anchovy TAC, subject to a maximum of 2,000 t.

Carryn de Moor

biophysical model encompasses physical processes (e.g. current movements and upwelling), biomass pools for lower trophic levels, age or size-based population structure for selected vertebrates of management interest, and habitat or trophic interactions. The economic model encompasses multiple fleets, impacts of discards and habitat modification, effort allocation, responses to mixed quotas, spatial management, and compliance. The management model includes various levels and combinations of regulatory tools including quotas, harvest strategies, season and area closures, gear and trip limitations, and days at sea limitations.

Sainsbury *et al.* (2000) suggest that a number of ecosystem objectives could benefit from multi-species fishery MSE: recovery of endangered species, limiting the effects of fishing on habitats and incidentally caught species, preserving food supply for marine predators, and maintaining biodiversity, ecosystem integrity, and resilience. It is within this multi-species context, specifically its expansion into ecosystem assessments or development of ecosystem management objectives, where ecosystem science can play a significant role in multi-species fishery MSE.

9.3.3.3 MSE and ecosystem models

The use and application of GLOBEC ecosystem science in the EAF is extensive. Much progress has been made in GLOBEC on determining tropho-dynamic pathways and their variability (see Moloney et al., Chapter 7, this volume), on coupled physical, biological, and chemical models to determine impacts of environmental variability on stock dynamics (see deYoung et al., Chapter 5, this volume), and on the capability to predict future impacts of global change (see Ito et al., Chapter 10, this volume). GLOBEC science provides the required process, modelling, and monitoring capabilities to develop ecosystem assessment and performance indicators of ecological objectives, hypothesis testing for selection of harvest strategies and decision rules, and insight into spatial processes required in the MSE implementation at ecosystem level.

Application of MSE in a full ecosystem context will, however, require evolution of the underlying operating models from a single resource and fishery system, to an ecosystem and multi-human use system. These models could range from simple to complex food web models, regional models linking physical forcing processes to lower trophic and higher trophic productivities, and global scale models that begin to account for global climate-change impacts on ecosystems. There will be a growing need for these models to be spatially explicit, perhaps linking fish stocks or sub-stocks as the management strategies may need to consider sequential closure of areas to fishing, or implementation of marine-protected areas (Kaplan et al. 2009). GLOBEC is poised to provide input since spatially explicit individual-based models became a de facto tool in GLOBEC modelling studies (deYoung et al., Chapter 5, this volume). There will likely be growing demand for management strategies that are robust to uncertainties in climate change. GLOBEC models such as NEMURO and NEMURO.FISH (Kishi et al. 2007c) have recently been used to predict the impacts of climate change on ecosystem dynamics (Hashioka and Yamanaka 2007; Ito et al., Chapter 10, this volume) and these approaches might provide hypothesis testing in an MSE arena. Finally, these models will need to address how an ecosystem will respond to natural and human-induced stress. The resilience of ecosystems is dependent on the structure of the ecosystem, particularly the presence or absence of structurally important species (Gifford et al., Chapter 4, this volume). An emerging paradigm in food web and ecosystem structure studies is the application of complexity theory. Using this theory, food webs have been illustrated to contain a few nodes (i.e. species) that have a large number of links (i.e. interactions) suggesting that food web structure and its maintenance is dependent on these key nodes (Borer et al. 2002). Complexity theory can be applied in understanding spatial patterns, migration and competition in ecosystems (Green and Sadedin 2005) and offers a new approach to understanding ecosystem structure and identifying key species whose conservation could serve as a key ecosystem objective within an MSE framework.

In conclusion, the application of the EAF will require an expanded empirical basis as well as novel approaches to modelling, bringing the ecosystem science developed by GLOBEC and similar programmes closer to management objectives. Murawski (2007) specifically notes an important new focus for science supporting EAF in terms of data integration across traditional disciplines at appropriate geographic scales, and in terms of understanding feedbacks and interactions among abiotic, biotic, and human parts of the ecosystem.

9.4 Communicating and increasing societal participation in ecosystem management

In the previous section, we illustrated the scientific basis for the transition from single-species management to the Ecosystem Approach, and the contribution that ecosystem science has, and continues to make, in this process. We have described a framework for the EAF and a process towards identifying management objectives that can address uncertainty and risk. Underlying this process is the recognition that scientific knowledge and management action requires full stakeholder participation.

Society's increasing concern over the sustainability of the common goods provided by marine ecosystems has broadened the concept of stakeholders to include not only fishing communities and management agencies but also conservation bodies, users of services conflictive with fisheries, and the public. This has brought fisheries management closer to the Sustainable Development framework, established to protect the ability of future generations to meet their needs. How marine ecosystem goods and services are to be sustainably developed depends on ecological, economic and social considerations, which have to be agreed to by all stakeholders. Communication, knowledge transfer, and meaningful consultation are thus crucial steps to ensure societal engagement in the new framework provided by the Ecosystem Approach.

9.4.1 Stakeholder engagement in resource management

If we want broad stakeholder engagement in resource management, we need to develop tools which can be understood by a broad section of our communities. There are different ways of ensuring this, as illustrated by two particular examples. On the one hand, we have the management of European fisheries, a classic example of top-down management with limited societal participation. A reduced number of individuals (Ministers of fisheries of the member states) negotiate decisions based on European Commission advice, which is in turn based on (but not necessarily follows) scientific advice from the International Council for the Exploration of the Sea (ICES). Stakeholder participation, be it from industry, conservation bodies, or public, has traditionally not been formalized and thus left to the informal contacts inside each member country. Stakeholder participation has been increased recently through the creation of Regional Advisory Councils (RAC), which is to advise the Commission on negotiated views based on the different position of stakeholder groups (Astorkiza *et al.* 2006). In contrast, we have the application of the Ecosystem Approach in South Africa, an area of long-term dedicated ecosystem research, substantial stakeholder involvement in management (Butterworth *et al.* 1997), and a firm commitment to EAF implementation (Shannon *et al.* 2004a, 2006a; Cochrane *et al.* 2007). As presented earlier, the latter included focused workshops with multiple stakeholders, providing preliminary lists of broad

operational objectives linked to management indicators (Cochrane *et al.* 2007; Nel *et al.* 2007). The second step is the development of tools to monitor performance in these fisheries (Paterson *et al.* 2007). Such tools, designed to structure discussions involving multiple, often conflicting objectives, are likely to lead to generally more acceptable solutions to stakeholders. Comparing prototype expert systems relevant to the management of small pelagic fisheries in the southern Benguela, Jarre *et al.* (2008) highlight that knowledge-based systems are worth developing to (1) synthesize complex problems in a logical and transparent framework, (2) help scientists to apply science to transdisciplinary issues, and (3) represent vehicles for delivering state-of-the-art science to users (Jarre *et al.* 2006, 2008). The conclusion to be drawn is that to implement EAF effectively, the engagement of multiple stakeholders in a collaborative process is essential, both to define the problems as well as to find solutions that will be transparent, credible, and owned by all. The development of this engagement, however, is specific to each resource and situation.

9.4.2 Communication tools: ecosystem status reports

A novel way to increase broad communication on marine ecosystem issues has been the publication of ecosystem status reports in various parts of the world, summarizing the status and trends of the marine ecosystem. Such reports are useful to advisory bodies and management agencies for communication and education purposes, even if the information is not formally incorporated in management systems. There are several examples of such assessment reports. On both its Atlantic and Pacific coasts, Canada produces State of the Ocean and Ecosystem reports (DFO 2003, 2007a), while the Alaska Fisheries Science Center includes an Ecosystem Considerations appendix (Boldt 2006) to its stock assessments which documents the state of the Gulf of Alaska ecosystem. In the United States, the National Oceanic and Atmospheric Administration (NOAA) Office of Marine Ecosystem Studies develops suites of indicators of changing ecosystem conditions based on a framework of ecosystem measurements, socio-economics, and governance of Large

Marine Ecosystems (LMEs). An important spin-off of the latter is that selected suites of indicators can be introduced into global observation and monitoring efforts through the Global Environment Observing System of Systems (GEOSS) and the Global Ocean Observing System (GOOS), linking monitoring to management needs. International scientific bodies also provide regional assessment reports. In 2004, the North Pacific Marine Science Organisation (PICES) published the first status report on the marine ecosystems of the North Pacific to identify, describe, and integrate observations of change in the North Pacific Ocean (PICES 2004; Box 9.5), with substantial input from GLOBEC-PICES research in the region. It reviewed climatic, oceanographic, and fisheries conditions for all major regions in the North Pacific, with a focus on 1999–2003, and identified some of the critical factors causing changes in these ecosystems. ICES and several organizations in the ICES area also produce reports of diverse periodicity and focus: the European Seas Quality Status Report (http://www.ices.dk/reports/germanqsr/23222_ICES_Report_samme.pdf), the Sir Alister Hardy Foundation for Ocean Science (http://www.sahfos.org) Ecological Status Reports, or the European Environment Agency assessment of environmental indicators (http://www.themes.eea.eu.int) are particular examples. To summarize the information on these reports and facilitate the provision of scientific or management advice, two approaches have been developed: a report card approach, based on matrices of ecosystem indicators (King *et al.* 2001; King 2005), and a traffic lights approach, which uses red, amber, and green lights to indicate qualitative levels of any indicator, and which has been suggested as a framework for management actions in response to multiple fishery limit reference points (Caddy 1999). The 2005 United Nations General Assembly endorsed the need for a regular process for global reporting and assessment of the state of the marine environment, including socio-economic aspects, based on existing regional mechanisms. Recently, the start-up phase of this process, the Assessment of Assessments, was launched as a preparatory stage towards the establishment of the Regular Process. The organizational arrangement for such AoA includes an Ad Hoc Steering Group to oversee its execution, two UN agencies to co-lead the process (UNEP and IOC), and a Group of Experts to undertake the Assessment of Assessments (see more information in http://www.unga-regular-process.org). The AoA remit includes:

- To cover relevant existing marine and coastal environmental assessments, including their ecological, social and economic aspects, at global and regional levels, complemented by available selected national assessments with broad geographical representation.
- To be implemented through a critical analysis of the assessments in order to evaluate their scientific credibility, policy relevance, legitimacy, and usefulness, in particular by identifying best practices and approaches, thematic and geographic assessment gaps and needs, uncertainties in scientific knowledge, data gaps and research needs, networking and capacity-building needs in developing countries and countries with economies in transition and a framework, and options for the regular process, based upon current relevant assessment process and practices.

While in general, regional and global reports increase awareness and facilitate understanding of the status of marine ecosystems, formalized procedures need be implemented if these are to be taken seriously in management. For that to happen, the quantitative relationship between ecosystem indicators and the stock(s) to be managed (in an EAF context) would need to be estimated. Simulations would be required to examine the extent to which uncertainties can be estimated and to assess the effect of management decisions. Ecosystem indicators with demonstrated quantitative relationships to the stock(s) in question, in contrast to those with qualitative relationships, are likely to provide direct quantitative assessment of the extent to which a management decision should be modified (Punt *et al.* 2001). This will be an important future development area in an EAF.

9.4.3 Future observational and scientific needs

If carried out successfully, the Ecosystem Approach process will highlight areas of uncertainty and show where further research is required, thus providing feedback to GLOBEC-like ecosystem research projects. It will identify the priority research needs

Box 9.5 Ecosystem status reports

An ecosystem approach to fisheries management accounts for ecosystem processes when providing management advice. Fundamental to this approach is the assessment of the state of the ecosystem, that is, knowledge of the status of the ecosystem. In an ecosystem management strategy evaluation context, ecosystem objectives are identified and are associated with quantified performance measures. These performance measures can be ecosystem indicators, and are included in an ecosystem assessment or status report. There are examples of ecosystem status reports that are typically produced annually. For groundfish stock assessments conducted for the Gulf of Alaska by the National Marine Fisheries Service, an appendix on Ecosystem Considerations (e.g. Boldt 2006) is produced annually. This appendix is a review of the current knowledge of ecosystem change or trends and encompasses ecosystem status indicators of the physical environment and of ecosystem or community indicators and also ecosystem-based management indices and information. Indicators presented include large-scale climate indices (e.g. Pacific Decadal Oscillation Index; Southern Oscillation Index), regional climate, and oceanography (e.g. Bering Sea ice cover; sea surface temperature); lower trophic level productivity (e.g. zooplankton biomass); invertebrate and fish abundance (e.g. species catch per unit effort from research surveys); and higher trophic level productivity (e.g. seabird breeding success). Other annual ecosystem status reports are simply summaries of the current physical and biological state of the marine ecosystems, for example the State of the Ocean reports produced by Fisheries and Oceans Canada (e.g. DFO 2007a,b).

The above examples provide executive summaries, but overall communication of ecosystem changes is via time series graphs and detailed descriptions, and for the most part is directed to the scientific community or resource managers. There are ecosystem status reports which focus on communicating observations in conceptual, graphic representations, directed to a broader audience which would include non-scientists such as policy makers, resource managers, and the general public. The State of Washington's Puget Sound Action Team periodically produces an ecosystem status report which presents over 24 environmental indicators that assess the status of marine life, habitats, water quality, and climate (PSAT 2007a,b). Detailed descriptions of each indicator are provided along with time series plots, however all indicators are summarized in a 'report card' in which the status of each indicator is represented by a dot on a gradient scale that moves from poor, critical, fair, good to excellent and each dot is associated with a trend arrow that indicates a negative trend, a positive trend, or no trend (no arrow). Another example of an ecosystem status report that attempts to synthesize the detailed scientific information into a simplified representation (Box 9.5, Table 1) is the Report of the Study Group on Fisheries and Ecosystem Responses to Recent Regime Shifts produced by the North Pacific Marine Science Organization (King 2005). That report looked at climate indices, physical oceanography, lower trophic level productivity, and invertebrate, fish and higher trophic level population dynamics for seven ecosystems in the North Pacific. The overall intent was to assess each parameter for previous known regime periods and to provide advice on the likelihood that a change in the state of the ecosystems occurred in 1998 (i.e. a regime shift year).

Jacquelynne R. King

Box 9.5, Table 1 An example of a synthesis of several ecosystem parameters used in an ecosystem status report intended to summarize a large number of time series to communicate to a broad audience: selected basin-wide climate-ocean indices and of selected physical oceanographic and fish population parameters in the California Current System. For each parameter, the extremes of the ranges of values are represented by ● and ○ and are explained in the comments column. In all cases, the symbol ● indicates moderate values, ↑ indicates a period of increasing trend, ↓ indicates a period of decreasing trend, and ⊕ indicates a period of variability in the parameter, that is, no apparent trend or persistent pattern. (Modified from table 2.1 of King 2005.)

Climate Ocean Indices

Parameter	Length of time series	1948–76	1977–88	1989–97	1998	1999	2000	2001	2002	2003	2004	Comments
Pacific Decadal Oscillation (winter)	1950–2004	●	○	○	○	◐	◐	◐	◐	○		● = negative ○ = positive
Victoria pattern (winter)	1950–2004	⊕	◐	●	◐	○	○	◐	○	○		● = negative ○ = positive
Arctic Oscillation	1950–2001	●	●	○	○	⊕	⊕	⊕	○			● = negative ○ = positive
Northern Oscillation Index	1950–2003	○	●	●	●	○	○	○	○	●		● = negative ○ = positive

California Current System

Parameter	Length of time series	1948–76	1977–88	1989–97	1998	1999	2000	2001	2002	2003	2004	Comments
Physical Oceanography												
Sea surface temperature	1900–2004	●	○	○	○	●	●	●	○	◐	○	● = cool; ○ = warm
Stratification intensity	1950–2003	●	↑	○	⊕	●	◐	◐	●	◐	◐	● = weak; ○ = strong
Upwelling strength	1946–2004		○	●	⊕	●	○	◐	●	●		● = weak; ○ = strong
Fish												
Coho salmon–British Columbia	1980–2003	○	○	↓	●	●	●	●	●	●		● = low abundance; ○ = high
Columbia River salmon	1940–2004	⊕	○	●	●	●	○	○	○	○		● = low returns; ○ = high
Pacific hake biomass	1948–2004	↑	○	↑	●	●	●	●	●	◐	◐	● = low; ○ = high

for fishery management and assist in guiding research investment.

Long-term monitoring of managed ecosystems, in particular, has to be an integral part of the performance evaluation of any management strategy, and is an area where ecosystem research could provide key information of management use. As technology develops, the monitoring of ocean systems is becoming more global; there will be a need to integrate spatial information on ocean processes and species' distribution into ecosystem models and ultimately into management strategies. Satellite remotely sensed oceanographic data is now being used to develop large-scale ecosystem indicators (Polovina and Howell 2005). Large-scale monitoring programmes, such as the international Argo float programme, which provides continuous and near real-time data on the physical properties of the upper ocean based on thousands of drifting profile floats, will provide a massive data set for management use. Developments in satellite and acoustic tags provide new opportunities to monitor species' associations with oceanic conditions or species' migratory behaviour at higher temporal and spatial resolutions than was previously possible. Several large-scale tagging programmes exist using either satellite tags (e.g. Tagging of Pacific Predators programme) or acoustic tags coupled with arrays of acoustic receivers on the ocean bottom (e.g. Pacific Ocean Shelf Tracking programme and the Ocean Tracking Network). These types of programmes provide information on physical parameters and detailed information on species' behaviour, providing the opportunity to address questions that relate to large-scale changes in ocean processes, to downscale large-scale changes to the spatial resolution of regional ecosystems and to investigate species' associations to conditions or movement in and out of ecosystems. The challenge will be to integrate this information into management, but questions might encompass residency extent, identification of critical habitat, and impacts on species that occur at life stages that are outside of a management zone.

As we move towards ecosystem-based approaches, ecosystem indicators will play a larger role in management evaluation, as discussed earlier. A large amount of work has already been conducted to review the properties of indicators, the current status

of indicator usage and approaches to the evaluation of performance of indicators (Cury and Christensen 2005). One of the difficulties that will need to be addressed in the future is the appropriate level of reporting and assessment given the multiple roles required (e.g. monitoring ecosystem status, impacts of human activities and success of management strategies). Adequate evaluation frameworks for selecting a suite of ecosystem indicators, and selection criteria, are needed (Rochet and Rice 2005).

Finally, ecosystem science has underpinned the development of the EAF through an increasingly strong multidisciplinary research strategy based on a combination of field studies, monitoring programmes and modelling efforts, and an effective use of comparative studies to extract broader lessons. Improvements in the effectiveness of EAF should result from the development of coupled climate-physical-biological *and social* models (Perry *et al.*, Chapter 8, this volume), demanding a broadening of the research agenda. Further, ideally an EAF strategy needs to understand the interactive and bidirectional nature of drivers of change, impacts, and responses between marine environments and human communities, recognizing humans as part of the ecosystem. The full incorporation of the human dimension in ecosystem research strategies recognizes the need to share ecosystem use and conservation with potentially conflicting users/stakeholders. Adequate tools to confront trade-offs, set broad objectives, specify societal preferences, etc. need to be developed in parallel to our natural science progress. Making the connection between ecology, economics, and human societies is an essential component of this process (Farber *et al.* 2006).

9.5 Summary

The management of marine fisheries is evolving rapidly in response to widespread evidence of commercial fish stocks overexploitation, climate-change impacts, and the need to align fisheries with the management of other uses of the marine ecosystem. In this contribution, we have reviewed this evolution, placing particular emphasis on the contribution of ecosystem science and scientific advice to management. In recent decades, it has been appreciated that fisheries management has not taken sufficient account of environmental and ecosystem issues. This was addressed in the 2001 Reykjavik Declaration, where signatory nations agreed to '*work on incorporating ecosystem considerations*' into management. A year later, the 2002 World Summit on Sustainable Development (WSSD) recognized that the management needs for the oceans have changed, needing integration of ocean management activities across sectors, and responding to the necessity of conservation objectives for the collective ocean use.

The ecosystem science community has contributed to this change in vision through increasingly focused research quantifying the role of climate variability and change on the dynamics of single species as well as ecosystem components, the role of habitat change in determining ecosystem productivity, the interactivity of target and non-target resources, and on patterns and cyclic fluctuations of ecosystem productivity, among others. This research has highlighted that good fisheries management alone may not be able to recover a depleted stock under unfavourable environmental conditions, and that poor management can prevent the recovery of a stock even if environmental conditions are favourable. As a result of interactions between the scientific, management, conservation and governance communities, a new framework for the implementation of an incremental, adaptive, and geographically specific ecosystem approach to fisheries has emerged. This approach is guided by the need to manage ecosystems for their long-term benefit, taking into consideration their multiple uses, the cumulative and long-term impacts of such uses, and the variable and sometimes non-stochastic role of the environment in supporting ecosystem processes. This approach is based on precautionary principles, taking full cognizance of uncertainty and risk, and supported by sound ecological understanding.

We have described how an Ecosystem Approach to Fisheries framework would require multiple-scale objective setting based on societal choices, including ecological, economic, and social considerations. Operational objectives need to be established, requiring the identification of indicators and reference points to mitigate sector impacts. Decision support and performance evaluation rules need to be set, including all uncertainties. We specifically consider the use of Management Strategy Evaluations (MSE), where the performance of alternative estimation and monitoring regimes, reference points, and management options can be evaluated via simulation tools before implementation. This approach can assist in resolving conflicting objectives and addressing management risks, and allows for the analysis of alternative mitigation measures with respect to the trade-offs. In this context, future GLOBEC-inspired ecosystem models can provide hypotheses testing in the MSE arena.

Ecosystem approaches to management worldwide require new and adequate communication and consultation to ensure full societal engagement in management supported by coordinated assessments of the state of the marine environment. In determining future steps, observational and scientific needs are identified, including the need to make better use of large-scale ecosystem monitoring tools, and the need to combine biological, economic, and social aspects of fisheries management in single multidisciplinary research efforts.

It is concluded that ecosystem management and ecosystem science have followed parallel concepts, but that increasing interaction between the two is essential to achieve marine resource sustainability. The EAF provides a foundation to operationalize this interaction, a new paradigm that weaves ecological, economic, and social threats into a tapestry of ecosystem health and sustainability (Lackey 1999). However, considerable work remains to be done to ensure an adequate balance between conservation and resource use, underpinned by appropriate and focused scientific research.

Acknowledgements

The comments and suggestions of two anonymous reviewers are greatly appreciated. They contributed appreciably to the final version.

PART IV

A way forward

The final section was the most challenging to write. Chapter 10 discusses the difficulties, approach, and state of the art in projecting marine ecosystem responses to future global change. It takes scenarios from the Intergovernmental Panel on Climate Change and projects likely future ocean changes in temperature, nutrient supply, plankton transport, acidification, etc. It then takes this further to postulate likely ecosystem responses to these scenarios—a completely new venture. This includes projections of future primary production, zooplankton, and higher trophic levels. The limitations of the statistical and scenario approaches are discussed. The authors then look towards the future with better understanding and modelling of mechanisms: physical processes, nutrients, phytoplankton, zooplankton, and higher trophic levels. They also discuss end-to-end food web models and future management approaches. Chapter 11 concludes the book by summarizing the major contributions of the Global Ocean Ecosystem Dynamics (GLOBEC) approach to marine ecosystem research. The incorporation of physical oceanography into biological studies, the links between modellers and observational scientists and the increasing interactivity between the natural and social sciences, have all resulted in new insights that are highlighted under various general themes. These themes are projected into the future of global change and an ecosystem approach to fisheries management.

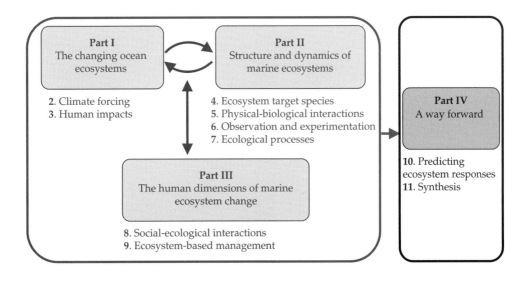

Part I
The changing ocean ecosystems

Part II
Structure and dynamics of marine ecosystems

2. Climate forcing
3. Human impacts

4. Ecosystem target species
5. Physical-biological interactions
6. Observation and experimentation
7. Ecological processes

Part III
The human dimensions of marine ecosystem change

8. Social-ecological interactions
9. Ecosystem-based management

Part IV
A way forward

10. Predicting ecosystem responses
11. Synthesis

CHAPTER 10

Ocean ecosystem responses to future global change scenarios: a way forward

Shin-ichi Ito, Kenneth A. Rose, Arthur J. Miller, Ken Drinkwater, Keith Brander, James E. Overland, Svein Sundby, Enrique Curchitser, James W. Hurrell, and Yasuhiro Yamanaka

10.1 Introduction

The overall aim of GLOBEC was 'To advance our understanding of the structure and functioning of the global ocean ecosystem, its major subsystems, and its response to physical forcing so that a capability can be developed to forecast the responses of the marine ecosystem to global change'. GLOBEC specified four objectives, and objective 3 was 'To determine the impacts of global change on stock dynamics using coupled physical, biological, and chemical models linked to appropriate observation systems and to develop the capability to project future impacts'. During the GLOBEC era, earth observational networks were developed such as the Global Climate Observing System (GCOS), which includes the Global Ocean Observing System (GOOS). Although imperfect, this global observational network is providing an unprecedented view of climate change in the earth system, and has increased our understanding tremendously over the past several decades. An increasing number of independent observations of the atmosphere, land, cryosphere, and ocean are providing a consistent picture of a warming world. Such multiple lines of evidence, the physical consistency among them, and the consistency of findings among multiple, independent analyses, form the basis for the iconic phrase of the observations chapter in the Fourth Assessment Report of the Intergovernmental Panel on Climate Change (IPCC-AR4, IPCC 2007a)

that the 'warming of the climate system is unequivocal'. Moreover, the evidence is strong that, especially in recent decades, human activities have contributed to the global warming. The IPCC-AR4 additionally cautioned that further warming and changes in the global climate system will very likely emerge over the next century.

The climate changes anticipated during the twenty-first century have the potential to greatly affect marine ecosystems. A major challenge facing the scientific community is to develop modelling and data analysis approaches for determining how climate change will affect the structure and functioning of marine ecosystems. During the GLOBEC era, our understanding of ecosystem structure and dynamics has improved greatly (deYoung et al., Chapter 5; Moloney et al., Chapter 7, both this volume), and new ecosystem modelling approaches have been developed and existing methods improved (see deYoung et al., Chapter 5, this volume). As we look forward, the next step is to use the knowledge gained from GLOBEC as a foundation, as we continue to develop data collection and modelling tools that can make sufficiently confident projections of marine ecosystem responses to future global climate change.

In this chapter, we summarize the available evidence for recent changes in climate effects in the oceans, and the status of our ability to project ecosystem responses to likely future global change.

We first present the evidence for changes in the physical and chemical properties of the oceans, including changes in water temperature, nutrient supply, mixing and circulation, micronutrient supply, transport of plankton, acidification, and sea-level rise. We then discuss the evidence for consequent responses of the marine ecosystem to the documented changes in the oceans, organized by trophic level (primary production, zooplankton, and higher trophic levels). Whenever possible, for each trophic level, we include the results of published examples of future projections of ecosystem

responses. With this information as the basis, we conclude with a discussion of our vision of the next steps that are needed to develop models capable of projecting ecosystem responses to global change.

To develop future climate projections, assumptions must be made on the levels of population increase and the rates of development (emission scenarios) that, in turn, determine the rate of CO_2 build-up in the atmosphere. Many examples of future projections cited in this chapter used emission scenarios taken from the IPCC (IPCC 2000), which are briefly described in Fig. 10.1. One exception is an

Projections of surface temperatures

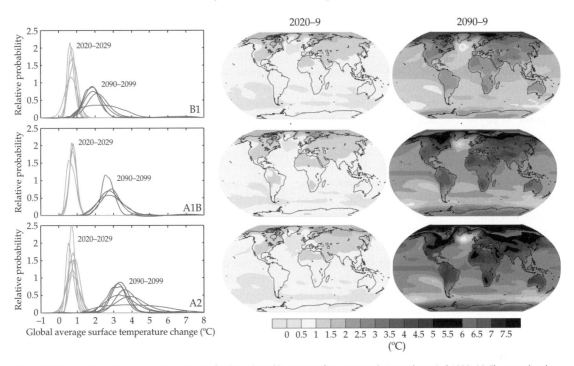

Figure 10.1 Projected surface temperature changes for the early and late twenty-first century relative to the period 1980–99. The central and right panels show the atmosphere-ocean coupled general circulation model (AOGCM) multi-model average projections for B1 (*top*), A1B (*middle*) and A2 (*bottom*) IPCC Special Report on Emissions Scenarios (SRES) scenarios averaged over the decades 2020–9 (*centre*) and 2090–9 (*right*). IPCC has assumed four storylines that encompass a marked portion of the underlying uncertainties in the main driving forces (demographic change, economic development, and technological change). Four qualitative storylines describe four sets of scenarios called families: A1 (world market), A2 (provincial enterprise), B1 (global sustainability), and B2 (local stewardship). The A1 family was divided into three groups characterizing alternative developments of energy technologies: A1FI (fossil fuel intensive), A1B (balanced), and A1T (predominantly non-fossil fuel). The scenario used historical data for atmospheric CO_2 between 1860 and 2000, and then assumed atmospheric CO_2 concentrations will reach 850 ppm (A2), 720 ppm (A1B), 500 ppm (B1) in the year 2100, respectively. The left panels show the corresponding uncertainties as the relative probabilities of estimated global average warming from several different AOGCM and Earth System Model of Intermediate Complexity studies for the same periods. (From IPCC 2007a). (See Plate 22).

earlier IPCC emission scenario, IS92, which was used by several cited studies. IS92a begins in 1990 and assumes an effective CO_2 concentration increase of 1% per year. IS92a differs from the IPCC Special Report on Emissions Scenarios (SRES) scenarios (Fig. 10.1) because it starts from 1990 and the many economic and political issues during the 1990s are not incorporated. IS92a can be roughly thought of as resulting in CO_2 concentrations in 2100 similar to those of A1B, although the emission scenario is very different and intermediate between the SRES B2 and A2 scenarios (IPCC 2000).

We note that we use the terms global climate change and global change throughout this chapter to refer to anthropogenic-induced climate changes. In many instances, and especially with the examination of historical data, we emphasize that the changes in the properties and biota of the oceans reflect various mixtures of local, regional, and global variation and trends in climate. We cannot attribute all of the documented changes or responses to global-scale phenomena, nor can we infer the role of anthropogenic versus natural influences (e.g., see Drinkwater *et al.*, Chapter 2, this volume, for a discussion of climate-induced effects on ocean ecosystems).

10.2 Emergence of global changes in the ocean environments and projected future ocean conditions

There has been extensive analysis of the physics and chemistry of the oceans to detect and quantify global change effects on ocean properties. We briefly summarize the changes in the ocean properties that have already emerged as signals of global change and that have clear links to ecosystem responses, and how these properties are projected to further change in the future. The properties we focus on are: water temperature, nutrient supply, mixing and circulation, micronutrient supply, transport of plankton, acidification, and sea-level rise.

10.2.1 Changes in sea water temperature

The global atmospheric concentrations of carbon dioxide, methane, and nitrous oxide have increased markedly as a result of human activities since 1750, owing mainly to combustion of fossil fuels and changes in land use. Together, the combined radiative forcing from these three greenhouse gases is a 2.3 watts per square metre (W m^{-2}) increase relative to 1750, which dominates the increase in the total net anthropogenic forcing (1.6 W m^{-2}) (IPCC 2007a). The total net anthropogenic forcing includes contributions from aerosols (a negative forcing) and several other sources, such as tropospheric ozone and halocarbons. Evidence for their important role is that climate model simulations, which include changes in greenhouse gases, better match the historical increase observed in water temperature than simulations without the increasing greenhouse gases (e.g. Barnett *et al.* 2005).

The increased heat is absorbed by the continents, the lower atmosphere, and the ocean, and is available to melt the glaciers, ice sheets on land, and sea ice. Since 1979, overall ocean surface water temperatures (SST) have warmed 0.133 ± 0.047°C (Rayner *et al.* 2006), and the warming is evident, to varying degrees, at most latitudes and in most ocean basins. A large part of the change in ocean heat content during the past 50 years has occurred in the upper 700 m of the world ocean, although warming has been detected to a depth of 3000 m (Levitus *et al.* 2005). The warming is not uniform; notable exceptions are the cooling observed in mid-latitude North Pacific and in some areas of the high-latitude North Atlantic. Current evidence indicates that the ocean is absorbing most of the heat being added to the climate system (Levitus *et al.* 2005). The global averaged surface temperature is projected to continue to increase in the future (Fig. 10.1, IPCC 2007a), although the magnitude depends on the specific emission scenario assumed.

SSTs in the oceans also show variation on the scale of years to decades related to the *El Niño*-Southern Oscillation (ENSO) and the Pacific Decadal Oscillation (PDO) for the Pacific Ocean (see Drinkwater *et al.*, Chapter 2, this volume). ENSO events are typically 2–3 years in duration that encompass the *El Niño* and *La Niña* phases. During the *La Niña* phase, the trade winds push warm water to the western equatorial Pacific. During the *El Niño* phase, the trade winds relax permitting the warm waters to propagate to the eastern Pacific.

Atmosphere-ocean coupled general circulation models (AOGCMs) project that ENSO events will continue to occur in the twenty-first century; however, general statements about how global change will affect the frequency and amplitude of ENSO events are difficult to make because the projections differ among the models.

There is also evidence for global change effects on SST variability reflected in changes in the PDO index. The PDO features an SST anomaly pattern nearly symmetric about the equator with larger amplitudes in the middle than in the low latitudes and an opposite signed anomaly along the eastern rim of the Pacific (Mantua *et al*. 1997). Overland and Wang (2007a) identified those AOGCMs that were able to capture the spatial pattern of the PDO (i.e. whose spatial correlation with the observed PDO exceeded 0.7) and compared their future SST projections. For example, the simulated spatial patterns of the first leading mode of North Pacific SST during the 2001–100 period under the A1B scenario showed a rather uniform and single-signed spatial pattern with linear trends in all of the examined models. The second leading mode of the simulated North Pacific SST during the 2001–100 (which corresponds to PDO) showed a pattern with projected amplitudes that did not change significantly. These projections also indicated that the change in the mean background SST field under anthropogenic influences will surpass the magnitude of natural variability in the North Pacific in less than 50 years (Fig. 10.2).

A sea surface height mode of variability is the North Pacific Gyre Oscillation (NPGO), which is likely to become more energetic under global warming (Di Lorenzo *et al*. 2008). The NPGO represents the second principal component of sea-level height (the breathing mode of Cummins and Freeland 2007; see also the Victoria mode of Bond *et al*. 2003). The NPGO is strongly correlated with salinity, nutrient supply, and biological variables in parts of the eastern North Pacific (Di Lorenzo *et al*. 2008). A more energetic NPGO could result in a shift in conditions in the eastern Pacific more associated with the NPGO than with the PDO.

Changes in temperature and precipitation are expected to cause a reduction in the meridional overturning circulation (MOC), with implications for general ocean circulation patterns and SST, especially in the North Atlantic (Clark *et al*. 2002). The latitudinal difference of ocean density gradients and the large-scale winds drive the MOC, together with the sinking of water in the high latitudes and broad upwelling in other regions. The MOC is connected to the Southern Ocean and the Indonesian Throughflow, yielding a global-scale pathway (Schmitz 1996), which has been called the Great Ocean Conveyor Belt (Broecker 1991). Under global warming conditions, there will be less cooling and increased precipitation at the high latitudes, which will weaken the MOC and affect broad oceanic areas. Schmittner *et al*. (2005), using 28 projections from nine different coupled global climate models under the A1B scenario, projected a gradual

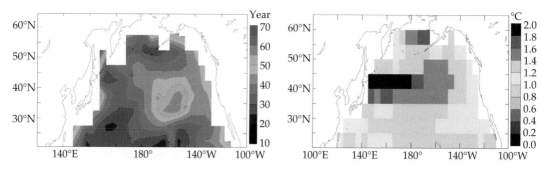

Figure 10.2 *Left*: Estimated year (+2000) when the net warming exceeds the magnitude of natural variability. This year for each location was defined by dividing twice the standard deviation of the observed sea surface temperature (SST) by the modelled temperature trend. This threshold is generally reached for the North Pacific before the middle of the twenty-first century. *Right*: Projected winter SST change for 2040–9 minus 1980–99. Changes are in the range of 1–2°C. (From Overland and Wang 2007a). (See Plate 23).

weakening of the North Atlantic MOC by 25% (±25%) through to 2100. The weakening of the MOC could cool the North American and European summer climate and SST in the North Atlantic (Sutton and Hodson 2005), and decrease the amplitude of a multi-decadal (~60 years) pattern in the SST covering the entire North Atlantic Ocean called the Atlantic Multidecadal Oscillation (AMO).

10.2.2 Changes in nutrient supply systems

Changes in mixing and circulation due to global warming will affect the nutrient supply to the euphotic zone, and can have major consequences for primary production and higher trophic levels. Warming increases stratification, thereby reducing vertical mixing by wind and tide, the timing and intensity of upwelling in coastal regions, and the strength of the MOC.

Several analyses suggest a general reduction of vertical mixing in recent decades over a wide region of the North Pacific. For example, surface layer stratification increased in intensity in the Oyashio region (subarctic north-western Pacific) from the 1960s to the 1990s (Ono et al. 2002). The apparent oxygen utilization (AOU) in the subsurface layer also increased during the same time period (e.g. Watanabe et al. 2001; Emerson et al. 2004), indicating reduced mixing between surface and subsurface layers. In the subtropical North Pacific, water density and nutrient concentrations decreased in the wintertime mixed layer (Watanabe et al. 2005).

In ocean areas associated with sea ice, melting water from sea ice could increase stratification due to the freshening of the oceanic waters and reduce nutrient supply into the surface layer. Sea-ice melting is one of the most visible results of global warming. The annual mean ice extent in the Arctic showed a significant decreasing trend in recent years, while sea-ice extent in the Antarctic showed a smaller decreasing trend (Comiso 2003; Stroeve et al. 2007). Arctic sea-ice extent at the end of the melt season in September has exhibited an especially sharp decline from 1953 to 2006. Although, future projections also indicate a decrease of Arctic ice cover (IPCC 2007a), the observational decline is much faster than model projections (Stroeve et al. 2007). Further global

warming will induce sea-ice melting and lead to surface freshening and increased stratification. A likely consequence of increased stratification is a reduction in nutrient supply to the surface layer. This effect will be especially strong in areas downstream of Antarctica and the Arctic seas.

Global warming will also affect stratification by changing the salinity of oceanic waters via changes in precipitation and the melting of water from other cryospheric (frozen water) sources besides sea ice. Changes in freshwater inputs affect salinity, which is a driver of vertical stratification. In high-latitude regions, salinity is the dominant source of vertical stratification (Carmack 2007). Salinity data in the ocean are more limited than temperature data. Boyer et al. (2005) attempted to calculate linear trends in salinity, averaged for 5-year periods from 1955–9 through 1994–8. Salinity decreased in the high latitudes and increased in the subtropics. Most of the Pacific is freshening with the exception of the subtropical South Pacific. Future projections under global warming scenarios by various AOGCMs suggest increased precipitation in the high latitudes and reduced precipitation in most subtropical land regions, which would tend to decrease surface salinity in high latitudes and increase surface salinity in subtropical areas (IPCC 2007a). The decrease of surface salinity in the high latitudes will increase the stratification and reduce the nutrient supply to the surface layer. Exceptions include those regions where advection carries the influence from one area to the other; for example, the Barents Sea where salinities are expected to increase due to advection of high-salinity waters from the southern regions (Bethke et al. 2006)

Upwelling is another important nutrient supply mechanism in coastal regions, especially in the eastern boundary of major ocean basins, such as the Benguela, Canary, Humboldt, and California Current systems. These areas support very high biological productivity and large populations of small pelagic fish, and were the focus of the Small Pelagic and Climate Change (SPACC) programme within GLOBEC (Fréon et al. 2009). Bakun (1990) proposed that, in eastern boundary regions, stronger warming on land compared to the sea leads to the enhancement of upwelling-favourable winds. He showed an increasing trend in upwelling in the eastern boundary regions, which was further supported by analyses of recent data for the California, Humboldt, and Benguela Current systems (Mendelssohn and Schwing 2002; McGregor *et al.* 2007). Auad *et al.* (2006) projected how global warming would affect upwelling in the California Current system using a high-resolution model forced by atmospheric fields derived from a coarse-resolution AOGCM. Their projection showed about a 30% increase in near-surface upwelling velocities during the key month of April, which cooled the upper-ocean (<70 m depth) region. The cooling effect was partially offset by heat fluxes from an unusually warm atmosphere during the time period of simulation, so that only a net mild cooling effect was projected.

However, other analyses provide contradictory results, suggesting that global warming would reduce upwelling in coastal systems. Diffenbaugh (2005) used an ensemble approach with coupled climate models and projected wind fields, and reported that some components of the ensemble showed a potential relaxation of the strength and variability of upwelling-favourable (equatorward) wind forcing in all four of the eastern boundary regions. Vecchi *et al.* (2006) used an AOGCM and projected a weakening of the Walker Circulation in the tropical Pacific. They found that the projected slowdown of the Walker Circulation was consistent with the slowdown observed historically since the mid-nineteenth century. Their model projections suggested further slowdown of the tropical Walker Circulation under global warming, which leads to a reduction in the trade winds and a concomitant reduction in upwelling-favourable winds. Perhaps these studies can be reconciled if one considers that global warming effects on local wind systems enhance upwelling, while effects on large-scale wind systems act to decrease upwelling. Regardless of the net effect, both local- and large-scale wind effects would act to change the timing (seasonality) of the upwelling.

The expected weakening of the MOC can have major effects on the nutrient supply to the euphotic zone. The sinking water in high latitudes drives deep-water circulation, and the deep water is rich in nutrients because decomposition of organic material takes a long time. These nutrient-rich waters

then supply nutrients to the euphotic zone in broad regions by upwelling. A weakening of the circulation could lower nutrient supplies in broad areas of the ocean. Schmittner (2005) estimated that a disruption of the Atlantic MOC would lead to a halving of plankton biomass in the North Atlantic, and a reduction in primary productivity in many other regions.

Recently, the concept of nutrients fueling primary productivity has been expanded from the classical focus on nitrogen and phosphorus to include micronutrients such as iron (Behrenfeld *et al.* 1996), and it is uncertain how global warming will affect the supply of these micronutrients. For example, one of the major sources of iron to the open ocean is atmospheric transport of dust by winds, such as the mid-latitude westerlies. Mid-latitude westerly winds have strengthened in both hemispheres since the 1960s (Gillett *et al.* 2005), and these winds are projected to be enhanced under global warming conditions. How such changes in wind patterns will affect the supply of iron to the ocean is not clear. One possibility is that more iron will be transported to the oceans; however, the net effect of changes in wind-driven transport on iron fluxes to the oceans is uncertain. Similar arguments can be made for other micronutrients; the best we can say, at this point, is that their delivery rates to the oceans might change substantially under global warming.

10.2.3 Transport of plankton

The possible changes in water temperature, salinity, and atmospheric circulation (wind patterns) would have major effects on the horizontal circulation in the ocean. Changes in horizontal circulation will affect the advective transport of phytoplankton, zooplankton, and early life stages of fish. Sakamoto *et al.* (2005) calculated the linear trend of the dynamic sea surface height between 1965 and 2003 in the north-western Pacific and detected an intensification of the Kuroshio Extension. The magnitude of the intensification was relatively small, especially in the context of the large decadal-scale variability observed in the north-western Pacific. Future projections of the ocean circulation under global warming scenarios also indicate an acceleration of the Kuroshio and Kuroshio Extension, accompanied by an enhancement of eddy activity in the Kuroshio Extension region (Fig. 10.3).

Another example of possible changes in horizontal circulation is how global change might affect circulation in the North Atlantic. The NAO (North Atlantic Oscillation) index is used to characterize the winter climate variability and is a rough meas-ure for the intensity of the Icelandic Low, which influences the intensity of ocean circulation between the northern North Atlantic and the Arctic (Hurrell 1995). A high NAO Index (i.e. a strong Icelandic Low) increases the flux of warm Atlantic water into the Nordic Seas and simultaneously increases the flux of cold Arctic water into the Labrador causing

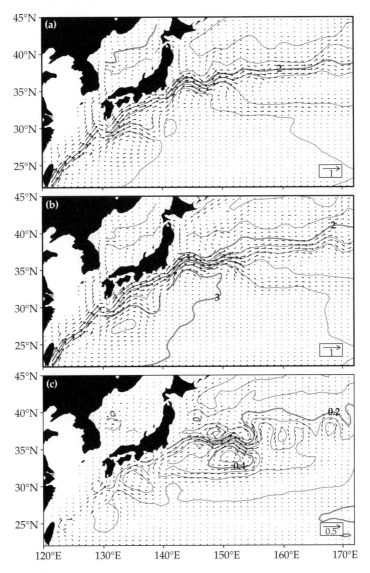

Figure 10.3 Long-term mean current velocities at 100 m depth (vectors, unit: m s^{-1}) and dynamic sea surface height (contours, unit: m) relative to 2048 m depth in (a) the simulation under pre-industrial conditions, (b) the simulation under double CO_2 conditions (1% increase of CO_2 concentration for 90 years), and (c) the difference between (b) and (a) (former minus latter). Contour intervals are 0.2 m in (a) and (b), and 0.05 m in (c). (Reprinted by permission of Macmillan Publishers Ltd: Geophysical Research Letters Sakamoto *et al.* Copyright 2005.)

opposite climate trends in the north-east Atlantic (e.g. the Nordic Seas) and the north-west Atlantic (e.g. the Labrador Sea) (Sundby and Drinkwater 2007). Historically, the NAO index has shown decadal-scale variation that creates temperature oscillations in the two North Atlantic regions with opposite phase. Miller *et al.* (2006) and Stephenson *et al.* (2006) used multiple AOGCMs to investigate how global warming would affect the NAO (or NAM: Northern Annular Mode). They found that none of the models projected a decrease in the NAO index and more than half of the models projected a positive increase in the NAO index; however, the magnitude of the increase was generally small. An increase in the NAO index potentially leads to an increase in the transport of warm Atlantic water and an associated influx of zooplankton into the Nordic Seas (Sundby 2000).

10.2.4 Acidification

The absorption of carbon in the ocean, including carbon from anthropogenic emissions, leads to increased acidification of oceanic waters. Over the period 1800 to 1994, the ocean was estimated to have absorbed 118 ± 19 petagrams (Pg) of carbon, or about 48% of the total emitted during the period (Sabine *et al.* 2004). Hydrolysis of CO_2 in water increases the hydrogen ion concentration, which can affect the ionic balance of many organisms and the ability of some organisms (e.g. corals and bivalves) to maintain their external calcium carbonate skeletons (Kurihara 2008; Vezina and Hoegh-Guldberg 2008).

Future projections using models of the ocean carbon cycle suggest that the Southern Ocean surface waters will become undersaturated with respect to aragonite, a metastable form of calcium carbonate, by the year 2050 (Orr *et al.* 2005). By 2100, this undersaturation could extend throughout the entire Southern Ocean and into the subarctic Pacific Ocean (Fig. 10.4). Ongoing studies show that the undersaturation in the Arctic Ocean may be even more severe than in the Southern Ocean.

Global-scale ocean acidification will also modify oceanic biogeochemical fluxes. Increased acidification reduces export fluxes related to $CaCO_3$ and hence weakens the strength of the biological carbon pump. A weakened biological pump is projected to increase oxidization of organic matter in shallow waters and can result in the expansion of hypoxic zones (Hofmann and Shellnhuber 2009).

10.2.5 Sea-level rise

Melting water from landlocked cryospheric fields and the thermal expansion of warmed sea water will contribute to a rise in sea level. Sea-level rise will influence the habitat quality and quantity of near-shore marine ecosystems and estuaries, which are nursery grounds for many higher trophic level species. Averaged sea-level rise by the thermal expansion was estimated as 0.33 mm year^{-1} during 1955 to 2003 (Antonov *et al.* 2005). The IPCC-AR4 (IPCC 2007a) suggested that global sea level will rise between 0.18 and 0.59 m during the twenty-first century. Thermal expansion is estimated to cause 70 to 75% of the rise in sea level; however, one of the major unknowns is what will happen to the Greenland ice sheet. The melting of the Greenland ice sheet would greatly increase sea-level rise. Indeed, melting at the edges of the Greenland ice sheet is presently more than compensated for by the accumulating mass in its interior regions due to increased precipitation (Rignot and Kanagaratnam 2006). Some investigators say that the Greenland ice sheet could respond to global warming on a timescale of hundreds of years, while other investigators suggest that the ice sheet could melt abruptly within this century (IPCC 2007a).

10.2.6 Feedback mechanisms

Changes in the physical or chemical properties of the oceans cause ecosystem responses that can, in turn, affect the physical or chemical changes. For example, phytoplankton can act to affect the distribution of heat within the water column by absorbing solar radiation in the surface layer. The heat absorption by phytoplankton influences both the mean and transient state of the equatorial climate (Murtugudde *et al.* 2002; Timmermann and Jin 2002), decadal variations of climate (Miller *et al.* 2003), and the global mean SST field (Frouin and Iacobellis 2002). Phytoplankton can also influence atmospheric radiation processes because they

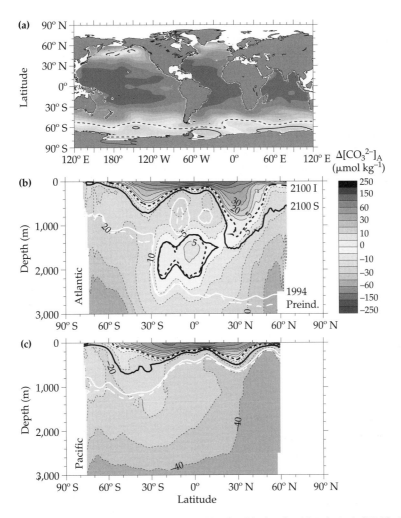

Figure 10.4 The aragonite saturation state in the year 2100 as indicated by $\Delta[CO_3^{2-}]_A$. The $\Delta[CO_3^{2-}]_A$ is the *in situ* $[CO_3^{2-}]$ minus that for aragonite-equilibrated sea water at the same salinity, temperature, and pressure. Shown are the OCMIP-2 median concentrations in the year 2100 under scenario IS92a: (a) surface map; (b) Atlantic; and (c) Pacific zonal averages. Thick lines indicate the aragonite saturation horizon in 1765 (pre-industrial; white dashed line), 1994 (white solid line) and 2100 (black solid line for S650; black dashed line for IS92a). Positive $\Delta[CO_3^{2-}]_A$ indicates supersaturation; negative $\Delta[CO_3^{2-}]_A$ indicates undersaturation. (After Orr *et al.* 2005. Reprinted by permission of Macmillan Publishers Ltd: Nature copyright 2005). (See Plate 24).

release dimethylsulfide (DMS), one of the major sources of cloud condensation nuclei over the ocean (Charlson *et al.* 1987). Another example is that absorbed carbon in the oceans can be transported to the subsurface layer by the sinking of particulate organic materials such as dead phytoplankton and zooplankton fecal pellets (Ducklow *et al.* 2001). Projecting ecosystem response to global warming may need to consider these and other feedbacks in order to ensure sufficient accuracy in the projected physical and chemical ocean properties.

10.3 Ecosystem responses to global change

Biota can respond in a variety of ways to the changes in the physical and chemical properties of the oceans that are expected under global warming. Analyses of ecosystem responses have their own uncertainties, and often use the outputs of the physical and biogeochemical models, and thus inherit the uncertainty associated with the physical and chemical property projections as well. High uncertainty in ecosystem responses results from the need to quantitatively represent how environmental conditions affect biota plus the ecological interactions that can substantially affect biological responses. Below, we summarize some of the available evidence for historical responses of oceanic ecosystems to variation in climate, and highlight some examples where ecosystem responses have been projected under future global change. We organize the results by trophic level (primary production, zooplankton and higher trophic levels), and to the extent possible, we relate ecosystem responses to changes in the physical and chemical properties discussed above.

10.3.1 Primary production

Primary production in the ocean will be strongly influenced by global warming through changes in the water temperature, nutrient supply, and transport. Understanding and projecting responses of primary production to global warming is critical because primary production acts as the starting point for much of the energy that is channelled into

biological pathways. Simple general statements about how primary production will respond to global warming are difficult because primary production, and the factors controlling it, vary spatially (e.g. latitude, upwelling versus mid-ocean, shelf seas versus deep-ocean). We must expect that the effects of climate change will therefore also vary spatially.

10.3.1.1 Emergence of primary production responses to global change

Several broad-scale analyses of ship-based observations have shown a decline in primary production in recent decades. Watanabe *et al.* (2005) noted a general decline in phosphate concentration in the subtropical region, and a decrease in net primary production of 0.26 gC m^{-2} year $^{-1}$ between winter and summer from 1971 to 2000. Ono *et al.* (2002) estimated an averaged decline in primary production of 0.51 ± 0.09 gC m^{-2} year^{-1} (corresponding to 0.9% of the primary production) in the Oyashio (subarctic) region (Fig. 10.5). Both trends were associated with a reduction of the wintertime

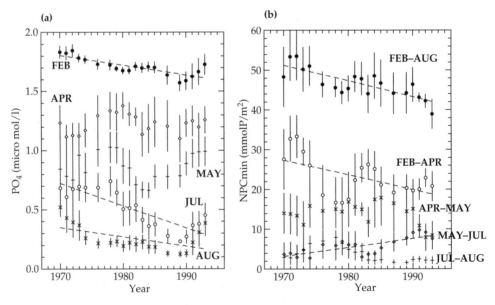

Figure 10.5 (a) Five-year running mean of monthly mean phosphate concentration within a mixed layer (μmol l^{-1}) for February, April, May, July, and August. (b) Five-year running mean of minimum net phosphate consumption within a mixed layer (mmol P m^{-2}) between the months: February to April, April to May, May to July, July to August, and the total (February to August). Regression lines are drawn only for the statistically significant case (95%). Error bars represent the 95% confidence limit of each average and are calculated from the original values rather than the 5-year running mean. (From Chiba *et al.* 2004 and redrawn from Ono *et al.* with permission of Macmillan Publishers Ltd: Geophysical Research Letters copyright 2002.)

mixed-layer depth, which was related to global change as a response to changes in the intensification of stratification (Ono *et al.* 2002; Chiba *et al.* 2004; Watanabe *et al.* 2005). Two caveats to these results are that Ono *et al.*'s analysis included phosphate only, even though nitrate and silicate are known to limit primary production in the Oyashio region, and the attribution of the changes to global warming in both analyses is tempered by the observed data being collected during certain PDO conditions that vary decadally and therefore the decrease in primary production may not purely reflect the effects of global warming.

Analyses of coastal regions using ocean colour satellite imagery suggest an increase in primary production over the past decade. Satellite imagery enables greater spatial and temporal coverage than ship-based observations; however, the imagery is for surface conditions and time series data are typically less than 10 years in duration. Gregg *et al.* (2005) used a 6 year (1998–2003) time series of Sea-viewing Wide Field-of-view Sensor (SeaWiFS) data, and detected a linear increase in ocean chlorophyll of 4.1% ($p < 0.05$), which corresponds to 0.00261 mg m^{-3} year^{-1}. However, this increase has occurred mostly in coastal regions where bottom depth is shallower than 200 m, whereas chlorophyll-*a* concentrations have decreased in the subtropical open ocean regions of both the Pacific and Atlantic. Gregg *et al.* (2005) showed that the decline in chlorophyll-*a* in the subtropics was associated with an increase in SST in the North and South Pacific and the South Atlantic. Behrenfeld *et al.* (2006) used a longer 10-year (1997–2006) SeaWiFS data set and found that chlorophyll and net primary production increased during 1997–9, corresponding to an *El Niño* to *La Niña* transition, and then subsequently decreased during 1999–2005. These changes occurred mainly in the permanently stratified regions of the open ocean (i.e. tropics and subtropics). Extrapolating these changes to the stratified areas of the ocean, results in changes in annual net primary production with an increase of 1.25 gC m^{-2} year^{-1} for the 1997–9 period and a decrease of 0.94 gC m^{-2} year^{-1} for the 1999–2005 period. Behrenfeld *et al.* (2006) related their estimated values to the Multivariate ENSO Index (MEI). The MEI (Wolter and Timlin 1998) is the

first leading mode in the tropical Pacific Ocean consisting of sea-level pressure, zonal and meridional components of the surface wind, SST, surface air temperature, and total cloudiness fraction of the sky. Behrenfeld *et al.* found that net primary production decreased with an increase in the MEI (warmer conditions), and that the MEI was positively correlated with the strength of stratification as indicated by the difference in water densities between the surface and the 200 m depth. These results suggest a possible mechanism whereby surface warming enhances stratification in the subtropical and tropical oceans, which then suppresses nutrient exchange, resulting in a decrease in net primary production. Recently, Polovina *et al.* (2008b) used a 9-year (1998–2006) SeaWiFS dataset to show an increasing trend in oligotrophic regions in the North and South Pacific and Atlantic (Fig. 10.6). In contrast, annual primary production in the Arctic increased between 2006 and 2007 with 30% of this increase attributable to decreased minimum summer ice extent and 70% due to a longer phytoplankton growing season (Arrigo *et al.* 2008). Assuming these trends continue, they estimated that the additional loss of ice during the Arctic spring could boost productivity more than threefold above 1998–2002 levels, potentially altering marine ecosystem structure and the degree of pelagic-benthic coupling.

Simultaneous with the enhanced stratification reducing nutrient supply in the North Pacific has been a freshening of the Okhotsk Sea, which could affect primary production by changing the supply of iron. The Okhotsk Sea Intermediate Water is an important source of iron for the North Pacific (Nishioka *et al.* 2007). Hill *et al.* (2003) suggested that the surface layer is freshening in the Okhotsk Sea, and Nakanowatari *et al.* (2007) reported warming of the Okhotsk Sea Intermediate Water and the consequent warming of the North Pacific Intermediate Water. These results suggest that the freshening of the Okhotsk Sea has reduced the formation of the Okhotsk Sea Intermediate Water, and hence reduced the iron supply from the Okhotsk Sea to the North Pacific. The relative contributions of reduced iron supply versus weakening of stratification to the reduction in primary production in the North Pacific are unknown.

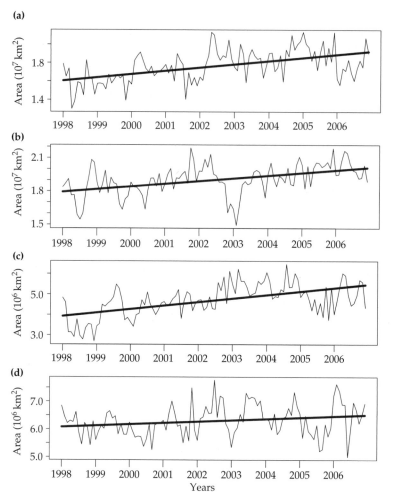

Figure 10.6 Time series of the monthly mean area (km²) of oligotrophic regions (surface chlorophyll less than or equal to 0.07 mg chlorophyll m⁻³) between 5–45°N/S latitude with the modelled seasonal cycle removed in (a) the North Pacific, (b) the South Pacific, (c) the North Atlantic, and (d) the South Atlantic. Straight lines show the linear trend. (Reprinted by permission of Macmillan Publishers Ltd: Geophysical Research Letters, Polovina *et al.* 2008b, copyright 2008.)

10.3.1.2 Future projections of primary production

Global-scale models project moderate reductions in global primary production. Cox *et al.* (2000) used a nutrient-phytoplankton-zooplankton-detritus (NPZD) coupled to an AOGCM with an ocean resolution of 2.5° × 3.75°. They projected a global average reduction of about 5% from 2000 to 2100 under global warming with the IS92a scenario. Boyd and Doney (2002) used a more complex ecosystem model that incorporated multi-nutrient limitation (N, P, Si, and Fe) and three groups of planktonic geochemically based functional groups: diatoms

(export flux and ballast), diazotrophs (nitrogen fixation), and calcifiers (alkalinity and ballast). They used the IPCC A1 scenario, and projected an average 5.5% decline in primary production. Bopp *et al.* (2005) used a similar type of multi-nutrient (NO_3, NH_4, PO_4, SiO_3, and Fe) and plankton community model (diatoms, nano-phytoplankton, microzooplankton, and mesozooplankton), named Pelagic Interactions Scheme for Carbon and Ecosystem Studies (PISCES), and projected a 15% decline with a scenario in which CO_2 started at pre-industrial conditions and increased at a rate of 1% year⁻¹

thereby reaching $4 \times CO_2$ after 140 years. All three models generated reduced primary production due to lower nutrient supply because of enhanced stratification, especially in lower latitudes, that exceeded the increased primary production projected to occur in higher latitudes due to an extended growing season.

Bopp *et al.* (2005) and Boyd and Doney (2002) also projected a shift in the phytoplankton community with a decrease in diatoms and an increase in nano-phytoplankton. The relative abundance of diatoms was projected to decline about 10% on a global scale and by up to 60% in the North Atlantic and in the subantarctic Pacific (Bopp *et al.* 2005). These shifts in diatoms showed good coincidence with the projected spatial distribution of the nutrient limitation factors (Fig. 10.7), with Fe limitation occurring south of 60°S (mainly in the South Pacific), NO_3 and Si limitation occurring between 30 and 40°S (mainly in the South Pacific and Indian Ocean), and NO_3 limitation occurring in the North Atlantic.

Cox *et al.* (2000) and Bopp *et al.* (2005) reported possible feedback effects. Cox *et al.* projected that the decrease of the ocean carbon uptake due to the reduction of primary production would be compensated for by an increase in natural CO_2 release due to reduced upwelling of CO_2 rich deep waters to the surface. Bopp *et al.* (2005) projected a 25% decrease in the export ratio, which was mainly explained by the recycling of nutrients and carbon in the surface layer by nano-phytoplankton that replaced the diatoms.

Hashioka and Yamanaka (2007) projected ecosystem responses in the western North Pacific for the end of the twenty-first century under the IS92a global warming scenario. Their analysis differed from that of Boyd and Doney (2002) and Bopp *et al.* (2005) by using a more complex food web representation (the North Pacific Ecosystem Model for Understanding Regional Oceanography or NEMURO, Kishi *et al.* 2007a) and by solving the physics with the biology rather than using off-line physics (i.e. physics is simulated first and the results are saved and read into the biology calculations). Hashioka and Yamanaka also projected that global warming would generally increase vertical stratification due to rising temperatures, decreasing nutrient and chlorophyll-*a* concentrations in the surface water,

and causing a shift in the dominant phytoplankton groups from diatoms to other smaller phytoplankton. They pointed out that these projected responses do not occur uniformly in all seasons, with the shift in phytoplankton groups occurring quite dramatically at the end of the spring, and that the onset of the diatom spring bloom was projected to occur about 2 weeks earlier (Fig. 10.8). However, the earlier onset of the diatom bloom is expected to be less at high latitudes where the Arctic summer day is exceeded by the Arctic winter night because the onset of the phytoplankton bloom would be

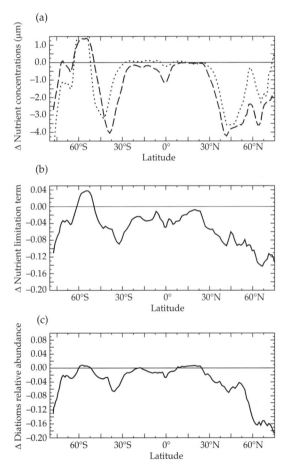

Figure 10.7 Zonal mean response to climate change ($4 \times CO_2 -$ $1 \times CO_2$) of (a) surface concentrations of nitrate (dashed line) and silicic acid (dotted line) in mM. (b) Nutrient limitation term of diatoms growth which is defined by four nutrient components. (c) The relative abundance of diatoms. (Reprinted by permission of Macmillan Publishers Ltd: Geophysical Research Letters, Bopp *et al.* 2005, copyright 2005.)

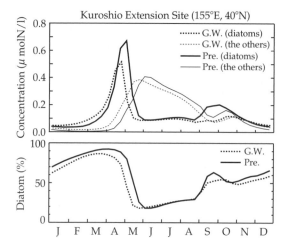

Figure 10.8 Seasonal variations of chlorophyll-*a* concentrations of diatoms (thick line) and other small phytoplankton (thin line) and biomass of diatoms as a percentage of total phytoplankton averaged over the surface 20 m at the Kuroshio Extension site (155°E, 40°N). Dotted and solid lines represent the global warming and present-day experiments, respectively. (Reprinted from Hashioka and Yamanaka 2007 with kind permission from Elsevier.)

determined by the seasonal change in light rather than a change in temperature (Eilertsen 1993).

10.3.2 Zooplankton

Zooplankton will be influenced by global warming through changes in water temperature, transport, and food availability. Shifts in the flora and timing of the bloom season can have especially important effects on the food available for zooplankton. The ship-based data available for zooplankton are more limited than for phytoplankton, and zooplankton is not monitored by satellite sensors. Furthermore, needing to know the trophodynamics between the phytoplankton and zooplankton adds considerable uncertainty to zooplankton modelling and projecting their population responses to global change. GLOBEC-related research has made progress on the understanding of phytoplankton to zooplankton trophodynamics (Moloney *et al.*, Chapter 7, this volume)

10.3.2.1 Emergence of zooplankton responses to global change

Richardson and Schoeman (2004) analysed Continuous Plankton Recorder (CPR) data for 1958 to 2002

for the north-east Atlantic and showed that zooplankton abundances were positively correlated to phytoplankton, with phytoplankton increasing during warm SST regimes in the cooler (northern) regions and decreasing in the warmer (southern) regions. The abundance of herbivorous copepods was positively related to phytoplankton, and the abundance of carnivorous zooplankton was positively related to the abundance of herbivorous copepods, but neither was directly correlated to SST. These results suggest climate-induced bottom-up forcing by which climate effects on phytoplankton propagate up the food web through herbivorous copepods to the carnivorous zooplankton. Richardson and Schoeman (2004) reported that SST during the period 1958–2002 decreased by about 0.1°C in the northern part of the study area and increased by about 0.5°C in the southern part. They also observed that plankton generally decreased in the southern area. Many AOGCMs project that SST will rise by 2–4°C in the north-eastern Atlantic by 2100, with an even higher increase in the northern area (IPCC 2007a). These results suggest that large responses of the zooplankton are possible for the north-east Atlantic, mediated through changes in their food. Additionally, CPR data from the Sir Alister Hardy Foundation for Ocean Science (SAHFOS) have shown a northward displacement of temperate species in the south and a northward displacement of boreal and arctic species in the north from the cool 1960s and 1970s to the present warm period (Beaugrand *et al.* 2002b; see Box 2.2, Fig. 1 in Drinkwater *et al.*, Chapter 2, this volume).

CPR data were also used to estimate changes in the phenology of plankton in the North Sea for 1958–90 (Edwards and Richardson 2004). The timing of species with summer peaks was earlier in the year, while the timing of species with blooms in the spring and autumn were unchanged (Fig. 10.9). The average shifts in bloom timing for the summer period were 23 days for dinoflagellates, 10 days for copepods, 10 days for non-copepod holozooplankton, and 27 days for meroplankton. SST during this same period increased by 0.9°C. The timing of the summer peaks was positively correlated with SST for all functional groups, with the highest correlations being for dinoflagellates ($r = 0.69$)

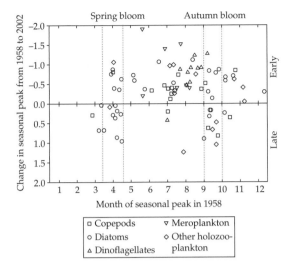

Figure 10.9 The change in the timing of the seasonal peaks (in months) for the 66 taxa over the 45-year period from 1958 to 2002 plotted against the timing of their seasonal peak in 1958. For each taxon, the linear regression was used to estimate the difference between the seasonal peak in 1958 and 2002. A negative difference between 1958 and 2002 indicates seasonal cycles are becoming earlier. (Reprinted by permission of Macmillan Publishers Ltd: Nature. Edwards and Richardson 2004. Copyright 2004.)

and meroplankton ($r = 0.70$). While the response of meroplankton can be explained by a direct species-specific effect of temperature, the response of the dinoflagellates is likely due to a mix of a direct temperature effect and indirect effects related to stratification and the responses of other taxa (Edwards and Richardson 2004).

A shift in the peak abundance of zooplankton has also been observed for the North Pacific. In the western North Pacific, Chiba *et al.* (2004) showed that the developmental timing of *Neocalanus flemingeri* has become earlier because of increased diatoms in the winter resulting from a shallower mixed layer improving light conditions. On the eastern side of the North Pacific, Mackas *et al.* (2007) reported a similar earlier shift in the peak day of *Neocalanus plumchrus* biomass since the mid-1970s that can be associated with warming (e.g. Miller *et al.* 1994), although Mackas *et al.* did not infer that the response was due to global warming. McGowan *et al.* (2003) found that the seasonal peak in bulk volumetric zooplankton biomass shifted 2 months earlier (July to May) after the mid-1970s in the southern part of the California

Current coinciding with warmer conditions and increased stratification (Kim and Miller 2007).

Wiltshire and Manly (2004) analysed a long-term data set from a single station in the North Sea for 1962 to 2002 and found the spring diatom bloom was actually delayed despite a general warming of about 1.1°C, perhaps due to responses of zooplankton to the temperature change. These data were collected three times a week (Greve *et al.* 2004). The delay in the spring bloom was hypothesized to be due to the warming trend during the autumn increasing the winter abundance of copepods, which resulted in higher grazing pressure that delayed the build-up of the diatom biomass.

In a recent review, Richardson (2008) summarized zooplankton responses to historical changes in climate. He also categorized responses of zooplankton to global warming as distributional shifts, phenology, and species composition and biomass changes. He found a general poleward movement of warm-water species and retreat of cold-water species, the opposite southward movement of cold-water species when physics pushed cooler water south (e.g. north-west Atlantic), earlier timing of life cycle events, and species abundances changes that varied in direction and mechanism by species and region.

10.3.2.2 Future projections of zooplankton
There are relatively few examples of mechanistic model projections (e.g. stage-based, NPZD) of zooplankton abundance and community composition under future climate conditions. Richardson's review (2008) cited Attrill *et al.* (2007) as an example of a correlative approach that related the NAO to jellyfish abundance and then projected jellyfish abundance under NAO conditions expected under global warming scenarios. Richardson (2008) presents some common criticisms of the empirical approach, and shows as an example how the *C. finmarchicus* abundance and NAO relationship broke down from 1996 (Reid *et al.* 2003a), likely because the mechanisms underlying the relationship had changed. The other example cited by Richardson (2008) was a version of an NPZD model (Bopp *et al.* 2004, 2005) but whose results under global warming were characterized as changes in primary production. We also were unable to locate clear examples of models being used to project zooplankton composition,

phenology, and abundances under global warming. Historically, models that included physics and zooplankton have focused on the geochemical cycling of nutrients (Le Quéré *et al.* 2005), and were not specifically designed for projecting zooplankton population dynamics. Forecasting zooplankton community and population dynamics was also not the focus of higher trophic level modelling where zooplankton was either ignored or represented very simply as food for fish. The exceptions generally involved models of larval fish feeding on a fairly detailed representation of the zooplankton (e.g. Letcher *et al.* 1996).

Ellingsen *et al.* (2008) applied a coupled model to simulate the change in the Barents Sea plankton during the period 1995–2059 (IPCC B2 scenario), and the model was validated with plankton observations for the period 1995–2004. The plankton model (Slagstad and Wassmann 1997; Slagstad *et al.* 1999; Wassmann *et al.* 2006) contained state variables for nitrate, ammonium, silicate, diatoms, flagellates, microzooplankton, dissolved organic carbon, heterotrophic flagellates, bacteria, and two groups of mesozooplankton representing Arctic and Atlantic (arcto-boreal) species. Phytoplankton production increased by 8%, and the major part of the increase happened in the eastern and northern Barents Sea (i.e. in the Arctic Water masses). The Atlantic mesozooplankton increased by 20%, but not enough to account for the 50% decrease in the Arctic zooplankton.

There are several efforts underway to use NPZD models coupled with three-dimensional physical models to project the population dynamics of key zooplankton species (e.g. *Calanus*) to global warming. These models are generally single species and use a stage-based or individual-based approach to simulating the population dynamics of the zooplankton species of interest. These efforts build upon some of the advances during the GLOBEC era in the modelling of key species of zooplankton coupled to the physics (see deYoung *et al.*, Chapter 5, this volume). For example, as part of the US GLOBEC synthesis efforts, population dynamics of key species of zooplankton will be simulated under global change using an individual-based model coupled to three-dimensional physical models of the north-west Atlantic and Arctic Ocean regions

(Cabell Davis, Woods Hole Oceanographic Institution, USA). Other models are being developed that could be used. For example, Slagstad and Tande (2007) recently developed a stage-based model of *Calanus finmarchicus* that was coupled to a nutrient, phytoplankton, and microzooplankton model and embedded into a three-dimensional hydrodynamic model for the Norwegian Sea that seems amenable to projecting zooplankton responses to global change.

10.3.3 Higher trophic levels

As we progress up the food chain, from phytoplankton to zooplankton to fish and other higher trophic level organisms, the possible interactions that can affect species responses increase, which puts enormous demands on the type and quantity of data needed. We focus on fish here but recognize that higher trophic level organisms also includes birds, marine mammals, and humans. Higher trophic level organisms such as fish are affected directly by changes in the physical and chemical properties of the ocean (temperature, salinity, and transport), but also by how these changes can directly and indirectly affect phytoplankton, zooplankton, benthos, and other fish that act as prey for piscivores. Transport and food availability during early life stages of fish are especially important because eggs and larval fish are subject to advective transport and their growth and survival are very sensitive to the timing, quantity, and quality of the available food (Lett *et al.* 2009). Mortality is high during the early life stages, and small changes become amplified when compounded over time and influence recruitment (e.g. cod in Gifford *et al.*, Chapter 4, this volume). The potential for indirect effects mediated through food web interactions to become important also presents a challenge. Our understanding of food web dynamics is limited in many situations. Higher trophic levels generally live for multiple years and can utilize a variety of habitats and prey. Thus, their monitoring is difficult but yet we need many years of data in order to be able to see their full responses at the population and community levels and to be able to isolate the responses due to global change versus other factors that vary and can directly and indirectly affect higher trophic level organisms (Rose 2000).

10.3.3.1 Emergence of higher trophic level responses to global change

While there are examples of single-species analysis of upper trophic level responses to global change (e.g. Drinkwater 2005; Vikebø et al. 2007b), there are far fewer studies that involve community responses over broad spatial scales. Perry et al. (2005) analysed 25 years of demersal fish species data for the North Sea and found shifts in some of the species. Of the 36 species that were analysed, about half (15 species) shifted their distributional range. These species included commercially exploited and non-exploited species. The distance of the shifts ranged from 48 to 403 km, with an average of 172.3 ± 98.8 km. They found that 13 out of 15 species that shifted went northward, and that many species were able to find cooler temperatures by moving into deeper waters. Species that shifted generally had a faster life history (small-bodied, short-lived, and smaller size at maturity) than non-shifting species (Fig. 10.10). However, individual growth rates did not differ significantly between shifting and non-shifting species. Greve et al. (2005) used a 10-year ichthyoplankton dataset for Helgoland Roads to investigate the phenological response of fish larvae to SST change. Eleven of 27 species showed a statistically significant negative correlation (i.e. earlier in the year) to SST (Fig. 10.11) and almost all of the species showed a negative correlation.

Mueter and Litzow (2008) showed that there was a northward shift in the summer distribution of the majority of the fish species on the eastern Barents Sea shelf in response to the pool of cold bottom water (<2°C) on the shelf retreating northward by over 200 km between the early 1980s and the early 2000s. This northward expansion resulted in an increase in total biomass, species richness, and average trophic level into the area formerly occupied by the cold pool. While the northward shifts were strongly related to temperature, they could not be explained by temperature alone and showed a non-linear, accelerating trend over time. Such non-linear effects suggested a reorganization of the fish community (i.e. an indirect effect) in response to a shift in average temperature. Understanding such indirect community-level responses of fish to a warming climate challenges our current state-of-knowledge (Mueter et al. 2009).

Intuitively, global warming effects on phytoplankton and zooplankton productivity and phenology will have large effects on higher trophic levels, although examples that document such responses are rare. We expect large responses to global change in some situations but it is difficult to find unequivocal evidence from historical analyses. Consideration of the match or mismatch between the timing of plankton blooms and the energy needs of young fish has a long history in fisheries science (Cury et al. 2008). Much attention has also been devoted to the mix of the species within the prey because very young fish often need certain sized

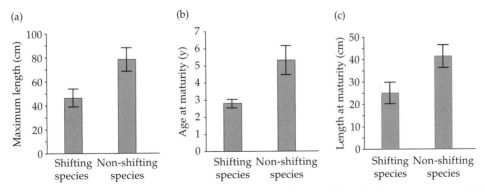

Figure 10.10 Differences in life history traits between shifting ($n = 15$) and non-shifting ($n = 21$) species with respect to centres of distribution (mean latitudes). (a) Maximum body size ($t = -2.41$, degrees of freedom (df) = 34, $P = 0.02$). (b) Age at maturity ($t = -2.86$, df = 27, $P = 0.01$). (c) Length at maturity ($t = -2.29$, df = 29, $P = 0.03$). Means are shown with standard errors. (From Perry et al. 2005.)

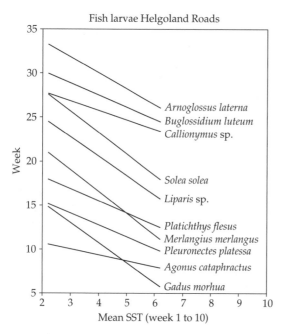

Figure 10.11 Regression lines of sea surface temperature (SST) to phenological timing of North Sea fish larvae. (From Greve *et al*. On the phenology of North Sea ichthyoplankton, ICES Journal of Marine Science, 2005, **62**, 1216–23. By permission of Oxford University Press.)

prey during their first days of exogenous feeding to ensure high enough capture rates of energy-rich food (Houde and Schekter 1980; Letcher *et al*. 1996). Beaugrand *et al*. (2003a) documented how a phenological response to warming in the plankton can lead to temporal mismatches between successive trophic levels, and influence commercially important fish species production such as the cod in the North Sea. Kasai *et al*. (1992) found that the advance of the spring bloom in the north-western Pacific has the potential to influence growth and survival of larval fishes. Grebmeier *et al*. (2006) discuss how warmer air temperatures and lower winter ice cover have resulted in a contraction of the summertime cool-water bottom pool and an associated movement of the more southern community dominated by pelagic fish into waters normally characterized by a benthic-dominated (arctic) community.

10.3.3.2 Future projections of higher trophic levels
Projected shifts in spatial distributions of higher trophic levels in response to global change have generally taken a species-centric approach of examining one or a few species in specific locations. Chavez *et al*. (2003) pointed out the existence of a periodicity of about 50 years in the landings of sardines and anchovies that coincided with variations in the productivity of coastal and open ocean ecosystems and temperature in the Pacific Ocean. They proposed a hypothesis that delineated a warm sardine regime (warm in the eastern boundary) as *El Viejo* (the old man) and a cool anchovy regime as *La Vieja* (the old woman) because of the similarity to *El Niño* and *La Niña* (Fig. 10.12). They drew attention to anthropogenic influences (e.g. global warming or overfishing) that can influence the character of regime shifts. Subsequently, Takasuka *et al*. (2007) proposed a simple hypothesis, called the optimal growth temperature hypothesis, to explain the species alternation between anchovy and sardine. The optimal growth temperature of Japanese sardine larvae is lower than that of Japanese anchovy; hence Japanese sardine will increase and anchovy will decrease during a cooler climate regime (Fig. 10.13). In contrast, the optimum temperature for California sardine is higher than that for California anchovy so that the temperature regimes have an opposite effect and cooler conditions that favour anchovy. The optimum growth temperature hypothesis is able to explain the paradox that the alternation of sardines and anchovies showed synchronicity on opposite sides of the Pacific, while the temperature anomaly showed asynchronicity. Takasuka *et al*. (2007) suggested that although fish might migrate to more favourable habitats, they cannot avoid basin-wide warming expected in the long term under global warming. Thus, historical data suggest that under global warming sardine may face collapse in the north-western Pacific while anchovy may face collapse in the north-eastern Pacific. Sardine and anchovy in upwelling systems (Peru-Humboldt, Benguela, and Canary) contribute greatly to the world's harvest (Schwartzlose *et al*. 1999). The optimal growth temperature hypothesis, developed for the Kuroshio-Oyashio and California Current, needs further evaluation, and whether the hypothesis can be applied to other upwelling regions is not known.

Vikebø *et al*. (2007b) used a temperature-dependent growth model for larval and juvenile cod coupled to a Regional Ocean Modelling System (ROMS; Shchepetkin and McWilliams 2005) circulation

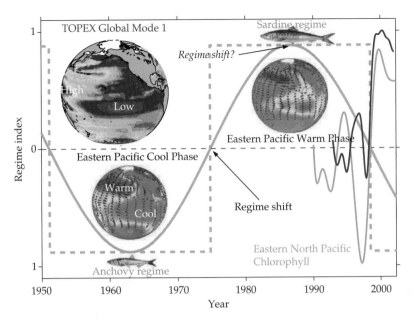

Figure 10.12 Hypothetical oscillation of a regime index with a period of 50 years. From the early 1950s to about 1975, the Pacific was cooler than average, and anchovies dominated. From about 1975 to the late 1990s, the Pacific was warmer, and sardines dominated. The spatial patterns of SST and atmospheric circulation anomalies are shown for each regime (Mantua *et al.* 1997). Some indices suggest that the shifts are rapid (dashed), whereas others suggest a more gradual shift (solid). The first empirical orthogonal function (EOF) of global TOPEX sea surface height (SSH) is shown above the cool, anchovy regime. The coefficient is shown in blue together with surface chlorophyll anomalies (green, mg m^{-3}) for the eastern margin of the California Current system from 1989 to 2001 (Pearcy 2005), also low-pass filtered. (From Chavez *et al.* 2003. Reprinted with permission from AAAS). (See Plate 25).

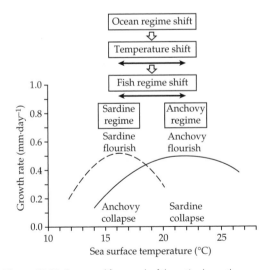

Figure 10.13 Conceptual framework of the optimal growth temperature hypothesis: a potential biological mechanism for the anchovy and sardine regime shifts under the assumption that both types of larvae experience similar temperatures during the same period. Quadratic functions for Japanese anchovy (*Engraulis japonicus*, solid curve) and Japanese sardine (*Sardinops melanostictus*, broken curve) were derived from observations. (From Takasuka *et al.* 2007 © 2008 NRC Canada or its licensors. Reproduced with permission.)

model to simulate growth and distribution of larval cod in the Nordic Seas under a climate scenario of reduced thermohaline circulation. The ROMS circulation model was forced by the Bergen Climate Model (BCM) to simulate the reduced thermohaline circulation by an increase in river run-off to the Nordic Seas and the Arctic Ocean. They found that fewer juvenile cod were transported eastwards in the Barents Sea, and those that were transported were considerably smaller in body size because of the cooler temperature. More juvenile cod were transported to the western parts of Spitsbergen, where today, and likely in the future, conditions are poorer and survival rate is lower than in the Barents Sea.

Drinkwater (2005) projected the expected responses of cod stocks throughout the North Atlantic to future temperature scenarios based upon observed responses of cod to temperature variability (Box 10.1). Population growth rates for many of the cod stocks were hypothesized to increase and lead to an overall rise in the total

Box 10.1 The response of Atlantic cod to future temperature changes

Atlantic cod are distributed throughout the continental shelves around the North Atlantic rim, which includes regions that are expected to experience some of the greatest anthropogenic climate changes. Multi-model scenarios suggest that by 2100 temperature increases of 2 to 4°C are likely in most regions of the North Atlantic occupied by cod, with a maximum increase of 6°C in the Barents Sea (IPCC 2007a). Drinkwater (2005) examined the response of cod to these modelled future warming scenarios. The responses included faster growth rates, larger weights-at-age, reduced age-of-maturity, and a northward spreading of their geographic distribution. He also suggested likely changes in recruitment based on the following two assumptions. First, if future annual mean bottom temperatures are projected to warm beyond 12°C, the cod are assumed to disappear from those regions. This is consistent with present observations that no cod stocks occupy waters with mean annual bottom temperatures exceeding approximately 12°C. Second, recruitment will increase for stocks whose bottom temperatures increase but remain below 5°C, and decrease for stocks whose bottom temperatures exceed 8.5°C. At temperatures between these two values, there would be little change in recruitment. These projected changes in temperature-dependent recruitment are consistent with observations (Planque and Frédou 1999; Drinkwater 2005).

The resulting projection showed decreased or collapsed stocks that depended on the magnitude of the temperature increase. Under a sustained 1°C change, several southern cod stocks would decrease or collapse (Box 10.1, Fig. 1a), while cold-water stocks, such as those off eastern Canada, Greenland, and in the Barents Sea, would benefit from increased recruitment. The net result would be an increase in the total production of Atlantic cod. Under greater temperature increases (2 and 3°C), the areas of decrease and collapse were projected to expand (Box 10.1, Fig. 1b and c), especially in the eastern Atlantic, while some stocks were still projected to increase in the western Atlantic. For a 4°C temperature change, however, even the Georges Bank stock in the western Atlantic was projected to likely disappear and regions exhibiting increased stocks were very limited (Box 10.1, Fig. 1d).

While many of these projections of possible changes to the Atlantic cod under future warming are consistent with past observations, the actual response remains uncertain. Indeed, Drinkwater (2005) noted that any projection will depend upon understanding in greater detail the physiological and behavioural responses of the cod to changes in environmental conditions, future fishing intensity, and the responses of other components of the marine ecosystem. Growth rate eventually will decrease at very warm temperatures, and cod movement is the result of a complicated behavioural response of individuals to multiple cues and to gradients and variation in temperature. Cod responses can also be dependent on the changes in primary (phytoplankton) and secondary (zooplankton) production, especially to changes to specific taxa that serve as food for different life stages (*C. finmarchicus* for larval and juvenile cod; capelin for adults).

continues

production of Atlantic cod in the North Atlantic. This was based on the relationship between recruitment anomalies and bottom water temperatures. The relationship derived between temperature and recruitment was consistent with how temperature affects larval growth rates (Otterlei *et al.* 1999; Bjornasson and Steinarsson 2002; Caldarone *et al.* 2003; Buckley *et al.* 2004), and hypothesized advection patterns of *C. finmarchicus* (Sundby 2000), an important prey item for early life stage cod. Buckley *et al.* (2004) suggested that the Atlantic cod, and also haddock, on Georges Bank will decline due to global warming unless accompanied by greater prey availability.

Box 10.1 *continued*

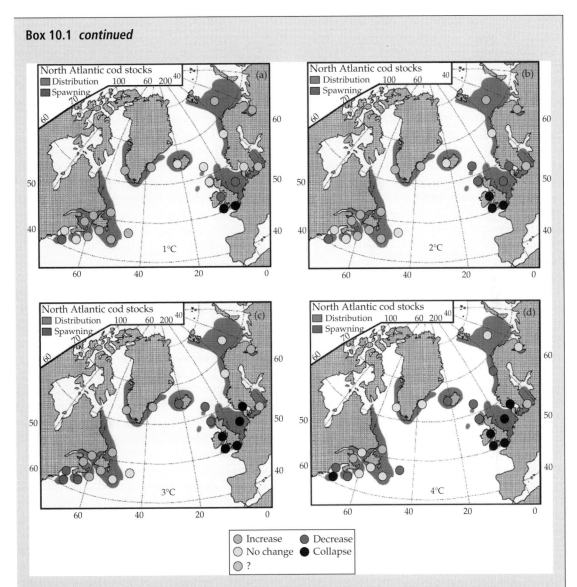

Box 10.1, Figure 1 The expected changes in the abundance of the cod stocks with a temperature increase of (a) 1°C, (b) 2°C, (c) 3°C, and (d) 4°C. Note the changes are relative to the previous change, that is, 4°C represents the changes from those at 3°C. (From Drinkwater. The response of Atlantic cod (*Gadus Morhua*) to future climate change. ICES journal of Marine Science, 2005, **62**, 1327–37. By permission of Oxford University). (See Plate 28).

Ken Drinkwater, Shin-ichi Ito, Kenneth Rose, Svein Sundby, Keith Brander

Huse and Ellingsen (2008) used an individual-based model to simulate how the Barents Sea capelin would change its spatial distribution related to its spawning migration and selection of feeding areas. They applied the same physical and plankton models as Ellingsen *et al.* (2008) described above, and used the surface layer values on a two-dimensional spatial grid from a simulation with the IPCC B2 scenario (Ellingsen *et al.* 2008). Movement of larvae was based on currents, while movement of adults was simulated as an evolving process using an artificial neural network with a genetic algorithm.

Multi-generational simulations used the information in the neural network from highest fitness adults to seed the neural network for next generation; movement patterns eventually stabilized. Three 300-year scenarios were simulated that repeated the 1996–2005, 2040–9, and 1996–2049 slices of output from the biophysical model. During warming, the feeding areas were projected to shift northwards as the ice edge retreated. The spawning areas were also projected to shift, from the southern border of the Barents Sea (Finnmark and Murman coasts) to the eastern border of the sea (Novaya Zemlya), and spawning was projected to occur earlier in the year. A similar analysis was also performed under the same IPCC B2 scenario that examined capelin distributions in the years 1990, 2020, 2050, and 2080 (Roderfeld *et al.* 2008).

Global change effects on temperature and horizontal circulation of zooplankton have been projected to cause a mix of effects on Norwegian fishes (Stenevik and Sundby 2007). With increased warming, the advection of *C. finmarchicus*, an important prey for many fish species, would be directed from the Norwegian Sea into the Barents Sea but would decrease in the North Sea (Sundby 2000). While *C. finmarchicus* abundance would decrease in the North Sea, the advection of southern copepod species, such as *C. helgolandicus*, from the Atlantic to the North Sea would increase. Combined effects of warmer temperature and changes in copepod composition would benefit the more southern fish species, such as sardine and anchovy, and decrease the productivity of the boreal fish species such as herring and cod. Stenevik and Sundby (2007) also projected that, under global warming (assuming warmer temperature, higher *C. finmarchicus*, and decreased sea ice), boreal fish production (e.g. cod) in the Arctic marine ecosystem of the Barents Sea would be expected to increase, with species distributions expanding northward and eastward. Fishery catches in the Russian and Norwegian sectors of the Barents Sea would therefore also likely increase.

Changes in upwelling would likely affect prey availability to fishes. Upwelling regions are areas of high fish production, and where increased upwelling and uplifted thermoclines are expected under global warming (Auad *et al.* 2006), we would expect enhanced primary production and a likely increase in the prey available to planktivorous fishes. However, offshore transport of larvae will also be modified under global warming. Auad *et al.* (2006) projected increased offshore surface transport of fish larvae, such as Pacific sardine, in the spring in the northern part of the California Current system, while onshore transport in winter and summer in the southern part increased. Offshore waters had lower prey densities.

Ito (2007) simulated the effects of changes in temperature and local vertical mixing on Pacific saury (*Cololabis saira*) growth and survival (Box 10.2). A simple three-box model, in which local vertical mixing strength was defined by the temperature difference between the surface and subsurface layer, was coupled to a nutrient-phytoplankton-zooplankton model (NPZ)-fish bioenergetics model called NEMURO.FISH (NEMURO for Including Saury and Herring; Ito *et al.* 2004, Megrey *et al.* 2007b). Saury growth rate, body size, and survival were followed as they seasonally migrated among the three spatial boxes. Changes in temperature affected the saury directly through bioenergetics (growth) and movement, and changes in temperature and vertical mixing affected the saury indirectly via their effects on primary and zooplankton production and thus prey availability to saury. Increases in temperature expected under future anthropogenic climate change caused the saury to alter their migration patterns, during which they experienced reduced food and slower growth.

Changes in horizontal transport would clearly affect egg and larval transport under future climate change, but general statements as to their effects on fish growth and survival depend on the species and ecosystem. One can envision a variety of future scenarios, but these need to be evaluated on a case-by-case basis. For example, Sakamoto *et al.* (2005) projected an acceleration of the Kuroshio Current system. Many fish species, including sardine, saury, and predatory fish (tuna) spawn in the upstream region of the Kuroshio, and their eggs and larvae are advected to the nursery grounds by the Kuroshio. Therefore, an intensification of the Kuroshio would likely cause greater distribution of the larvae. What is uncertain is how such changes in distribution would translate into changes in growth and survival.

Multi-species approaches that attempt to include some of the interspecific interactions in their projections of global change effects on higher trophic levels are relatively rare in marine systems. Spatial Environmental Population Dynamics Model (SEAPODYM; Lehodey *et al.* 2003) is an example of a multi-species model that simulates the population dynamics of three tuna species (skipjack, *Katsuwonus pelamis*; yellowfin, *Thunnus albacares*; and bigeye, *T. obesus*) and their forage fish species. SEAPODYM combines physical, biochemical, forage, and fish population dynamics. Lehodey *et al.* (2007) used an updated version of SEAPODYM, which includes an

enhanced definition of habitat indices, movement, and accessibility of tuna predators to different vertically migrant and non-migrant micronekton functional groups, to project the potential impact of climate change on the bigeye tuna population. The simulation was driven by physical-biogeochemical fields projected from a global earth system simulation (Bopp *et al.* 2001) with parameterizations obtained for bigeye tuna in the Pacific Ocean using optimization techniques (Senina *et al.* 2008) and the historical catch data for the last 50 years. Projections

---

Box 10.2 Projection of Pacific saury response

The response of Pacific saury to future climate change was explored using the NEMURO.FISH model by Ito *et al.* (2007). A simple three-box spatial model including physical, biochemical, and plankton dynamics was constructed with saury migrating among the boxes. Fish growth was calculated according to a bioenergetics equation using the simulated plankton from the lower trophic model as food. The SST input to the model was obtained from a global warming scenario (IPCC A2) simulation of an atmosphere-ocean coupled general circulation models (AOGCM; Sakamoto *et al.* 2005). The simulated SST in years 2001, 2050, and 2099 was averaged for the Oyashio (subarctic) region, the mixed water region, and the Kuroshio (subtropical) region. SST anomalies were calculated by subtracting monthly mean SSTs in 2001 from those in 2050 and in 2099 (Box 10.2, Fig. 1a). These anomalies were added to observational climatological SST and used as input to the saury version of the NEMURO. FISH model.

The model suggested that the body weight of saury will decrease with increased temperature (Box 10.2, Fig. 1b), but it was likely that there would be increased survival. Reduced body weight resulted from lowered zooplankton densities due to reduced primary production in a shallower mixed layer. The decrease in nutrient supply from

the subsurface layer to the surface due to warmer winter conditions exceeded any positive direct effect of temperature on phytoplankton production. Reduced food and higher metabolic rate resulted in slowed growth of adult saury, which triggered a change in their size-based migration patterns. Larger saury migrate to the north earlier than smaller saury because they can tolerate colder water, and the critical temperature below which saury start to migrate southward is higher for larger fish. The warmer temperatures in the mixed water region under global warming allowed smaller saury (young-of-the-year) to remain in the north and not migrate back to the south (Box 10.2, Fig. 1c). The higher prey density in the mixed water region compared to the Kuroshio region enabled faster growth and saury body sizes that were similar to baseline values by February of the first winter when saury started to spawn. Similar body size and higher food resulted in higher egg production. However, the life span of Pacific saury is 2 years and when age-1 fish repeated the northward migration in their second year, their body weight decreased by about 10 g (about 8% of final body weight) because they experienced reduced prey densities during their migration. The net effect of these changes would likely be an increase in saury abundance but with reduced individual weights.

continues

Box 10.2 *continued*

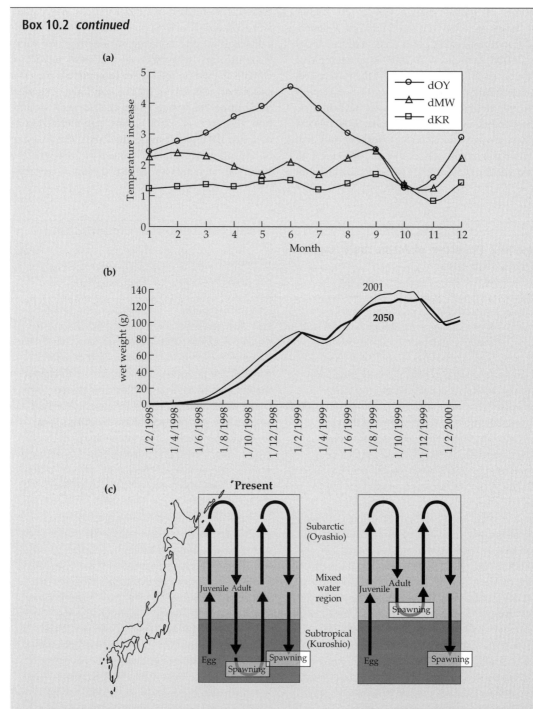

Box 10.2, Figure 1 (a) The temperature anomaly (°C) in 2050, relative to 2001, used as the forcing of the NEMURO model (circles for the Oyashio region, triangles for the mixed water region, squares for the Kuroshio region, respectively). (b) Wet weight (g) change of Pacific saury calculated by NEMURO.FISH (thin line for 2001 and thick line for 2050). (c) Schematic picture of simulated migration pattern of Pacific saury in the present (*left*) and in future (*right*). The saury migrate from the subtropical to the subarctic for feeding during spring-autumn and migrate back to the south during autumn-winter and spawn during autumn-winter in the mixed water region and subtropical. However, under global warming conditions, the migration pattern was modified especially in the spawning migration in the first winter. (From Ito 2007.)

Shin-ichi Ito, Kenneth A. Rose, and Arthur J. Miller

of future changes in bigeye tuna population dynamics were developed based on the IPCC A2 emission scenario. The model suggests an increase in the bigeye tuna population in the eastern Pacific because increased temperatures approached the tuna's preferred temperature range, while the western bigeye tuna population would decrease (Fig. 10.14).

Salmon utilize freshwater, coastal areas, and the open ocean, and thus provide an exceptionally difficult challenge for projecting their population responses to global change (Box 10.3). In order to fully express global change effects, models that simulate the entire life cycle are needed. Representing the full life cycle (birth to death, with survivors producing the new young each year) allows for effects to accumulate through the different life stages. For example, warmer temperatures could have mixed effects on sockeye salmon during their lake phase, and positive effects in the short term but negative

effects in the long term during their migration to the ocean (Box 10.3).

We have taken a narrow view of higher trophic levels by focusing on fish, and some models consider other higher trophic organisms including humans (see Perry *et al.*, Chapter 8, this volume). In the context of modelling fish responses, we may need to take into account the economic impacts of global warming and its feedback on fisheries. Eide and Heen (2002) used a coupled fishery-economics model for northern Norway and projected how the fishery and cod and herring populations would respond to global warming scenarios. They used a biologically simple multi-species population model (focused on recruitment) coupled with a model of the multiple fishing fleets (ECONMULT); the economic factors that were simulated used an input-output approach. Warm and cold future states were simulated. The warm state was created by changing the atmospheric

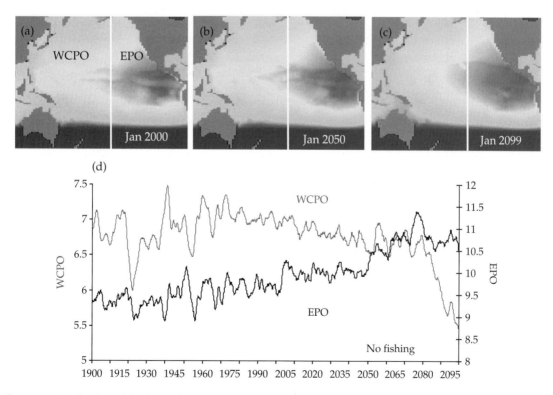

Figure 10.14 Predicted spatial distribution of bigeye tuna biomass in the Pacific Ocean (a) current, (b) in 2050 and (c) in 2099 under a global warming scenario, and (d) comparison of predicted biomass (million tons) of bigeye tuna in the western central (WCPO) and eastern (EPO) Pacific Ocean under a climate change scenario (without fishing). (From Lehodey *et al.* 2007). (See Plate 26).

Box 10.3 Salmon migration and bioenergetics

Pacific salmon *Oncorhynchus* spp. live for 1–2 years (and can live for up to 4 years) in freshwater before migrating to the ocean, where they spend the next 1–3 years. Welch *et al.* (1998) showed a sharp boundary of sockeye salmon distribution defined by temperature. They suggested that food was always limiting, and that salmon growth was very sensitive to temperature. Based on the observational results, they projected the future distribution of sockeye salmon in the North Pacific (Box 10.3, Fig. 1). Warmer temperatures in the North Pacific are expected to reduce salmon growth in the ocean. Weaker winter convection and increased glacial run-off to lakes and rivers will increase turbidity and thereby decrease prey production in freshwater lakes. These two negative effects will lead to a decline in the productivity of sockeye salmon (Melack *et al.* 1997).

Schindler *et al.* (2005) suggested that warming could have positive effects on sockeye salmon during their lake phase. They suggested that warmer conditions would bring an earlier spring ice breakup and induce a more rapid seasonal temperature rise, thereby enhancing prey for juvenile sockeye in lakes. Therefore, they projected that global warming may enhance growth conditions for juvenile salmonids in large lakes of Alaska.

Rand *et al.* (2006) focused on the energy costs of the spawning migration and applied a bioenergetics model. Their model successfully reproduced the observed energy loss of sockeye during the spawning migration from ocean to the spawning area (about 1200 km upstream). They projected that higher temperatures and higher discharge would enhance energy loss of salmon during freshets. As a result, expected future reductions in

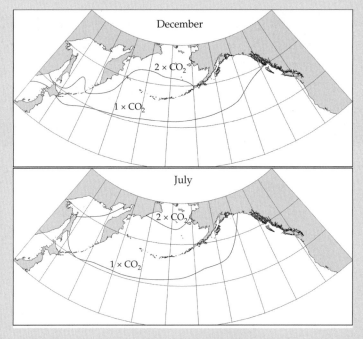

Box 10.3, Figure 1 Comparison of the predicted winter (7°C) and summer (12°C) positions of the sockeye salmon distribution under current and future climates (Albers equal area projection). Under a doubling of atmospheric CO_2 the area of acceptable thermal habitat in the North Pacific is predicted to decrease to 0 in summer and decline sharply in winter. The predictions are based on the Canadian Climate Centre's coupled ocean-atmosphere general climate model (Boer *et al.* 1992; McFarlane *et al.* 1992). (From Welch *et al.* 1998. © 2008 NRC Canada or its licensors. Reproduced with permission.)

continues

Box 10.3 *continued*

peak flows during freshets under global warming would reduce transit time to the spawning ground, resulting in an energy savings that compensated for the increased metabolic rates under warmer temperatures. However, if the body mass of sockeye salmon at the beginning of the spawning migration was smaller (Welch *et al.* 1998), then their model suggests a decreased fitness during the river migration that could jeopardize their sustainability (Box 10.3, Fig. 2).

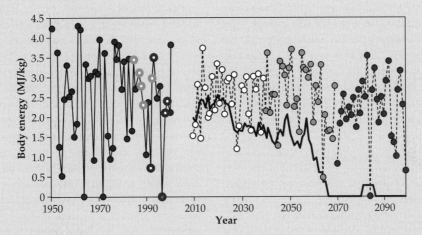

Box 10.3, Figure 2 Simulated time series of body energy for early Stuart sockeye salmon (Fraser River, British Columbia) at the spawning ground during 1950–2001 and 2010–99. With respect to the first period, note the 10 years identified with open circles, of which the five with heavy black borders represent years of high *en route* mortality and the five with heavy grey borders represent years of low *en route* mortality. Three climate stanzas are shown for the second period. Simulation results from the future scenario involving a long-term decline in the body mass of sockeye salmon at the beginning of the migration are shown with a solid black line (plotted as a 5-year running average of model output). (From Rand *et al.* 2006.)

Shin-ichi Ito, Kenneth A. Rose,
and Arthur J. Miller

forcing and increasing prey production and cod, and herring growth and recruitment, while the cold state future was assumed to occur from a reduction of warm Atlantic water masses inflowing to the Barents Sea with negative effects on cod and herring growth and recruitment. Model projections showed increased stock, catch, profit (wage paying ability), and flow of profits for the warm state, and decreased values for the cold state. Without fisheries management, the increase in catch in the warm state was more than the decrease in catch in the cold state. Under some fisheries management strategies, however, the opposite was predicted: the decrease in the cold state exceeded the increase in the warm state. The range of responses was generally larger among different management strategies than between different environmental scenarios. Thus, projections of fish population responses to global change can be affected by the fishery and by economics.

10.4 The future and challenges

The development of modelling tools capable of projecting the effects of global climate change on zooplankton and higher trophic levels has lagged behind those for nutrients (biogeochemical cycling) and physics. As documented above, the effects of global change are already being observed and thus there is a need for more and improved forecasting tools. GLOBEC and related studies have laid a sound foundation for how to proceed to project marine ecosystem responses to likely future climate conditions. The shift from trying to understand past and current conditions to projecting ecosystem responses to previously unobserved, and somewhat uncertain, future conditions is a major challenge.

In this section, we look at the big picture issues of what is needed in the long term to develop models for the general problem of predicting ecosystem responses to future global change. We grouped the issues into the categories of mechanistic approaches, physical modelling, nutrients and phytoplankton, zooplankton and higher trophic levels, and resource management. We also focus here on fish as the higher trophic level organisms, but acknowledge that birds and marine mammals are also of great interest.

10.4.1 Mechanistic approaches

Our discussion of the next steps centres on issues related to advancing mechanistic approaches over scenario and empirical approaches (see Box 10.4). All three approaches will be required to tackle the prob-

lem of developing credible projections of ecosystem responses to global change. We envision scenario and empirical approaches as being especially useful for certain specific situations (e.g. the short-term response of a very well-studied process or species in a well-monitored location), and for broad-scale semi-quantitative analyses of processes. For example, Cheung *et al.* (2009) combined an empirical approach with simple logistic population models to project how invertebrate and fish biodiversity would change worldwide under global warming. Scenario and empirical approaches will be useful for constraining the problems that then should be investigated using mechanistic approaches; what processes and variables appear to be important and need to be included in the mechanistic models. However, as the problem requires projecting ecosystem responses on basin or similar spatial scales and for decades or longer, scenario and empirical approaches are often pushed beyond their domain of applicability, leading to the increasing reliance on mechanistic approaches that combine physical and biological simulation models. The use of mechanistic approaches has its costs. Mechanistic approaches

---

Box 10.4 Empirical modelling approaches for projecting primary production

We can categorize the approaches to projecting future marine ecosystem responses into three groups (Sarmiento *et al.* 2004): (1) mechanistic models of marine ecosystems, (2) scenario projections based on phenomena observed in the past, and (3) empirical model development based on statistical descriptions of observational data. Mechanistic models offer the best opportunity for projecting ecosystem responses under previously unobserved conditions, and these models have been improved greatly during the GLOBEC era (Chapter 5, deYoung *et al.*, this volume). The models of Boyd and Doney (2002), Bopp *et al.* (2005), and Hashioka and Yamanaka (2007) summarized in the text are examples of what we consider to be mechanistic approaches. Scenario projections are tempting because they rely on empirical observations but often generate

projections with little or unknown confidence because of the many possible differences between historical phenomena and future conditions. Here we focus on the empirical modelling approach.

Sarmiento *et al.* (2004) developed an empirical model to project chlorophyll and primary production using SST, surface salinity, surface density, upwelling, vertical density gradient at 50 m depth, mixed-layer depth, and sea-ice cover. They divided the world's ocean ecosystems into seven biomes and developed separate statistical models for each biome. Satellite-based chlorophyll data were used to estimate the parameters of the statistical models. They applied the statistical models to projected future conditions under global warming generated by atmosphere-ocean coupled general circulation models (AOGCMs). Major results were:

continues

Box 10.4 *continued*

Chlorophyll change (warming-control) (mg Chlorophyll m⁻³)

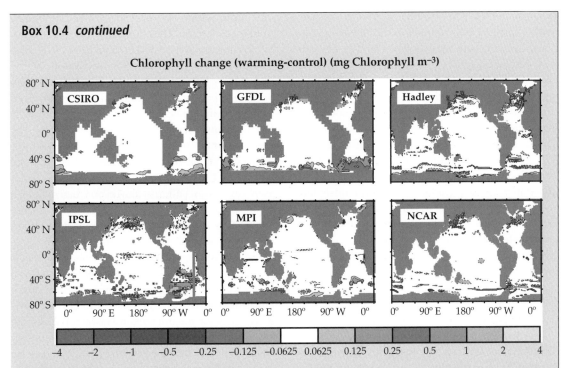

Box 10.4, Figure 1 Simulations of the impact of global warming on the chlorophyll concentration in mg m⁻³. Chlorophyll is calculated for both the control and the warming simulations using the empirical model. The figure shows the difference between the warming simulation and the control simulation averaged over the period 2040 to 2060 (except for Max Planck Institut (MPI), which is for the period 2040 to 2049). Areas in white are those for which the chlorophyll change is smaller than ±0.0625 mg m⁻³. (From Sarmiento *et al.* 2004). (See Plate 27).

(1) Decreased chlorophyll in the North Pacific primarily due to the retreat of the marginal sea-ice biome. (2) Possible increased chlorophyll in the North Atlantic due to a complex combination of factors. (3) Increased chlorophyll in the Southern Ocean due to the retreat of the marginal sea-ice zone and to changes in its northern boundary. (4) Possibly decreased chlorophyll adjacent to the Antarctic continent due to freshening within the marginal sea-ice zone (Box 10.4, Fig. 1). The net effects of these changes were that global warming was projected to increase primary production by 0.7 to 8.1%, although the uncertainty was large due to differences among the various AOGCM models in their projected future conditions.

Shin-ichi Ito, Arthur J. Miller, Kenneth A. Rose

require extensive data, can be computationally challenging, difficult to validate, and involve trying to mesh together processes that operate on many different spatial and temporal scales.

10.4.2 Physical modelling

Physical models are continuously evolving towards a higher fidelity representation of the climate sys-

tem (Trenberth 1997). Some of the improvements stem from more accurate and efficient numerical solution schemes (e.g. time-stepping and advection; Shchepetkin and McWilliams 2005; Haidvogel *et al.* 2008), improved sub-grid-scale parameterizations (e.g. mixed-layer-depth computations, meso- and sub-mesoscale features; Gent and McWilliams 1990; Large *et al.* 1994; Fox-Kemper and Ferrari 2008), and a better understanding of the climate system feedbacks (e.g. clouds and aerosols;

Trenberth 1997). An ensemble approach is often used to quantify uncertainties in the climate projections. In the ocean, models have tended towards finer resolution grids, and the nesting of grids to allow for coverage of broad spatial areas while having finer resolution where it is needed. Highly skilled oceanographic models are now also being developed for near-shore areas (e.g. estuaries and upwelling regions) where fish productivity is high (e.g. Houde and Rutherford 1993; Cury *et al.* 2008). Even with their sometimes high degree of uncertainty, oceanographic physical models are ahead of the biological models in attempting to project responses to climate change. As summarized above, physical models are using global climate change scenarios as inputs (e.g. IPCC outputs) and projecting subsequent changes in ocean properties such as vertical and horizontal circulation patterns, stratification, temperature and salinity distributions, and sea-ice cover.

The challenge for physical models is to evolve in directions that lead not only to improvements in the representation of the physics but also to account for the needs of the biological models that rely on the outputs of the physical models. From a physical ocean point of view, this means an explicit representation of a wider range of relevant spatio-temporal scales. For example, for physical models to provide useful information for the estimation of near-shore fish production, the physical model needs to resolve the shelf and coastal morphology (bathymetry and coastline) on relatively fine temporal and spatial scales. Physical models also need to be linked to high-resolution atmospheric and hydrological models in order to resolve variation in biologically important mesoscale processes related to ocean fronts, coastal upwelling, and river run-off. Furthermore, this fine-scale variation needs to be part of decadal-long integrations. Decadal simulations are necessary in order to explore low-frequency environmental conditions and their effects on higher trophic levels. Finally, we need to anticipate the inclusion of new processes (e.g. acidification) and 'new' micronutrients (e.g. iron), and others yet to be determined.

One approach that is being used to achieve high-resolution climate-scale simulations in a given domain is the nesting of a high-resolution limited area grid within a lower resolution large-scale numerical domain. Traditionally, for hindcast simulations, information is downscaled from the coarse-resolution region to the fine resolution region through an overlap in the domains. Downscaling works well when the forcing data are constrained by observations, such as in the reanalysis products (e.g. Curchitser *et al.* 2005; Hermann *et al.* 2009). The high-resolution nest can explicitly resolve features missing from the large-scale model, though it is still constrained through the boundaries by the large-scale circulation patterns. However, when making a future projection, the forcing functions are unconstrained by data and can be expected to respond differently to an alternative (high-resolution) ocean in the coupled model. One of the challenges, then, is to not only downscale information to the local scales, but also to understand how regional changes affect the global climate (i.e. upscaling). Furthermore, if we consider the interaction between the climate system and higher trophic levels, high-resolution nests need to be designed so as to permit large-scale spatial shifts in the distribution of biota as the climate changes.

Another challenge that needs to be addressed by global and downscaled model projections of future conditions is the estimation of uncertainties in the computations. The IPCC approach for global models is to create a super-ensemble of all the available models; that is, each individual model creates an ensemble of runs which are then assembled to a multi-model super-ensemble. Based on the super-ensemble, the assessment report assigns uncertainties to the projections (e.g. highly likely, very likely, etc.). In general, confidence limits given by ensemble modelling may be artificially narrow if the models all use similar assumptions and parameterizations, which do not reflect the full underlying uncertainty. An important question is how to use this information on uncertainty to initialize and force optimal downscaled regional projections, whose output is very relevant to biological modelling. Ådlandsvik (2008) suggested that initialization for regional studies should be derived from more than a single model. Overland and Wang (2007b) evaluated the collection of IPCC models for their suitability to force regional models of the North Pacific and the Barents Sea. They used criteria based on each model's ability to reproduce known patterns of the current climate in a particular geographic region and on the assumption that present-day skill is indicative of future performance. They concluded that some models are considerably more skillful in the North Pacific and the Barents Sea and thus their individual ensembles should be favoured for forcing regional models. The corollary is that many of the models, when evaluated regionally, were not suitable for forcing downscaled models of specific regions.

Considerations of using models for coupled ecosystem-physical projections may also influence the choice of the large-scale models employed for downscaling. For example, in the North Pacific, some biological activity is related to the PDO, which is based on the dominant mode of variability in SST anomalies (Mantua *et al.* 1997). In other regions, variability in stratification may correlate better with biological activity than with SST, and indeed global models may show different skill levels for different variables or statistical modes. Determining which large-scale models to use for specific downscaling situations, and quantifying their associated uncertainty, will require observational networks that allow for rigorous evaluation of the models.

10.4.3 Nutrients and phytoplankton

Several issues emerge as mechanistic models of nutrients and phytoplankton dynamics are pushed towards simulating global climate change. To date, the focus has been on biogeochemical cycling of nutrients and primary production (Anderson 2005; Le Quéré *et al.* 2005). Some suggest that commonly used formulations for nutrient limitation and nutrient ratios are not always appropriate for biogeochemical cycling and predator-prey interactions (Flynn 2005). Ecological stoichiometry (Andersen *et al.* 2004) will become especially important as we push the models towards conditions (global climate change) well outside of their domain of calibration and validation. In addition, the inclusion of micronutrients, such as iron, and new biogeochemical considerations such as acidity, have stretched the available data (Hood *et al.* 2006). A long-standing issue is how to represent the effects of multiple

nutrients on primary production (Tilman *et al.* 1982; Flynn 2003; Vallina and Le Quere 2008); adding micronutrients to the possible limiting factors highlights the uncertainty on how to simulate their cycling and how to combine multiple nutrients into an overall effect on photosynthesis. For some quantities, inclusion of more detailed phytoplankton (and zooplankton) may be needed (Hood *et al.* 2006), and even inclusion of higher trophic level effects may be needed to ensure realistic simulation of biogeochemical cycling (Hood *et al.* 2006; Wilson *et al.* 2009).

For some short-lived species (e.g. phytoplankton) with rapid turnover times, adaptation via evolutionary processes can become important to accurately predict their distributions and productivity (Fussmann *et al.* 2003; Yoshida *et al.* 2003). The trade-offs among species traits as environmental conditions change affect community structure (Litchman and Klausmeier 2008). Follows *et al.* (2007) used a typical NPZD model formulation, applied globally by coupling with a three-dimensional physics model, but with key parameters of the phytoplankton species related to nutrient, light, and temperature randomly assigned within allometric constraints. They then did 10-year simulations with repeating annual environmental conditions to determine which phytoplankton types (defined by their parameter values) would comprise the phytoplankton community and contribute to primary production. Patterns of phytoplankton types were robust to different initial conditions and other variation, and roughly corresponded to the types of phytoplankton communities found in different regions. The Follows *et al.* (2007) approach allowed for self-organization among a wide range of fixed phytoplankton types, but did not allow for adaptation during the simulation. Global change will occur over long time scales relative to the generation times of phytoplankton, therefore evolutionary adaptation may need to be considered in multi-decadal projections.

Representation of the phytoplankton community with a few functional groups may need to be reconsidered as more nutrients and processes are added to the models and the models are asked to simulate a variety of systems. With broad questions that do not focus too closely on specific regions, a single formulation of a NPZD can be used for a global analysis (e.g. Moore *et al.* 2001; Aita *et al.* 2003; Le Quéré *et al.* 2005), including very simple models that use adjustments to approximate a more complex community structure (e.g. Denman 2003). However, recent experience with the NEMURO NPZD model illustrates the fundamental trade-off we must make between generality and accuracy (*sensu* Levins 1966) when building ecological models. As the NEMURO model was applied to different areas and systems, adjustments were made to accommodate regional conditions, resulting in a dozen or more versions of the NEMURO NPZD model being in use (e.g. Yoshie *et al.* 2005; Wainwright *et al.* 2007; Werner *et al.* 2007a). Friedrichs *et al.* (2007) compared NPZD models of differing complexity for a standard water column configured for two different sites (equatorial Pacific and Arabian Sea) with all models using the same physics and concluded that simple models performed well when applied to single sites but that complex models were more portable.

We also expect the classical formulation of NPZD models to be challenged. A major overhaul of how the phytoplankton community is represented may be needed to accommodate what we anticipate to be changes in the formulation of the zooplankton. As discussed below, we envision the development of end-to-end models that combine the physics, nutrients, phytoplankton, zooplankton, fish, other higher tropic levels, and people into a single model or series of tightly coupled models. The role of zooplankton will change from primarily ensuring realistic chlorophyll to having to ensure realistic pathways of carbon transfer from the primary producers to food for fish and higher trophic levels. This will involve more detailed representation or alternative formulations of the zooplankton (e.g. more groups; size, stage, or individual-based approaches), and thus will likely require substantial changes in how the food of the zooplankton (i.e. phytoplankton) is represented.

10.4.4 Zooplankton, higher trophic levels, and end-to-end models

We expect that the mechanistic approach taken to date, and exemplified by GLOBEC, of focusing on target species of zooplankton and fish in certain locations (deYoung *et al.* 2004a; Gifford *et al.*, Chapter 4, this volume) will continue into the future. The target species approach has been used for decades in fisheries and ecology, and has been further championed by GLOBEC. Analyses of field and long-term monitoring data, and the development of coupled biophysical models, will continue. Such approaches are needed because there will always be global change questions that are best addressed with single species and life-stage-specific models. There will be gradual improvements in the physics and a slow expansion of lower food web complexity and further inclusion of higher trophic levels, and these analyses will continue to serve a vital role in increasing our understanding of ecosystem responses to climate change.

A challenge always associated with the target species approach is how to extrapolate the methods and results to other species and to other locations. Can the results be transported or are the methods general and thus data collection and models must be replicated at each new site? The issue of extrapolation has a long history as part of how results, whether modelling or empirical, can be generalized, and we expect progress will continue in this area.

While the target species approach has been successful and extrapolation questions will continue to be addressed, this approach is better designed for examining past and current conditions rather than future conditions. Past and current conditions can be used to constrain what is needed in mechanistic models and what aspects can be reasonably treated as fixed versus dynamic. Working within the past and current conditions allows for a much more constrained problem, and for more confidence that the model structure is realistic for the new conditions in model simulations of the future. Using models to project for global change puts enormous demands on the models for a greater range of possible effects and responses. We foresee a new generation of 'end-to-end' or 'physics to fish to fishers' models (Travers *et al.* 2007) that allow for food web interactions, community composition shifts, feedbacks, and thresholds responses, all of which can be beyond the capabilities of the single-species approach.

Several things have converged making the development of end-to-end models a possibility. Ever increasing computing power is continuing to relax computing limitations. Recent advances in physical modelling now allow for simulation of mesoscale features within broad geographic regions, while at the same time permitting multi-year simulations. We are quickly approaching the situation of being able to generate sufficient years of simulation of the high-resolution physics to allow for full life cycle models of higher trophic levels. There have also been advances in higher trophic level modelling, such as size-based and individual-based approaches that permit more realistic representation of multi-species population dynamics than the classical biomass-based approach (DeAngelis and Mooij 2005; Rose and Sable 2009). On the data side, the availability of more and more detailed data (e.g. genetics, spatially resolved measurements), especially on movement of animals (e.g. acoustic and sonic tagging) is rapidly expanding. Finally, concerns about declining fish populations, demands for ecosystem-based management, and the uncertainty of ecosystem health under global warming add a sense of urgency to the development of end-to-end models.

Understanding feedbacks and thresholds within the ecosystem is necessary for accurate projections of future ecosystem states. Short-term forecasts may avoid the necessity for inclusion of slow-reacting feedbacks, but the long-term projections needed for assessing responses to global change require that the appropriate feedbacks are included in order to ensure realistic projections. Feedbacks often become important drivers, and sometimes can dominate ecosystem responses when complex systems are integrated over long time scales. Feedbacks can be purely physical (e.g. air-water exchanges), density-dependence processes of organisms (populations hitting life cycle bottlenecks that result in non-proportional responses; deYoung et al., Chapter 5, this volume), indirect responses due to changes in interspecific interactions (food web effects), coupling of pelagic and benthic pathways, evolutionary adaptations of rapid turnover biota (e.g. Follows et al. 2007), and treating humans as part of the ecosystem so that the environment and humans affect each other dynamically (Perry et al., Chapter 8, this volume).

Density-dependent responses of growth, mortality, and reproduction are critical for projecting fish and other higher trophic level population responses because they can operate as a feedback response

and modulate the more direct effects of global change. For example, an increase in egg mortality due to warmer temperatures can be offset, or compensated for, by less crowded conditions in a later life stage leading to a decrease in mortality in that stage. Despite the importance of density-dependence for understanding and managing fish stocks and its intense study for decades, it remains a major unknown and source of controversy (Rose et al. 2001). Inclusion of density-dependent processes in higher trophic level organisms can also create computing issues because density-dependence requires simultaneously solving the equations of state of the food, fish, and their predators. Knowing the relative roles of bottom-up, top-down, and wasp-waist controls (Field et al. 2006; Bakun 2006b; Cury et al. 2008) is also vital for making and understanding accurate projections. The feeding interactions among a large suite of members of the food web are complex but potentially may need to be represented for long-term projections of higher trophic level responses.

We also need to include humans in the models as fully fledged members of the ecosystem (Brander et al., Chapter 3; Perry et al., Chapter 8, both this volume); we cannot assume a fixed fishing mortality rate or that people will maintain their identical behaviour under global change. Models are needed that can simulate density-dependent responses when they are important and allow for threshold responses in order to minimize surprises under previously unobserved future conditions. Threshold responses include abrupt changes in the ecosystem state (e.g. regime shifts), and previously rare species blooming under the new environmental conditions (e.g. jellyfish).

A major challenge will be how to represent behavioural movement of higher-level organisms (Tyler and Rose 1994). The spatial dynamics of higher trophic level organisms determine their exposure to environmental conditions, and we need to model changes in movement behaviour in response to global change in order to determine how the changes in physical and chemical properties and in the other biota will affect the growth, mortality, and reproduction of the higher trophic level organisms (Lett et al. 2009). The passive particle approach used for eggs and larvae is not appropriate in many situations for juveniles and adults. The modelling approach must accommodate ontogenetic (larvae to juvenile to adult) changes in movement abilities, preferences, and tolerances. How to realistically simulate the movement-related responses of higher trophic level organisms on a mechanistic, rather then prescriptive basis, should be the focus of a concerted effort (Railsback *et al.* 1999; Humston *et al.* 2004); such algorithms are needed for realistic spatial models involving non-passive life stages of higher trophic level organisms.

Zooplankton are clearly a focal point for future model development, especially as we attempt to include food web interactions and higher trophic level organisms. GLOBEC focused on bottom-up approaches of climate forcing the physics that forces the nutrient and phytoplankton dynamics and then more vague, but still one-way upward, linkages to zooplankton and higher trophic levels (Moloney *et al.*, Chapter 7, this volume). Fisheries science, as a discipline, usually starts at the fish level and looks down to 'food' and up to predators and fishing. There are very few examples of projected zooplankton population and community responses to global change. Zooplankton have generally been included in large-scale models as biomass in order to simulate biogeochemical cycling and not included in way conducive to projecting their population and community dynamics in response to climate change (Le Quéré *et al.* 2005). Models that include key zooplankton species with structured (stage- or individual-based) approaches are presently being geared up for projecting global change effects, but there are few examples reported to date in the literature. How to include zooplankton in ecosystem models to allow for projecting their population and community responses to global change and accurate rates of energy flows up pathways appropriate to 'feed' the higher trophic level organisms, while still ensuring biogeochemical cycling, will be a major challenge. We do not think it is a simple division of the presently used functional groups into more groups, but rather a major retooling of how zooplankton should be represented for use in end-to-end models. This may require changes in how phytoplankton is represented to accommodate the new zooplankton formulations. Such restructuring of the functional groups in the mod-

els should be pursued carefully but with skepticism (Anderson 2005).

There are several approaches that show promise for end-to-end food web modelling (see review in Travers *et al.* 2007). We briefly summarize some of them here. Coupling age-structured single-species higher trophic level models to an NPZD model was used to simulate global change effects on saury growth and migration (Box 10.2), and climate regime effects on herring population dynamics (Rose *et al.* 2007). Both analyses were versions of models developed as part of the NEMURO family of models (Megrey *et al.* 2007b), and used a very simple set of spatial boxes. The NEMURO effort is continuing in this direction by converting the age-structured approach to an individual-based approach, expanding from single species to a limited multi-species food web, and by embedding the models in the ROMS physical model. Shin and Cury (2001) also used an individual-based approach but they simulated a many-species food web (called OSMOSE: Objected-oriented Simulator of Marine ecOSystems Exploitation) on a two-dimensional spatial grid of cells and coupled the higher trophic level with an NPZ-like model. The model was used to examine various aspects of fishing on the food web (e.g. Shin and Cury 2004; Travers *et al.* 2007), and a new version is being developed that is driven by the output of a three-dimensional ROMS physical model. Movement of higher trophic levels was achieved by forcing spatial distributions to match historical distributions. An alternative to representing the community at the species level is size-based models (e.g. Baird and Suthers 2007; Maury *et al.* 2007a), whereby the state variables represent a progression of size classes rather than association with any particular species. Other approaches being developed include the Integrated Generic Bay Ecosystem Model (IGBEM) and Bay Model 2 (BM2) coupled models that follow biomass for the lower trophic levels and age-structured cohort approach for fish in a coarse three-dimensional spatial grid (Fulton *et al.* 2004c), and the use of outputs from an NPZ as inputs to the Ecopath with Ecosim (EwE; Christensen and Walters 2004) modelling package (Aydin *et al.* 2005). Models being developed in parallel efforts are beneficial to the overall process of model development, and we expect eventually that hybrid versions that combine features across these approaches will be developed as we make progress towards the goal of developing end-to-end models for projecting ecosystem responses to global change.

Two issues not yet addressed by the ongoing end-to-end modelling efforts are large-scale spatial distributional shifts and emergence of new dominants. Predicting large-scale spatial redistributions of phytoplankton, zooplankton, and fish and other predator communities, and the subsequent reorganization of the local food webs (e.g. previously rare species becoming important), have not been a focus of modelling to date. A fundamental issue is how to represent rare species that become dominants, explosions of problematic species (e.g. jellyfish—Purcell 2005, harmful algal blooms—Van Dolah 2000), and broad enough spatial areas to allow for changes in community structure and alteration of migration routes and new migration patterns to emerge.

The development of end-to-end models not only stretches our biological knowledge but also presents some practical considerations such as to how to solve coupled models that operate on different scales. Solving the zooplankton with the physics and using stored results is computationally efficient but prevents feedback between the upper trophic levels and their zooplankton food (i.e. prevents density-dependence as an emergent phenomenon). Solving the zooplankton with the fish allows for feedback between the predators and their prey but can complicate the coding, and then shifts the decision up the food chain to how to solve the zooplankton and fish with other higher trophic levels (e.g. marine mammals) and with any human components (e.g. dynamic fishing; human movement patterns). It seems clear that solving the entire modelling system simultaneously is computationally inefficient, but how to parse up the solution to allow for feasibility and computational speed while allowing for some feedbacks but ignoring others remains a challenge. Furthermore, as with any analysis that uses multiple coupled models on different scales, quantifying the uncertainty of projections is critical but difficult. One becomes limited by the number of simulations that are possible, with the challenge of specifying the uncertainty in so many variables

and parameters, and the large number of possible of combinations of factors and conditions that could be analysed. Computational issues still remain with the development and exercising of comprehensive end-to-end models.

10.4.5 Resource management

Conservatively, the translation of these multi-species coupled biophysical models and end-to-end models into direct management decision making seems a long way off. These models can inform management in terms of cautionary warnings about possible changes in the population(s) of interest resulting from changed environmental conditions, and can suggest key variables to monitor. However, until these models can demonstrate that their accuracy and precision are sufficient to be defensible for management decisions, managers will continue to use their simpler population-centric approaches. Declining fish populations, increasing levels of conflict associated with multiple resource management and increasing numbers of endangered species, has made the managers less willing to change their models. Few of the currently available target species models developed during GLOBEC have been incorporated into management decision making because of their complexity, potential conflicts among stakeholders, and general unproven skill at projection (see also Barange *et al.*, Chapter 9 this volume).

Establishing the accuracy and precision of coupled physical-food web and end-to-end models will likely remain an issue into the foreseeable future. Extensive data on the physical environment, coupled with emerging methods of data assimilation, will result in ongoing improvements to the physical models. Uncertainty will remain too high with the higher trophic level components of these coupled and end-to-end models for direct management decision making. Simply, high-quality data for calibration and skill assessment data for fish population dynamics and productivity remains limited. Yet, fish are usually of greatest interest to managers. This problem of highly uncertain fish data has plagued fisheries for generations and will continue to do so, although some progress is being made.

10.5 Prognosis

We have documented how changes are already occurring in marine ecosystems, some of which are related to global climate change. While the development of climate and physical models has progressed to the point of making projections, the biological models related to zooplankton and higher trophic levels have not kept pace. There is a lack of examples of projections of zooplankton responses to global change, and the projections of higher trophic levels tend to be site-specific, species-specific, and even life-stage-specific. We anticipate a surge in mechanistic modelling that will expand the existing efforts, continuing in the GLOBEC approach, and a concerted effort for the development of end-to-end models.

Maintaining long-term monitoring, and expanding it in critical areas and dimensions, provides the empirical information for defining the questions and locations ripe for mechanistic modelling. We cannot model everything everywhere and responses detected by scenario and empirical analyses of long-term monitoring data lead to us being able to focus on specific types of ecosystems, selected species and biological communities, and certain processes in the mechanistic models. A challenge to monitoring and empirical data collection is to know to measure some variables now, which may appear unimportant under current conditions, but will become important under future conditions (e.g. acidification, iron). Development and implementation of ocean observing systems (Harris *et al.*, Chapter 6, this volume) should continue, with particular emphasis on how the observing data will mesh with quantifying changes in ocean properties directly relevant to zooplankton and higher trophic levels (i.e. fish). Further development of sophisticated data analysis methods is needed to make sense of the ever increasing amounts and types of data becoming available.

With a sound empirical foundation, we are optimistic that many of the issues surrounding the development of mechanistic end-to-end forecasting models can be solved with a concerted effort that continues and expands on the GLOBEC legacy.

GLOBEC-style analyses that focus on specific species in particular locations will continue to make significant contributions. However, when we take the long-term view, we see a need for investment in monitoring, data analysis, and model development now so that results and models that allow for broad-scale analyses and ecosystem-based analyses with end-to-end models are available in the near-future. The onus is on the oceanographic community. It is becoming a cliché to say so, but diverse groups must work together to address these issues. GLOBEC is an excellent example of such multidisciplinary, cross-cutting research. We now need to move even further from multidisciplinarity to truly interdisciplinarity. We view the difference between multidisciplinarity and interdisciplinarity as operational in terms of the degree of communication and feedback that occurs between the various research groups. We view multidisciplinary as different groups with specific skills working together on the same problem but exchanging information periodically at, say annual or semi-annual, team meetings. Interdisciplinary research has the groups together, ideally co-located physically near each other, with constant exchange of information and adjustments to plans in response to the needs of the other groups. Hopefully, after the next decade of effort, one of the legacies of the GLOBEC programme will be that it laid down the foundation for the development of methods for forecasting the full ecosystem responses of marine systems to global climate change.

Acknowledgements

The authors gratefully acknowledge the constructive comments and suggestions of two anonymous reviewers which improved the quality of our chapter. The editors Manuel Barange, Francisco Werner, John Field, and Eileen Hofmann gave us many helpful suggestions. The help and assistance of Dawn Ashby is also gratefully acknowledged. We thank Patrick Lehodey for providing the tuna prediction result. We are also grateful to many scientists who provided high-quality figures.

CHAPTER 11

Marine ecosystems and global change: a synthesis

Eileen E. Hofmann, Manuel Barange, John G. Field, Roger P. Harris, R. Ian Perry, and Francisco E. Werner

11.1 Introduction

The 1990s and 2000s have brought increasing awareness of the vulnerability of marine animal populations, especially of commercially important species (Pauly *et al.* 2002), and awareness of significant changes in the composition of marine ecosystems (Frank *et al.* 2005; Heath 2005). These two decades have also brought increasing recognition of the effects of climate change and overexploitation on marine ecosystems (Perry *et al.* 2010a), but have also underscored the critical importance and often lack of understanding of the mechanisms that control marine ecosystems and their constituent populations. While correlations between biological and physical variables provide some insights into potential controls, these are generally unsuccessful *predictors* of marine ecosystem variability. The development of the capability to provide predictions of marine ecosystem variability has necessitated understanding couplings between the natural and human systems, development of interdisciplinary research programmes, and development of models based on first principles. The desire to develop predictive capabilities of marine ecosystem and population variability laid the foundation (NAS 1987; US GLOBEC 1988) for development of the Global Ocean Ecosystems Dynamics (GLOBEC) programme (GLOBEC 1997, 1999).

Prior to GLOBEC, recruitment to marine fish populations was believed to depend on variations in larval feeding and nutrition (e.g. Lasker 1978; Lasker and Sherman 1981; Sherman *et al.* 1983) or on the vagaries of advection of larvae into favourable/unfavourable environmental conditions (e.g. Iles and Sinclair 1982). In the first case, larval feeding was viewed in terms of a direct link from zooplankton to the consumer. In the latter, the scales of the physical environment were known but the details of the scales that were relevant to the planktonic organism were essentially unexplored (Chapter 5). These disconnects were partly the result of the fragmented approach that characterized research into marine ecosystems. Physical oceanography and biological studies were essentially separate disciplines. Studies that focused on coupling between the physical environment and biological populations were not typically integral parts of research programmes, resulting in difficulties in attributing causes, effects, and mechanisms. This in turn contributed to a lack of linkage and cohesion between the observational and modelling communities.

The research programmes that did attempt interdisciplinary studies were often restricted to a limited location and to a single season and were too short to provide insights into the effects of longer-term environmental variability on marine populations. The decadal and longer patterns of variability that are now known to be significant contributors to marine ecosystem and population variability (Chapters 2 and 10) were unresolved and hence overlooked as driving forces. A consequence of this is that, with the exception of the two hypotheses mentioned above, recruitment in marine animal populations was considered to be random variability around a constant mean.

The need for better understanding of environmental and biological controls on marine populations is of increasing relevance for the management of these natural resources during a period of increasing human and climate impacts. Management of marine natural resources since the 1980s has been generally structured around maintaining the biomass or maximizing the fishery yields of single species and as a result has disregarded interactions between species and has not explicitly included environmental effects on these interactions or on the target species. The trend now is towards ecosystem-based approaches with a focus on ecosystem stability (including maintenance of biodiversity), resilience to multiple impacts and protection of multiple ecosystem services and multiple species (Chapter 9). The role of human and societal needs was not, until recently, a component of the scientific advice underpinning marine resource management (Chapter 8), but such a need is now clear (Perry and Ommer 2003; Ommer *et al.* 2008, 2009).

The scientific research that preceded the GLOBEC programme clearly demonstrated the need to understand the dependence of population dynamics on the physical structure of the ocean and to link this to ecosystem dynamics. In turn, studies of ecosystem dynamics needed to focus on important components of particular ecosystems while retaining links to processes at basin and global scales (e.g. climate). New technologies for sampling and monitoring were needed which could be integrated with existing technologies and could facilitate linking physical and biological ecosystem components in a new generation of models (Chapters 5 and 7). The vision and perceived needs articulated by these earlier studies provided the basis for development of the GLOBEC programme in the mid-1990s. GLOBEC marked a fundamental change in research into marine ecosystem variability, which is reflected in the programme objective 'of advancing understanding of the structure and functioning of the global ocean ecosystem, its major subsystems, and its response to physical forcing so that a capability can be developed to forecast the responses of the marine ecosystem to global change'.

The preceding chapters provide detailed overviews of GLOBEC science that address aspects of the overall programme goal. This chapter summarizes these advances into general themes and indicates fundamental changes in conceptual views of the marine ecosystem that have resulted. The lessons learnt from the GLOBEC programme will result in significant changes in how to approach marine research programmes, thereby extending the legacy of GLOBEC science into future interdisciplinary research.

11.2 What has been learnt?

11.2.1 Understanding scales of processes

The view of marine ecosystems as operating along a continuum defined by space and time underpinned much of the research that was undertaken during GLOBEC (Chapter 5) because this organization of physical and biological interactions provided a convenient framework for developing research programmes and modelling efforts. However, the view of how these space- and time-scales interact to structure marine ecosystems has been significantly altered during GLOBEC. The early ideas presented a view of continuous interaction in which energy cascaded through the food web from larger to smaller scales. The scales at which organisms associated with different trophic levels accessed other parts of the space-time continuum were limited, as were their feedbacks to other parts of the food web. This view has changed to one in which marine ecosystem variability and population recruitment result from the integration of processes across all scales and includes direct as well as indirect interactions (Chapter 5). One of the most important advances has been the integration of humans into these scale-dependent processes (Chapters 3 and 8).

The initial approach used in GLOBEC was to focus on the physical environment as the major driver affecting changes in marine population abundances and variability, that is, bottom-up control of marine ecosystems. As GLOBEC science developed, this bottom-up control approach was replaced by one in which the effects of trophic level interactions and human effects (top-down controls) were accorded equal importance in controlling marine ecosystems (Chapters 4 and 7). As a result, the view of either bottom-up or top-down control of marine food webs has been modified to recognize that both types of control operate simultaneously and that the relative strength of these controls produces the marine ecosystem that is observed at any time or location. Thus, the two-dimensional space-time continuum view of marine ecosystems has been modified to reflect a multi-dimensional system that includes inputs and feedbacks among and between trophic or organizational levels at multiple scales (Chapters 5, 7, and 8).

The realization that marine ecosystems exhibit different scales of response to environmental forcing and perturbations highlighted the need for coupled physical-biological models that could explore the range of possible outcomes. As a result, most GLOBEC research programmes incorporated strong modelling components from the outset. However, much of the coupled modelling done in GLOBEC focused on the responses of key target species (Chapter 4) and included limited resolution of food web trophic structure in specific locations (Chapters 5 and 7). The focus on specific trophic levels (e.g. deYoung *et al.* 2004a) advanced food web understanding but clearly pointed to the importance of including processes and interactions at shorter and longer scales.

Understanding the scales of connectivity of marine populations and how these intersect trophic levels that apparently operate at different scales is now known to be critical in determining marine population variability (Cowen *et al.* 2006). The interactions between zooplankton species distribution and abundance, fisheries, and climate provide one example of how processes at multiple scales interact to produce variability in marine ecosystems (Lehodey *et al.* 2006; Chapter 2). Variability in recruitment to marine populations is now viewed in terms of circulation processes at multiple scales and growth and mortality processes that result from food web dynamics at multiple trophic levels (Werner *et al.* 1996; Chapters 2, 5, and 7). Thus, the two-dimensional space-time framework for viewing interactions in marine ecosystems is being replaced with one in which food web components evolve and intersect with one another and the environment at different scales.

An emergent issue from GLOBEC science is the importance of threshold responses in marine ecosystems to both climate and human-induced variability. An awareness of the ranges and scales of natural climate variability has resulted in a new understanding that ecosystem response to climate variability occurs over multi-decadal scales (Chapters 2 and 10). In its broadest sense, scales include the organizational and space- and time-scales that typically range from climate to plankton to fish and occasionally other top predators, such as seabirds, seals, and cetaceans. Such a definition excludes an important additional consideration, which is that human activities are changing how marine ecosystems respond to changes in climate (Chapter 3). For example, fishing that is focused on older, larger individuals and that targets specific stocks or sub-populations reduces the age structure and variability of fish populations and increases their sensitivity to climate variability and change (Perry *et al.* 2010a; Planque *et al.* 2010). The inclusion of the effects of human activities adds yet another dimension to the space-time framework used to view interactions in marine ecosystems. The scales chosen for studies of marine systems and human interactions typically match better across large geographical scales (e.g. global), less so for small-scale qualitative social and cross-scale studies, and can be sufficiently mismatched to prevent the recognition of the drivers and responses to global changes (Perry and Ommer 2003).

The recognition that changes in marine ecosystems can occur over short time-scales resulted in acceptance for the concept of regime shifts during GLOBEC. The view of marine ecosystems has developed from one that was characterized by variability about a constant state to one that considers a relatively rapid transition (regime shift) from one mean state to another mean state (Chapter 9). Studies during GLOBEC focused on identifying the causes of regime shifts and developing an understanding of the consequences for marine ecosystem structure and productivity. The importance of regional regime shifts for marine resources has led to improved understanding of coupling between coastal, open ocean, and atmospheric systems (Chapter 2). The feedbacks of these shifts into the larger climate system and into social and economic structures are now the subject of ongoing research programmes (e.g. Allison *et al.* 2009). Understanding of the causes and effects may make it possible to manage stocks in certain types of regime shifts (deYoung *et al.* 2008).

11.2.2 Alternative views of marine food webs

Prior to GLOBEC much of the understanding of marine food webs was from studies focused on quantifying primary production and microbial processes in the cycling and export of carbon (e.g. the Joint Global Ocean Flux Study (JGOFS, Fasham 2003). As the need for understanding of larval recruitment and marine population variability in a fisheries context developed it became apparent that lack of understanding of the dynamics by which lower and higher trophic levels are connected limited progress in this area. The role of zooplankton as intermediaries between primary producers and higher trophic levels was recognized as a critical but relatively unknown connection. For this reason, GLOBEC undertook targeted studies that focused on key zooplankton species (Chapter 4). Species-specific studies were not a new approach developed under GLOBEC. Rather, the contribution made by GLOBEC was to study the key species in conjunction with studies of their habitat, competitors, and predators, that is, in an ecosystem context. From these targeted studies emerged a view of the key species in relation to the environment and other components of the food web (Chapter 7).

The regional programmes undertaken during GLOBEC (Table 11.1) focused on understanding food web dynamics and their modifications due to environmental variability. A significant advancement from these studies was the recognition of the importance of functional diversity in marine food webs. The view of food webs is now one of the linked systems of alternative food webs that change in importance in response to environmental conditions and/or changes in species assemblages and abundance, that is, top-down and bottom-up controls (Frank *et al.* 2005; Chapter 7). The changed view of food webs includes the concept of dual pathways, which are composed of slow and fast response paths, which has provided insights into ecosystem resiliency in response to natural and anthropogenic forcings (Chapter 7).

Many examples arising from GLOBEC studies of this new understanding of what controls marine food webs are discussed in the previous chapters. Notable examples that were significantly advanced by GLOBEC science are articulations of the wasp-waist control idea (Rice 1995; Cury *et al.* 2000, 2001; Bakun 2006b; Chapter 7), observations of top-down control of zooplankton by fish and gelatinous zooplankton (Chapter 7), and observations of the effects of habitat change, such as reductions in sea ice extent and concentration in southern and northern polar regions, on interannual variability in energy flow through food webs (Chapter 7).

The importance of top predators, including humans, in controlling marine food web structure and function has been clearly demonstrated in all GLOBEC science programmes (Chapters 3, 7, and 8). Explicit inclusion of upper trophic levels allowed advancement of many of the ideas of the importance of trophic controls and alternative trophic pathways. That GLOBEC successfully incorporated higher trophic levels in all of the regional programmes (see discussion in Chapter 7) sets a precedent for future marine programmes focused on food web dynamics.

GLOBEC has advanced the concept that humans are integral components of marine ecosystems, with dual roles both as predators in marine food webs and as recipients of marine goods and services that enhance human livelihoods and well-being (Chapters 3, 8, and 9). The awareness that fisheries impact marine ecosystems and the status and productivity of their constituent fishery resources make it necessary to include ecosystem considerations in management strategies and policies, that is, an ecosystem-based approach to management. The revised views of food webs and the understanding of the effects of habitat change, trophic and species interactions, and responses to climate change (e.g. regime shifts) that have come from GLOBEC science programmes help to provide a foundation for moving from ecosystem science to ecosystem-based management of marine living resources (Chapter 9).

Table 11.1 Objectives, locations, and time frame of GLOBEC regional programmes[a].

Programme	Objective	Location	Time frame[a]
CLimate **I**mpacts on **O**ceanic **TO**p **P**redators (CLIOTOP)	Identify impact of both climate variability and fishing on the structure and function of open ocean pelagic ecosystems and their top-predator species	Global	2004–14
Ecosystem **S**tudies of **S**ub-Arctic **S**eas (ESSAS)	Understand the impact of global change on the sub-Arctic Seas, and to suggest innovative approaches to manage these important regions	Atlantic and Pacific sub-Arctic Seas	2005–15
ICES-GLOBEC **C**od and **C**limate **C**hange (CCC)	Understand how climate variability affects the productivity and distribution of cod stocks	North Atlantic Ocean	1992–2009
PICES-GLOBEC **C**limate **C**hange and **C**arrying **C**apacity (CCCC)	Understand and link climate change and variability to physical conditions and ecosystem structure and function in the North Pacific and its adjacent marginal seas	North Pacific Ocean and adjacent marginal seas	1995–2007
Southern **O**cean GLOBEC (SO GLOBEC)	Understand the physical and biological factors that contribute to enhanced Antarctic krill growth, reproduction, recruitment, and survivorship throughout the year and to link these to climate variability	West Antarctic Peninsula, Bransfield Strait, South Georgia and Scotia Sea, east Antarctica, Lazarev Sea	1993–2010
Small **P**elagic **F**ish and **C**limate **C**hange (SPACC)	Understand and predict climate-induced changes in the fish production of marine ecosystems	Eastern boundary current regions, Japan seas, Baltic Sea, northwest Atlantic Ocean	1997–2009

[a] In addition, science plans for two new regional programmes have been published in GLOBEC reports: ICED (Integrating Climate and Ecosystem Dynamics, Murphy *et al.* 2008, focusing on the Southern Ocean ecosystem) and BASIN (Basin-scale Analysis, Synthesis and INtegration, Wiebe *et al.* 2009, investigating the North Atlantic ecosystem). These programmes are considered the follow-on to previous GLOBEC programmes and have commenced their activities in 2008/09.

11.2.3 Marine ecosystems as social-ecological systems

Social-ecological systems are systems that have marine (including physical-biological subsystems) and human (including cultural, management, economic, and socio-political subsystems) components, which are highly interconnected and interactive. Traditionally, natural physical-biological marine ecosystems have been studied independently from their human components, and by different scientific disciplines with largely different scientific traditions (Chapter 8). GLOBEC was one of the new generation of programmes that attempted to collaborate with other disciplines to correctly interpret the causes and deal with the consequences of global changes in marine social-ecological systems.

Changes in marine ecosystems can have dramatic impacts on human societies that depend on marine resources and how these societies' responses affect marine ecosystems. Responses of the marine physical-biological and human subsystems to changes and impacts occur over a range of scales (Chapter 8). For example, the physical-biological marine system can adjust migration and distribution patterns (short-scale response) or food web dynamics and structure (longer-scale response). Human systems can adjust through changes in intensification of fishing (short-scale response) or diversification outside of fishing (longer scale response). At shorter time-scales, coping responses by both natural and human subsystems have common elements, but at longer time-scales, many of the adaptive capacities of the human subsystem have no analogues in the natural subsystem (Chapter 8). A result of the latter is to create a divergence in the long-term consequences of natural subsystem adaptations to environmental alterations compared with those of the human subsystem. The challenge is to develop governance systems for coupled marine social-ecological systems which account for these short and long-term consequences of significant environmental changes.

11.2.4 Paradigm shifts in the study of marine ecosystems

Many of the coupled numerical model systems now considered as standards for studying marine ecosystems underwent significant development and refinement in GLOBEC science programmes. The field of interdisciplinary modelling advanced as a result of GLOBEC studies showing that coupled models could be used to integrate physical and biological processes across a range of scales to produce realistic simulations of circulation and ecosystem processes for environments that ranged from polar to tropical and from coastal to open ocean. The merger of observational and modelling systems facilitated development of data assimilative marine ecosystem models. As a result, GLOBEC fostered an important shift in how research programmes are planned and implemented by including integration of physical-biological modelling and observations from the initial phases of research programmes (Chapter 5). However, this integration is still a work-in-progress because of the difficulties of

including all biological components above zooplankton, which has limited the integration to fish larvae (e.g. Vikebø *et al.* 2005) or particular fish species (Megrey *et al.* 2007b).

The GLOBEC science programmes explicitly included studies of top predators (fish, seals, seabirds, and cetaceans) and upper trophic level dynamics. The organisms associated with upper trophic levels are typically long-lived and are likely to be most impacted by decadal and perhaps longer climate variability (Chapter 2). Alterations in upper trophic level dynamics potentially can alter controls on ecosystem processes and result in the emergence of alternative pathways in marine food webs (Chapter 7). Thus, understanding the responses of these organisms and including these in coupled models will allow the development of projections of ecosystem responses to climate-scale processes to be developed. The response of top predators in the context of environmental and lower trophic level variability is now an integral consideration in studies of marine ecosystems' responses to climate change (Chapters 2 and 10).

GLOBEC science has provided concepts, data, and models that have strengthened ecosystem-based management approaches for marine living resources, thereby providing a direct transfer of basic science results to applications (Chapters 3, 8, and 9). The ecosystem approach to management, for example, requires the productivity of an individual exploited fish stock to be considered in the context of the productivity of the entire ecosystem. How changes in the ecosystem and repartition of energy between components lead to different exploitation scenarios is an area that has gained exposure as a result of GLOBEC studies (Chapter 9), the consequences of which are still relatively unexplored and unknown (Collie *et al.* 2008).

11.2.5 Importance of the comparative approach

Species-level studies were central to addressing the GLOBEC objective of understanding marine ecosystem and population variability. The target species were chosen for their ecological (e.g. Antarctic krill, *Euphausia superba*) or economic (e.g. Atlantic cod, *Gadus morhua*) importance in their regional ecosystem. Through the use of target species it was possible to isolate the direct effects of environmental processes and food web interactions on critical components of the life cycle, such as larval growth and recruitment processes. Most of the GLOBEC programmes included focused studies of specific zooplankton species or the early pelagic life stages of fisheries species, the results of which were synthesized through individual-based models that included growth and behavioural dynamics. Embedding these ecological models into physical circulation models allowed simulation of the growth and development of these organisms during transport by the ocean currents (Chapter 5). The results of these detailed studies provided enhanced understanding of the spatial and temporal dynamics of plankton and fish larvae and provided answers to questions of dispersal, retention, and connectivity of marine populations in a range of environments. The results from these different ecosystems can serve as proxies for manipulations of the same system, particularly when climatic change and over-fishing provide perturbations at the bottom and top of the food webs (Travers *et al.* 2007). Thus, similar approaches used to study the different target species provide a basis for comparisons of properties of different ecosystems across trophic levels and/or trophic function (Chapter 4).

Another important aspect of GLOBEC was the comparative approach among similar (and different) marine ecosystems across the globe (Chapters 4, 5, and 7) and the use of standard approaches and methodologies across systems (Chapter 6). Through the GLOBEC regional programmes (Table 11.1), comprehensive data sets were collected and a range of models was developed. Development of methods for using these data and models to make comparisons from one end of the food web to the other (e.g. end-to-end comparisons, Chapter 7) both within and between these systems is a goal of GLOBEC synthesis and will provide a long-term legacy for the programme. Through comparisons of diverse ecosystems, a new understanding is emerging of the potential impacts of climate and human-induced changes on important ecosystems and fisheries in eastern boundary current upwelling

ecosystems (SPACC: Checkley *et al.* 2009), ecosystems that produce Atlantic cod (CCC: Drinkwater *et al.* 2005), the pelagic and demersal ecosystems across the North Pacific (CCCC: Batchelder and Kim 2008), and the ecosystems of northern (ESSAS: Hunt *et al.* 2007) and southern (SO GLOBEC: Hofmann *et al.* 2004c, 2008) polar ecosystems. The use of coupled physical-biological models as the means for effective integration and synthesis across these programmes (Steele *et al.* 2007; Chapters 5 and 7) is a clear result of GLOBEC.

Comparative studies of the responses of human systems to environmental and/or socio-economic forcing to show the important consequences of these for marine food webs are also part of the approach used in GLOBEC (Chapters 3, 8, and 9). For example, the susceptibilities of Atlantic cod populations to environmental and climate variability (Chapter 2) and the consequences of variability in these populations to the human communities which depend on this species (Chapter 8) are becoming well known.

The comparative approach favoured in GLOBEC often required the use of new technologies, which were perfected or newly developed during GLOBEC. For example, the Continuous Underway Fish Egg Sampler (CUFES) was used to provide high-resolution distributional maps of fish eggs in several regions of the world's oceans to establish generic principles of fish-spawning strategies and habitat (Checkley *et al.* 2000) and genetic techniques were implemented to analyse the distribution of congeneric copepod species in the North Atlantic (Lindeque *et al.* 2006). New technologies were also refined and improved through use in GLOBEC field programmes that addressed questions that could not be resolved with traditional methodologies. For example, satellite relay data loggers that measure ocean temperature were deployed on crabeater seals (*Lobodon carcinophagus*) to resolve upper ocean variability in Antarctic continental shelf waters (Costa *et al.* 2008) and multi-frequency acoustics were used to map the distribution of Antarctic krill (Lawson *et al.* 2008a). These technologies improved field observations and provided data sets that influenced the development of coupled physical-biological models (Chapter 5) and the understanding of marine ecosystems (Chapter 7).

11.3 Emerging scientific themes

11.3.1 Scales and thresholds

Understanding the scales of processes and interactions in marine ecosystems gained during GLOBEC has provided the basis for new ideas about how the results of change are manifested in these systems. For example, linear changes, such as increasing temperature, can result in non-linear responses in ecosystems (Chapters 2, 5, and 8). An important result is the recognition that threshold effects, which arise from system non-linearities, are likely to have unanticipated outcomes in marine ecosystems. There are likely many such thresholds in the complex marine physical-biological and social-ecological systems. The adjustments of these systems to a changing climate may compound or increase the number of unexpected system changes. While it may be possible to identify, characterize, and locate some thresholds, others can only be anticipated as sources of risk. It is the latter that may in fact provide the response that shifts a marine ecosystem from one state to another (deYoung et al. 2008).

The unknown nature of threshold responses renders critical the understanding of alternative pathways in ecosystem energy flows. The potential for alternative states needs explicit inclusion in marine ecosystem research programmes and in coupled models. Without this the ability to recognize and understand fundamental changes in environmental structure will be compromised. These ecological regime shifts can result in a new 'base state' for the ecosystem. A potential consequence of such shifts is that the ecosystem may no longer support the food web or the fishery that is of commercial or ecological interest. For example, the reduction in sea ice extent and concentration in the South Atlantic sector of the Southern Ocean may be contributing to reduction in Antarctic krill abundance (Atkinson et al. 2004), which in turn has severe consequences for the top-predator community (Chapters 2 and 7). Alternative pathways also arise where fishing has induced a restructuring of the ecosystem in response to removal of particular species. For these systems, the cessation of fishing does not necessarily result in stock recovery because of fundamental changes in the ecosystem (Chapter 9).

11.3.2 Non-climate drivers of ecosystem change

The non-climate drivers of change, such as intensive fishing, are now recognized as important factors that result in ecosystem change (Pauly et al. 2002). This relatively recent understanding, which was recognized in the GLOBEC Implementation Plan (GLOBEC 1999), was not activated in the original focus of the programme. Recent research has demonstrated an interaction between fishing and environmental factors that compounds their direct impacts on marine ecosystem structure and functioning (Anderson et al. 2008). Specifically, the selectivity of fishing operations removes individuals with particular characteristics from the gene pool, thereby affecting structure and function at higher levels of organization. Fishing also leads to a loss of older age classes, reduces the mean size of individuals and mean trophic level of communities, and results in spatial contraction, all of which make marine ecosystems more sensitive to climate variability at interannual to interdecadal scales (Perry et al. 2010a). A consequence is that marine ecosystems under intense exploitation evolve towards stronger bottom-up control, greater sensitivity to climate forcing (Perry et al. 2010a; Planque et al. 2010), and less dynamic stability (Anderson et al. 2008). Research into the interactions and feedbacks of fisheries and ecosystems will continue to evolve in coming years. In particular, understanding how selective harvesting can lead to the unstable booms and busts that can precede systematic declines in stock levels and the implications of this for resource management will be of importance.

The current view of the Earth is of a system that functions in a fully integrated and coordinated manner, where human activities have become so pervasive and profound that they affect the earth at a global scale in complex, interactive, and accelerating ways (Crutzen and Stoermer 2000). Humans must be considered as significant drivers of change (Chapter 3) as well as being affected by these changes (Ommer et al. 2008, 2009). It is therefore essential that future research programmes embrace the need for multidisciplinary research in their design and implementation, if they are to respond to societal challenges. Multi- and interdisciplinary research programmes, in this context, include economic and social science components in order to bridge the multidisciplinary gap to ensure the sustainability of our marine ecosystems.

11.3.3 Advancements in modelling capability

An emerging area of marine modelling is the development of nested models that allow transfer of information from global to regional-scale models (downscaling) and from regional to global-scale models (upscaling). Initial implementation of nested ocean circulation models has shown marked improvement in model skill and fidelity (Chapter 5). Continued advancements in two-way scaling approaches, especially for regionally based models, are needed to develop projections of oceanic responses to climate change. Incorporation of upscaling/downscaling into the marine ecosystem module of coupled physical-biological models remains a challenge and will require development of realistic marine ecosystem models at basin and global scales (Chapter 10).

Data assimilative marine ecosystem models have advanced as a result of GLOBEC science (Chapters 5 and 10) and these models are now considered integral parts of marine ecosystem research (e.g. Friedrichs *et al.* 2007; Stow *et al.* 2009). The development of these models is such that real-time predictions of standing stock up to the level of zooplankton are now being attempted (see www.ncof.co.uk/ Ecosystem-Model-Forecast.html) and conceptual frameworks for modelling fish and shellfish responses to climate change are being developed (Hollowed *et al.* 2009). The challenge now is to develop these models to provide the equivalent of weather forecasts for components of the marine ecosystem, such as fishery stocks. These projections will require climate projections as well as statistics of the uncertainties associated with the projections (Chapter 10). Integration across the wide range of space- and time-scales involved in predicting from ocean to fish is a major challenge.

The issues and stresses facing marine systems require consideration and inclusion of human and social systems in more traditional marine ecosystem models; a significant and daunting challenge. A potential approach for meeting this challenge is

to continue development and expansion of community-based models, such as the Regional Ocean Modelling System (Haidvogel *et al.* 2008) and the North Pacific Ecosystem Model for Understanding Regional Oceanography (NEMURO; Kishi et al. 2007a; Werner *et al.* 2007a; Chapter 5). The value of community-based versus individual models is a matter of debate. Multiple individual models, such as promoted by the Intergovenmental Panel on Climate Change approach of ensemble climate models, potentially provide a range of results, but community-based models can include the results of multiple models. However, the focus of an interdisciplinary research community on a particular model structure allows advances to be made more quickly than would happen with resources distributed over several models. GLOBEC fostered collaborations between empiricists, social scientists, and modellers and continued development of this community is needed as much as the emergence of new model structures to address issues of climate change.

The modelling studies initiated under GLOBEC made significant advances in linking scales across environmental, food web, and climate processes. These models provide realistic simulations at large space- (e.g. basin) and time-scales (e.g. annual and greater) (Chapters 5 and 10). However, continued advances in model skill and the availability of realistic projections will require inclusion of disparate information at multiple scales. A large unknown at present is how to include human effects in these models. These areas are where new efforts in observational and experimental work, conceptual understanding, model development, and scientific capacity building need to be directed.

11.3.4 New directions in modelling marine ecosystems

The realistic simulation of the life history and physiology of individual species was an original goal of the modelling efforts in GLOBEC (GLOBEC 1993a,b; US GLOBEC 1995). Numerous models were developed as part of GLOBEC that simulated growth and development of individual zooplankton species and fish larvae and some were embedded in circulation models and/or lower trophic level models (Chapters 5 and 7). Thus, the integration of species-based observational and experimental studies in detailed and realistic models for a limited number of marine zooplankton and fish larvae (Chapters 5 and 7) is a success of the GLOBEC programme. The challenge now is to incorporate detailed species-level models into food web and biogeochemical models and to extend to scales that are larger than those of individual organisms. The melding of species, food web, and biogeochemical models is needed to link marine animal populations, ecosystem, and climate variability. The quantification of direct climate impacts on the production of marine ecosystems has been historically hampered by difficulties of downscaling global climate models to the scales of biological relevance, lack of adequate ecosystem models capable of scaling-up biological production and transfers across the food web, and difficulties in separating the multiple additional biological and socio-ecological stressors affecting marine ecosystem production. These issues remain as challenges to be addressed for the next generation of marine ecosystem models (Barange *et al.* in press).

The development of models that use mechanistic approaches based on metabolic and bioenergetic theory and/or size-based structures provide a starting point for moving in a direction that will potentially result in a major advancement for marine ecosystem modelling (Chapter 7). These models raise the understanding of ecological interactions to the level of communities, rather than species and use macro-ecological theory to predict the steady-state and dynamic properties of marine food webs (e.g. Brown *et al.* 2004). These approaches assume that the fundamental size-based processes that determine the use and transfer of energy in communities respond to changes in temperature and primary production in consistent and predictable ways. Despite offering greater potential to give accurate predictions of total production, the disadvantage of a community approach is that it focuses on aggregate production, so their predictions are particularly valuable in countries where fisheries yields primarily meet subsistence needs and/or are converted to fishmeal (Barange *et al.* in press a).

Recent advances in adaptive ecosystem models with emergent properties (Follows *et al.* 2007) provide a new direction for marine ecosystem model development. Explicit inclusion of the inherent variability in marine organisms (genetic diversity) provides the possibility of exploring ranges of responses to climate forcing and hence insights into the consequences of threshold responses. However, these models will require the ability to map genotype to phenotype, which in turn requires integration of the developing area of marine genomics, as well as the more traditional experimental, field and laboratory-based studies. Experimentalists and modellers will still need to continue to work together in the manner that has developed during GLOBEC.

11.4 The future

11.4.1 Key lessons learnt from GLOBEC

The initial GLOBEC programme focus on marine zooplankton populations and their predators resulted in advances in understanding of population dynamics, community interactions, animal behaviour, and animal physiology (Chapters 4, 6, and 7). However, studies of the dynamics of the trophic level(s) that support zooplankton populations (e.g. food) and related carbon and nutrient cycling, and flux export were not explicitly included in the research programmes that were developed around GLOBEC science objectives. Thus, potential changes in marine primary production and biogeochemical cycling that could arise from the same climate and anthropogenic factors influencing marine zooplankton populations were not explicitly part of GLOBEC science. Moreover, within the larger food web, attention was focused on a limited number of key species (Chapter 4). This placed restrictions on the science programmes and ultimately the understanding gained from them. The interconnectivity of marine ecosystems requires a view of these systems as the product of integrated environmental, biogeochemical, and food web interactions that are part of a larger earth system.

The GLOBEC programme evolved to include human, social, and cultural issues as part of its research agenda (Chapter 3, 8, and 9). However, these issues were introduced during the mid-term of GLOBEC and full integration of these into all research programmes did not occur. The field studies associated with regional and national research programmes were already ongoing and resources to add additional studies and components were limited or non-existent. Also, some of the GLOBEC programmes were developed around science questions and for regions where inclusion of human and societal factors was difficult to implement after the fact (e.g. Southern Ocean). Inclusion of social and human considerations and close collaboration with social scientists must be included in future marine ecosystem research programmes from the start (e.g. Ommer *et al.* 2009).

Marine ecosystems respond to changes that occur at seasonal and interannual scales and at local and regional scales, which are sufficiently short to make their study feasible within the lifetime of a typical research programme. The benefit of GLOBEC is that it sustained decade-long (and longer for some national programmes) research programmes for some regions. As a result, some of the GLOBEC field studies focused on regional ecosystems that were actually changing during the lifetime of the programme. For example, the Southern Ocean GLOBEC programme factored climate-related changes in the sea ice environment into the field study design (Hofmann *et al.* 2002).

The GLOBEC programme provided a long-term structure that maintained the focus of a research community on a specific set of science questions. The advances in understanding that accrue from the individual regional and national programmes developed under GLOBEC provide a measure of success for the overall programme. As evidenced by the previous chapters, these were significant. Without the infrastructure that GLOBEC provided, it is unlikely that a research community that included participants from multiple countries and ocean-wide field programmes would have directed resources and facilities towards studying marine ecosystem variability would have directed resources and facilities towards studying marine ecosystem variability. Thus, GLOBEC showed the importance and need for long-term coordinated programmes to address long-term issues, such as the consequences of climate change for marine ecosystems.

11.4.2 GLOBEC benefits to society

GLOBEC was fundamentally a natural science programme focused on understanding the effects and impacts of climate variability and climate change on marine ecosystems. The science programmes developed under GLOBEC were designed to provide descriptions of the current state of marine ecosystems rather than descriptions of what could exist. As a programme, GLOBEC provided scientific information to advance understanding of marine ecosystems and some of this eventually was incorporated into policy and decision making. However, direct interaction with policy and decision makers was limited and was typically one way. Through GLOBEC science strategies for engagement of scientific research in policy and decision making were eventually developed and the extent to which these continue to be used provides a measure of how society has benefited from GLOBEC research.

During the GLOBEC programme it has become clear that both the atmospheric climate and the marine environment are dynamic, unstable, and vary over a wide range of space scale and timescale. A simple but powerful outcome of ecosystem science following from this, in the context of exploited resources, is the understanding that an unfavourable environment may prevent resources to continue delivering services despite best management efforts, and that bad management will in turn prevent ecosystem service delivery even if the environment is adequate (Chapters 2, 3, 8, and 9). This duality has provided a scientific basis for two of the most powerful management principles now well entrenched in natural resource management: (1) the precautionary approach, which states that if an action or policy might cause severe or irreversible harm, and in the absence of scientific consensus that harm would not ensue, the burden of proof falls on those who would advocate taking the action; and (2) the ecosystem approach, a strategy that assumes that resource management needs to take into consideration the effects of actions on every element of the ecosystem of which the resource is part of, recognizing that all elements of the ecosystem are linked (Chapter 9). In this way GLOBEC has been particularly effective in providing conceptual understanding of ecosystem functioning for policy development, and tools and methodologies to incorporate climate-driven variability for resource management applications.

In the second phase the GLOBEC programme was influenced by its interactions with the social sciences. Issues of ecosystem sustainability, resilience, and adaptive capacity (Chapters 3, 8, and 9) became of interest and were incorporated into GLOBEC research programmes and outputs from these have been incorporated into sustainable development resolutions (e.g. the World Summit for Sustainable Development of 2002) and social and economic fishery policies. Thus, GLOBEC science has informed policy and improved decision making at all levels. Marine ecosystem research programmes that follow GLOBEC will have to continue developing science-based principles in support of this evolving policy that expects all ecosystem components and all human actions to be addressed by comprehensive and consistent policy strategies.

The GLOBEC programme facilitated training of a generation of scientists who will form the core for future research into ocean, climate, and societal issues related to marine systems. GLOBEC had a policy to particularly engage with developing countries, facilitating their participation in programme meetings, workshops and symposia, and with younger scientists through summer schools and travel grants. This next generation will take ownership of approaches, contributions, innovative technologies, and connections to social and economic systems that form the legacy of the GLOBEC programme (Table 11.2) and extend these to address the next set of issues and questions.

11.4.3 Importance of infrastructure

The existence of infrastructure that will support and sustain long-term integrated programmes remains a significant challenge. International collaboration and coordination has been the hallmark of GLOBEC. The regional programmes developed under the auspices of GLOBEC (Table 11.1) typically consisted of studies proposed by individual nations that were developed around GLOBEC science goals and objectives. The international nature of these programmes came from coordination of the nationally based programmes through the GLOBEC International Scientific Steering Committee (SSC) and by coordination bodies with a regional focus, such as the Southern Ocean or North Pacific (e.g. the North

Table 11.2 Topics to which GLOBEC research has made significant contributions.

Topic	Contribution
Approaches to science	Multidisciplinary/interdisciplinary international collaboration
	Coupled models as integrative tools
	Multi-scale (time, space, and institutional) analysis
	Enhanced understanding of high trophic levels and top-down controls
Contributions to knowledge	Ecosystem structure and function
	Forcing of ecosystems
	Physical-biological-human interactions and feedbacks
Innovative technologies	Sampling and technological advances
	Coupled models (trophic, scale, and time)
	Retrospective studies on past ecosystem states
Information transfer to management bodies	Policy (providing conceptual understanding of ecosystem function)
	Managers (providing tools to incorporate climate-driven variability)
	Communities (enhancing communication on global environmental change and marine sustainability)
Education and outreach	Curriculum development
	Web-based instruction and data availability, including animations (scenarios)
	Lessons learnt transfer to future marine ecosystem research programmes
	Training of graduate students and postdoctoral researchers

Pacific Marine Science Organisation, PICES). These other coordination bodies had representation on the GLOBEC SSC, which ensured communication and coordination. The GLOBEC SSC provided a central stable structure that allowed evaluation of scientific ideas and programmes through the broader scientific community and through international organizations.

The establishment of a stable International Project Office (IPO) with a supported staff provided a central focus for the GLOBEC programme, which allowed coordination and implementation at national, regional, and international levels. The IPO provided fund-raising and management for the international GLOBEC programme and centrally managed regional activities, which allowed the regional programmes to focus resources on science activities. Much of the funding and fund-raising activities were in support of workshops and conferences that led to development of regional programmes (Table 11.1), dissemination of GLOBEC science results (e.g. Open Science Meetings), and development of new foci within the programme (e.g. human dimensions). The IPO also provided infrastructure and support for publication of GLOBEC reports, special contributions, and newsletters, which facilitated communication within the GLOBEC programme and with the broader scientific community. An important role filled by the GLOBEC IPO was as a meta-data archive and repository for GLOBEC programmes. The central organizing structure provided by the GLOBEC IPO was critical to the success of the overall programme. The GLOBEC IPO set a high standard for future international multidisciplinary research programmes.

A particular achievement of GLOBEC has been the success of resolving processes at geographical scales larger than traditional science programmes. For example, the combination of several national efforts led to a basin-scale understanding of the overwintering dynamics of the copepod *Calanus finmarchicus* in the North Atlantic from Labrador to Norway (Heath *et al.* 2004). This focus on large-scale processes required that research funding, resources, and logistics be coordinated across funding agencies, institutions, and scientific communities from many countries. Although programme timing and level of funding varied among the countries that participated in GLOBEC regional programmes (Table 11.1), the clear message was that it is possible to coordinate science at an international level to address common science questions. GLOBEC also showed that synthesis of programme results at an international level provides far more beyond what is gained through individual programme synthesis efforts. This achievement has inspired new research programmes that are no longer limited by the field sampling capabilities of a single country (e.g. BASIN, Wiebe *et al.* 2009; ICED, Murphy *et al.* 2008; Table 11.1). The continuation and success of large-scale programmes will require development of international science infrastructure that facilitates funding and resources for multinational research programmes.

The globalization of science and the issues now facing society, such as environmental change, requires programmatic infrastructure with the flexibility to allow international collaborations to go forward in a truly coordinated manner. The scientific community is now faced with a shift in focus from demonstrating that global environmental change is occurring to determining approaches for mitigation and adaptation. This will open new and complex research opportunities, although many important basic natural science questions about responses of marine ecosystems to the dual effects of climate and human activities will remain.

11.4.4 Next steps

A legacy from GLOBEC is the demonstrated need to develop research programmes with a whole ecosystem approach that connects across all trophic levels and encompasses regional-scale to basin/global-scale studies and models in order to understand processes and impacts of global change. The Integrated Marine Biogeochemistry and Ecosystem Research (IMBER) programme is an attempt to build upon what has been learnt in GLOBEC and JGOFS. For example, the IMBER programme goal of understanding food webs end to end recognizes from the outset that more complete characterizations of dominant energy pathways and potential alternative pathways is required to develop scenarios of responses of marine ecosystems to climate and human-induced changes.

The ecosystem-based approaches to management of marine resources must include strategies for incorporating changing human behaviour and prediction of consequences on marine ecosystems (Chapter 8). Many potential structures exist for development of ecosystem-based management, for example, keystone species, social needs, cultural target species, regional connectivity and marine protected areas, and which of these is most appropriate will be determined by the understanding of the system.

A major objective of GLOBEC was the development of new technologies to sample the ocean and integrating these with existing systems (Chapter 6). The role of monitoring the environment to establish baselines and to detect change was emphasized early in the development of GLOBEC. The early emphasis in GLOBEC on modelling resulted in a natural connectivity to observing and experimental systems. GLOBEC fostered the development of Observational System Simulation Experiments (OSSEs; Chapter 5) that included nested arrays, multi-platform instrumentation, and scale-based physical and biological sampling. However, the institutional support, infrastructure, and user community needed to implement and sustain such systems was lacking. Monitoring is now considered an integral part of science and is the basis for many important science discoveries. Global observational networks, such as Argo (Wilson 2000), the Global

Ocean Observing System (GOOS: Alverson 2005; Alverson and Baker 2006), and the Global Earth Observation System of Systems (GEOSS: Alverson 2007), now exist and data from these systems are readily available. Technological changes are moving towards sampling systems with autonomous sensors, although much remains to be done for biological sensors. The science results and new technology, including advances in modelling, arising from GLOBEC perhaps now can provide transitions to the development of long-term sustained observing systems that include biological as well as physical measurements (Chapters 5 and 6).

Many challenges will have to be overcome, including the development of methodologies to underpin integrative research between the natural, social, and economic sciences. The multiple-scale nature of the problems will require zooming from local to global impacts and response, highlighting policy needs across the same scales and potentially illuminating the path to global stewardship principles. Finally, at the start the GLOBEC science programmes were designed to advance understanding of an overarching science objective. However, this traditional view of science is changing. Science is required more and more to engage in public issues where facts are uncertain, values are in dispute, stakes are high, and decisions are urgent (Funtowicz and Ravetz 1993). Future research programmes must be designed to incorporate this new approach, which will make them challenging to develop but potentially rewarding for society.

References

Acheson, J. (1981). Anthropology of fishing. *Annual Review of Anthropology*, **10**, 275–316.

Adams, P.B., Botsford, L.W., Gobalet, K.W., *et al*. (2007). Coho salmon are native south of San Francisco Bay: a reexamination of North American coho salmon's southern range limit. *Fisheries*, **32**, 441–51.

Adger, W.N., Hughes, T.P., Folke, C., *et al*. (2005). Social-ecological resilience to coastal disasters. *Science*, **309**, 1036–9.

Ådlandsvik, B. (2008). Marine downscaling of a future climate scenario for the North Sea. *Tellus A*, **60**, 451–8.

Aebischer, N.J., Coulson, J.C., and Colebrook, J.M. (1990). Parallel long-term trends across four marine trophic levels and weather. *Nature*, **347**, 753–5.

Agostini, V.N., Bakun, A., and Francis, R.C. (2007). Larval stage controls on Pacific sardine recruitment variability: high zooplankton abundance linked to poor reproductive success. *Marine Ecology Progress Series*, **345**, 237–44.

Ainley, D.G. (2003). *The Adelie penguin: bellwether of climate change*. Columbia University Press, New York.

Ainley, D.G. and Jacobs, S.S. (1981). Sea bird affinities for ocean and ice boundaries in the Antarctic. *Deep-Sea Research I*, **28**, 1173–86.

Ainley, D.G. and MeMaster, D.P. (1990). The upper trophic levels in polar marine ecosystem. In: Dayton, P.K., ed. *Polar oceanography, Part B: chemistry, biology, and geology*. Academic Press, La Jolla, San Diego, CA, pp. 599–629.

Ainley, D.G. Spear, L.B., Tynan C.T., *et al*. (2005). Physical and biological variables affecting seabird distributions during the upwelling season of the northern California Current. *Deep-Sea Research II*, **52**, 123–43.

Aita, M.N., Yamanaka, Y., and Kishi, M.J. (2003). Effects of ontogenetic vertical migration of zooplankton on annual primary production-using NEMURO embedded in a general circulation model. *Fisheries Oceanography*, **12**, 284–90.

Aksnes, D.L. (2007). Evidence for visual constraints in large marine fish stocks. *Limnology and Oceanography*, **52**, 198–203.

Aksnes, D.L. and Blindheim, J. (1996). Circulation patterns in the North Atlantic and possible. impact on population dynamics of *Calanus finmarchicus*. *Ophelia*, **44**, 7–28.

Aksnes, D.L. and Giske, J. (1990). Habitat profitability in pelagic environments. *Marine Ecology Progress Series*, **64**, 209–15.

Aksnes, D.L. and Ohman, M.D. (1996). A vertical life table approach to zooplankton mortality estimation. *Limnology and Oceanography*, **41**, 1461–9.

Aksnes, D.L., Miller, C.B., Ohman, M.D., *et al*. (1997). Estimation techniques used in studies of copepod population dynamics- a review of underlying assumptions. *Sarsia*, **82**, 279–96.

Aksnes, D.L. Nejstgaard, J., Soedberg, E., *et al*. (2004). Optical control of fish and zooplankton populations. *Limnology and Oceanography*, **49**, 233–8.

Aksnes, D.L., Troedsson, C., and Thompson, E.M. (2006). A model of developmental time applied to planktonic embryos. *Marine Ecology Progress Series*, **318**, 75–80.

Alcaraz, M., Paffenhöfer, G.-A., and Strickler J.R. (1980). Catching the algae: a first account of visual observations on filter-feeding calanoids. In: Kerfoot, W.C., ed. *Evolution and ecology of zooplankton communities*. University Press of New England, Hanover, New Hampshire and London, pp. 241–8.

Alcaraz, M., Saiz, E., and Calbet, A. (1994). Small-scale turbulence and zooplankton metabolism: effects of turbulence on hearthbeat rates of planktonic crustaceans. *Limnology and Oceanography*, **39**, 1465–70.

Alder, J. and Sumaila, U.R. (2004). Western Africa: a fish basket of Europe past and present. *Journal of Environment and Development*, **13**, 156–78.

Alexander, V. and Niebauer, H.J. (1989). Recent studies of phytoplankton blooms at the ice edge in the Southeast Bering Sea. *Rapports et Proces-Verbaux des Reunions du Conseil International pour l'Exploration de la Mer*, **188**, 98–107.

Alexander, V., Henrichs, S.M., and Niebauer, H.J. (1996). Bering Sea ice dynamics and primary production. *Polar Biology*, **9**, 13–25.

Alheit, J. and Bakun, A. (2009). History of international co-operation in research. In: Checkley, D., Alheit, J., Oozeki, Y. *et al.* eds. *Climate change and small pelagic fish*. Cambridge University Press, Cambridge.

Alheit, J. and Bakun, A. (2009). Population synchronies within and between ocean basins: Apparent teleconnections and implications as to physical–biological linkage mechanisms. *Journal of Marine Systems*, **79**, 267–285.

Alheit, J. and Hagen, E. (1997). Long-term climatic forcing of European herring and sardine populations. *Fisheries Oceanography*, **6(2)**, 130–9.

Alheit, J. and Hagen, E. (2001). The effect of climatic variation on pelagic fish and fisheries. In: Jones, P.D., Ogilvie, A.E.J., Davies, T.D. *et al. History and climate: memories of the future.* Kluwer Academic/Plenum Publishers, New York, pp. 247–65.

Alheit, J. and Niquen, M. (2004). Regime shifts in the Humboldt current ecosystem. *Progress in Oceanography*, **60**, 201–22.

Alheit, J., Möllmann, C., Dutz, J., *et al.* (2005). Synchronous ecological regime shifts in the central Baltic and the North Sea in the late 1980s. *ICES Journal of Marine Science*, **62**, 1205–15.

Alheit, J., Roy, C., and Kifani, S. (2009). Decadal-scale variability in populations. In: Checkley, D., Alheit, J., Oozeki, Y., *et al.* eds. *Climate change and small pelagic fish.* Cambridge University Press, Cambridge.

Allain, G., Petitgas, P., and Lazure, P. (2001). The influence of mesoscale ocean processes on anchovy (*Engraulis encrasicolus*) recruitment in the Bay of Biscay estimated with a three-dimensional hydrodynamic model. *Fisheries Oceanography*, **10**, 151–63.

Allain, G., Petitgas, P., and Lazure, P. (2004). Use of a biophysical larval drift growth and survival model to explore the interaction between a stock and its environment: anchovy recruitment in Biscay. *ICES Council Meeting Papers*, **CM 2004/J:14**.

Allison, E.H. (2001). Big laws, small catches: global ocean governance and the fisheries crisis. *Journal of International Development*, **13**, 933–50.

Allison, E.H. and Ellis, F. (2001). The livelihoods approach and management of small-scale fisheries. *Marine Policy*, **25**, 377–88.

Allison, E.H., Adger, W.N., Badjeck, M.-C., *et al.* (2005). *Effects of climate change on the sustainability of capture and enhancement fisheries important to the poor: analysis of the vulnerability and adaptability of fisherfolk living in poverty.* Final Technical Report. Project No. R4778J Fisheries Management Science Programme. Department for International Development, London.

Allison, E.H., Perry, A.L., Badjeck, M.-C., *et al.* (2009). Vulnerability of national economies to potential impacts of climate change on their fisheries. *Fish and Fisheries*, **10**, 173–96.

Alverson, K. (2005). Watching over the world's oceans. *Nature*, **434**, 19–20.

Alverson, K. (2007). Why the world needs a global ocean observing system. In: *The full picture*. Tudor Rose, London, pp. 76–8.

Alverson, K. and Baker, D.J. (2006). Taking the pulse of the oceans. *Science*, **314**, 1657.

Ambler, J.W. (2002). Zooplankton swarms: characteristics, proximal cues and proposed advantages. *Hydrobiologia*, **480**, 155–64.

Andersen, T., Elser, J.J., and Hessen, D.O. (2004). Stoichiometry and population dynamics. *Ecology Letters*, **7**, 884–900.

Andersen, T., Faeerovig, P.J., and Hessen, D.O. (2007). Growth rate versus biomass accumulation: different roles of food quality and quantity for consumers. *Limnology and Oceanography*, **52**, 2128–34.

Anderson, C.N.K., Hsieh, C.-H., Sandin, S.A., *et al.* (2008). Why fishing magnifies fluctuations in fish abundance. *Nature*, **452**, 835–9.

Anderson, P.J. and Piatt, J.F. (1999). Community reorganization in the Gulf of Alaska following ocean climate regime shift. *Marine Ecology Progress Series*, **189**, 117–23.

Anderson, T.R. (2005). Plankton functional type modeling: running before we can walk?. *Journal of Plankton Research*, **27**, 1073–81.

Antonov, J.I., Levitus, S., and Boyer, T.P. (2005). Thermosteric sea level rise, 1955–2003. *Geophysical Research Letters*, **32(12)**, L12602, doi:10.1029/2005GL023112.

Arai, M.N. (1997). *A functional biology of Scyphozoa*. Chapman & Hall, London, 316pp.

Arai, M.N. (2005). Predation on pelagic coelenterates: a review. *Journal of the Marine Biological Association of the United Kingdom*, **85**, 523–36.

Arcos, D.F., Cubillos, L.A., and Núñez, S.P. (2001). The jack mackerel fishery and *El Niño* 1997–98 effects off Chile. *Progress in Oceanography*, **49**, 597–617.

Argo Science Team. (2001). ARGO: the global array of profiling floats, In: Koblinksy, C.J. and Smith, N.R., eds. *Observing the ocean for climate in the 21st Century*. GODAE Bureau of Meteorology, Melbourne, Australia, pp. 248–58.

Arnason, R. (2006). Global warming, small pelagic fisheries and risk. In: Hannesson, R., Barange, M., and Herrick, S., Jr., eds. *Climate change and the economics of the world's fisheries: examples of small pelagic fish stocks*. Edward Elgar, Cheltenham, UK, pp. 1–32.

Arrigo, K.R., van Dijken, G., and Pabi, S. (2008). Impact of a shrinking Arctic ice cover on marine primary production. *Geophysical Research Letters*, **35**, L19603, doi:10.1029/2008GL035028.

Arrow, K., Solow, R., Portney, P.R., *et al.* (1993). Report of NOAA panel on contingent valuation. *Federal Register*, **58(10)**, 4601–14.

Arts, M.T., Ackman, R.G., and Holub, B.J. (2001). Essential fatty acids in aquatic ecosystems: a crucial link between diet and human health and evolution. *Canadian Journal of Fisheries and Aquatic Sciences*, **58**, 122–37.

Ashby, D.M. (2004). Update of the GLOBEC national, multinational and regional programme activities. *GLOBEC Special Contribution*, **7**, 195pp.

Ashjian, C.J., Davis, C.S., Gallager, S.M., *et al.* (2001). Distribution of plankton, particles, and hydrographic features across Georges Bank described using the Video Plankton Recorder. *Deep-Sea Research II*, **48**, 245–82.

Ashjian, C.J., Davis, C.S., Gallager, S.M., *et al.* (2008). Distribution of larval krill and zooplankton in association with hydrography in Marguerite Bay, Antarctic Peninsula, in austral fall and winter 2001 described using the Video Plankton Recorder. *Deep-Sea Research II*, **55**(3–4), 455–71.

Astorkiza, K., Del Valle, I., Astorkiza, I., *et al.* (2006). Participation. In: Motos, L. and Wilson, D.C., eds. *The knowledge base for fisheries management. Developments in Aquaculture and Fisheries Science Series*, **36**, 239–65. Elsevier.

Ástthórsson, O.S. and Gislason, A. (1995). Long-term changes in zooplankton biomass in Icelandic waters in spring. *ICES Journal of Marine Science*, **52**, 657–68.

Ástthórsson, O.S. and Gislason, A., Hallgrimsson, I., *et al.* (1983). Variations in zooplankton densities in Icelandic waters in spring during the years 1961–1982. *Journal of the Marine Research Institute*, **7**(2), 73–113.

Atkinson, A., Siegel, V., Pakhomov, E., *et al.* (2004). Long-term decline in krill stock and increase in salps within the Southern Ocean. *Nature*, **432**, 100–3.

Atkinson, A., Siegel, V., Pakhomov, E.A. *et al.* (2008). Oceanic circumpolar habitats of Antarctic krill. *Marine Ecology Progress Series*, **362**, 1–23.

Atkinson, A., Whitehouse, M.J., Priddle, J., *et al.* (2001). South Georgia, Antarctica: a productive, coldwater, pelagic ecosystem. *Marine Ecology Progress Series*, **216**, 279–308.

Atkinson, D.B., Rose, G.A., Murphy, E., *et al.* (1997). Distribution changes and abundance of northern cod (*Gadus morhua*), 1981–1993. *Canadian Journal of Fisheries and Aquatic Sciences*, **54**, 132–8.

Attrill, M.J. and Power, M. (2002). Climatic influence on a marine fish assemblage. *Nature*, **417**, 275–8.

Attrill, M.J., Wright, J., and Edwards, M. (2007). Climate-related increases in jellyfish frequency suggest a more gelatinous future for the North Sea. *Limnology and Oceanography*, **52**, 480–5.

Auad, G., Miller, A., and Di Lorenzo, E. (2006). Long term forecast of oceanic conditions off California and biological implications. *Journal of Geophysical Research*, **111**, C09008, doi:10.1029/2005JC003219.

Augustin, C.B. and Boersma, M. (2006). Effects of nitrogen stressed algae on different *Acartia* species. *Journal of Plankton Research*, **28**, 429–36.

Aydin, K.Y., Boldt., J., Gaichas, S., *et al.* (2006). Ecosystem assessment of the Bering Sea/Aleutian Islands and Gulf of Alaska management regions. In: *Stock assessment and fishery evaluation report for the groundfish resources or the Bering Sea/Aleutian Islands regions*. Northeast Pacific Fisheries Management Council, Anchorage, AK, pp. 24–89.

Aydin, K.Y., McFarlane, G.A., King, J.R., *et al.* (2005). Linking oceanic food webs to coastal production and growth rates of Pacific salmon (*Oncorhynchus* spp.), using models on three scales. *Deep-Sea Research II*, **52**, 757–80.

Backhaus, J.O., Harms, I.H., Krause, M., *et al.* (1994). An hypothesis concerning the space-time succession of *Calanus finmarchicus* in the northern North Sea. *ICES Journal of Marine Science*, **51**, 169–80.

Badjeck, M.-C. (2008). *Vulnerability of coastal fishing communities to climate variability and change: implications for fisheries livelihoods and management in Peru*. Ph.D. Thesis. Center for Tropical Marine Ecology, University of Bremen, Germany, 220pp.

Badjeck, M.-C., Mendo, J., Wolff, M., *et al.* (2009). Climate variability and the Peruvian scallop fishery: the role of formal institutions in resilience building. *Climatic Change*, **94**(1–2), 211–32.

Bagoien, E. and Kiørboe, T. (2005a). Blind dating-mate finding in planktonic copepods, I: tracking the pheromone trail of *Centropages typicus*. *Marine Ecology Progress Series*, **300**, 105–15.

Bagoien, E. and Kiørboe, T. (2005b). Blind dating-mate finding in planktonic copepods, III: hydromechanical communication in *Acartia tonsa*. *Marine Ecology Progress Series*, **300**, 129–33.

Bagoien, E., Kaartvedt, S., Aksnes, D.L., *et al.* (2001). Vertical distribution and mortality of overwintering *Calanus*. *Limnology and Oceanography*, **46**, 1494–510.

Baier, C.T. and Napp, J.M. (2003). Climate-induced variability in *Calanus marshallae* populations. *Journal of Plankton Research*, **25**, 771–82.

Bailey, K.M. and Macklin, S.A. (1994). Analysis of patterns in larval walleye pollock *Theragra chalcogramma* survival and wind mixing events in Shelikof Strait, Gulf of Alaska. *Marine Ecology Progress Series*, **113**, 1–12.

Baird, M. and Suthers, I.M. (2007). A size-resolved pelagic ecosystem model. *Ecological Modelling*, **203**, 185–203.

Bakun, A. (1990). Global climate change and intensification of coastal ocean upwelling. *Science*, **247**, 198–201.

Bakun, A. (1996). *Patterns in the ocean: ocean processes and marine population dynamics*. California Sea Grant College System, University of California Press, La Jolla, CA, 323pp.

Bakun, A. (2001). 'School-mix feedback': a different way to think about low frequency variability in large mobile fish populations. *Progress in Oceanography*, **49**, 485–511.

Bakun, A. (2005). Regime shifts. In: Robinson, A.R. and Brink, K., eds. *The sea*, vol. 13. Harvard University Press, Cambridge, MA, pp. 971–1026.

Bakun, A. (2006a). Fronts and eddies as key structures in the habitat of marine fish larvae: opportunity, adaptive response and competitive advantage. In: Olivar, M.P. and Govoni, J.J., eds. *Recent advances in the study of fish eggs and larvae. Scientia Marina*, **70(Suppl. 2)**, 105–22.

Bakun, A. (2006b). Wasp-waist populations and marine ecosystem dynamics: navigating the 'predator pit' topographies. *Progress in Oceanography*, **68**, 271–88.

Bakun, A. (2009). Research challenges in the twenty-first century. In: Checkley, D., Alheit, J., Oozeki, Y., et al. eds. *Climate change and small pelagic fish*. Cambridge University Press, Cambridge.

Bakun, A. and Broad, K., eds. (2002). *Climate and fisheries: interacting paradigms, scales, and policy approaches*. Columbia Earth Institute and International Research Institute for Climate Prediction, New York. IRI Publication-IRI-CW/02/1, 70pp. Available online at http://iri.columbia.edu/outreach/publication/irireport/FisheriesWS2001.pdf

Bakun, A. and Broad, K., (2003). Environmental 'loopholes' and fish population dynamics: comparative pattern recognition with focus on *El Niño* effects in the Pacific. *Fisheries Oceanography*, **12**, 458–73.

Bakun, A. and Cury, P. (1999). The school trap: a mechanism promoting large-amplitude out-of-phase population oscillations of small pelagic fish species. *Ecology Letters*, **2**, 349–51.

Bakun, A. and Weeks, S. (2004). Greenhouse gas buildup, sardines, submarine eruptions and the possibility of abrupt degradation of intense marine upwelling systems. *Ecology Letters*, **7**, 1015–23.

Bakun, A. and Weeks, S. (2006). Adverse feedback sequences in exploited marine systems: are deliberate interruptive actions warranted? *Fish and Fisheries*, **7**, 316–33.

Baldursson, F.M., Danielsson, A., and Stefansson, G. (1996). On the rational utilization of the Icelandic cod stock. *ICES Journal of Marine Science*, **53(4)**, 643–58.

Balvanera, P., Pfisterer, A.B., Buchmann, N., et al. (2006). Quantifying the evidence for biodiversity effects on ecosystem functioning and services. *Ecology Letters*, **9**, 1146–56.

Bamstedt, U., Gifford, D.J., Irigoien, X., et al. (2000). Feeding. In: Harris, R.P., Wiebe, P.H., Lenz, J. et al. eds. *ICES zooplankton methodology manual*. Academic Press, London, pp. 297–400.

Ban, S., Burns, C., Castel, J., et al. (1997). The paradox of diatom-copepod interactions. *Marine Ecology Progress Series*, **157**, 287–93.

Barange, M. and Perry, R.I. (2009). Physical and ecological impacts of climate change relevant to marine and inland capture fisheries and aquaculture. In: Cochrane, K., De Young, C., Soto, D., et al. eds. *Climate change implications for fisheries and aquaculture. Overview of current scientific knowledge. FAO Fisheries and Aquaculture Technical Paper*, **530**. FAO, Rome.

Barange, M., Coetzee, J.C., and Twatwa, N.M. (2005). Strategies of space occupation by anchovy and sardine in the southern Benguela: the role of stock size and intra-species competition. *ICES Journal of Marine Science*, **62(4)**, 645–54.

Barange, M., Bernal, M., Cercole, M.C., et al. (2009). Current trends in the assessment and management of small pelagic fish stocks. In: Checkley, D., Alheit, J., Oozeki, Y., et al. eds. *Climate change and small pelagic fish*. Cambridge University Press, Cambridge.

Barange, M., O'Boyle, R., Cochrane, K.L., et al. (2010). Marine resources management in the face of change: from ecosystem science to ecosystem-based management. *This volume*. Oxford University Press, Oxford.

Barange, M. Allen, I., Allison, E., et al. (in press). Predicting the impacts and socio-economic consequences of climate change on global marine ecosystems and fisheries: the QUEST_Fish framework. In: Ommer, R., Perry, R.I., Cury, P., et al. eds. *Coping with climate change in marine socio-ecological systems*. Blackwell Fisheries and Aquaculture Research Series.

Barange, M., Coetzee, J., Takasuka, A., et al. (2009b). Habitat expansion and contraction in anchovy and sardine populations. *Progress in Oceanography* **83**, 251–60.

Barber, R.T and Chavez, F.P. (1986). Ocean variability in relation to living resources during the 1982/83 *El Niño*. *Nature*, **319**, 279–85.

Barnard, R., Batten, S., Beaugrand, G., et al. (2004). Continuous plankton records: plankton atlas of the North Atlantic Ocean (1958–1999), II: biogeographical charts. *Marine Ecology Progress Series*, **Suppl.**, 11–75.

Barnett, T., Pierce, D., Rao, K., et al. (2005). Penetration of human-induced warming into the world's oceans. *Science*, **309**, 284–7.

Barnston, A.G. and Livezey, R.E. (1987). Classification, seasonality, and persistence of low-frequency atmospheric circulation patterns. *Monthly Weather Review*, **115**, 1083–126.

Barth, J.A., Menge, B.A., Lubchenco, J., et al. (2007). Delayed upwelling alters nearshore coastal ocean ecosystems in the northern California current. *Proceedings of the National Academy of Sciences of the United States of America*, **104**, 3719–24.

Barth, J.A., Pierce, S.D., and Cowles, T.J. (2005). Mesoscale structure and its seasonal evolution in the northern California Current System. *Deep-Sea Research II*, **5**, 5–28.

Bartsch, J., Brander, K., Heath, M., *et al.* (1989). Modelling the advection of herring larvae in the North Sea. Nature, **340**, 632–6.

Batchelder, H.P. and Kashiwai, M. (2007). Ecosystem modeling with NEMURO within the PICES Climate Change and Carrying Capacity Program. *Ecological Modelling*, **202**, 7–11.

Batchelder, H.P. and Kim, S. (2008). Lessons learned from the PICES/GLOBEC Climate Change and Carrying Capacity (CCCC) Program and Synthesis Symposium. *Progress in Oceanography*, doi:10.1016/j.pocean.2008.03.003.

Batchelder, H.P., Barth, J.A., Kosro, P.M., *et al.* (2002a). The GLOBEC Northeast Pacific California Current System program. *Oceanography*, **15**, 36–47.

Batchelder, H.P., Edwards, C.A., and Powell, T.M. (2002b). Individual-based models of copepod populations in coastal upwelling regions: implications of physiologically and environmental influenced diel vertical migration on demographic success and nearshore retention. *Progress in Oceanography*, **53**, 307–33.

Batchelder, H.P., Lessard, E.J., Strub, P.T., *et al.* (2005). US GLOBEC biological and physical studies of plankton, fish and higher trophic level production, distribution and variability in the northeast Pacific. *Deep-Sea Research II*, **52**, 1–4.

Batchelder, H.P., 2006. Forward-in-Time/Backward-in-Time Trajectory (FITT/BITT) modeling of particles and organisms in the coastal ocean. *Journal of Atmospheric and Oceanic Technology*, **23**, 727–741.

Baumann, H., Hinrichsen, H.-H., Möllmann, C., *et al.* (2006). Recruitment variability in Baltic Sea sprat (*Sprattus sprattus*) is tightly coupled to temperature and transport patterns affecting the larval and early juvenile stages. *Canadian Journal of Fisheries and Aquatic Sciences*, **63**, 2191–201.

Baumgartner, T.R., Soutar, A., and Ferreira-Bartrina, V. (1992). Reconstructions of the history of Pacific sardine and northern anchovy populations over the past two millennia from sediments of the Santa Barbara Basin, California. *California Cooperative Oceanic Fisheries Investigations Report*, **33**, 24–40.

Bavinck, M., Chuenpagdee, R., Diallo, M., *et al.* (2005). *Interactive fisheries governance.* Eburon Publishers, Delft. 72pp.

Bax, N., Williamson, A., Agüero, M., *et al.* (2003). Marine invasive alien species: a threat to global biodiversity. *Marine Policy*, **27**, 313–23.

Beamish, R.J., Mahnken, C and Neville, C.V.M. (2004). Evidence that reduced early marine growth is associated with lower marine survival of coho salmon. *Transactions of the American Fisheries Society*, **133**, 26–33.

Beare, D.J. and McKenzie, E. (1999a). Connecting ecological and physical time-series: the potential role of changing seasonality. *Marine Ecology Progress Series*, **178**, 307–9.

Beare, D.J. and McKenzie, E. (1999b). The multinominal logit model: a new tool for exploring Continuous Plankton Recorder data. *Fisheries Oceanography*, **8**, 25–39.

Beare, D.J., Batten, S., Edwards, M., *et al.* (2002). Prevalence of boreal Atlantic, temperature Atlantic and neritic zooplankton in the North Sea between 1958 and 1998 in relation to temperature, salinity, stratification intensity and Atlantic inflow. *Journal of Sea Research*, **48**, 29–49.

Beare, D.J., Burns, F., Greig, A., *et al.* (2004). An increase in the abundance of anchovies and sardines in the northwestern North Sea since 1995. *Global Change Biology*, **10(7)**, 1209–13.

Beaugrand, G. (2004). The North Sea regime shift: evidence, causes, mechanisms and consequences. *Progress in Oceanography*, **60**, 245–62.

Beaugrand, G. and Ibanez, F. (2002). Spatial dependence of calanoid copepod diversity in the North Atlantic Ocean. *Marine Ecology Progress Series*, **232**, 197–211.

Beaugrand, G., Reid, P.C., Ibanez, F., *et al.* (2000). Biodiversity of North Atlantic and North Sea calanoid copepods. *Marine Ecology Progress Series*, **204**, 299–303.

Beaugrand, G., Ibañez, F., Lindley, J.A., *et al.* (2002a). Diversity of calanoid copepods in the North Atlantic and adjacent seas: species associations and biogeography. *Marine Ecology Progress Series*, **232**, 179–95.

Beaugrand, G., Reid, P.C., Ibanez, F., *et al.* (2002b). Reorganization of North Atlantic marine copepod biodiversity and climate. *Science*, **296**, 1692–4.

Beaugrand, G., Brander, K.M., Lindley, J.A., *et al.* (2003a). Plankton effect on cod recruitment in the North Sea. *Nature*, **426**, 661–4.

Beaugrand, G., Ibañez, F., and Lindley, J.A. (2003b). An overview of statistical methods applied to CPR data. *Progress in Oceanography*, **58**, 235–62.

Beaugrand, G., Lindley, J.A., Helaouët, P., *et al.* (2007). Macroecological study of *Centropages typicus* in the North Atlantic Ocean. *Progress in Oceanography*, **72**, 259–73.

Beaugrand, G., Edwards, M., Brander, K., *et al.* (2008). Causes and projections of abrupt climate-driven ecosystem shifts in the North Atlantic. *Ecology Letters*, **11**, 1157–68.

Beaugrand, G., Luczak, C., and Edwards, M. (2009). Rapid biogeographical plankton shifts in the North Atlantic Ocean. *Global Change Biology*, doi: 10.1111/j.1365-2486.2009.01848.x.

Begon, M., Townsend, C.R., and Harper, J.L. (2005). *Ecology: from individuals to organisms*, 4th edition. Blackwell Publishing, Boston, MA, 752pp.

Behrenfeld, M.J., Bale, A.J., Kolber, Z.S., *et al.* (1996). Confirmation of iron limitation of phytoplankton photosynthesis in the equatiorial Pacific Ocean. *Nature*, **383**, 508–11.

Behrenfeld, M.J. and Kolber, Z. (1999). Widespread iron limitation of phytoplankton in the south Pacific Ocean. *Science*, **283(5403)**, 840–3.

Behrenfeld, M.J., O'Malley, R.T., Siegel, D.A., *et al.* (2006). Climate-driven trends in contemporary ocean productivity. *Nature*, **444**, 752–5.

Bell, M.V., Dick, J.R., Anderson, T.R., *et al.* (2007). Application of liposome and stable isotope tracer techniques to study polyunsaturated fatty acid biosynthesis in marine zooplankton. *Journal of Plankton Research*, **29**, 417–22.

Bellier, E., Planque, B., and Petitgas, P. (2007). Historical fluctuations in spawning location of anchovy (*Engraulis encrasicolus*) and sardine (*Sardina pilchardus*) in the Bay of Biscay during 1967–73 and 2000–2004. *Fisheries Oceanography*, **16**, 1–15.

Benaka, L.R. (1999). Fish habitat: essential fish habitat and rehabilitation. *American Fisheries Society Symposium*, **22**. Bethesda, MD, 459pp.

Béné, C. (2003). When fishery rhymes with poverty: a first step beyond the old paradigm on poverty and small-scale fisheries. *World Development*, **31**, 949–75.

Benfield, M.C., Davis, C.S., Wiebe, P.H., *et al.* (1996). Comparative distributions of calanoid copepods, pteropods, and larvaceans estimated from concurrent video plankton recorder and MOCNESS tows in the stratified region of Georges Bank. *Deep-Sea Research II*, **43**, 1925–45.

Benfield, M.C., Grosjean, P., Culverhouse, P.G., *et al.* (2007). RAPID: research on automated plankton identification. *Oceanography*, **20**, 172–87.

Benfield, M.C., Lavery, A.C., Wiebe, P.H., *et al.* (2003). Distributions of *Physonect siphonulae* in the Gulf of Maine and their potential as important sources of acoustic scattering. *Canadian Journal of Fisheries and Aquatic Sciences*, **60**, 759–72.

Berkeley, S.A., Chapman, C., and Sogard, S.M. (2004). Maternal age as a determinant of larval growth and survival in marine fish, *Sebastes melanops*. *Ecology*, **85**, 1258–64.

Berkes, F., Colding, J., and Folke, C. (2003). Introduction. In: Berkes, F., Colding, J., and Folke, C., eds. *Navigating social-ecological systems: building resilience for complexity and change*. Cambridge University Press, Cambridge, pp. 1–30.

Berkes, F., and Folke, C., eds. (1998). *Linking social and ecological systems: management practices and social mechanisms for building resilience*. Cambridge University Press, Cambridge.

Berman, M. and Sumaila, U.R. (2006). Discounting, amenity values and marine ecosystem restoration. *Marine Resource Economics*, **21**, 211–19.

Bertram, D.F., Mackas, D.L., and Mckinnell, S.M. (2001). The seasonal cycle revisited: interannual variation and ecosystem consequences. *Progress in Oceanography*, **49**, 283–307.

Bethke, I., Furevik, T., and Drange, H. (2006). Towards a more saline North Atlantic and a fresher Arctic under global warming. *Geophysical Research Letters*, **33**, L21712.

Beverton, R.J.H. and Holt, J. (1957). On the dynamics of exploited fish populations. *Fish and fisheries series 11*. Chapman & Hall, London, 533pp.

Bianchi, G., Gislason, H., Graham, K., *et al.* (2000). Impact of fishing on size composition and diversity of demersal fish communities. *ICES Journal of Marine Science*, **57**, 558–71.

Bindoff, N.L., Willebrand, J., Artale, V., *et al.* (2007). Observations: oceanic climate change and sea level. In: Solomon, S., Qin, D., Manning, M., *et al.*, eds. *Climate change 2007: the physical science basis. Contribution of Working Group I to the Fourth Assessment Report of the Intergovernmental Panel on Climate Change*. Cambridge University Press, Cambridge and New York.

Bishop, D., Gammel, P., and Giles, C.R. (2001). The little machines that are making it big. *Physics Today*, **54**, 38–44.

Bishop, J.K.B., Davis, R.E., and Sherman, J.T. (2002). Robotic observations of dust-storm enhancement of carbon biomass in the North Pacific. *Science*, **298**, 817–21.

Biuw, M., Boehme, L., Guinet, C., *et al.* (2007). Variations in behavior and condition of a Southern Ocean top predator in relation to *in situ* oceanographic conditions. *Proceedings of the National Academy of Sciences of the United States of America*, **104**, 13705–10.

Björnsson, B. and Steinarsson, A. (2002). The food-unlimited growth rate of Atlantic cod *Gadus morhua*. *Canadian Journal of Fisheries and Aquatic Sciences*, **59**, 494–502.

Björnsson, B. and Steinarsson, A., and Oddgeirsson, M. (2001). Optimal temperature for growth and feed conversion of immature cod (*Gadus morhua* L.). *ICES Journal of Marine Science*, **58**, 29–38.

Bjornstad, O.N., Fromentin, F.-M., Stenseth, N.C., *et al.* (1999). Cycles and trends in cod populations. *Proceedings of the National Academy of Sciences of the United States of America*, **96**, 5066–71.

Bjornstad, O.N., Nisbet, R.M., and Fromentin, J.-M. (2004). Trends and cohort resonance in age-structured populations. *Journal of Animal Ecology*, **73**, 1157–67.

Blacker, R.W. (1957). Benthic animals as indicators of hydrographic conditions and climatic change in

Svalbard waters. *Fisheries Investigations (Series 2)*, **20**, 1–49.

Blankenship, L.E. and Yayanos, A.A. (2005). Universal primers and PCR of gut contents to study marine invertebrate diets. *Molecular Ecology*, **14**, 891–9.

Block, B.A. (2005). Physiological ecology in the 21st century: advancements in biological science. *Integrative and Comparative Biology*, **45**, 305–20.

Block, B.A., Dewar, H., Williams, T., *et al.* (1998). Archival tagging of Atlantic bluefin tuna (*Thunnus thynnus thynnus*). *Marine Technology Society Journal*, **32**, 37–46.

Block, B.A., Dewar, H., Blackwell, S.B., *et al.* (2001). Migratory movements, depth preferences, and thermal biology of Atlantic bluefin tuna. *Science*, **293**, 1310–4.

Block, B.A., Costa, D.P., Boehlert, G.W., *et al.* Revealing pelagic habitat use: the tagging of Pacific pelagics program. *Oceanologica Acta*, **25**, 255–66.

Bluhm, B.A., Coyle, K.O., Konar, B., *et al.* (2007). High gray whale relative abundances associated with an oceanographic front in the south-central Chukchi Sea. *Deep-Sea Research II*, **54**, 2919–33.

Bochdansky, A.B., and Bollens, S.M. (2004). Relevant scales in zooplankton ecology: distribution, feeding, and reproduction of the copepod *Acartia hudsonica* in response to thin layers of the diatom *Skeletonema costatum*. *Limnology and Oceanography*, **49**, 625–36.

Bochenek, E.A., Klinck, J.M., Powell, E.N., *et al.* (2001). Biochemically based model of the growth and development of *Crassostrea gigas* larvae. *Journal of Shellfish Research*, **20**, 243–65.

Bode, A., Alvarez-Ossorio, M.T., Cunha, M.E., *et al.* (2007). Stable nitrogen isotope studies of the pelagic food web on the Atlantic shelf of the Iberian Peninsula. *Progress in Oceanography*, **74**, 115–31.

Bode, A., Carrera, P., and Lens, S. (2003). The pelagic food-web in the upwelling ecosystem of Galicia (NW Spain) during spring: natural abundance of stable carbon and nitrogen isotopes. *ICES Journal of Marine Science*, **60**, 11–22.

Boehlert, G.W. (1997). *Application of acoustic and archival tags to assess estuarine, nearshore and offshore habitat utilization and movement by salmonids. NOAA Technical Memorandum, NMFS NOAA-TM-NMFS-SWFSC-236*, U.S. Department of Commerce, NOAA, NMFS, SWFSC.

Boehlert, G.W., Costa, D.P., Crocker, D.E., *et al.* (2001). Autonomous pinniped environmental samplers: using instrumented animals as oceanographic data collectors. *Journal of Atmospheric and Oceanic Technology*, **18**, 1882–93.

Boer, G.J., McFarlane, N.A., and Lazare, M. (1992). Greenhouse gas induced climate change simulated with

the CCC second-generation general circulation model. *Journal of Climate*, **5**, 1045–77.

Boero, F., Bouillon, J., Gravili, C., *et al.* (2008). Gelatinous plankton: irregularities rule the world (sometimes). *Marine Ecology Progress Series*, **356**, 299–310.

Bogstad, B., Hiis Hauge, K., and Ultang, Ø. (1997). MULTSPEC-a multispecies model for fish and marine mammals in the Barents Sea. *Journal of Northwest Atlantic Fisheries Science*, **22**, 317–41.

Boldt, J., ed. (2006). Ecosystem considerations for 2007. *North Pacific Fisheries Management Council, Bering Sea/ Aleutian Islands and Gulf of Alaska Stock Assessment and Fisheries Evaluation Report 2006*. North Pacific Fisheries Management Council, Anchorage, AK, 360pp.

Bolster, W.J. (2008). Putting the ocean in Atlantic history: maritime communities and marine ecology in the northwest Atlantic, 1500–1800. *American Historical Review*, **113**, 19–47.

Bonardelli, J.C., Himmelman, J.H., and Drinkwater, K. (1996). Relation of spawning of the giant scallop, *Placopecten magellanicus*, to temperature fluctuations during downwelling events. *Marine Biology*, **124**, 637–49.

Bond, N.A., Overland, J.E., Spillane, M., *et al.* (2003). Recent shifts in the state of the North Pacific, *Geophysical Research Letters*, **30**, 2183, doi:10.1029/2003GL018597.

Bonnet, D., Titelman, J., and Harris, R. (2004). *Calanus* the cannibal. *Journal of Plankton Research*, **26**, 937–48.

Bonnet, D., Richardson, A., Harris, R., *et al.* (2005). An overview of *Calanus helgolandicus* ecology in European waters. *Progress in Oceanography*, **65**, 1–53.

Bopp, L.P., Monfray, P., Aumont, O., *et al.* (2001). Potential impact of climate change on marine export production. *Global Biogeochemical Cycles*, 15, 81–99.

Bopp, L.P., Boucher, O., Aumont, O., *et al.* (2004). Will marine dimethylsulfide emissions amplify or alleviate global warming? A model study. *Canadian Journal of Fisheries and Aquatic Sciences*, **61**, 826–35.

Bopp, L.P., Aumont, O., Cadule, P., *et al.* (2005). Response of diatoms distribution to global warming and potential implications: a global model study. *Geophysical Research Letters*, **32**, L19606, doi:10.1029/2005GL023653.

Borer, E.T., Anderson, K., Blanchette, C.A., *et al.* (2002). Topological approaches to food web analyses: a few modifications may improve our insights. *Oikos*, **99**, 398–403.

Borja, A., Uriarte, A., Egaña, J., *et al.* (1998). Relationship between anchovy (*Engraulis encrasicolus* L.) recruitment and environment in the Bay of Biscay. *Fisheries Oceanography*, **7**, 375–80.

Borja, A., Uriarte, A., Egaña, J., *et al.* (1996). Relationships between anchovy (*Engraulis encrasicolus* L.) recruitment and the environment in the Bay of Biscay. *Scientia Marina*, **60(Suppl. 2)**, 179–92.

Botsford, L.W. (1981). The effects of increased individual growth rates on depressed population size. *American Naturalist*, **117**, 38–63.

Botsford, L.W. (1986). Effects of environmental forcing on age-structured populations: northern California Dungeness crab (*Cancer magister*) as an example. *Canadian Journal of Fisheries and Aquatic Sciences*, **43**, 2345–52.

Botsford, L.W. (1997). Dynamics of populations with density-dependent recruitment and age structure. Chapter 12. In: Tuljapurkar, S., and Caswell, H. eds. *Structured population models in marine, terrestrial, and freshwater systems*. Chapman & Hall, New York.

Botsford, L.W. and Wickham, D.E. (1978). Behavior of age-specific, density-dependent models and the Northern California Dungeness crab (*Cancer magister*) fishery. *Journal of the Fisheries Research Board of Canada*, **35**, 833–43.

Botsford, L.W. and Lawrence, C.A. (2002). Patterns of co-variability among California current chinook salmon, coho salmon, Dungeness crab, and physical oceanographic conditions. *Progress in Oceanography*, **53**, 283–305.

Botsford, L.W., Moloney, C.L., Hastings, A., *et al.* (1994). The influence of spatially and temporally varying oceanographic conditions on meroplanktonic metapopulations. *Deep-Sea Research II*, **41**, 107–45.

Botsford, L.W., Castilla, J.C., and Peterson, C.H. (1997). The management of fisheries and marine ecosystems. *Science*, **277**, 509–15.

Botsford, L.W., Lawrence, C.A., and Hill, M.F. (2005). Differences in dynamic response of California Current salmon species to changes in ocean conditions. *Deep-Sea Research II*, **52**, 331–45.

Bouillon, J., Medel, M.D., Pages, F., *et al.* (2004). Fauna of the Mediterranean Hydrozoa. *Scientia Marina*, **68**, 5–449.

Boustany, A.M., Davis, S.F., Pyle, P., *et al.* (2002). Expanded niche for white sharks. *Nature*, **415**, 35–6.

Boyd, P.W. and Doney, S.C. (2002). Modelling regional responses by marine pelagic ecosystems to global climate change. *Geophysical Research Letters*, **29(16)**, 1806, doi:10.1029/2001GL014130.

Boyer, D.C., Boyer, H.J., Fossen, I., *et al.* (2001). Changes in abundance of the northern Benguela sardine stock during the decade 1990 to 2000 with comments on the relative importance of fishing and the environment. *South African Journal of Marine Science*, **23**, 67–84.

Boyer, D.C., Levitus, S., Antonov, J.I., *et al.* (2005). Linear trends in salinity for the world ocean, 1955–1998, *Geophysical Research Letters*, **32**, L01604, doi:10.1029/2004GL021791.

Braccini, J.M., Bronwyn, G.M., and Terence, W.I. (2007). Hierarchical approach to the assessment of fishing effects on non-target chondrichthyans: case study of *Squalus megalops* in southeastern Australia. *Canadian Journal of Fisheries and Aquatic Sciences*, **63(11)**, 2456–66.

Brander, K.M. (1981). Disappearance of common skate *Raia batis* from Irish Sea. *Nature*, **290**, 48–9.

Brander, K.M. ed. (1993). Cod and climate change. Report of the first meeting of an ICES/International GLOBEC Working Group, Lowestoft, England, 7–11 June 1993. *GLOBEC Report*, **4**, 67pp.

Brander, K.M. (1994). Patterns of distribution, spawning, and growth in North Atlantic cod: the utility of inter-regional comparisons. *ICES Marine Science Symposia*, **198**, 406–13.

Brander, K.M. (1995). The effect of temperature on growth of Atlantic cod (*Gadus morhua* L.) *ICES Journal of Marine Science*, **52**, 1–10.

Brander, K.M. (2003a). Fisheries and climate. In: G. Wefer, F. Lamy and F. Mantoura, eds. *Marine science frontiers for Europe*. Springer-Verlag, Berlin, pp. 29–38.

Brander, K.M. (2003b). What kinds of fish stock predictions do we need and what kinds of information will help us to make better predictions?. *Scientia Marina*, **67(Suppl. 1)**, 21–33.

Brander, K.M. (2005). Cod recruitment is strongly affected by climate when stock biomass is low. *ICES Journal of Marine Science*, **62**, 339–43.

Brander, K.M. (2007a). Global fish production and climate change. *Proceedings of the National Academy of Sciences of the United States of America*, **104**, 19709–14.

Brander, K.M. (2007b). The role of growth changes in the decline and recovery of North Atlantic cod stocks since 1970. *ICES Journal of Marine Science*, **64**, 211–17.

Brander, K.M. and Mohn, R. (2004). Effect of the North Atlantic Oscillation on recruitment of Atlantic cod (*Gadus morhua*). *Canadian Journal of Fisheries and Aquatic Sciences*, **61**, 1558–64.

Brander, K.M., Blom, G., Borges, M.F., *et al.* (2003a). Changes in fish distribution in the eastern North Atlantic: are we seeing a coherent response to changing temperature. *ICES Journal of Marine Science*, **219**, 261–70.

Brander, K.M., Dickson, R.R., and Edwards, M. (2003b). Use of Continuous Plankton Recorder information in support of marine management: applications in fisheries, environmental protection, and in the study of ecosystem response to environmental change. *Progress in Oceanography*, **58**, 175–91.

Brander, K.M., Botsford, L., Ciannelli, L., *et al.* (2010). Human impacts on marine ecosystems. *This volume.* Oxford University Press, Oxford.

Brashares, J.S., Arcese, P., Sam, M.K., *et al.* (2004). Bushmeat hunting, wildlife declines, and fish supply in West Africa. *Science*, **306**, 1180–3.

Bremner, J. and Perez, J. (2002). A case study of human migration and the sea cucumber crisis in the Galapagos Islands. *Ambio*, **31**, 306–10.

Brentnall, S.J., Richards, K.J., Brindley, J., *et al.* (2003). Plankton patchiness and its effect on larger-scale productivity. *Journal of Plankton Research*, **25**, 121–40.

Brierley, A.S. and Thomas, D.N. (2002). Ecology of southern pack ice. *Advances in Marine Biology: An Annual Review*, **43**, 171–276.

Broad, K., Pfaff, A.S.P., and Glantz, M.H. (1999). Climate information and conflicting goals: *El Niño* 1997–98 and the Peruvian fishery. In: *Public philosophy, environment, and social justice.* Carnegie Council on Ethics and International Affairs, New York.

Brodeur, R.D. and Ware, D.M. (1992). Long-term variability in zooplankton biomass in the subarctic Pacific Ocean. *Fisheries Oceanography*, **1**, 32–8.

Brodeur, R.D., Fisher, J.P., Teel, D.J., *et al.* (2004). Juvenile salmonid distribution, growth, condition, origin, environmental and species associations in the Northern California Current. *Fishery Bulletin*, **102**, 25–46.

Brodeur, R.D., Daly, E.A., Schabetsberger, R.A., *et al.* (2007). Interannual and interdecadal variability in juvenile coho salmon (*Oncorhynchus kisutch*) diets in relation to environmental changes in the northern California Current. *Fisheries Oceanography*, **16**, 395–408.

Brodeur, R.D., Decker, M.B., Ciannelli, L., *et al.* (2008a). Rise and fall of jellyfish in the eastern Bering Sea in relation to climate regime shifts. *Progress in Oceanography*, **77**, 103–11.

Brodeur, R.D., Suchman, C.L., Reese, D.C., *et al.* (2008b). Spatial overlap and trophic interactions between pelagic fish and large jellyfish in the northern california Current. *Marine Biology*, **154**, 649–59.

Broecker, W.S. (1991). The great ocean conveyor. *Oceanography*, **4**, 79–89.

Broglio, E., Johansson, M., and Jonsson, P.R. (2001). Trophic interaction between copepods and ciliates: effects of prey swimming behavior on predation risk. *Marine Ecology Progress Series*, **220**, 179–86.

Brookfield, K., Gray, T., and Hatchard, J. (2005). The concept of fisheries-dependent communities. A comparative analysis of four UK case studies: Shetland, Peterhead, North Shields and Lowestoft. *Fisheries Research*, **72**, 55–69.

Brothers, E.B., Mathews, C.P., and Lasker, R. (1976). Daily growth increments in otoliths from larval and adult fishes. *Fishery Bulletin*, **74**, 1–8.

Broughton, E.A. and Lough, R.G. (2006). A direct comparison of MOCNESS and Video Plankton Recorder zooplankton abundance estimates: possible applications for augmenting net sampling with video systems. *Deep-Sea Research II*, **53**, 2789–807.

Browman, H. and Boyd, P.W. (2008). Implications of large-scale iron fertilization of the oceans. *Marine Ecology Progress Series*, **364**, 213–18.

Browman, I. (1996). Predator-prey interaction in the sea: commentaries on the role of turbulence. *Marine Ecology Progress Series*, **139**, 301–2.

Browman, I., and Skiftesvik, A.B. (1996). The effects of turbulence on the predation cycle of fish larvae: comments on some of the issues. *Marine Ecology Progress Series*, 139, 309–12.

Brown, B.E., Brennan, J.A., Grosslein, M.D., *et al.* (1976). The effect of fishing on the marine finfish biomass in the Northwest Altantic from the Gulf of Maine to Cape Hatteras. *International Commission for the Northwest Atlantic Fisheries Research Bulletin*, **12**, 49–68.

Brown, J.H., Gillooly, J.F., Allen, A.P., *et al.* (2004). Toward a metabolic theory of ecology. *Ecology*, **85(7)**, 1771–89.

Brown, T.C. (1984). The concept of value in resource allocation. *Land Economics*, **60**, 231–46.

Bruggeman, J. and Kooijman, S.A.L.M. (2007). A biodiversity-inspired approach to aquatic ecosystem modeling. *Limnology and Oceanography*, **52**, 1533–44.

Bryant, A.D., Hainbucher, D., and Heath, M. (1998). Basin-scale advection and population persistence of *Calanus finmarchicus*. *Fisheries Oceanography*, **7**, 235–44.

Buckley, L.J. and Durbin, E.G. (2006). Seasonal and interannual trends in the zooplankton prey and growth rate of Atlantic cod (*Gadus morhua*) and haddock (*Melanogrammus aeglefinus*) larvae on Georges Bank. *Deep-Sea Research II*, **53**, 2758–70.

Buckley, L.J., Caldarone, E., and Ong, T.-L. (1999). RNA-DNA ratio and other nucleic acid-based indicators for growth and condition of marine fishes. *Hydrobiologia*, **401**, 265–77.

Buckley, L.J., Caldarone, E., and Lough, R.G. (2004). Optimum temperature and food-limited growth of larvae cod (*Gadus morhua*) and haddock (*Melanogrammus aeglefinus*) on Georges Bank. *Fisheries Oceanography*, **13**, 134–40.

Buckley, L.J., Caldarone, E., and Lough, R.G., *et al.* (2006). Ontogenetic and seasonal trends in recent growth rates of Atlantic cod and haddock larvae on Georges Bank: effects of photoperiod and temperature. *Marine Ecology Progress Series*, **325**, 205–26.

Buckley, L.J., Caldarone, E.M., and Clemmesen, C. (2008). Multi-species larval fish growth model based on temperature and fluorometrically derived RNA/DNA ratios: results from a meta-analysis *Marine Ecology Progress Series* 371:221–232.

Bucklin, A. and Wiebe, P.H. (1998). Low mitochondrial diversity and small effective population sizes of the copepods *Calanus finmarchicus* and *Nannocalanus minor*: possible impact of climatic variation during recent glaciation. *Journal of Heredity*, **89**, 383–92.

Bullard, S.G., Lambert, G., Carman, M.R., *et al.* (2007). The colonial ascidian *Didemnum* sp. A: current distribution, basic biology and potential threat to marine communities of the northeast and west coasts of North America. *Journal of Experimental Marine Biology and Ecology*, **342**, 99–108.

Bundy, A. and Fanning, L.P. (2005). Can Atlantic cod (*Gadus morhua*) recover? Exploring trophic explanations for the non-recovery of the cod stock on the eastern Scotian Shelf, Canada. *Canadian Journal of Fisheries and Aquatic Sciences*, **62(7)**, 1471–89.

Bundy, M.H. and Paffenhöfer, G.-A. (1993). Innervation of copepod antennules investigated using laser scanning confocal microscopy. *Marine Ecology Progress Series*, **102**, 1–14.

Bundy, M.H., Gross, T.F., Coughlin, D.J., *et al.* (1993). Quantifying copepod searching efficiency using swimming pattern and perceptive ability. *Bulletin of Marine Science*, **53**, 15–28.

Bunker, A.J. and Hirst, A.G. (2004). Fecundity of marine planktonic copepods: global rates and patterns in relation to chlorophyll *a*, temperature and body weight. *Marine Ecology Progress Series*, **279**, 161–81.

Burns, J.M. and Kooyman, G.L. (2001). Habitat use by Weddell seals and emperor penguins foraging in the Ross Sea, Antarctica. *American Zoologist*, **41**, 90–8.

Burns, J.M., Costa, D.P., Fedak, M.A., *et al.* (2004). Winter habitat use and foraging behavior of crabeater seals along the Western Antarctic Peninsula. *Deep-Sea Research II*, **51**, 2279–303.

Burns, J.M., Hindell, M.A., Bradshaw, C.J.A., *et al.* (2008). Fine-scale habitat selection of crabeater seals as determined by diving behavior. *Deep-Sea Research II*, **55(3–4)**, 500–14.

Buskey, E.J. (1984). Swimming pattern as an indicator of the roles of copepod sensory systems in the recognition of food. *Marine Biology*, **79**, 165–76.

Buskey, E.J., and Hartline, D.K. (2003). High-speed video analysis of the escape responses of the copepod *Acartia tonsa* to shadows. *Biological Bulletin*, **204**, 28–37.

Buskey, E.J., Coulter, C., and Strom, S. (1993). Locomotory patterns of microzooplankton: potential effects on food

selectivity of larval fish. *Bulletin of Marine Science*, **53**, 29–43.

Buskey, E.J., Peterson, J.O., and Ambler, J.W. (1996). The swarming behavior of the copepod *Dioithona oculata*: *in situ* and laboratory studies. *Limnology and Oceanography*, **41**, 513–21.

Butterworth, D.S. (2007). Why a management procedure approach? Some positives and negatives. *ICES Journal of Marine Science*, **64**, 613–17.

Butterworth, D.S., and Bergh, M.O. (1993). The development of a management procedure for the South African anchovy resource. In: Smith, S.J., Hunt, J.J., and Rivard, D., eds. *Risk evaluation and biological reference points for fisheries management. Canadian Special Publication of Fisheries and Aquatic Sciences*, **120**, 83–90.

Butterworth, D.S., Cochrane, K.L., and De Oliveria, J.A.A. (1997). Management procedures: a better way to manage fisheries? The South African experience. In: Pikitch, E.K., Huppert, D.D., and Sissensine, M.P., eds. *Global trends: fisheries management. American Fisheries Society Symposium*, **20**, 83–90.

Buttino, I., Santo, M.D., Ionora, A., *et al.* (2004). Rapid assessment of copepod (*Calanus helgolandicus*) embryo viability using fluorescent probes. *Marine Biology*, **145**, 393–9.

Caddy, J.F. (1999). Deciding on precautionary management measures for a stock based on a suite of Limit Reference Points (LRPs) as a basis for a multi-LRP harvest law. *NAFO Science Council Studies*, **32**, 55–68.

Calambokidis, J., Steiger, G.H., Rasmussen, K., *et al.* (2000). Migratory destinations of humpback whales that feed off California, Oregon and Washington. *Marine Ecology Progress Series*, **192**, 295–304.

Calbet, A and Saiz, E. (2005). The ciliate-copepod link in marine ecosystems. *Aquatic Microbial Ecology*, **38**, 157–67.

Caldarone, E.M., Onge-Burns, J.M., and Buckley, L.J. (2003). Relationship of RNA-DNA ratio and temperature to growth in larvae of Atlantic cod *Gadus morhua*. *Marine Ecology Progress Series*, **262**, 229–40.

Caley, M.J., Carr, M.H., Hixon, M.A., *et al.* (1996). Recruitment and the local dynamics of open marine populations. *Annual Review of Ecology and Systematics*, **27**, 477–500.

Campagna, C., Piola, A.R., Marin, M.R., *et al.* (2006). Southern elephant seal trajectories, fronts and eddies in the Brazil/Malvinas Confluence. *Deep-Sea Research I*, **53**, 1907–24.

Campana, S.E. (2005). Otolith science entering the 21st century. *Marine and Freshwater Research*, **56**, 485–95.

Campana, S.E., and Hurley, P.C.F. (1989). An age- and temperature-mediated growth model for cod (*Gadus morhua*) and haddock (*Melanogrammus aeglefinus*) larvae in the

Gulf of Maine. *Canadian Journal of Fisheries and Aquatic Sciences*, **46**, 603–13.

Campbell, R.G., Runge, J.A., and Durbin, E.G. (2001a). Evidence for food limitation of *Calanus finmarchicus* production rates on the southern flank of Georges Bank during April 1997. *Deep-Sea Research II*, **48**, 531–49.

Campbell, R.G., Wagner, M.M., Teegarden, G.J., *et al.* (2001b). Growth and development rates of the copepod *Calanus finmarchicus* reared in the laboratory. *Marine Ecology Progress Series*, **221**, 161–83.

Campbell, R.W. and Head, E.J.H. (2000). Egg production rates of *Calanus finmarchicus* in the western North Atlantic: effect of gonad maturity, female size, chlorophyll concentration, and temperature. *Canadian Journal of Fisheries and Aquatic Sciences*, **57**, 518–29.

Caparroy, P., Perez, M.T., and Carlotti, F. (1998). Feeding behaviour of *Centropages typicus* in calm and turbulent conditions. *Marine Ecology Progress Series*, **168**, 109–18.

Carlotti, F. and Wolf, K.-U. (1998). A Lagrangian ensemble model of *Calanus finmarchicus* coupled with a 1-D ecosystem model. *Fisheries Oceanography*, **7**, 191–204.

Carlotti, F., Giske, J., and Werner, F. (2000). Modelling zooplankton dynamics. In Harris, R.P., Wiebe P.H., Lenz J., *et al.* eds. *Zooplankton methodology manual*. Academic Press, San Diego, CA, pp. 571–667.

Carlton, J.T. (1989). Man's role in changing the face of the ocean: biological invasions and implications for conservation of near-shore environments. *Conservation Biology*, **3**, 265–73.

Carlton, J.T., and Geller, J.B. (1993). Ecological roulette: the global transport of nonindigenous marine organisms. *Science*, **261**, 78–82.

Carmack, E.C. (2007). The alpha/beta ocean distinction: a perspective on freshwater fluxes, convection, nutrients and productivity in high-latitutde seas. *Deep-Sea Research II*, **54**, 2578–98.

Carpenter S.R., Kitchell, J.F., and Hodgson, J.R. (1985). Cascading trophic interactions and lake productivity. *Bioscience*, **35**, 634–9.

Carr, M.E. and Kearns, E.J. (2003). Production regimes in four eastern boundary current systems. *Deep-Sea Research II*, **30**, 3199–221.

Carr, M.E., Friedrichs, M.A.M., Schmeltz, M., *et al.* (2006). A comparison of global estimates of marine primary production from ocean color. *Deep-Sea Research II*, **53**, 741–70.

Carrion, J.S., Fuentes, N., González-Sampériz, P., *et al.* (2007). Holocene environmental change in a montane region of southern Europe with a long history of human settlement. *Quaternary Science Reviews*, **26**, 1455–75.

Carroll, M.L. and Carroll, J. (2003). The Arctic Ocean. In: Black, K.D. and Shimmield, G., eds. *Biogeochemistry of marine systems*. Blackwell Publishing, Oxford, pp. 127–56.

Cassie, R.M. (1959). Micro-distribution of plankton. *New Zealand Journal of Science*, **2**, 398–409.

Castellani, C., Irigoien, X., Harris, R.P., *et al.* (2005). Feeding and egg production of *Oithona similis* in the North Atlantic. *Marine Ecology Progress Series*, **288**, 173–82.

Castellani, C., Irigoien, X., Harris, R.P., *et al.* (2007). Regional and temporal variation of *Oithona* spp. biomass, stage structure and productivity in the Irminger Sea, North Atlantic. *Journal of Plankton Research*, **29**, 1051–70.

Caswell, H. (2001). *Matrix population models: construction, analysis, and interpretation*, 2nd edition. Sinauer Associates, Sunderland, MA.

Cayré, P. and Marsac, F. (1993). Modelling the yellowfin tuna (*Thunnus albacares*) vertical distribution using sonic tagging results and local environmental parameters. *Aquatic Living Resources*, **6**, 1–14.

CCAMLR. (2001). *Report of the twentieth meeting of the commission*. CCAMLR, Hobart, Australia.

CCAMLR. (2002). *Report of the twenty-first meeting of the commission*. CCAMLR, Hobart, Australia.

CCAMLR. (2005). Text of the convention on the conservation of antarctic marine living resources. *Basic Documents*, CCAMLR, Hobart, Australia.

Cerrano, C., Bavestrello, G., Bianchi, C.N., *et al.* (2000). A catastrophic mass-mortality episode of gorgonians and other organisms in the Ligurian Sea (north-western Mediterranean), summer (1999). *Ecology Letters*, **3**, 284–93.

Chai, F., Jiang, M., Barber, R.T., *et al.* (2003). Interdecadal variation of the transition zone chlorophyll front, a physical-biological model simulation between 1960 and 1990. *Journal of Oceanography*, **59**, 461–75.

Chapin, F.S., III, Walker, B.H., Hobbs, R.J., *et al.* (1997). Biotic control over the functioning of ecosystems. *Science*, **277**, 500–4.

Chapman, E.W., Ribic, C.A., and Fraser, W.R. (2004). The distribution of seabirds and pinnipeds in Marguerite Bay and their relationship to physical features during austral winter 2001. *Deep-Sea Research II*, **51**, 2261–78.

Charles, A.T. (2001). *Sustainable fishery systems*. Blackwell Science, Oxford, 370pp.

Charlson, R.J., Lovelock, J.E., Andreae, M.O., *et al.* (1987). Oceanic phytoplankton, atmospheric sulphur, cloud albedo and climate. *Nature*, **326**, 655–61.

Charrassin, J.-B., Park, Y.-H., Le Maho, Y., *et al.* (2002). Penguins as oceanographers unravel hidden mechanisms of marine productivity. *Ecology Letters*, **5**, 317–19.

Chassignet, E.P., Hurlburt, H.E., Smedstad, O.M., *et al.* (2006). Generalized vertical coordinates for eddy-resolving global and coastal ocean forecasts. *Oceanography*, **19**, 20–31.

Chavez, F.P., Collins, C.A., Huyer, A., *et al. El Niño* along the west coast of North America. *Progress in Oceanography*, **54**, 1–6.

Chavez, F.P., Ryan, J., Lluch-Cota, S.E., *et al.* (2003). From anchovies to sardines and back: multidecadal change in the Pacific Ocean. *Science*, **299**, 217–21.

Checkley, D.M., Jr., Ortner, P.B., Cummings, S.R., *et al.* (1997). A continuous, underway fish egg sampler. *Fisheries Oceanography*, **6**, 58–73.

Checkley, D.M., Motos, L., Uriarte, A., *et al.* (1999). Continuous, underway sampling of pelagic fish eggs and the environment: the Bay of Biscay anchovy and machine vision research. *ICES Council Meeting Papers*, **CM 1999/M:05**, 15pp.

Checkley, D.M., Hunter, J.R., Motos, L., *et al.* (2000). Report of a workshop on the use of the Continuous Underway Fish Egg Sampler (CUFES) for mapping spawning habitats of pelagic fish. *GLOBEC Report*, **14**, 65pp.

Checkley, D.M., Davis, R.E., Herman, A.W., *et al.* (2008). Assessing plankton and other particles *in situ* with the SOLOPC. *Limnology and Oceanography*, **53(5)**, 2123–36.

Checkley, D.M., Alheit, J., Oozeki, Y., *et al.* (2009). *Climate change and small pelagic fish*. Cambridge University Press, Cambridge, 408pp.

Cheng, J.-H., Ding, F.-Y., Li, S.-F., *et al.* (2005). A study on the quantity distribution of macro-jellyfish and its relationship to seawater temperature and salinity in the East China Sea region. *Acta Ecologica Sinica*, **25**, 440–5.

Cheung, W., Alder, J., Karpouzi, V., *et al.* (2005). Patterns of species richness in the high seas. *CBD Technical Series*, **20**, 31pp. Secretariat of the Convention on Biological Diversity.

Cheung, W.W.L., Lam, V.W.Y., Sarmiento, J.L., *et al.* (2009). Projecting global marine biodiversity impacts under climate change scenarios. *Fish and Fisheries*, doi: 10.1111/j.1467–2979.2008.00315.x

Chiba, S., Aita, M.N., Tadokoro, K., *et al.* (2008). From climate regime shifts to lower-trophic level phenology: synthesis of recent progress in retrospective studies of the western North Pacific. *Progress in Oceanography*, **77**, 112–26.

Chiba, S., Ono, T., Tadokoro, K., *et al.* (2004). Increased stratification and decreased lower trophic level productivity in the Oyashio region of the North Pacific: a 30-year retrospective study. *Journal of Oceanography*, **60**, 149–62.

Childers, A.R., Whitledge, T.E., and Stockwell, D.A. (2005). Seasonal and interannual variability in the distribution of nutrients and chlorophyll *a* across the Gulf of Alaska shelf: 1998–2000. *Deep-Sea Research II*, **52**, 193–216.

Choi, J.S., Frank, K.T., Leggett, W.C., *et al.* (2004). Transition to an alternate state in a continental shelf ecosystem. *Canadian Journal of Fisheries and Aquatic Sciences*, **61**, 505–10.

Choi, K., Lee, C.I., Hwang, C., *et al.* (2008). Distribution and migration of Japanese common squid, *Todarodes pacificus*, in the southwestern part of the East (Japan) Sea. *Fisheries Research*, **91**, 281–90.

Choi, K.-H. and Kimmerer, W.J. (2008). Mate limitation in an estuarine copepod. *Limnology and Oceanography*, **53**, 1656–64.

Christensen, A., Daewel, U., Jensen, H., *et al.* (2007). Hydrodynamic backtracking of fish larvae by individual-based modelling. *Marine Ecology Progress Series*, **347**, 221–32.

Christensen, V. and McClean, J.L., eds. (2004). Placing fisheries in their ecosystem context. *Ecological Modelling*, **172(2–4)**, 103–440.

Christensen, V., and Pauly, D. (1992). ECOPATH II – A software for balancing steady-state ecosystem models and calculating network characteristics. *Ecological Modelling*, **61**, 169–85.

Christensen, V., and Walters, C.J. (2004). Ecopath with ecosim: methods, capabilities, and limitations. *Ecological Modelling*, **172**, 109–39.

Christensen, V., Guenette, S, Heymanns, J.J., *et al.* (2003). Hundred-year decline of North Atlantic predatory fishes. *Fish and Fisheries*, **4**, 1–24.

Chu, D. and Wiebe, P.H. (2003). Application of sonar techniques to oceanographic-biological surveys. *Current Topics in Acoustical Research*, **3**, 1–25.

Ciannelli, L., Bailey, K., Chan, K.S., *et al.* (2007). Phenological and geographical patterns of walleye pollock spawning in the Gulf of Alaska. *Canadian Journal of Fisheries and Aquatic Sciences*, **64**, 713–22.

Ciannelli, L., Hjermann, D.Ø., Lehodey, P., *et al.* (2004). Climate forcing, food web structure, and community dynamics in pelagic marine ecosystems. In: Stenseth, N.C., Ottersen, G., Hurrell, J.W., *et al.* eds. *Marine ecosystems and climate variation: the North Atlantic, a comparative perspective*. Oxford University Press, Oxford, pp. 143–69.

Cipollini, P., Cromwell, D., Chellenor, P.G., *et al.* (2001). Rossby waves detected in ocean colour data. *Geophysical Research Letters*, **28**, 323–6.

Clapham, P.J. and Link, J.S. (2006). Whales, whaling and ecosystems in the North Atlantic Ocean. In: Estes, J.A., DeMaster, D.P., Doak, D.F., *et al.* eds. *Whales, whaling and ecosystems*. University of Chicago Press, Berkeley, CA, pp. 314–23.

Clark, B. (2006). Climate change: a looming challenge for fisheries management in southern Africa. *Marine Policy*, **30(1)**, 84–95.

Clark, P.U., Pisias, N.G., Stocker, T.F., *et al.* (2002). The role of the thermohaline circulation in abrupt climate change. *Nature*, **415**, 863–9.

Clark, S.H. and Brown, B.E. (1977). Changes in biomass of finfishes and squids from the Gulf of Maine to Cape Hatteras, 1963–74, as determined from research vessel survey data. *Fishery Bulletin*, **75**, 1–21.

Clark, S.H., and Tyler, P.A. (2008). Adult Antarctic krill feeding at abyssal depths. *Current Biology*, **18**, 282–5.

Clarke, A., Murphy, E.J., Meredith, M.P., *et al.* (2007). Climate change and the marine ecosystem of the western Antarctic Peninsula. *Philosophical Transactions of the Royal Society of London, Series B, Biological Sciences*, **362**, 149–66.

Clay, C.S. (1983). Deconvolution of the fish scattering PDF from the echo PDF for a single transducer sonar. *Journal of the Acoustical Society of America*, **73**, 1989–94.

Cochrane, K. and Starfield, A. (1992). The potential use of predictions of recruitment success in the management of the south african anchovy resource. In: Payne, A.I.L., Brink, K.H., Mann, K.H., *et al.* eds. *Benguela trophic functioning. South African Journal of Marine Science*, **12**, 891–902.

Cochrane, K., Butterworth, D.S., De Oliveira, J.A., *et al.* (1998). Management procedures in a fishery based on highly variable stocks and with conflicting objectives: experiernces in the South African pelagic fishery. *Reviews in Fish Biology and Fisheries*, **8**, 177–214.

Cochrane, K., Augustyn, C.J., Bianchi, G., *et al.* (2007). Ecosystem approaches for fisheries (EAF) management in the BCLME (Project LMR/EAF/03/01). *FAO Fisheries Circular*, **1026**, 167pp.

Colebrook, J.M. (1964). Continuous plankton records: a principal component analysis of the geographical distribution of zooplankton. *Bulletin of Marine Ecology*, **6**, 78–100.

Colebrook, J.M. (1969). Variability in plankton. *Progress in Oceanography*, **5**, 115–25.

Colebrook, J.M. and Robinson, G.A. (1964). Continuous plankton records: annual variations of abundance of plankton, 1948–1960. *Bulletin of Marine Ecology*, **6**, 52–69.

Coll, M., Palomera, I., Tudela, S., *et al.* (2006a). Trophic flows, ecosystem structure and fishing impacts in the south Catalan Sea, northwestern Mediterranean. *Journal of Marine Systems*, **59**, 63–96.

Coll, M., Shannon, L.J., Moloney, C.L., *et al.* (2006b). Comparing trophic flows and fishing impacts of a NW Mediterranean ecosystem with coastal upwelling

systems by means of standardized models and indicators. *Ecological Modelling*, **198**, 53–70.

Collie, J.S. and Spencer, P.D. (1994). Modeling predator-prey dynamics in a fluctuating environment. *Canadian Journal of Fisheries and Aquatic Sciences*, **51**, 2665–72.

Collie, J.S., Richardson, K., and Steele, J.H. (2004). Regime shifts: can ecological theory illuminate the mechanisms? *Progress in Oceanography*, **60**, 281–302.

Collie, J.S., Wood, A.D., and Jeffries, H.P. (2008). Long-term shifts in the species composition of a coastal fish community. *Canadian Journal of Fisheries and Aquatic Sciences*, **65**, 1352–65.

Colton, J.B., Jr. (1959). A field observation of mortality of marine fish larvae due to warming. *Limnology and Oceanography*, **4**, 219–22.

Comiso, J.C. (2003). Warming trends in the Arctic from clear sky satellite observations. *Journal of Climate*, **16(21)**, 3498–510.

Concelman, S., Bollens, S.M., Sullivan, B.K., *et al.* (2001). Distribution, abundance and benthic-pelagic coupling of suspended hydroids on Georges Bank. *Deep-Sea Research II*, **48**, 645–58.

Conover, D.O. (2007). Nets versus nature. *Nature*, **450(8)**, 179–80.

Conover, R.J. (1988). Comparative life histories in the genera *Calanus* and *Neocalanus* in high latitudes of the northern hemisphere. *Hydrobiologia*, **167/168**, 127–42.

Constable, A.J. (2005). Implementing plausible ecosystem models for the Southern Ocean: an ecosystem, productivity, ocean, climate (EPOC) model. *CCAMLR WG-EMM 05/33*. CCAMLR, Hobart, Australia.

Constable, A.J., de la Mare, W.K., Agnew, D.J., *et al.* (2000). Managing fisheries to conserve the Antarctic marine ecosystem: practical implementation of the Convention on the Conservation of Antarctic Marine Living Resources (CCAMLR). *ICES Journal of Marine Science*, **57**, 778–91.

Cook, K.B., Bunker, A., Hay, S., *et al.* (2007). Naupliar development times and survival of the copepods *Calanus helgolandicus* and *Calanus finmarchicus* in relation to food and temperature. *Journal of Plankton Research*, **29**, 757–67.

Cooney, R.T., Allen, J.R., Bishop, M.A., *et al.* (2001). Ecosystem control of pink salmon (*Oncorhynchus gorbuscha*) and Pacific herring (*Clupea pallasi*) populations in Prince William Sound, Alaska. *Fisheries Oceanography*, **10**, 1–13.

Corbett, M. (2007). *Learning to leave: The irony of schooling in a coastal community*. Fernwood, Black Point, Nova Scotia, Canada.

Corkett, C. and McLaren, I.A. (1969). Egg production and oil storage by the copepod *Pseudocalanus* in the laboratory.

Journal of Experimental Marine Biology and Ecology, **3**, 90–105.

Corten, A. (2001). Northern distribution of North Sea herring as a response to high water temperatures and/or low food abundance. *Fisheries Research*, **50(1–2)**, 189–204.

Corten, A. (2002). The role of 'conservatism' in herring migrations. *Reviews in Fish Biology and Fisheries*, **11**, 339–61.

Costa, D.P. (1993). The secret life of marine mammals: novel tools for studying their behavior and biology at sea. *Oceanography*, **6**, 120–8.

Costa, D.P., and Sinervo, B. (2004). Field physiology: physiological insights from animals in nature. *Annual Review of Physiology*, **66**, 209–38.

Costa, D.P., Croxall, J.P., and Duck, C.D. (1989). Foraging energetics of Antarctic fur seals in relation to changes in prey availability. *Ecology*, **70**, 596–606.

Costa, D.P., Klinck, J.M., Hofmann, E.E., *et al.* (2008). Upper ocean variability in West Antarctic Peninsula continental shelf waters as measured using instrumented seals. *Deep-Sea Research II*, **55(3–4)**, 323–37.

Costanza, R., d'Arge, R., de Groot, R., *et al.* (1997). The value of the world's ecosystem services and natural capital. *Nature*, **387**, 253–60.

Costello, J.H., Strickler, J.R., Marrasé, C., *et al.* (1990). Grazing in a turbulent environment: behavioral response of a calanoid copepod, *Centropages hamatus*. *Proceedings of the National Academy of Sciences of the United States of America*, **87**, 1648–52.

Costello, J.H., Sullivan, B.K., and Gifford, D.J. (2006). A bio-physical interaction underlying variable phenological responses to climate change by coastal zooplankton. *Journal of Plankton Research*, **28**, 1099–105.

Couzin, I.D., Krause, J., Franks, N.R., *et al.* (2006). Effective leadership and decision-making in animal groups on the move. *Nature*, **433**, 513–6.

Cowan, J.H., Jr., Rose, K.A., and DeVries, D.R. (2000). Is density-dependent growth in young-of-the-year fishes a question of critical weight? *Reviews in Fish Biology and Fisheries*, **10**, 61–89.

Cowen, R.K., Lwiza, K.M.M., Sponaugle, S., *et al.* (2000). Connectivity of marine populations: open or closed? *Science*, **287**, 857–9.

Cowen, R.K., Paris, C.B., and Srinivasan, A. (2006). Scaling of connectivity in marine populations. *Science*, **311**, 522–7.

Cowen, R.K., Thorrold, S., Pineda, J., *et al.* (2007). Population connectivity in marine systems: an overview. *Oceanography*, **20(3)**, 16–23.

Cox, P.M., Betts, R.A., Jones, C.D., *et al.* (2000). Acceleration of global warming due to carbon cycle feedbacks in a coupled climate model. *Nature*, **408**, 184–7.

Coyle, K.O., Pinchuk, A.L., Eisner, L.B., *et al.* (2008). Zooplankton species composition, abundance and biomass on the eastern Bering Sea shelf during summer: the potential role of water column stability and nutrients in structuring the zooplankton community. *Deep-Sea Research II*, **55**, 1775–91.

Crawford, R.J.M. and Shelton, P.A. (1978). Pelagic fish and seabird interrelationships off the coasts of south west and South Africa. *Biological Conservation*, **14**, 85–109.

Crawford, R.J.M. and Shelton, P.A., and Hutchings, L. (1980). Implications of availability, distribution and movements of pilchard (*Sardinops ocellata*) and anchovy (*Engraulis capensis*) for assessment and management of the South African purse-seine fishery. *Rapports et Proces-verbaux des Reunions. Conseil International pour l'Exploration de la Mer*, **177**, 355–73.

Crocker, D.E., Costa, D.P., Le Boeuf, B.J., *et al.* (2006). Impact of *El Niño* on the foraging behavior of female northern elephant seals. *Marine Ecology Progress Series*, **309**, 1–10.

Croll, D.A., Tershy, B.R., Hewitt, R.P., *et al.* (1998). An integrated approach to the foraging ecology of marine birds and mammals. *Deep-Sea Research II*, **45**, 1353–7.

Croll, D.A., Marinovic, B., Benson, S., *et al.* (2005). From wind to whales: trophic links in a coastal upwelling system. *Marine Ecology Progress Series*, **289**, 117–30.

Croxall, J.P., Prince, P.A., and Ricketts, C. (1985). Relationships between prey life-cycles and the extent, nature and timing of seal and seabird predation in the Scotia Sea. In: Siegfried, W.R., Condy, P.R., and Laws, R.M., eds. *Antarctic nutrient cycles and food webs*. Springer-Verlag, Berlin.

Crutzen, P.J. and Stoermer, E.F. (2000). The Anthropocene. *Global Change Newsletter*, **41**, 17–18.

Culverhouse, P.F., Williams, R., Reguera, B., *et al.* (2003). Do experts make mistakes? A comparison of human and machine identification of dinoflagellates. *Marine Ecology Progress Series*, **247**, 17–25.

Culverhouse, P.F., Williams, R., Benfield, M., *et al.* (2006). Automatic image analysis of plankton: future perspectives. *Marine Ecology Progress Series*, **312**, 297–309.

Cummins, P.F. and Freeland, H.J. (2007). Variability of the north Pacific current and its bifurcation. *Progress in Oceanography*, **75**, 253–65.

Cunningham, C.L. and Butterworth, D.S. (2005). Re-revised OMP-04. *Unpublished Marine and Coastal Management Document*, **SWG/DEC2005/PEL/05**, 14pp.

Curchitser, E.N., Haidvogel, D.B., Hermann, A.J., *et al.* (2005). Multi-scale modeling of the North Pacific Ocean: assessment of simulated basin-scale variability (1996–2003).

Journal of Geophysical Research-Oceans, **110**, C11021, doi:101029/2005JC002902.

Cury, P. and Christensen, V. (2005). Quantitative ecosystem indicators for fisheries management. *ICES Journal of Marine Science*, **62(3)**, 307–10.

Cury, P., and Pauly, D. (2000). Patterns and propensities in reproduction and growth in marine fishes. *Ecological Research*, **15**, 101–6.

Cury, P., and Roy, C. (1989). Optimal environmental window and pelagic fish recruitment success in upwelling areas. *Canadian Journal of Fisheries and Aquatic Sciences*, **46**, 670–80.

Cury, P., and Shannon, L.J. (2004). Regime shifts in upwelling ecosystems: observed changes and possible mechanisms in the northern and southern Benguela. *Progress in Oceanography*, **60**, 223–43.

Cury, P., Bakun, A., Crawford, R.J.M., *et al.* (2000). Small pelagics in upwelling systems: patterns of interaction and structural changes in "wasp-waist" ecosystems. *ICES Journal of Marine Science*, **57**, 603–18.

Cury, P., Shannon, L., and Shin, Y.-J., Shannon, L., (2001). *The functioning of marine ecosystems*. Reykjavik Conference on Responsible Fisheries in the Marine Ecosystem, 22pp.

Cury, P., Shannon, L., and Shin, Y.-J. (2003). The functioning of marine ecosystems: a fisheries perspective. In: Sinclair, M. and Valdimarsson, G., eds. *Responsible fisheries in the marine ecosystem*, pp. 103–23. FAO and CABI Publishing, Rome, and Wallingford, UK. 426pp.

Cury, P., Shannon, L., Roux, J.-P., *et al.* (2005). Trophodynamic indicators for an ecosystem approach to fisheries. *ICES Journal of Marine Science*, **62**, 430–42.

Cury, P., Shin, Y.-J., Planque, B., *et al.* (2008). Ecosystem oceanography for global change in fisheries. *Trends in Ecology and Evolution*, **23**, 338–46.

Cushing, D.H. (1975). *Marine ecology and fisheries*. Cambridge University Press, Cambridge, 278pp.

Cushing, D.H. (1982). *Climate and fisheries*. Academic Press, London, 373pp.

Cushing, D.H., and Dickson, R.R. (1976). The biological response in the sea to climatic changes. *Advances in Marine Biology: An Annual Review*, **14**, 1–122.

Daan, N., Christensen, V., and Cury, P.M., eds. (2005). Quantitative ecosystem indicators for fisheries management. Proceedings of a symposium held in Paris, France, 31 March-3 April 2004. *ICES Journal of Marine Science*, **62(3)**, 613pp.

Dagg, M.J. (1978). Estimated, *in situ*, rates of egg production for the copepod *Centropages typicus* (Kroyer) in the New York Bight. *Journal of Experimental Marine Biology and Ecology*, **34**, 183–96.

Dagg, M.J., Liu, H., and Thomas, A.C. (2006). Effects of mesoscale phytoplankton variability on the copepods

Neocalanus flemingeri and *N. plumchrus* in the coastal Gulf of Alaska. *Deep-Sea Research I*, **53**, 321–32.

Dale, T., Bagoien, E., Melle, W., *et al.* (1999). Can predator avoidance explain varying overwintering depth of *Calanus* in different oceanic water masses? *Marine Ecology Progress Series*, **179**, 113–21.

Dale, T., Kaartvedt, S., Ellertsen, B., *et al.* (2001). Large-scale oceanic distribution and population structure of *Calanus finmarchicus*, in relation to physical environment, food and predators. *Marine Biology*, **139**, 561–74.

Dalpadado, P. and Skjoldal, H.R. (1996). Abundance, maturity and growth of krill species *Thysanoessa inermis* and *T. longicaudata* in the Barents Sea. *Marine Ecology Progress Series*, **144**, 175–83.

Dalsgaard, J., St. John, M., Kattner, G., *et al.* (2003). Fatty acid trophic markers in the pelagic marine environment. *Advances in Marine Biology: An Annual Review*, **46**, 225–340.

Daly, K.L. (2004). Overwintering growth and development of larval *Euphausia superba*: an interannual comparison under varying environmental conditions west of the Antarctic Peninsula. *Deep-Sea Research II*, **51**, 2139–68.

Dam, H.G. and Tang, K.W. (2001). Affordable egg mortality: constraining copepod egg mortality with life history traits. *Journal of Plankton Research*, **23**, 633–40.

Dann, P., Norman, F.I., Cullen, J.M., *et al.* (2000). Mortality and breeding failure of little penguins, *Eudyptula minor*, in Victoria 1995–1996, following a widespread mortality of pilchard *Sardinops sagax*. *Marine and Freshwater Research*, **51**, 355–62.

D'Arrigo, R.D., Wilson, R.J., Deser, C., *et al.* (2005). Tropical-north Pacific climate linkages over the past four centuries. *Journal of Climate*, **18**, 5253–65.

Daskalov, G.M. (2000). Mass-balance modelling and network analysis of the Black Sea pelagic ecosystem. *Izvestiya IRR Varna*, **25**, 49–62.

Daskalov, G.M. (2002). Overfishing drives a trophic cascade in the Black Sea. *Marine Ecology Progress Series*, **225**, 53–63.

Daskalov, G.M. (2003). Long-term changes in fish abundance and environmental indices in the Black Sea. *Marine Ecology Progress Series*, **255**, 259–70.

Daskalov, G.M., Grishin, A.N., Rodionov, S., *et al.* (2007) Trophic cascades triggered by overfishing reveal possible mechanisms of ecosystem regime shifts. *Proceedings of the National Academy of Sciences of the United States of America*, **104(25)**, 10518–23.

Davis, C.S. (1987). Components of the zooplankton production cycle in the temperate ocean. *Journal of Marine Research*, **45**, 947–83.

Davis, C.S., Gallager, S.M., Berman, M.S., *et al.* (1992). The video plankton recorder (VPR): design and initial results. *Archiv für Hydrobiologie, Beiheft: Ergebnisse der Limnologie*, **36**, 67–81.

Davis, C.S., Gallager, S.M., Marra, M., *et al.* (1996). Rapid visualization of plankton abundance and taxonomic composition using the Video Plankton Recorder. *Deep-Sea Research II*, **43**, 1947–70.

Davis, C.S., Hu, Q., Gallager, S.M., *et al.* (2004). Real-time observation of taxa-specific plankton distributions: an optical sampling method. *Marine Ecology Progress Series*, **284**, 77–96.

Davis, C.S., Thwaites, F.T., Gallager, S.M., *et al.* (2005). A three-axis fast-tow digital Video Plankton Recorder for rapid surveys of plankton taxa and hydrography. *Limnology and Oceanography Methods*, **3**, 59–74.

Davis, R.E., Sherman, J.T., and Dufour, J. (2001). Profiling ALACEs and other advances in autonomous subsurface floats. *Journal of Atmospheric and Oceanic Technology*, **18**, 982–93.

Davis, R.E., Eriksen, C.C., and Jones, C.P. (2003). Autonomous buoyancy-driven underwater gliders. Chapter 3. In: Griffiths, C., ed. *Technology and application of autonomous underwater vehicles*. Taylor & Francis, London, pp. 37–58.

Dawson, M.N., Martin, L.E., and Penland, L.K. (2001). Jellyfish swarms, tourists, and the Christ-child. *Hydrobiologia*, **451**, 131–44.

de Baar, H.J.W. and Boyd, P.W. (2000). The role of iron in plankton ecology and carbon dioxide transfer of the global oceans. In: Hanson, R.B., Ducklow, H.W., and Field, J.G., eds. *The changing ocean carbon cycle. A midterm synthesis of the Joint Global Ocean Flux Study*. Cambridge University Press, Cambridge, pp. 61–140.

de Figueiredo, G.M., Nash, R.D.M., and Montagnes, D.J.S. (2005). The role of the generally unrecognised microprey source as food for larval fish in the Irish Sea. *Marine Biology*, **148**, 395–404.

de Figueiredo, G.M., Nash, R.D.M., and Montagnes, D.J.S. (2007). Do protozoa contribute significantly to the diet of larval fish in the Irish Sea? *Journal of the Marine Biological Association of the United Kingdom*, **87**, 843–50.

De la Mare, W.K. (1996). Some recent developments in the management of marine living resources. In: Floyd, R.B., Shepherd, A.W., and De Barro, P.J., eds. *Frontiers of population ecology*. CSIRO Publishing, Melbourne, Australia, pp. 599–616.

De Oliveira, J.A.A. (2006). Long-term harvest strategies for small pelagic fisheries under regime shifts: the South African fishery for pilchard and anchovy. In: Hannesson, R., Barange, M., and Herrick, S., Jr., eds. *Climate change and the economics of the world's fisheries: examples of small pelagic fish stocks*. Edward Elgar, Cheltenham, UK, pp. 151–204.

De Oliveira, J.A.A. and Butterworth, D.S. (2004). Developing and refining a joint management procedure for the mul-

tispecies South African pelagic fishery. *ICES Journal of Marine Science*, **61**, 1432–42.

De Oliveira, J.A.A., and Butterworth, D.S. (2005). Limits to the use of environmental indices to reduce risk and/or increase yield in the South African anchovy fishery. *African Journal of Marine Science*, **27(1)**, 191–203.

De Oliveira, J.A.A., Uriarte, A., and Roel, B.A. (2005). Potential improvements in the management of Bay of Biscay anchovy by incorporating environmental indices as recruitment predictors. *Fisheries Research*, **75**, 2–14.

De Oliveira, J.A.A., Kell, L.T., Punt, A.E., *et al.* (2008). Managing without best predictions: the management strategy evaluation framework. In: Payne, A., Cotter, J., and Potter, T., eds. *Advances in fisheries science: 50 years on from Beverton and Holt*. Blackwell Publishing, Oxford, pp. 104–34.

De Robertis, A. (2002). Size-dependent visual predation risk and the timing of vertical migration: an optimization model. *Limnology and Oceanography*, **47**, 925–33.

DeAngelis, D.L. and Mooij, W.M. (2005). Individual-based modeling of ecological and evolutionary processes. *Annual Review of Ecology, Evolution and Systematics*, **36**, 147–68.

Degnbol, P. (2001). Science and the user perspective-the scale gap and the need for shared indicators. Paper presented at people and the sea. *Inaugural Conference Center for Maritime Research (MARE), 30 August – 1 September 2001*. Amsterdam, The Netherlands.

Degnbol, P., and Jarre, A. (2004). Review of indicators in fisheries management: a development perspective. In: Shannon, L.J., Cochrane, K.L., and Pillar, S.C., eds. *Ecosystem approaches to fisheries in the Southern Benguela*. *African Journal of Marine Science*, **26**, 303–26.

Delgado, C.L., Wada, N., Rosegrant, M.W., *et al.* (2003). *Outlook for fish to 2020: meeting global demand*. International Food Policy Research Institute, Washington, DC and WorldFish Center, Penang, Malaysia.

DeLong, E.F., Taylor, L.T., Marsh, T.L., *et al.* (1999). Visualisation and enumeration of marine planktonic archaea and bacteria by using polyribonucleotide acid probes and fluorescent *in situ* hybridization. *Applied Environmental Microbiology*, **65**, 5554–63.

Delong, R.L., Stewart, B.S., and Hill, R.D. (1992). Documenting migrations of northern elephant seals using day length. *Marine Mammal Science*, **8**, 155–9.

Delworth, T.L. and Mann, M.E. (2000). Observed and simulated multidecadal variability in the Northern Hemisphere. *Climate Dynamics*, **16**, 661–76.

Denman, K.L. (2003). Modelling planktonic ecosystems: parameterizing complexity. *Progress in Oceanography*, **57**, 429–52.

Deser, C., Walsh, J.E., and Timlin, M.S. (2000). Arctic sea ice variability in the context of recent atmospheric circulation trends. *Journal of Climate*, **13**, 617–33.

deVeen, J.F. (1978). On selective tidal transport in the migration of North Sea plaice and other flatfish species. *Netherlands Journal of Sea Research*, **12**, 115–47.

Devine, J.A., Baker, J.D., and Haedrich, R.L. (2006). Fisheries: deep-sea fishes qualify as endangered. *Nature*, **439(7072)**, 29.

DeWailly, E. and Knap, A. (2006). Food from the oceans and human health: balancing risks and benefits. *Oceanography*, **19(2)**, 84–93.

Dewar, W.K., Bingham, R.J., Iverson, R.L., *et al*. (2006). Does the marine biosphere mix the ocean? *Journal of Marine Research*, **64**, 541–61.

deYoung, B., Heath, M., Werner, F., *et al*. (2004a). Challenges of modeling ocean basin ecosystems. *Science*, **304**, 1463–6.

deYoung, B., Harris, R., Alheit, J., *et al*. (2004b). Detecting regime shifts in the ocean: data considerations. *Progress in Oceanography*, **60**, 143–64.

deYoung, B., Barange, M., Beaugrand, G., *et al*. (2008). Regime shifts in marine ecosystems: detection, prediction and management. *Trends in Ecology and Evolution*, **23**, 402–9.

deYoung, B., Werner, F., Batchelder, *et al*. (2010). Physical-biological interactions: integration and modelling. *This volume*. Oxford University Press, Oxford.

DFO (Fisheries and Oceans Canada). (2003). State of the eastern Scotian Shelf ecosystem. *Canadian Science Advisory Secretariat Science Advisory Report*, **2003/004**.

DFO (2007a). State of the Pacific Ocean 2006. *Ocean Status Report*, **2007/019**, 85pp.

DFO (2007b). 2006 State of the ocean: physical oceanographic conditions in the Newfoundland and Labrador region. *Canadian Science Advisory Secretariat Science Advisory Report*, **2007/025**, 11pp.

Di Lorenzo, E., Schneider, N., Cobb, K.M., *et al*. (2008). North Pacific Gyre oscillation links ocean climate and ecosystem change. *Geophysical Research Letters*, **35**, L08607, doi:10.1029/2007GL032838.

Diaz, R.J. (2001). Overview of hypoxia around the world. *Journal of Environmental Quality*, **30**, 275–81.

Diaz, R.J., and Rosenburg, R. (2008). Spreading dead zones and consequences for marine ecosystems. *Science*, **321**, 926–9.

Diaz, R.J., and Solow, A. (1999). Ecological and economic consequences of hypoxia. Topic 2. Gulf of Mexico hypoxia assessment. *NOAA Coastal Ocean Program Decision Analysis Series*. NOAA COP, Silver Springs, MD.

Dickey, T.D. (1988). Recent advances and future directions in multi-disciplinary *in situ* oceanographic measurement systems. In: Rothschild, B.J., ed. *Toward a theory on biological-physical interactions in the world ocean*. Springer, Berlin, pp. 555–98.

Dickey, T.D. (1990). Physical-optical-biological scales relevant to recruitment in large marine ecosystems. In: Sherman, K., Alexander, L.M., and Gold, B.D., eds. *Large marine ecosystems: patterns, processes, and yields*. AAAS Press, Washington, DC, pp. 82–98.

Dickey, T.D. ed. (1993). Report of the first meeting of the International GLOBEC working group on sampling and observation systems, Paris, France, March 30–April 2, 1993. *GLOBEC Report*, **3**, 99pp.

Dickey, T.D. (2003). Emerging ocean observations for interdisciplinary data assimilation systems, *Journal of Marine Systems*, **40/41**, 5–48.

Dickey, T.D. (2004). Studies of coastal ocean dynamics and processes using emerging optical technologies. *Oceanography*, **17**, 9–13.

Dickey, T.D. and Bidigare, R.R. (2005). Interdisciplinary oceanographic observations: the wave of the future. *Scientia Marina*, **69**, 23–42.

Dickey, T.D., Lewis, M.R., and Chang, G.C. (2006). Optical oceanography: recent advances and future directions using global remote sensing and *in situ* observations. *Reviews of Geophysics*, **44**, RG 1001, doi:10.1029/2003RG000148.

Dickmann, M., Möllmann, C., and Voss, R. (2007). Feeding ecology of central Baltic sprat *Sprattus sprattus* larvae in relation to zooplankton dynamics: implications for survival. *Marine Ecology Progress Series*, **342**, 277–89.

Dickson, R.R., Meincke, J., Malmberg, S.-A., *et al*. (1988). 'The "great salinity anomaly" in the northern North Atlantic 1968–82. *Progress in Oceanography*, **20**, 103–51.

Diffenbaugh, N.S. (2005). Atmosphere-land cover feedbacks alter the response of surface temperature to CO_2 forcing in the western United State. *Climate Dynamics*, **24**, 237–51.

Dingle, H. (1996). Migration. *The biology of life on the move*. Oxford University Press, Oxford, 474pp.

Doall, M.H., Colin, S.P., Strickler, J.R., *et al*. (1998). Locating a mate in 3D: the case of *Temora longicornis*. *Philosophical Transactions of the Royal Society of London, Series B, Biological Sciences*, **353**, 681–9.

Doall, M.H., Strickler, J.R., Fields, D.M., *et al*. (2002). Mapping the free-swimming attack volume of a planktonic copepod, *Euchaeta rimana*. *Marine Biology*, **140**, 871–9.

Doherty, P.J., Planes, S., and Mather, P. (1995). Gene flow and larval duration in seven species of fish from the Great Barrier Reef. *Ecology*, **76**, 2373–91.

Dolan, A.H., Taylor, M., Neis, B., *et al*. (2005). Restructuring and health in Canadian coastal communities. *EcoHealth*, **2**, 1–14.

Domingues, C.M., Church, J.A., White, N.J., *et al.* (2008). Improved estimates of upper-ocean warming and multi-decadal sea-level rise. *Nature*, **453**, 1090–3.

Doney, S.C., Lindsay, K., Caldeira, K., *et al.* (2004). Evaluating global ocean carbon models: the importance of physics. *Global Biogeochemical Cycles.* 18, GB3017, doi:10.1029/2003GB002150.

Dower, J.F., Miller, T.J., and Leggett, W.C. (1997). The role of microscale turbulence in the feeding ecology of larval fish. *Advances in Marine Biology: An Annual Review*, **31**, 169–220.

Drinkwater, K.F. (2000). Changes in ocean climate and its general effect on fisheries: examples from the Northwest Atlantic. In: D. Mills, ed. *The ocean life of Atlantic salmon-environmental and biological factors influencing survival.* Fishing News Books, Oxford, pp. 116–36.

Drinkwater, K.F. (2002). A review of the role of climate variability in the decline of northern cod. *American Fisheries Society Symposium Series*, **32**, 113–30.

Drinkwater, K.F. (2005). The response of Atlantic cod (*Gadus morhua*) to future climate change. *ICES Journal of Marine Science*, **62**, 1327–37.

Drinkwater, K.F. (2006). The regime shift of the 1920s and 1930s in the North Atlantic. *Progress in Oceanography*, **68**, 134–51.

Drinkwater, K.F., and Myers, R.A. (1987). Testing predictions of marine fish and shellfish landings from environmental variables. *Canadian Journal of Fisheries and Aquatic Sciences*, **44**, 1568–73.

Drinkwater, K.F., Belgrano, A., Borja, A., *et al.* (2003). The response of marine ecosystems to climate variability associated with the North Atlantic Oscillation. In: Hurrell, J., Kushnir, Y., Ottersen, G., *et al.* eds. *The north Atlantic oscillation, climatic significance and environmental impact. Geophysical Monograph Series* 134. AGU Press, Washington, DC, pp. 211–34.

Drinkwater, K.F., Loeng, H., Megrey, B., *et al.* eds. (2005). The influence of climate change on north Atlantic fish stocks. *ICES Journal of Marine Science*, **62**, 1203–542.

Drinkwater, K.F., Beaugrand, G., Kaeriyama, M., *et al.* (2009a). On the processes linking climate to ecosystem changes. *Journal of Marine Systems*, **79**, 374–88.

Drinkwater, K.F., Hunt, G.L., Lehodey, P., *et al.* (2010b). Climate forcing on marine ecosystems. *This volume.* Oxford University Press, Oxford.

Ducklow, H. (2003). Biogeochemical provinces: towards a JGOFS synthesis. In: Fasham, M.J.R., ed. *Ocean biogeochemistry: the role of the ocean carbon cycle in global change. Global change-the IGBP Series*, pp. 3–17. Springer, Heidelberg, 297pp.

Ducklow, H., Steinberg, D.K., and Buesseler, K.O. (2001). Upper ocean carbon export and the biological pump. *Oceanography*, **14**, 50–8.

Ducklow, H., Baker, K., Martinson, D.G., *et al.* (2007). Marine pelagic ecosystems: the west Antarctic peninsula. *Philosophical Transactions of the Royal Society of London, Series B, Biological Sciences*, **362**, 67–94.

Dufrêne, M. and Legendre, P. (1997). Species assemblages and indicator species: the need for a flexible asymmetrical approach. *Ecological Monographs*, **67**, 345–66.

Duhamel du Monceau, M. (1771). *Traite general des pesches: et histoire des poissons qu'elles fournissent, tant pour la subsistance des hommes, que pour plusieurs autres usages qui ont rapport aux arts et au commerce.* Chez Saillant and Nyon, Paris.

Dulvy, N.K., Sadovy, Y., and Reynolds, J.D. (2003). Extinction vulnerability in marine populations. *Fish and Fisheries*, **4**, 25–64.

Dulvy, N.K., Ellis, J.R., Goodwin, N.B., *et al.* (2004). Methods of assessing extinction risk in marine fishes. *Fish and Fisheries*, **5**, 255–75.

Dulvy, N.K., Rogers, S.I., Jennings, S., *et al.* (2008). Climate change and deepening of the North Sea fish assemblage: a biotic indicator of warming seas. *Journal of Applied Ecology*, **45**, 1029–39.

Dunn, J., Mitchell, B., Urquhart, G.G., *et al.* (1989). New opening and closing nets for sampling fish larvae. *The early life history of fish, the third ICES Symposium. Rapports et Proces-verbaux des Reunions. Conseil International pour l'Exploration de la Mer*, **191**, p. 450.

Dunn, J., Hall, C.D., Heath, M.R., *et al.* (1993). ARIES-a system for concurrent physical, biological, and chemical sampling at sea. *Deep-Sea Research I*, **40**, 867–78.

Durant, J.M., Hjermann, D.O., Ottersen, G., *et al.* (2007). Climate and the match or mismatch between predator requirements and resource availability. *Climate Research*, **33**, 271–83.

Durbin, E.G. and Campbell, R.G. (2007). Reassessment of the gut pigment method for estimating *in situ* zooplankton ingestion. *Marine Ecology Progress Series*, **331**, 305–7.

Durbin, E.G. and Casas, M.C. (2006). Abundance and spatial distribution of copepods on Georges Bank during the winter/spring period. *Deep-Sea Research II*, **53**, 2537–69.

Durbin, E.G., Durbin, A.G., Smayda, T.J., *et al.* (1983). Food limitation of production by adult *Acartia tonsa* in Narragansett Bay, Rhode Island. *Limnology and Oceanography*, **28**, 1199–213.

Durbin, E.G., Campbell, R.G., Gilman, S.L., *et al.* (1995). Diel feeding-behavior and ingestion rate in the copepod *Calanus finmarchicus* in the southern Gulf of Maine during late spring. *Continental Shelf Research*, **15(4–5)**, 539–70.

Durbin, E.G., Runge, J.A., Campbell, R., *et al.* (1997). Late fall-early winter recruitment of *Calanus finmarchicus* on

Georges Bank. *Marine Ecology Progress Series*, **151**, 103–14.

Durbin, E.G., Campbell, R.G., Casas, M.C., *et al.* (2003). Interannual variation in phytoplankton blooms and zooplankton productivity and abundance in the Gulf of Maine during winter. *Marine Ecology Progress Series*, **254**, 81–100.

Durbin, E.G., Casas, M., Rynearson, T.A., *et al.* (2008). Measurement of copepod predation on nauplii using qPCR of the cytochrome oxidase I gene. *Marine Biology*, **153**, 699–707.

Duxbury, J. and Dickinsen, S. (2007). Principles for sustainable governance of the coastal zone: in the context of natural disasters. *Ecological Economics*, **63**, 319–30.

Eckert, S.A., Dolar, L.L., Kooyman, G.L., *et al.* (2002). Movements of whale sharks (*Rhincodon typus*) in southeast Asian waters as determined by satellite telemetry. *Journal of Zoology*, **257**, 111–5.

Edeline E., Carlson S.M., Stige L.C., *et al.* (2007). Trait changes in a harvested population are driven by a dynamic tug-of-war between natural and harvest selection. *Proceedings of the National Academy of Sciences of the United States of America*, **104(40)**, 15799–804.

Edmondson, T. (1960). Reproductive rates of rotifers in natural populations. *Memorie dell'Istituto Italiano di Idrobiologia*, **12**, 21–77.

Edvardsen, A., Zhou, M., Tande, K.S., *et al.* (2002). Zooplankton population dynamics: measuring *in situ* growth and mortality rates using an optical plankton counter. *Marine Ecology Progress Series*, **227**, 205–19.

Edwards, M. and Richardson, A.J. (2004). Impact of climate change on marine pelagic phenology and trophic mismatch. *Nature*, **430**, 881–4.

Edwards, M., Beaugrand, G., Reid, P.C., *et al.* (2002). Ocean climate anomalies and the ecology of the North Sea. *Marine Ecology Progress Series*, **239**, 1–10.

Ehrenberg, J.E. (1974). Two applications for a dual beam transducer in hydroacoustic fish assessment systems. In: *IEEE international conference on engineering in the ocean environment*. Halifax, Nova Scotia, Canada, pp. 152–5.

Ehrenberg, J.E. (1979). A comparative analysis of *in situ* methods for directly measuring the acoustic target strength of individual fish. *IEEE Journal of Oceanic Engineering*, **OE-4**, 141–52.

Eiane, K. and Ohman, M.D. (2004). Stage-specific mortality of *Calanus finmarchicus*, *Pseudocalanus elongatus* and *Oithona similis* on Fladen Ground, North Sea, during a spring bloom. *Marine Ecology Progress Series*, **268**, 183–93.

Eiane, K., Aksnes, D.L., Bagoien, E., *et al.* (1999). Fish or jellies-a question of visibility? *Limnology and Oceanography*, **44**, 1352–7.

Eiane, K., Aksnes, D.L., Ohman, M.D., *et al.* (2002). Stage-specific mortality of *Calanus* spp. under different predation regimes. *Limnology and Oceanography*, **47**, 636–45.

Eide, A. and Heen, K. (2002). Economic impacts of global warming: a study of the fishing industry in North Norway. *Fisheries Research*, **56**, 261–74.

Eilertsen, H.C. (1993). Spring blooms and stratification. *Nature*, **363(6424)**, 24.

Ekstrom, P. (2004). An advance in geolocation by light. *Memoirs of the National Institute of Polar Research*, **58**, 210–26.

Ellertsen, B., Fossum, P., Solemdal, P., *et al.* (1989). Relation between temperature and survival of eggs and first-feeding larvae of Northeast Arctic cod. *Rapports et Proces-Verbaux des Reunions du Conseil International pour l'Exploration de la Mer*, **191**, 209–19.

Ellingsen, I.H., Dalpadado, D., Slagstad, D., *et al.* (2008). Impact of climatic change on the biological production in the Barents Sea. *Climatic Change*, **87**, 155–75.

Elmgren, R. (1989). Man's impact on the ecosystem of the Baltic Sea: energy flows today and at the turn of the century. *Ambio*, **18**, 326–32.

Elser, J.J. and Hessen, D.O. (2005). Biosimplicity via stociometry: the evolution of food-web structure and processes. In: Belgrano, A., Scharler, U.M., Dunne J., *et al.* eds. *Aquatic food webs: an ecosystem approach*. Oxford University Press, Oxford, pp. 7–18.

Elton, C. (1927). *Animal ecology*. Sidgwick and Jackson, London, 207pp.

Emerson, S., Watanabe, Y.W., Ono, T., *et al.* (2004). Temporal trends in apparent oxygen utilization in the upper pycnocline of the North Pacific. *Journal of Oceanography*, **60**, 139–48.

Enghoff, I.B., MacKenzie, B.R., and Nielsen, E.E. (2007). The Danish fish fauna during the warm Atlantic period (*ca.* 7000–3900 B.C.): forerunner of future changes? *Fisheries Research*, **87**, 167–80.

Enquist, B.J., Economo, E.P., Huxman, T.E., *et al.* (2003). Scaling metabolism from organisms to ecosystems. *Nature*, **423**, 639–42.

EPICA (2006). One-to-one coupling of glacial climate variability in Greenland and Antarctica. *Nature*, **444**, 195–8.

Escribano, R., Daneri, G., Farias, L., *et al.* (2004). Biological and chemical consequences of the 1997–1998 *El Niño* in the Chilean coastal upwelling system: a synthesis. *Deep-Sea Research II*, **51**, 2389–411.

Eslinger, D.L. and Iverson, R.L. (2001). The effects of convective and wind-driven mixing on spring phytoplankton dynamics in the southeastern Bering Sea middle shelf domain. *Continental Shelf Research*, **21**, 627–50.

Essington, T.E., Beaudreau, A.H., and Wiedenmann, J. (2006). Fishing through marine food webs. *Proceedings of* the National Academy of Sciences of the United States of America, **103**, 3171–5.

Estes, J.A. and Palmisano, J.F. (1974). Sea otters: their role in structuring nearshore communities. *Science*, **185**, 1058–60.

Estes, J.A., DeMaster, D.P., Doak, D.F., *et al.* eds. (2007). Whales, whaling and ecosystems. In: *The state of world fisheries and aquaculture*. University of Chicago Press and Food and Agriculture Organization of the United Nations, Berkeley, CA and Rome.

Ezer, T., Arango, H., and Shchepetkin, A.F. (2002). Developments in terrain-following ocean models: intercomparisons of numerical aspects. *Ocean Modelling*, 4, 249–67.

Fabricius, C., Folke, C., Cundill, G., *et al.* (2007). Powerless spectators, coping actors, and adaptive co-managers: a synthesis of the role of communities in ecosystem management. *Ecology and Society*, **12(1)**, art:29.

Fach, B.A. and Klinck, J.M. (2006). Transport of Antarctic krill (*Euphausia superba*) across the Scotia Sea, Part I: circulation and particle tracking simulations. *Deep-Sea Research I*, 53, 987–1010.

Fach, B.A., Hofmann, E.E., and Murphy, E.J. (2006). Transport of Antarctic krill (*Euphausia superba*) across the Scotia Sea, Part II: krill growth and survival. *Deep-Sea Research I*, 53, 1011–43.

Fairweather, T.P., van der Lingen, C.D., Booth, A.J., *et al.* (2006). Indicators of sustainable fishing for South African sardine *Sardinops sagax* and anchovy *Engraulis encrasicolus*. *African Journal of Marine Science*, **28**, 661–80.

FAO. (1995a). Precautionary approach to fisheries, Part 1: guidelines on the precautionary approach to capture fisheries and species introductions. *FAO Fisheries Technical Paper*, **350(1)**, 52pp.

FAO. (1995b). *Code of conduct for responsible fisheries*. FAO, Rome, 41pp.

FAO. (2001). *Fisheries governance*. FIGIS Topics and Issues Fact Sheet. Fishery Policy and Planning Division, FAO, Rome.

FAO. (2003). Fisheries management 2: the ecosystem approach to fisheries. *FAO Technical Guidelines for Responsible Fisheries*, **4(Suppl. 2)**, 112pp.

FAO. (2005). Increasing the contribution of small-scale fisheries to poverty alleviation and food security. *FAO Technical Guidelines for Responsible Fisheries*, **10**, 79pp.

FAO. (2006). Contribution of fisheries to national economies in West and Central Africa-policies to increase the wealth generated by small-sale fisheries. *New Directions in Fisheries-A Series of Policy Briefs on Development Issues*, **3**, 12pp.

FAO. (2007). *The state of world fisheries and aquaculture (SOFIA) 2006*. FAO, Rome. 162pp.

Farber, S., Costanza, R., Childers, D.L., *et al.* (2006). Linking ecology and economics for ecosystem management. *BioScience*, **56(2)**, 117–29.

Fasham, M.J.R., ed. (1984). *Flows of energy and materials in marine ecosystems: theory and practice*. Plenum Press, New York, 733pp.

Fasham, M.J.R. (1993). Modelling the marine biota. In: Heimann, M., ed. *The global carbon cycle*. Springer-Verlag, Berlin, pp. 457–504.

Fasham, M.J.R. ed. (2003). *Ocean biogeochemistry: the role of the ocean carbon cycle in global change*. Global change-the IGBP Series. Springer-Verlag, Heidelberg, 297pp.

Fedak, M.A., Lovell, P., and Grant, S.M. (2001). Two approaches to compressing and interpreting time-depth information as collected by time-depth recorders and satellite-linked data recorders. *Marine Mammal Science*, **17**, 94–110.

Feigenbaum, D.L. and Kelly, M. (1984). Changes in the lower Chesapeake Bay food chain in the presence of the sea nettle *Chrysaora quinquecirrha* (Scyphomedusa). *Marine Ecology Progress Series*, **19**, 39–47.

Feinberg, L.R. and Peterson, W.T. (2003). Variability and duration of euphausiid spawning off central Oregon, 1996–2001. *Progress in Oceanography*, **57**, 363–79.

Feinberg, L.R., Shaw, C.T., and Peterson, W.T. (2007). Long-term laboratory observations of *Euphausia pacifica* fecundity: comparison of two geographic regions. *Marine Ecology Progress Series*, **341**, 141–52.

Fennel, K., Wilkin, J., Levin, J., *et al.* (2006). Nitrogen cycling in the Middle Atlantic Bight: results from a three-dimensional model and implications for the North Atlantic nitrogen budget. *Global Biogeochemical Cycles*, **20**, GB3007, doi:10.1029/2005GB002456.

Fernö, A., Giske, J., and Sundby, S. (1997). Foreword. Mare Cognitum. *Sarsia*, **82**, p.79.

Ferron, A. and Leggett, W.C. (1994). An appraisal of condition measures for marine fish larvae. *Advances in Marine Biology: An Annual Review*, **30**, 217–303.

Fiedler, P.C., Reilly, S.B., Hewitt, R.P., *et al.* (1998). Blue whale habitat and prey in the California Channel Islands. *Deep-Sea Research II*, **45**, 1781–801.

Field, D.B., Baumgartner, T.R., Ferreira, V., *et al.* (2009). Variability from scales in marine sediments and other historical records. In: Checkley, D., Alheit, J., Oozeki, Y., *et al.* eds. *Climate change and small pelagic fish*. Cambridge University Press, Cambridge.

Field, J.C., Francis, R.C., and Aydin, K.Y. (2006). Top-down and bottom-up dynamics: linking a fisheries-based ecosystem model with climate hypotheses in the Northern California Current. *Progress in Oceanography*, **68**, 238–70.

Fielding, S., Griffiths, G., and Roe, H.S.J. (2004). The biological validation of ADCP acoustic backscatter through direct comparison with net samples and model predictions based on acoustic-scattering models. *ICES Journal of Marine Science*, **61**, 184–200.

Fields, D.M. and Weissburg, M.J. (2004). Rapid firing rates from mechanosensory neurons in copepod antennules. *Journal of Comparative Physiology A*, **190**, 877–82.

Fields, D.M., and Yen, J. (2002). Fluid mechanosensory stimulation of behaviour from a planktonic marine copepod, *Euchaeta rimana* Bradford. *Journal of Plankton Research*, **24**, 747–55.

Fields, D.M., Shaeffer, D.S., and Weissburg, M.J. (2002). Mechanical and neural responses from the mechanosensory hairs on the antennule of *Gaussia princeps*. *Marine Ecology Progress Series*, **227**, 173–86.

Fiksen, Ø. (2000). The adaptive timing of diapause-a search for evolutionarily robust strategies in *Calanus finmarchicus*. *ICES Journal of Marine Science*, **57**, 1825–33.

Fiksen, Ø., and Mackenzie, B.R. (2002). Process-based models of feeding and prey selection in larval fish. *Marine Ecology Progress Series*, **243**, 151–64.

Fiksen, Ø., Utne, A.C.W., Aksnes, D.L., *et al.* (1998). Modelling the influence of light, turbulence and ontogeny on ingestion rates in larval cod and herring. *Fisheries Oceanography*, **7**, 355–63.

Fiksen, Ø., Jørgensen, C., Kristiansen, T., *et al.* (2007). Linking behavioural ecology and oceanography: larval behaviour determines growth, mortality and dispersal. *Marine Ecology Progress Series*, **347**, 195–205.

Finlayson, A.C. (1994). *Fishing for truth: a sociological analysis of northern cod stock assessments from 1977–1990.* Institute of Social and Economic Research, Memorial University of Newfoundland, St. John's, NF.

Finney, B.P. (1998). Long-term variability in Alaskan sockeye salmon abundance determined by analysis of sediment cores. *North Pacific Anadromous Fish Commission Bulletin*, **1**, 388–95.

Finney, B.P., Gregory-Eaves, I., Douglas, M.S.V., *et al.* (2002). Fisheries productivity in the northeast Pacific Ocean over the past 2,200 years. *Nature*, **416**, 729–33.

Fisher, J.A.D., Frank, K.T., Petrie, B., *et al.* (2008). Temporal dynamics within a contemporary latitudinal diversity gradient. *Ecology Letters*, **11**, doi 10.1111/j.1461–0248.2008.01216.x

Fisher, R., Sogard, S.M., and Berkley, S.A. (2007). Trade-offs between size and energy reserves reflect alternative strategies for optimizing larval survival potential in rockfish. *Marine Ecology Progress Series*, **344**, 257–70.

Fleminger, A. and Hulsemann, K. (1977). Geographical range and taxonomic divergence in North Atlantic *Calanus* (*Calanus helgolandicus*, *Calanus finmarchicus*, and *Calanus glacialis*). *Marine Biology*, **40**, 233–48.

Fletcher, W.J. (2005). The application of qualitative risk assessment methodolgy to prioritize issues for fisheries management. *ICES Journal of Marine Science*, **62**, 1576–87.

Flierl, G.R. and McGillicuddy, D.J. (2002). Mesoscale and submesoscale physical-biological interactions. In: Robinson, A.R., McCarthy, J.J., and Rothschild, B.J., eds, *The sea, vol. 12: biological-physical interactions in the sea*. Wiley, New York, pp. 113–85.

Flinkman, J., Aro, E., Vuorinen, I., *et al.* (1998). Changes in northern Baltic zooplankton and herring nutrition form 1980s to 1990s: top-down and bottom-up processes at work. *Marine Ecology Progress Series*, **165**, 127–36.

Flynn, B.A. and Gibbons, M.J. (2007). A note on the diet and feeding of *Chrysaora hysoscella* in Walvis Bay lagoon, Namibia, during September 2003. *African Journal of Marine Science*, **29**, 303–7.

Flynn, K.J. (2003). Modelling multi-nutrient interactions in phytoplankton; balancing simplicity and realism. *Progress in Oceanography*, **56**, 249–79.

Flynn, K.J. (2005). Modelling marine phytoplankton growth under eutrophic conditions. *Journal of Plankton Research*, **54**, 92–103.

Fogarty, M.J. (2005). Impacts of fishing activities on benthic habitat and carrying capacity: approaches to assessing and managing risk. *American Fisheries Society Symposia*, **41**, 769–84.

Fogarty, M.J. and Botsford, L.W. (2007). Population connectivity and spatial management of marine fisheries. *Oceanography*, **20(3)**, 112–23.

Fogarty, M.J. and Murawski, S.A. (1998). Large-scale disturbance and the structure of marine systems: fishery impacts on Georges Bank. *Ecological Applications*, **8(S1)**, S6–S22.

Folke, C., Carpenter, S., Walker, B., *et al.* (2004). Regime shifts, resilience and biodiversity in ecosystem management. *Annual Review of Ecology and Evolution*, **35**, 557–81.

Folkvord, A. (2005). Comparison of size-at-age of larval Atlantic cod (*Gadus morhua*) from different populations based on size- and temperature-dependent growth models. *Canadian Journal of Fisheries and Aquatic Sciences*, **62**, 1037–52.

Follows, M.J., Dutkiewicz, S., Grant, S., *et al.* (2007). Emergent biogeography of microbial communities in a model ocean. *Science*, **315**, 1843–6.

Fonteneau, A., Lucas, V., Tewkai, E., *et al.* (2008). Mesoscale exploitation of a major tuna concentration in the Indian Ocean. *Aquatic Living Resources*, **21(2)**, 109–21.

Foote, K.G. and Stanton, T.K. (2000). Acoustical methods. In: Harris, R.P., Wiebe, P.H., Lenz, J., *et al.* eds. *ICES zooplankton methodology manual*. Academic Press, London, pp. 223–58.

Forchhammer, M.C. and Post, E. (2004). Using large-scale climate indices in climate change ecology studies. *Population Ecology*, **46**, 1–12.

Ford, J.D., Smit, B., and Wandel, J. (2006). Vulnerability to climate change in the Arctic: a case study from Arctic Bay, Canada. *Global Environmental Change*, **16**, 145–60.

Fox-Kemper, B and Ferrari, R. (2008). Parameterization of mixed layer eddies, Part II: prognosis and impact. *Journal of Physical Oceanography*, **38**, 1166–79.

Francis, R.C. and Shotton, R. (1997). Risk in fisheries management: a review. *Canadian Journal of Fisheries and Aquatic Sciences*, **54**, 1699–715.

Francis, R.C., Hare, S.R., Hollowed, A., *et al.* (1998). Effects of interdecadal climate variability on the oceanic ecosystems of the NE Pacific. *Fisheries Oceanography*, **7**, 1–21.

Frank, K.T., Perry, R.I., and Drinkwater, K.F. (1990). Predicted response of Northwest Atlantic invertebrate and fish stocks to CO_2 induced climate change. *Transactions of the American Fisheries Society*, **119**, 353–65.

Frank, K.T., Carscadden, J.E., and Simon, J.E. (1996). Recent excursions of capelin (*Mallotus villosus*) to the Scotian Shelf and Flemish Cap during anomalous hydrographic conditions. *Canadian Journal of Fisheries and Aquatic Sciences*, **53**, 1473–86.

Frank, K.T., Petrie, B., Choi, J.S., *et al.* (2005). Trophic cascades in a formerly cod-dominated ecosystem. *Science*, **308**, 1621–3.

Frank, K.T., Petrie, B., Shackell, N.L., *et al.* (2006). Reconciling differences in trophic control in mid-latitude marine ecosystems. *Ecology Letters*, **9**, 1096–105.

Frank, K.T., Petrie, B., Shackell, N.L. (2007). The ups and downs of trophic control in continental shelf ecosystems. *Trends in Ecology and Evolution*, **22(5)**, 236–42.

Franks, P.J.S. (1997). New models for the exploration of biological processes at fronts. *ICES Journal of Marine Science*, **54**, 161–7.

Fraser, W.R. and Hofmann, E.E. (2003). A predator's perspective on causal links between climate change, physical forcing and ecosystem response. *Marine Ecology Progress Series*, **265**, 1–15.

Fraser, W.R. and Trivelpiece, W.Z. (1996). Factors controlling the distribution of seabirds: winter-summer heterogeneity in the distribution of Adelie penguin populations. In: Ross, R.M., Hoffman, E.E., and Quetin, L.B., eds. *Foundations for ecological research west of the Antarctic peninsula*. American Geophysical Union, Washington, DC, pp. 257–72.

Fraser, W.R., Pitman, R.L., and Ainley, D.G. (1989). Seabird and fur seal responses to vertically migrating winter krill swarms in Antarctica. *Polar Biology*, **10**, 37–41.

Fraser, W.R., Trivelpiece, W.Z., Ainley, D.G., *et al.* (1992). Increases in Antarctic penguin populations-reduced competition with whales or a loss of sea ice due to environmental warming. *Polar Biology*, **11**, 525–31.

Frederiksen, M., Edwards, M., Richardson, A.J., *et al.* (2006). From plankton to top predators: bottom-up control of a marine food web across four trophic levels. *Journal of Animal Ecology*, **75**, 1259–68.

Freeland, H.J. and Cummins, P.F. (2005). Argo: A new tool for environmental monitoring and assessment of the world's ocean, an example from the NE Pacific. *Progress in Oceanography*, **64(1)**, 31–44.

Freeland, H.J., Gatien, G., Huyer, A., *et al.* (2003). Cold halocline in the northern California current: an invasion of subarctic water. *Geophysical Research Letters*, **30**, art:1141.

Fréon, P., Mullon, C., and Voisin, B. (2003). Investigating remote synchronous patterns in fisheries. *Fisheries Oceanography*, **12(4–5)**, 443–57.

Fréon, P., Bouchon, M., Mullon, C., *et al.* (2008). Interdecadal variability of anchoveta abundance and overcapacity of the fishery in Peru. *Progress in Oceanography*, **79(2–4)**, 401–12.

Fréon, P., Werner, F., and Chavez, F. (2009). Conjectures on future climate effects on marine ecosystems dominated by small pelagic fish. In: Checkley, D., Alheit, J., Oozeki, Y., *et al.* eds. *Climate change and small pelagic fish*. Cambridge University Press, Cambridge.

Friedlaender, A.S., Halpin, P.N., Qian, S.S., *et al.* (2006). Whale distribution in relation to prey abundance and oceanographic processes in shelf waters of the western Antarctic peninsula. *Marine Ecology Progress Series*, **317**, 297–310.

Friedland, K.D., Hare, J.A., Wood, G.B., *et al.* (2008). Does the fall phytoplankton bloom control recruitment of Georges Bank haddock, *Melanogrammus aeglefinus*, through parental condition? *Canadian Journal of Fisheries and Aquatic Sciences*, **65(6)**, 1076–86.

Friedlingstein, P., Cox, P., Betts, R., *et al.* (2006). Climate-carbon cycle feedback analysis: results from the (CMIP)-M-4 model intercomparison. *Journal of Climate*, **19**, 3337–53.

Friedrichs, M.A.M., Dusenberry, J.A., Anderson, L.A., *et al.* (2007). Assessment of skill and portability in regional marine biogeochemical models: role of multiple planktonic groups. *Journal of Geophysical Research – Oceans*, **112**, C08001, doi:10.1029/2006JC003852.

Fringer, O.B., Gerritsen, M., and Street, R.L. (2006). An unstructured-grid, finite-volume, nonhydrostatic parallel coastal ocean simulator. *Ocean Modelling*, **14**, 1139–73.

Frischer, M.E., Lee, R.F., Sheppard, M.A., *et al.* (2006). Evidence for a free-living life stage of the blue crab parasitic dinoflagelate, *Hematodinium* sp. *Harmful Algae*, **5**, 548–57.

Fromentin, J.-M. and Kell, L.T. (2007). Consequences of variations in carrying capacity or migration for the perception of Atlantic bluefin tuna (*Thunnus thynnus*) population dynamics. *Canadian Journal of Fisheries and Aquatic Sciences*, **64**, 827–36.

Fromentin, J.-M. and Planque, B. (1996). *Calanus* and environment in the eastern North Atlantic, II: influence of the North Atlantic Oscillation on *C. finmarchicus* and *C. hegolandicus*. *Marine Ecology Progress Series*, **134**, 111–8.

Frouin R. and Iacobellis, S.F. (2002). Influence of phytoplankton on the global radiation budget, *Journal of Geophysical Research – Atmosphere*, **107**, 4377.

Fry, B. and Quiñones, R.B. (1994). Biomass spectra and stable isotope indicators of trophic level in zooplankton of the northwest Atlantic. *Marine Ecology Progress Series*, **112**, 201–4.

Fuiman, L.A., Cowan, J.H., Jr., Smith, M.E., *et al.* (2005). Behavior and recruitment success in fish larvae: variation with growth rate and the batch effect. *Canadian Journal of Fisheries and Aquatic Sciences*, **62**, 1337–49.

Fuiman, L.A., Rose, K.A., Cowan, J.H., Jr., *et al.* (2006). Survival skills required for predator evasion by fish larvae and their relation to laboratory measures of performance. *Animal Behaviour*, **71**, 1389–99.

Fukami, K., Watanabe, A., Fujita, S., *et al.* (1999). Predation on naked protzoan microzooplankton by fish larvae. *Marine Ecology Progress Series*, **185**, 285–91.

Fulton, E.A., Parslow, J.S., Smith, A.D.M., *et al.* (2004a). Biogeochemical marine ecosystem models. 2. The effect of physiological detail on model performance. *Ecological Modelling*, **173**, 371–406.

Fulton, E.A., Smith, A.D.M., and Punt, A.E. (2004b). Ecological indicators of the ecosystem effects of fishing: *Final Report*, **R99/1546**, Australian Fisheries Management Authority, Canberra.

Fulton, E.A., Smith, A.D.M., and Johnson, C.R. (2004c). Biogeochemical marine ecosystem models I: IGBEM-a model of marine bay ecosystems. *Ecological Modelling*, **174**, 267–307.

Fulton, E.A., Smith, A.D.M., and Punt, A.E. (2005). Which ecological indicators can robustly detect effects of fishing? *ICES Journal of Marine Science*, **62**, 540–51.

Funtowicz, S.O. and Ravetz, J.R. (1993). Science for the post-normal age. *Futures*, **25**, 735–55.

Fussmann, G.F., Ellner, S.P., and Hairston, N.G. (2003). Evolution as a critical component of plankton dynamics. *Proceedings of the Royal Society of London, Series B, Biological Sciences*, **270**, 1015–22.

Gaard, E. (1996). Life cycle, abundance and transport of *Calanus finmarchicus* in Faroese waters. *Ophelia*, **44**, 59–70.

Gaines, S.D., Gaylord, B., Gerber, L.R., *et al.* (2007). Connecting places: the ecologicall consequences of dispersal in the sea. *Oceanography*, **20(3)**, 90–9.

Galbraith, P.S., Browman, H.I., Racca, R.G., *et al.* (2004). Effect of turbulence on the energetics of foraging in Atlantic cod *Gadus morhua* larvae. *Marine Ecology Progress Series*, **281**, 241–57.

Gallager, S.M., Davis, C.S., Epstein, A.W., *et al.* (1996). High-resolution observations of plankton spatial distributions correlated with hydrography in the Great South Channel, Georges Bank. *Deep-Sea Research II*, **43**, 1627–64.

Gallager, S.M., Daly, K., Fisher, K., *et al.* (2001). Seasonal changes in the association of larval krill with its potential microplankton food resource along the Western Antarctic Peninsula. *EOS, Transactions, American Geophysical Union*, **82(suppl.)**, np.

Gallager, S.M., Yamazaki, H., and Davis, C.S. (2004). Contribution of fine-scale vertical structure and swimming behavior to formation of plankton layers on Georges Bank. *Marine Ecology Progress Series*, **267**, 27–43.

Gallego, A., Heath, M.R., Basfrod, D.J., *et al.* (1999). Variability in growth rates of larval haddock in the northern North Sea. *Fisheries Oceanography*, **8**, 77–92.

Gallego, A., North, E.W., and Petitgas, P. (2007). Introduction: status and future of modelling physical-biological interactions during the early life of fishes. *Marine Ecology Progress Series*, **347**, 122–6.

Gallienne, C.P. and Robins, D.B. (2001). Is *Oithona* the most important copepod in the world's ocean? *Journal of Plankton Research*, **23(12)**, 1421–32.

Galopin, G.C. (2006). Linkages between vulnerability, resilience, and adaptive capacity. *Global Environmental Change*, **16**, 293–303.

Garcia, S.M. and Cochrane, K.L. (2005). Ecosystem approach to fisheries: a review of implementation guidelines. *ICES Journal of Marine Science*, **62(3)**, 311–8.

Garcia, S.M., Zerbi, A., Aliaume, C., *et al.* (2003). The ecosystem approach to fisheries. Issues, terminology, principles, institutional foundations, implementation and outlook. *FAO Fisheries Technical Paper*, **443**, 71pp.

Garçon, V.C., Oschlies, A., Doney, S.C., *et al.* (2001). The role of mesoscale variability on plankton dynamics in the North Atlantic. *Deep-Sea Research II*, **48**, 2199–226.

Gargett, A.E. (1997). The optimal stability "window": a mechanism underlying decadal fluctuations in North Pacific salmon stocks? *Fisheries Oceanography*, **6**, 109–17.

Gaylord, B. and Gaines, S.D. (2000). Temperature or transport: species ranges mediated solely by flow. *American Naturalist*, **155**, 769–89.

Gedalof, Z., Mantua, N.J., and Peterson, D.L. (2002). A multi-century perspective of variability in the Pacific Decadal Oscillation: new insights from tree rings and coral. *Geophysical Research Letters*, **29(2204)**, doi:10.1029/2002GL015824.

Gell, F.R. and Roberts, C.M. (2003). Benefits beyond boundaries: the fishery effects of marine reserves. *Trends in Ecology and Evolution*, **18**, 448–55.

Gent, P.R. and McWilliams, J.C. (1990). Isopycnal mixing in ocean circulation models. *Journal of Physical Oceanography*, **20**, 150–5.

Gentleman, W., Leising, A., Frost, B., *et al.* (2003). Functional responses for zooplankton feeding on multiple resources: a review of assumptions and biological dynamics. *Deep-Sea Research II*, **50**, 2847–75.

German, A.W. (1987). History of the early fisheries, 1720–1930. Chapter 40. In: Backus, R.H., ed. *Georges Bank*. MIT Press, Cambridge, MA, pp. 409–24.

Gerritsen, J. (1980). Adaptative responses to encounter problems. In: Kerfoot, C.W., ed. *Evolution and ecology of zooplankton communities*. University Press of New England, Hanover and London, pp. 52–62.

Gerritsen, J. and Strickler, J.R. (1977). Encounter probabilities and community structure in zooplankton. *Journal of the Fisheries Research Board of Canada*, **35**, 1370–3.

Gibbons, M.J. and Richardson, A.J. (2009). Patterns of jellyfish abundance in the North Atlantic. *Hydrobiologia*, **616**, 51–64.

Gifford, D.J., Harris, R.P., McKinnel, S.M., *et al.* (2010). Target species. *This volume.* Oxford University Press, Oxford.

Gill, P.G. (2002). A blue whale (*Balaenoptera musculus*) feeding ground in a southern Australian coastal upwelling zone. *Journal of Cetacean Research and Management*, **4**, 179–84.

Gillett, N.P., Allan, R.J., and Ansell, T.J. (2005). Detection of external influence on sea level pressure with a multi-model ensemble. *Geophysical Research Letters*, **32**, L19714, doi:10.1029/2005GL023640.

Gillett, N.P., Huse, G., and Fiksen, Ø. (1998). Modelling spatial dynamics of fish. *Reviews of Fish Biology and Fisheries*, **8**, 57–91.

Gillett, N.P., Huse, G., and Berntsen, J. (2001). Spatial modelling for marine resource management, with a focus on fish. *Sarsia*, **86**, 405–10.

Gillett, N.P., Mangel, M., Jakobsen, P., *et al.* (2003). Explicit trade-off rules in proximate adaptive agents. *Evolutionary Ecology Research*, **5**, 835–65.

Gislason, A. (2005). Seasonal and spatial variability in egg production and biomass of *Calanus finmarchicus* around Iceland. *Marine Ecology Progress Series*, **286**, 177–92.

Gislason, A., Eiane, K., and Reynisson, P. (2007). Vertical distribution and mortality of *Calanus finmarchicus* during overwintering in oceanic waters southwest of Iceland. *Marine Biology*, **150**, 1253–63.

Gismervik, I. and Andersen, T. (1997). Prey switching by *Acartia clausi* experimental evidence and implications of intraguild predation assessed by a model. *Marine Ecology Progress Series*, **157**, 247–59.

Gjøsæter, H., Bogstad, B., and Tjelmeland, S. (2002). Assessment methodology for Barents Sea capelin (*Mallotus villosus* Müller). *ICES Journal of Marine Science*, **59**, 1086–95.

Glantz, M.H. and Feingold, L.E. (1992). Climate variability, climate change, and fisheries: a summary. In: Glantz,

M.H., ed. *Climate variability, climate change and fisheries*. Cambridge University Press, Cambridge.

GLOBEC. (1993a). Population dynamics and physical variability. *GLOBEC Report*, **2**, 104pp.

GLOBEC. (1993b). Numerical modeling. Report of the first meeting of an international GLOBEC working group, Villefranche-sur-mer, France, 12–14 July 1993. *GLOBEC Report*, **6**, 60pp.

GLOBEC. (1994). An Advanced Modeling/Observation System (AMOS) for physical-biological-chemical ecosystem research and monitoring (concepts and methodology). *GLOBEC Special Contribution*, **2**, 149pp.

GLOBEC. (1997). Global Ocean Ecosystem Dynamics science plan. *GLOBEC Report 9 and IGBP Report 40*, 82pp.

GLOBEC. (1999). Global Ocean Ecosystem Dynamics implementation plan. *GLOBEC Report 13 and IGBP Report 47*, 207pp.

Gómez-Gutiérrez, J., Peterson, W.T., DeRobertis, A., *et al.* (2003). Mass mortality of krill cased by parasitoid ciliates. *Science*, **301**, 339.

Gómez-Gutiérrez, J., and Miller, C.B. (2005). Cross-shelf life-stage segregation and community structure of the euphausiids off central Oregon (1970–1972). *Deep-Sea Research II*, **52**, 289–315.

Goodyear, C.P. (1993). Spawning stock biomass per recruit in fisheries management: foundation and current use. In: Smith, S.J., Hunt, J.J., and Rivard, D., eds. *Risk evaluation and biological reference points for fisheries management. Canadian Special Publication of Fisheries and Aquatic Sciences*, **120**, 67–81.

Goulder, H. and Kennedy, D. (1997). Valuing ecosystem services: philosophical bases and empirical methods. In: Daily, G.C., ed. *Nature's services: societal dependence on natural ecosystems*. Island Press, Washington, DC, pp. 23–48.

Goy, J., Morand, P., and Etienne, M. (1989). Long-term fluctuations of *Pelagia noctiluca* (Cnidaria, Scyphomedusa) in the western Mediterranean sea-prediction by climatic variables. *Deep-Sea Research I*, **36**, 269–79.

Graham, W.M. (2001). Numerical increases and distributional shifts of *Chrysaora quinquecirrha* (Desor) and *Aurelia aurita* (Linne) (Cnidaria: Scyphozoa) in the northern Gulf of Mexico. *Hydrobiologia*, **451**, 97–111.

Graham, W.M. and Bayha, K.M. (2007). Biological Invasions by marine jellyfish. In: Nentwig, W., ed. *Ecological studies*, **193**. Springer-Verlag, Berlin and Heidelberg, pp. 239–56.

Graham, W.M., Pages, F., and Hamner, W.M. (2001). A physical context for gelatinous zooplankton aggregations: a review. *Hydrobiologia*, **451**, 199–212.

Graybill, M.R. and Hodder, J. (1985). Effects of the 1982–83 El Niño on reproduction of six species of seabirds in

Oregon. In: W.S. Wooster and D.L. Fluharty, eds. *El Niño north: El Niño effects in the eastern subarctic Pacific Ocean.* University of Washington, Seattle, WA, pp. 205–10.

Grebmeier, J.M., Overland, J.E., Moore, S.E., *et al.* (2006). A major ecosystem shift in the northern Bering Sea. *Science,* **311,** 1461–4.

Grechina, A. (1998). Historia de investigaciones y aspectos básicos de la ecología del jurel (*Trachurus symmetricus murphyi* (Nichols)) en alta mar del Pacífico Sur. In: Arcos, D., ed. *Biología y ecología del jurel en aguas Chilenas.* Editorial Anibal Pinto, Concepción, Chile, pp. 11–34.

Green, D.G. and Sadedin, S. (2005). Interactions matter-complexity in landscapes and ecosystems. *Ecological Complexity,* **2(2),** 117–30.

Greenberg, D.A., Dupont, F., Lyard, F.H., *et al.* (2007). Resolution issues in numerical models of oceanic and coastal circulation. *Continental Shelf Research,* **27,** 1317–43.

Greenlaw, C.F. and Johnson, R.K. (1983). Multiple-frequency acoustical estimation. *Biological Oceanography,* **2,** 227–52.

Greenstreet, S.P. and Rogers, S.I. (2006). Indicators of the health of the North Sea fish community: identifying reference levels for an ecosystem approach to management. *ICES Journal of Marine Science,* **63,** 573–93.

Greenstreet, S.P., Spence, F.E., and McMillan, J.A. (1999). Fishing effects in northeast Atlantic shelf seas: patterns in fishing effort, diversity and community structure, V: changes in structure of the North Sea groundfish species assemblage between 1925 and 1996. *Fisheries Research,* **40,** 153–83.

Gregg, W.W., Casey, N.W., and McClain, C.R. (2005). Recent trends in global ocean chlorophyll, *Geophysical Research Letters,* **32,** L03606, doi:10.1029/2004GL021808.

Gregg, W.W., Friedrichs, M.A.M., Robinson, A.R., *et al.* (2008). Skill assessment in ocean biological data assimilation. *Journal of Marine Systems,* **76,** 16–33.

Grémillet, D., Lewis, S., Drapeau, L., *et al.* (2008). Spatial match-mismatch in the Benguela upwelling zone: should we expect chlorophyll and sea-surface temperature to predict marine predator distributions? *Journal of Applied Ecology,* **45,** 610–21.

Greve, W., Lange, U., Reiners, F., *et al.* (2001). Predicting the seasonality of North Sea zooplankton. *Senckenbergiana Maritima,* **31,** 263–8.

Greve, W., Prinage, S., Zidowitz, H., *et al.* (2005). On the phenology of North Sea ichthyoplankton. *ICES Journal of Marine Science,* **62,** 1216–23.

Greve, W., Reiners, F., Nast, J., *et al.* (2004). Helgoland Roads time-series meso- and macrozooplankton 1975 to 2004: lessons from 30 years of single spot high frequency sampling at the only off-shore island of the North Sea. *Helgoland Marine Research,* **58,** 274–88.

Griffiths, C.L., van Stittert, L., Best, P.B., *et al.* (2004). Impacts of human activities on marine animal life in the Benguela: a historical overview. *Oceanography and Marine Biology: An Annual Review,* **42,** 303–92.

Griffiths, G., ed. (2003). *Technology and applications of autonomous underwater vehicles.* Taylor & Francis, London, 342pp.

Grimes, C.B., Brodeur, R.D., Halderson, L.J., *et al.,* eds. (2007). *The ecology of juvenile salmon in the northeast Pacific Ocean.* American Fisheries Society Symposium 87, Bethseda, MD.

Grimm, V. and Railsback, S.F. (2005). *Individual-based modelling and ecology.* Princeton University Press, Princeton, NJ, 480pp.

Grishin, A.N., Kovalenko, L.A., and Sorokolit, L.K. (1994). Trophic relations in plankton communities in the Black Sea before and after *Mnemiopsis leidyi* invasion. In: *The main results of YugNIRO complex researches in the Azov-Black Sea region and the World ocean in 1993.* pp. 38–44. Trudi YugNIRO, Kerch, 40.

Grosholz, E. (2002). Ecological and evolutionary consequences of coastal invasions. *Trends in Ecology and Evolution,* **17(1),** 22–7.

Grosjean, P., Picheral, M., Warembourg, C., *et al.* (2004). Enumeration, measurement, and identification of net zooplankton samples using the ZOOSCAN digital imaging system. *ICES Journal of Marine Science,* **61,** 518–25.

Grove, M. and Breitburg, D.L. (2005). Growth and reproduction of gelatinous zooplankton exposed to low dissolved oxygen. *Marine Ecology Progress Series,* **301,** 185–98.

Gücü, A.C. (2002). Can overfishing be responsible for the successful development of Mnemiopsis? *Estuarine, Coastal and Shelf Science,* **54,** 439–51.

Guinet, C., Khoudil, M., Bost, C.A., *et al.* (1997). Foraging behaviour of satellite-tracked king penguins in relation to sea-surface temperatures obtained by satellite telemetry at Crozet Archipelago, a study during three austral summers. *Marine Ecology Progress Series,* **150,** 11–20.

Gulland, J.A. and Garcia, S.M. (1984). Observed patterns in multispecies fisheries. In: May, R.M., ed. *Exploitation of marine communities.* Springer, Berlin, pp. 155–90.

Gunderson, L.H. and Holling, C.S. (2001). *Panarchy: understanding transformations in systems of humans and nature.* Island Press, Washington, DC, 450pp.

Gunn, J. and Block, B.A. (2001). Acoustic, archival and pop-up satellite tagging of tunas. In: Block, B.A. and Stevens E.D., eds. *Tunas: ecological physiology and evolution.* Academic Press, San Diego, CA, pp. 167–224.

Gurevitch, J. and Padilla, D.K. (2004). Are invasive species a major cause of extinctions? *Trends in Ecology and Evolution,* **19(9),** 470–4.

Haag, A.L. (2007). Algae bloom again. *Nature*, **447**, 520–1.

Haedrich, R.L. and Hamilton, L.C. (2000). The fall and future of Newfoundland's cod fishery. *Society and Natural Resources*, **13**, 359–72.

Haidvogel, D.B., Arango, H., Budgell, W.P., *et al.* (2008). Ocean forecasting in terrain-following coordinates: formulation and skill assessment of the Regional Ocean Modeling System. *Journal of Computational Physics*, **227**, 3595–624.

Haidvogel, D.B. and Beckmann, A. (1999). *Numerical ocean circulation modelling*. Imperial College Press, London, 300pp.

Hall, S.J. and Harding, M.J.C. (1997). Physical disturbance and marine benthic communities: the effects of mechanical harvesting of cockles on non-target benthic infauna. *Journal of Applied Ecology*, **34(2)**, 497–517.

Halley, J.M. and Stergiou, K. (2005). The implications of increasing variability of fish landings. *Fish and Fisheries*, **6**, 266–76.

Halpern, B.S., Walbridge, S., Selkoe, K.A., *et al.* (2008). A global map of human impact on marine ecosystems. *Science*, **319**, 948–52.

Hamilton, L.C. (2003). Fisheries dependent communities: propositions about ecological and social change. In: Duhaime, G. and Bernard, N., eds. *Arctic economic development and self-government*. GÉTIC, Université Laval, Québec, Canada, pp. 49–61.

Hamilton, L.C. (2007). Climate, fishery and society interactions: observations from the North Atlantic. *Deep-Sea Research II*, **54**, 2958–69.

Hamilton, L.C. and Butler, M.J. (2001). Outport adaptations: social indicators through Newfoundland's cod crisis. *Human Ecology Review*, **8(2)**, 1–11.

Hamilton, L.C., Lyster, P., and Otterstad, O. (2000). Social change, ecology and climate in 20th century Greenland. *Climatic Change*, **47**, 193–211.

Hamilton, L.C., Brown, B.C., and Rasmussen, R.O. (2003). West Greenland's cod-to-shrimp transition: local dimensions of climatic change. *Arctic*, **56**, 271–82.

Hamilton, L.C., Colocousis, C.R., and Johansen, S.T.F. (2004a). Migration from resource depletion: the case of the Faroe Islands. *Society and Natural Resources*, **17**, 443–53.

Hamilton, L.C., Haedrich, R.L., and Duncan, C.M. (2004b). Above and below the water: social-ecological transformation in northwest Newfoundland. *Population and Environment*, **25**, 195–215.

Hamilton, L.C., Jónsson, S., Ögmundardóttir, H., *et al.* (2004c). Sea changes ashore: the ocean and Iceland's herring capital. *Arctic*, **57**, 325–35.

Hamilton, L.C., Otterstad, O., and Ögmundardóttir, H. (2006). Rise and fall of the herring towns: impacts of climate and human teleconnections. In: Hannesson, R., Barange, M., and Herrick, S., Jr., eds. *Climate change and the economics of the world's fisheries: examples of small pelagic fish stocks*. Edward Elgar, Cheltenham, UK, pp. 100–25.

Hammond, P.S. and Donovan, G.P. (in press). Development of the IWC revised management procedure. *Journal of Cetacean Research and Management, Special Edition* **4**.

Hamner, W.M. and Jenssen, R.M. (1974). Growth, degrowth, and irreversible cell-differentiation in *Aurelia aurita. American Zoologist*, **14**, 833–49.

Hampton, I., Sweijd, N., and Barange, M., eds. (2009). Benguela Environment Fisheries Interaction and Training Programme (BENEFIT) research projects. *GLOBEC Report*, **25**, 108pp.

Handy, S.M., Hutchins, D.A., Cary, S.C., *et al.* (2006). Simultaneous enumeration of multiple raphidophyte species by quantitative real-time PCR: capabilities and limitations. *Limnology and Oceanography Methods*, **4**, 193–204.

Haney, J.C. (1986). Seabird patchiness in tropical oceanic waters: the influence of *Sargassum* 'reefs'. *Auk*, **103**, 141–51.

Hannah, C.G., Naimie C.E., Loder J.W., *et al.* (1998). Upper-ocean transport mechanisms from the Gulf of Maine to Georges Bank, with implications for *Calanus* Supply. *Continental Shelf Research*, **15**, 1887–911.

Hannesson, R. (2006). Sharing the herring: fish migrations, strategic advantage and climate change. In: Hannesson, R., Barange, M., and Herrick, S., Jr., eds. *Climate change and the economics of the world's fisheries: examples of small pelagic fish stocks*. Edward Elgar, Cheltenam, UK, pp. 66–99.

Hannesson, R., Barange, M., and Herrick, S., Jr., eds. (2006). *Climate change and the economics of the world's fisheries: examples of small pelagic fish stocks*. Edward Elgar, Cheltenham, UK.

Hansen, B., Gaard, E., and Reinert, J. (1994). Physical effects on recruitment of Faroe Plateau cod. *ICES Marine Science Symposia*, **198**, 520–8.

Hansson, H.G. (2006). Ctenophores of the Baltic and adjacent Seas-the invader *Mnemiopsis* is here!. *Aquatic Invasions*, **1(4)**, 295–8.

Harden-Jones, F.R. (1968). *Fish migration*. Edward Arnold, London, 325pp.

Hardman-Mountford, N.J., Richardson, A.J., Boyer, D.C., *et al.* (2003). Relating sardine recruitment in the northern Benguela to satellite-derived sea surface height using a neural network pattern recognition approach. *Progress in Oceanography*, **59**, 241–55.

Hardy, A.C. (1924). The herring in relation to its animate environment, Part I: the food and feeding habits of the herring. *Fishery Investigations London, Series 2*, **7(3)**, 53pp.

Hardy, A.C. (1926). A new method of plankton research. *Nature*, **118**, 630–2.

Hare, J.A. and Whitfield, P.E. (2003). An integrated assessment of the introduction of lionfish (*Pterois volitans/miles complex*) to the Western Atlantic Ocean. *NOAA Technical Memorandum NOS NCCOS*, **2**, 31pp.

Hare, S.R. and Mantua, N.J. (2000). Empirical evidence for North Pacific regime shifts in 1977 and 1989. *Progress in Oceanography*, **47**, 103–45.

Harley, C.D.G., Hughes, R.A., Hultgren, K.M., *et al.* (2006). The impacts of climate change in coastal marine systems. *Ecology Letters*, **9**, 228–41.

Harris, M. (1998). *Lament for an ocean: the collapse of the Atlantic cod fishery*. McClelland and Stewart, Toronto.

Harris, R.P., Irigoien, X., Head, R.N., *et al.* (2000). Feeding, growth, and reproduction in the genus *Calanus*. *ICES Journal of Marine Science*, **57**, 1708–26.

Harris, R.P., Buckley, L.J., Campbell, R., *et al.* (2010). Dynamics of marine ecosystems: observation and experimentation. *This volume*. Oxford University Press, Oxford.

Harris, R.P. and Steele, J.H., eds. (2004). Regime shifts in the ocean: reconciling observations and theory. *Progress in Oceanography*, **60(2–4)**, 402pp.

Harrison, D.E. and Carson, M. (2007). Is the world ocean warming? Upper-ocean temperature trends: 1950–2000. *Journal of Physical Oceanography*, **37**, 174–87.

Hashioka, T. and Yamanaka, Y. (2007). Ecosystem change in the western North Pacific associated with global warming using 3D-NEMURO. *Ecological Modelling*, **202(1–2)**, 95–104.

Hassel, A., Skjoldal, H.R., Gjøsæter, H., *et al.* (1991). Impact of grazing from capelin (*Mallotus villosus*) on zooplankton: a case study in the northern Barents Sea in August 1985. *Polar Research*, **10**, 371–88.

Hastings, A. and Botsford, L.W. (2006). Persistence of spatial populations depends on returning home. *Proceedings of the National Academy of Sciences of the United States of America*, **103**, 6067–72.

Hauck, M. and Sowman, M. (2003). Waves of change: coastal and fisheries co-management in South Africa. UCT Press, Cape Town, South Africa.

Haury, L.R. and Kenyon, D.E. (1980). Experimental evaluation of the avoidance reaction of *Calanus finmarchicus*. *Journal of Plankton Research*, **2**, 187–202.

Haury, L.R., McGowan, J.A., and Wiebe, P.H. (1978). Patterns and process in the time-space scales of plankton distribution. In: J.H. Steele, J.H., ed. *Spatial pattern in plankton communities*. Plenum Press, New York, pp. 277–327.

Haury, L.R., Yamazaki H., and Itsweire H.E. (1990). Effects of turbulent shear flow on zooplankton distributions. *Deep-Sea Research I*, **37**, 447–61.

Hawkins, S.J., Southward, A.J., and Genner, M.J. (2003). Detection of environmental change in a marine ecosystem-evidence from the western English Channel. *Science of the Total Environment*, **310**, 245–56.

Hay, S. (2006). Marine ecology: gelatinous bells may ring change in marine ecosystems. *Current Biology*, **16(17)**, doi:10.1016/j.cub.2006.08.010.

Hayes, D.B., Ferreri, C.P., and Taylor, W.W. (1996). Linking fish habitat to their population dynamics. *Canadian Journal of Fisheries and Aquatic Sciences*, **53(Suppl. 1)**, 383–90.

Hays, G.C., Robinson, C., and Richardson, A.J. (2005). Climate change and marine plankton. *Trends in Ecology and Evolution*, **20**, 337–44.

Heath, M.R. (1995). An holistic analysis of the coupling between physical and biological processes in the coastal zone. *Ophelia*, **42**, 95–125.

Heath, M.R. (2005). Changes in the structure and function of the North Sea fish foodweb, 1973–2000, and the impacts of fishing and climate. *ICES Journal of Marine Science*, **62(5)**, 847–68.

Heath, M.R. (2007). The consumption of zooplankton by early life stages of fish in the North Sea. *ICES Journal of Marine Science*, **64**, 1650–63.

Heath, M.R. and Gallego, A. (1998). Bio-physical modelling of the early life stages of haddock, *Melanogrammus aeglefinus*, in the North Sea. *Fisheries Oceanography*, **7(2)**, 110–25.

Heath, M.R. and Lough, R.G. (2007). A synthesis of large-scale patterns in the planktonic prey of larval and juvenile cod (*Gadus morhua*). *Fisheries Oceanography*, **16**, 169–85.

Heath, M.R., Henderson, E.W., and Baird, D.L. (1988). Vertical distribution of herring larvae in relation to physical mixing and illumination. *Marine Ecology Progress Series*, **47**, 211–28.

Heath, M.R., Backhaus, J.O., Richardson, K., *et al.* (1999a). Climate fluctuations and the spring invasion of the North Sea by *Calanus finmarchicus*. *Fisheries Oceanography*, **8(Suppl. 1)**, 163–76.

Heath, M.R., Dunn, J., Fraser, J.G., *et al.* (1999b). Field calibration of the Optical Plankton Counter with respect to *Calanus finmarchicus*. *Fisheries Oceanography*, **8(Suppl. 1)**, 13–24.

Heath, M.R., Astthorsson, O.S., Dunn, J., *et al.* (2000). Comparative analysis of *Calanus finmarchicus* demography at locations around the northeast Atlantic. *ICES Journal of Marine Science*, **57**, 1562–80.

Heath, M.R., Boyle, P.R., Gislason, A., *et al.* (2004). Comparative ecology of over-wintering *Calanus finmarchicus* in the northern North Atlantic, and implications for life-cycle patterns. *ICES Journal of Marine Science*, **61**, 698–708.

Heath, M.R., Kunzlik, P.A., Gallego, A., *et al.* (2008a). A model of meta-population dynamics for North Sea and

West of Scotland cod-the dynamic consequences of natal fidelity. *Fisheries Research*, **93**, 92–116.

Heath, M.R., Rasmussen, J., Ahmed, Y., *et al.* (2008b). Spatial demography of *Calanus finmarchicus* in the Irminger Sea. *Progress in Oceanography*, **76**, 39–88.

Hedd, A., Ryder, J.L., Cowen, L.L., *et al.* (2002). Inter-annual variation in the diet, provisioning and growth of Cassin's auklet at Triangle Island, British Columbia: responses to variation in ocean climate. *Marine Ecology Progress Series*, **229**, 221–32.

Heinle, D.R. (1969). Temperature and zooplankton. *Chesapeake Science*, **10**, 186–209.

Heino, M. and Godø, O.R. (2002). Fisheries-induced selection pressures in the context of sustainable fisheries. *Bulletin of Marine Science*, **70**, 639–56.

Heithaus M.R., Frid, A., Wirsing, A.J., *et al.* (2008). Predicting ecological consequences of marine top predator declines. *Trends in Ecology and Evolution*, **23(4)**, 202–10.

Helaouët, P. and Beaugrand, G. (2007). Macroecology of *Calanus finmarchicus* and *C. helgolandicus* in the North Atlantic Ocean and adjacent seas. *Marine Ecology Progress Series*, **345**, 147–65.

Helbig, J.A. and Pepin, P. (1998a). Partitioning the influence of physical processes on the estimation of ichthyoplankton mortality rates, I: theory. *Canadian Journal of Fisheries and Aquatic Sciences*, **55**, 2189–205.

Helbig, J.A. and Pepin, P. (1998b). Partitioning the influence of physical processes on the estimation of ichthyoplankton mortality rates, II: application to simulated and field data. *Canadian Journal of Fisheries and Aquatic Sciences*, **55**, 2206–20.

HELCOM. (2002). Oxygen deficiency in Danish marine water 2002. Helsinki Commission, Monitoring and Assessment Group Fourth Meeting Warnemünde, Germany, 21–25 October 2002. Agenda Item 3 Information by the Contracting Parties and the Observer Organizations. *HELCOM MONAS 4/2002*, **3/9**, 8pp.

Helland, S., Nejstgaard, J.C., Humlen, R., *et al.* (2003). Effects of season and maternal food on *Calanus finmarchicus* reproduction, with emphasis on free amino acids. *Marine Biology*, **142**, 1141–51.

Helland-Hansen, B. and Nansen, F. (1909). The Norwegian Sea. Its physical oceanography based upon the Norwegian researches 1900–1904. *Report on Norwegian Fishery and Marine Investigations*, **2(2)**, 1–360.

Helmuth, B., Broitman, B.R., Blanchette, C.A., *et al.* (2006). Mosaic patterns of thermal stress in the rocky intertidal zone: implications for climate change. *Ecological Monographs*, **76**, 461–79.

Hennemuth, R.C. and S. Rockwell. (1987). History of fisheries conservation and management. In R.H. Baukus, ed. *Georges Bank*. MIT Press, Cambridge, MA, pp. 430–6.

Henriksen, C.I., Saiz, E., Calbet, A., *et al.* (2007). Feeding activity and swimming patterns of *Acartia grani* and *Oithona davisae* nauplii in the presence of motile and non-motile prey. *Marine Ecology Progress Series*, **331**, 119–29.

Herman, A.W. (1988). Simultaneous measurement of zooplankton and light attenuance with a new optical plankton counter. *Continental Shelf Research*, **8**, 205–21.

Herman, A.W. (1992). Design and calibration of a new optical plankton counter capable of sizing small zooplankton. *Deep-Sea Research II*, **39**, 395–415.

Herman, A.W., Beanlands, B., Chin-Yee, M., *et al.* (1998). Moving Vessel Profiler (MVP): *in-situ* sampling of plankton and physical parameters at 12 kts and the integration of a new laser optical plankton counter. In: *Oceanology international 98. The global ocean, 10–13 March 1998, Brighton, UK. Conference proceedings, vol. 2.* Spearhead Exhibitions Limited, Surrey, pp. 123–35.

Hermann, A.J., Rugen, W.C., Stabeno, P.J., *et al.* (1996a). Physical transport of young pollock larvae (*Theragra chalcogramma*) near Shelikof Strait as inferred from a hydrodynamic model. *Fisheries Oceanography*, **5**, 58–70.

Hermann, A.J., Hinckley, S., Megrey, B.A., *et al.* (1996b). Interannual variability of the early life history of walleye pollock near Shelikof Strait, as inferred from a spatially explicit, individual-based model. *Fisheries Oceanography*, **5(Suppl. 1)**, 39–57.

Hermann, A.J., Hinckley, S., Megrey, B.A., *et al.* (2001). Applied and theoretical considerations for constructing spatially explicit individual-based models of marine fish early life history which include multiple trophic levels. *ICES Journal of Marine Science*, **58**, 1030–41.

Hermann, A.J., Haidvogel, D.B., Dobbins, E.L., *et al.* (2002). Coupling global and regional circulation models in the coastal Gulf of Alaska. *Progress in Oceanography*, **53**, 335–67.

Hermann, A.J., Curchitser, E.N., Dobbins, E.L., *et al.* (2009). A comparison of remote versus local influence of El Nino on the coastal circulation of the Northeast Pacific. *Deep-Sea Research II*, doi: 10.1016/j.dsr2.2009.02.005.

Hewitt, R.P., Watkins, J.L., Naganobu, M., *et al.* (2004). Biomass of Antarctic krill in the Scotia Sea in January/February 2000 and its use in revising an estimate of precautionary yield. *Deep-Sea Research II*, **51**, 1215–36.

Heymans, J.J., Shannon, L.J., and Jarre, A. (2004). Changes in the northern Benguela ecosystem over three decades: 1970s, 1980s and 1990s. *Ecological Modelling*, **172**, 175–95.

Hickey, B.M. (1998). Coastal oceanography of western North America from the tip of Baja California to Vancouver Island. In: Robinson, A.R. and Brink, K., eds. *The sea, vol. 11*. Harvard University Press, Cambridge, pp. 345–93.

Hidalgo, P. and Escribano, R. (2001). Succession of pelagic copepod species during the period 1996/1998 in northern Chile: the influence of the 1997–98 *El Niño*. *Hydrobiologia*, **453**, 153–60.

Hidalgo, P. and Escribano, R. (2007). Coupling of life cycles of the copepods *Calanus chilensis* and *Centropages brachiatus* to upwelling induced variability in the central-southern region of Chile. *Progress in Oceanography*, **75**, 501–17.

Hilborn, R. (2003). The state of the art in stock assessment: where we are and where we are going. *Scientia Marina*, **67(Suppl. 1)**, 15–20.

Hilborn, R. and Walters, C.J. (1992). *Quantitative fisheries stock assessment: choice dynamics and uncertainty.* Chapman and Hall, New York, p. xv + 570pp.

Hilborn, R., Maguire, J.J., Parma, A.M., *et al.* (2001). The precautionary approach and risk management: can they increase the probability of success in fisheries management. *Canadian Journal of Fisheries and Aquatic Sciences*, **58**, 99–107.

Hilborn, R., Quinn, T.P., Schindler, D.E., *et al.* (2003). Biocomplexity and fisheries sustainability. *Proceedings of the National Academy of Sciences of the United States of America*, **100(11)**, 6564–8.

Hill, K.L., Weaver, A.J., Freeland, H.J., *et al.* (2003). Evidence of change in the Sea of Okhotsk: implications for the North Pacific. *Atmosphere-Ocean*, **41**, 49–63.

Hill, S.L., Murphy, E.J., Reid, K., *et al.* (2006). Modelling Southern Ocean ecosystems: krill, the food-web, and the impacts of harvesting. *Biological Reviews*, **81**, 581–608.

Hinckley, S., Hermann, A.J., and Megrey, B.A. (1996). Development of a spatially explicit, individual-based model of marine fish early life history. *Marine Ecology Progress Series*, **139**, 47–68.

Hind, A., Gurney, W.S.C., Heath, M., *et al.* (2000). Overwintering strategies in *Calanus finmarchicus*. *Marine Ecology Progress Series*, **193**, 95–107.

Hinrichsen, H.H., Bottcher, U., Oberst, R., *et al.* (2001). The potential for advective exchange of the early life stages between western and eastern Baltic cod (*Gadus morhua*). *Fisheries Oceanography*, **10**, 249–58.

Hinrichsen, H.H., St. John, M.A., Lehmann, A., *et al.* (2002a). Resolving the impact of physical forcing variations on the eastern Baltic cod spawning environment. *Journal of Marine Systems*, **32**, 281–94.

Hinrichsen, H.H., Möllmann, C., Voss, R., *et al.* (2002b). Biophysical modelling of larval Baltic cod (*Gadus morhua*) survival and growth. *Canadian Journal of Fisheries and Aquatic Sciences*, **59**, 1858–73.

Hinrichsen, H.H., Bottcher, U., Koster, F.W., *et al.* (2003). Modeling the influences of atmospheric forcing conditions on Baltic cod early life stages: distributions and drift. *Journal of Sea Research*, **49**, 187–201.

Hirche, H.J., Meyer, U., and Niehoff, B. (1997). Egg production of *Calanus finmarchicus* – effect of temperature, food and season. *Marine Biology*, **127**, 609–20.

Hirst, A.G. and Bunker, A.J. (2003). Growth of marine planktonic copepods: global rates and patterns in relation to chlorophyll *a*, temperature, and body weight. *Limnology and Oceanography*, **48**, 1988–2010.

Hirst, A.G. and Kiørboe, T. (2002). Mortality of marine planktonic copepods: global rates and patterns. *Marine Ecology Progress Series*, **230**, 195–209.

Hirst, A.G., Peterson, W.T., and Rothery, P. (2005). Errors in juvenile growth rate estimates are widespread: problems with the moult rate method. *Marine Ecology Progress Series*, **296**, 263–79.

Hirst, A.G., Bonnet, D., and Harris, R.P. (2007). Seasonal dynamics and mortality rates of *Calanus helgolandicus* over two years at a station in the English Channel. *Marine Ecology Progress Series*, **340**, 189–205.

Hitchcock, G.L., Lane, P., Smith, S., *et al.* (2002). Zooplankton spatial distributions in coastal waters of the northern Arabian Sea, August, 1995. *Deep-Sea Research II*, **49**, 2403–23.

Hjermann, D.Ø., Ottersen, G., and Stenseth, N.C. (2004). Competition among fishermen and fish causes the collapse of Barents Sea capelin. *Proceedings of the National Academy of Sciences of the United States of America*, **101**, 11679–84. doi: 10.1073/pnas.0402904101.

Hjermann, D.Ø., Melsom, A., Dingsør, G.E., *et al.* (2007). Fish and oil in the Lofoten-Barents Sea system: synoptic review of the effect of oil spills on fish populations. *Marine Ecology Progress Series*, **339**, 283–99.

Hjort, J. (1914). Fluctuations in the great fisheries of northern Europe. *Rapports et Proces-verbaux des Reunions. Conseil International pour l'Exploration de la Mer*, **20**, 1–228.

Hoag, H. (2003). Atlantic cod meet icy death. *Nature*, **422**, 792.

Hoegh-Guldberg, O. (1999). Climate change, coral bleaching and the future of the world's coral reefs. *Marine and Freshwater Research*, **50**, 839–66.

Hoerling, M.P., Kumar, A., and Zhong, M. (1997). *El Niño*, *La Niña*, and the nonlinearity of their teleconnections. *Journal of Climate*, **10**, 1769–86.

Hofmann, E.E. and Friedrichs, M.A.M. (2002). Predictive modeling for marine ecosystems. In: Robinson, A.R., McCarthy, J.J., and Rothschild, B.J., eds. *The sea, vol. 12.* Wiley, New York, pp. 537–65.

Hofmann, E.E. and Hüsrevoglu, Y.S. (2003). A circumpolar modeling study of habitat control of Antarctic krill (*Euphausia superba*) reproductive success. *Deep-Sea Research II*, **50**, 3121–42.

Hofmann, E.E. and Murphy, E.J. (2004). Advection, krill and Antarctic marine ecosystems. *Antarctic Science*, **16**, 487–99.

Hofmann, E.E. and Powell, T.M. (1998). Environmental variability effects on marine fisheries: four case histories. *Ecological Applications*, **8(1)**, S23–32.

Hofmann, E.E. and Schellnhuber, H.J. (2009). Oceanic acidification affects marine carbon pump and triggers extended marine oxygen holes. *Proceedings of the National Academy of Sciences of the United States of America*, **106**, 3017–22.

Hofmann, E.E., Costa, D.P., Daly, K.L., *et al.* (2002). US Southern Ocean Global Ocean Ecosystems Dynamics Program. *Oceanography*, **15**, 64–74.

Hofmann, E.E., Klinck, J.M., Costa, D.P., *et al.* (2004a). U.S. Southern Ocean Global Ocean Ecosystems Dynamics Program. *Oceanography*, **15**, 64–74.

Hofmann, E.E., Wiebe, P.H., Costa, D.P., *et al.* (2004b). An overview of the Southern Ocean Global Ocean Ecosystems Dynamics program. *Deep-Sea Research II*, **51(17–19)**, 1921–4.

Hofmann, E.E., Wiebe P.H., Costa, D.P. *et al.* eds. (2004c). Integrated ecosystem studies of western Antarctic Peninsula continental shelf waters and related Southern Ocean regions. *Deep-Sea Research II*, **51**, 1921–2344.

Hofmann, E.E., Wiebe P.H., Costa, D.P. *et al.* eds. (2008). Dynamics of plankton, krill, and predators, in predators in relation to environmental features of the western Antarctic Peninsula and related areas: SO GLOBEC Part II. *Deep-Sea Research II*, **55**, 269–558.

Holliday, D.V. and Pieper, R.E. (1989). Determination of zooplankton size and distribution with multifrequency acoustic technology. *Journal du Conseil*, **46**, 52–61.

Holliday, D.V. and Pieper, R.E., (1995). Bioacoustical oceanography at high frequencies. *ICES Journal of Marine Science*, **52**, 279–96.

Holling, C.S. (1973). Resilience and stability of ecological systems. *Annual Reviews of Ecology and Systematics*, **4**, 1–23.

Hollingworth, C.E., ed. (2000). Ecosystem effects of fishing. Proceedings of an ICES/SCOR Symposium held in Montpellier, France, March 1999. *ICES Journal of Marine Science*, **57**, 465–791.

Hollowed, A.B., Ianelli, J.N., and Livingston, P.A. (2000). Including predation mortality in stock assessments: a case study involving Gulf of Alaska walleye pollock. *ICES Journal of Marine Science*, **57**, 279–93.

Hollowed, A.B., Bond, N.A., Wilderbuer, T.K., *et al.* (2009). A framework for modelling fish and shellfish responses to future climate change. *ICES Journal of Marine Science*, doi:10.1093/icesjms/fsp057.

Honda, M., Kushnir, Y., Nakamura, H., *et al.* (2005). Formation, mechanisms and predictability of the Aleutian-Icelandic low seesaw in ensemble AGCM simulations. *Journal of Climate*, **18**, 1423–34.

Hood, R.R., Lwas, E.A., Armstrong, R.A., *et al.* (2006). Pelagic functional group modeling: progress, challenges and prospects. *Deep-Sea Research II*, **53**, 459–512.

Hooff, R.C. and Peterson, W.T. (2006a). Copepod biodiversity as an indicator of changes in ocean and climate conditions of the northern California current ecosystem. *Limnology and Oceanography*, **51**, 2607–20.

Hooff, R.C. and Peterson, W.T. (2006b). Recent increases in copepod biodiversity as an indicator of changes in ocean and climate conditions in the California Current ecosystem. *Limnology and Oceanography*, **51**, 2042–51.

Hooker, S.K. and Boyd, I.L. (2003). Salinity sensors on seals: use of marine predators to carry CTD data loggers. *Deep-Sea Research I*, **50**, 927–39.

Hopcroft, R.R., Clarke, C., Byrd, A.G., *et al.* (2005). The paradox of *Metridia* spp. egg production rates: a new technique and measurements from the coastal Gulf of Alaska. *Marine Ecology Progress Series*, **286**, 193–201.

Horne, E.P.W., Loder, J.W., Naimie, C.E., *et al.* (1996). Turbulence dissipation rates and nitrate supply in the upper water column on Georges Bank. *Deep-Sea Research II*, **43**, 1683–712.

Horsted, S.A. (1991). Advice and management of major fisheries in Greenland waters. *NAFO Scientific Council Studies*, **16**, 79–94.

Horsted, S.A. (2000). A review of the cod fisheries at Greenland 1910–1995. *Journal of Northwest Atlantic Fishery Science*, **28**, 1–112.

Houde, E.D. (1987). Fish early life dynamics and recruitment variability. *American Fisheries Society, Symposium*, **2**, 17–29.

Houde, E.D. (1989). Comparative growth, mortality, and energetics of marine fish larvae: temperature and implied latitudinal effects. *Fishery Bulletin*, **87**, 471–95.

Houde, E.D. (1996). Evaluating stage-specific survival during the early life of fish. In: Watanabe, Y., Yamashita, Y., and Oozeki, Y., eds. *Survival strategies in early life stages of marine resource*. Balkema, Rotterdam, the Netherlands, pp. 51–66.

Houde, E.D. (2008). Emerging from Hjort's shadow. *Journal of Northwest Atlantic Fishery Science*, **41**, 53–70.

Houde, E.D. and Rutherford, E. (1993). Recent trends in estuarine fisheries: predictions of fish production and yield. *Estuaries and Coasts*, **16**, 161–76.

Houde, E.D. and Schekter, R.C. (1980). Feeding by marine fish larvae: developmental and functional response. *Environmental Biology of Fishes*, **5**, 315–34.

Hovgård, H. and Wieland, K. (2008). Fishery and environmental aspects relevant for the emergence and decline of Atlantic cod (*Gadus morhua*) in West Greenland waters. In: Kruse, G.H., Drinkwater, K., Ianelli, J.N., *et al.*, eds. *Resilience of gadid stocks to fishing and climate change*. University of Alaska, Fairbanks, AK, Alaska Sea Grant, pp. 89–110.

Howard, J.A.E., Jarre, A., Clarke, A., *et al.* (2007). Application of the sequential t-test algorithm for analysing regime shifts to the southern Benguela ecosystem. *African Journal of Marine Science*, **29**, 437–51.

Hsieh, C.-H., Glaser, S., Lucas, A., *et al.* (2005). Distinguishing noise from catastrophic shifts in the North Pacific. *Nature*, **435**, 1–26.

Hsieh, C.-H., Reiss, C.S., Hunter, J.R., *et al.* (2006). Fishing elevates variability in the abundance of exploited species. *Science*, **443(19)**, 859–62.

Hu, Q. and Davis, C.S. (2006). Automatic plankton image recognition with co-occurrence matrices and support vector machine. *Marine Ecology Progress Series*, **306**, 51–61.

Hughes, T.J.R., Mazzei, L., and Jansen, K.E. (2000). Large eddy simulation and the variational multiscale method. *Computing and Visualization in Science*, **3**, 47–59.

Hughes, T.P., Bellwood, D.R., Folke, C., *et al.* (2005). New paradigms for supporting the resilience of marine ecosystems. *Trends in Ecology and Evolution*, **20(7)**, 380–6.

Hui, C.A. (1979). Undersea topography and distributions of dolphins of the genus *Delphinus* in the southern California bight. *Journal of Mammalogy*, **60**, 521–7.

Humston, R., Olson, D.B., and Ault, J.S. (2004). Behavioral assumptions in models of fish movement and their influence on population dynamics. *Transactions of the American Fisheries Society*, **133**, 1304–28.

Hunt von Herbing, I. and Gallager, S.M. (2000). Foraging behavior in early Atlantic cod larvae (*Gadus morhua*) feeding on a protozoan (*Balanion* sp.) and a copepod nauplius (*Pseudodiaptomus* sp.). *Marine Biology*, **136**, 591–602.

Hunt von Herbing, I., Gallager, S.M., and Halteman, W. (2001). Metabolic costs of pursuit and attack in early larval Atlantic cod. *Marine Ecology Progress Series*, **216**, 201–12.

Hunt, G.L., Jr. (1997). Physics, zooplankton, and the distribution of least auklets in the Bering Sea: a review. *ICES Journal of Marine Science*, **54**, 600–7.

Hunt, G.L., Jr., and McKinnell, S. (2006). Interplay between top-down, bottom-up, and wasp-waist control in marine ecosystems. *Progress in Oceanography*, **68**, 115–24.

Hunt, G.L., Jr., Heinemann, D., and Everson, I. (1992). Distributions and predator-prey interactions of macaroni penguins, Antarctic fur seals, and Antarctic krill near Bird Island, South Georgia. *Marine Ecology Progress Series*, **86**, 15–30.

Hunt, G.L., Stabeno, P., Walters, G., *et al.* (2002). Climate change and control of the southeastern Bering Sea pelagic ecosystem. *Deep-Sea Research II*, **49(26)**, 5821–53.

Hunt, G.L., Jr. Drinkwater, K., McKinnell, S.M., *et al.*, eds. (2007). Effects of climate variability on sub-arctic marine ecosystems-A GLOBEC symposium, GLOBEC-ESSAS Symposium on 'Effects of climate variability on sub-arctic marine ecosystems'. *Deep-Sea Research II*, **54**, 2453–970.

Hunt, G.L., Stabeno, P.J., Strom, S., *et al.* (2008). Patterns of spatial and temporal variation in the marine ecosystem of the southeastern Bering Sea, with special reference to the Pribilof Domain. *Deep-Sea Research II*, **55**, 1919–44.

Hunter, J.R. and Alheit, J. (1995). International GLOBEC Small Pelagic Fish and Climate Change Program. Report of the first planning meeting, La Paz, Mexico, June 20–24, 1994. *GLOBEC Report*, **8**, 72pp.

Hunter, J.R. and Alheit, J. eds. (1997). International GLOBEC Small Pelagic Fishes and Climate Change Program. Implementation Plan. *GLOBEC Report*, **11**, 36pp.

Hunter, M.D. and Price, P.W. (1992). Playing chutes and ladders: heterogeneity and the relative roles of bottom-up and top-down forces in natural communities. *Ecology*, **73(3)**, 724–32.

Huntley, M.E. and Lopez, M.D.G. (1992). Temperature-dependent production of marine copepods: a global synthesis. *American Naturalist*, **140**, 201–42.

Huntley, M.E. and Zhou, M. (2004). Influence of animals on turbulence in the sea. *Marine Ecology Progress Series*, **273**, 65–79.

Hurrell, J. (1995). Decadal trends in the North Atlantic oscillation: regional temperatures and precipitation. *Science*, **269**, 676–9.

Huse, G. (2001). Modelling habitat choice in fish using adapted random walk. *Sarsia*, **86**, 477–83.

Huse, G. and Ellingsen, I. (2008). Capelin migrations and climate change-a modelling analysis. *Climatic Change*, **87**, 177–97.

Hutchings, J.A. (2000). Collapse and recovery of marine fishes. *Nature*, **406**, 882–5.

Hutchings, J.A. and Myers, R.A. (1994). Timing of cod reproduction: interannual variability and the influence of temperature. *Marine Ecology Progress Series*, **108**, 21–31.

Hwang, J.-S., Costello, J.H., and Strickler, J.R. (1994). Copepod grazing in turbulent flow: elevated foraging behavior and habituation of escape responses. *Journal of Plankton Research*, **16**, 421–43.

Hygum, B.H., Rey, C., Hansen, B.W., *et al.* (2000). Importance of food quantity to structural growth rate and neutral lipid reserves accumulated in *Calanus finmarchicus*. *Marine Biology*, **136**, 1057–73.

Ianelli, J.N., Barbeaux, S., Honkalehto, T., *et al.* (2007). Eastern Bering Sea walleye pollock. In: the plan team for the groundfish fisheries of the Bering Sea and Aleutian Islands, compilers. *Stock assessment and fisheries evaluation report for the groundfish resources of the Bering Sea/Aleutian Islands regions*. NOAA, Seattle.

Ianora, A., Poulet, S.A., and Miralto, A. (2003). The effects of diatoms on copepod reproduction: a review. *Phycologia*, **42**, 351–63.

Ianora, A., Miralto, A., Poulet, S.A., *et al.* (2004). Aldehyde suppression of copepod recruitment in blooms of a ubiquitous planktonic diatom. *Nature*, **429**, 403–7.

ICES. (1949). Climatic changes in the Arctic in relation to plants and animals. *Rapports et Proces-Verbaux des Reunions du Conseil International pour l'Exploration de la Mer*, **125**, 5–52.

ICES. (1999). Report of the workshop on gadoid stocks in the North Sea during the 1960s and 1970s. *ICES Cooperative Research Report*, **244**, 55pp.

ICES. (2005). Guidance on the application of the ecosystem approach to management of human activities in the european marine environment. *ICES Cooperative Research Report*, **273**, 22pp.

ICES. (2006). Incorporation of process information into stock-recruitment models. *ICES Cooperative Research Report*, **282**, 154pp.

ICES. (2007a). Report of the Baltic Fisheries Assessment Working Group (WGBFAS). 17–26 April 2007, ICES Headquarters. *ICES Council Meeting Papers*, **CM ACFM/15**.

ICES. (2007b). Report of the Study Group on Management Strategies (SGMAS), 22–26 January 2007, ICES Headquarters. *ICES Council Meeting Papers*, **CM 2007/ ACFM:04**. 59pp.

ICES. (2007c). Report of the Working Group on the Assessment of Mackerel, Horse Mackerel, Sardine and Anchovy (WGMHSA), 3–4 September 2007, ICES Headquarters. *ICES Council Meeting Papers*, **CM 2007/ ACFM:31**, 712pp.

ICES. (2007d). Report of the Workshop on Limit and Target Reference Points (WKREF), 29 January–2 February 2007, Gdynia, Poland. *ICES Council Meeting Papers*, **CM 2007/ ACFM:05**, 89pp.

ICES. (2007e). Report of the Workshop on the Integration of Environmental Information into Fisheries Management Strategies and Advice (WKEFA), 18–22 June 2007, ICES Headquarters, Copenhagen, Denmark. *ICES Council Meeting Papers*, **CM 2007/ACFM:25**, 182pp.

ICES. (2007f). Status of introductions of non-indigenous marine species to the North Atlantic and adjacent waters 1992–2002. *ICES Cooperative Research Report*, **284**, 149pp.

Ikeda, T. (1985). Metabolic rates of epipelagic marine zooplankton as a function of body mass and temperature. *Marine Biology*, **85**, 1–11.

Iles, T.D. and Sinclair, M. (1982). Atlantic herring: stock discreteness and abundance. *Science*, **215**, 627–33.

IMARPE. (2008). *Resultado de la Segunda Encuesta Estructural de la Pesquería Artesanal en el Litoral Peruano" II ENEPA 2004–2005*. Unidad de Estadistica y Pesca Artesanal, Instituto del Mar del Peru, Callao.

Incze, L.S., Hebert, D., Wolff, N., *et al.* (2001). Changes in copepod distributions associated with increased turbulence from wind stress. *Marine Ecology Progress Series*, **213**, 229–40.

IPCC. (2000). Special report on emissions scenarios: a special report of Working Group III of the Intergovernmental Panel on Climate Change. Cambridge University Press, Cambridge, 570pp.

IPCC. (2001). *Climate change 2001: the scientific basis. Contribution of Working Group I to The Third Assessment Report of the IPCC*. Cambridge University Press, Cambridge.

IPCC. (2007a). *Climate change 2007: the physical science basis. Contribution of Working Group I to the Fourth Assessment Report of the Intergovernmental Panel on Climate Change*. Cambridge University Press, Cambridge and New York, 966pp.

IPCC. (2007b). Summary for policymakers. In: Parry, M.L., Canziani, O.F., Palutikof, J.P., *et al.* eds. *Climate change 2007: impacts, adaptation and vulnerability. Contribution of Working Group II to the Fourth Assessment Report of the Intergovernmental Panel on Climate Change*. Cambridge University Press, Cambridge, pp. 7–22.

Irigoien, X. (2004). Some ideas about the role of lipids in the life cycle of *Calanus finmarchicus*. *Journal of Plankton Research*, **26(3)**, 259–63.

Irigoien, X., and Harris, R.P. (2006). Comparative population structure, abundance and vertical distribution of six copepod species in the North Atlantic (Station India, 59°N, 19°W, 1971–1974). *Marine Biology Research*, **2**, 276–90.

Irigoien, X., Head, R., Klenke, U., *et al.* (1998). A high frequency time series at weathership M, Norwegian Sea, during the 1997 spring bloom: feeding of adult female *Calanus finmarchicus*. *Marine Ecology Progress Series*, **172**, 127–37.

Irigoien, X., Head, R., Harris, R.P., *et al.* (2000). Feeding selectivity and egg production of *Calanus helgolandicus* in the English Channel. *Limnology and Oceanography*, **45**, 44–54.

Irigoien, X., Head, R., Harris, R.P., *et al.* (2002). Copepod hatching success in marine ecosystems with high diatom concentrations. *Nature*, **419**, 387–9.

Irigoien, X., Huisman, J., and Harris, R.P. (2004). Global biodiversity patterns of marine phytoplankton and zooplankton. *Nature*, **429**, 864–7.

Irigoien, X., Verheye, H.M., Harris, R.P., *et al.* (2005). Effect of food composition on egg production and hatching success rate of two copepod species (*Calanoides carinatus* and *Rhincalanus nasutus*) in the Benguela upwelling system. *Journal of Plankton Research*, **27**, 735–42.

Irigoien, X., Fiksen, O., Cotano, U., *et al.* (2007). Could Biscay Bay anchovy recruit through a spatial loophole? *Progress in Oceanography*, **74**, 132–48.

Isaacs, M. (2006). Small-scale fisheries transform: expectations, hopes and dreams for a "better life for all". *Marine Policy*, **30**, 51–9.

Ishii, H. and Tanaka, F. (2001). Food and feeding of *Aurelia aurita* in Tokyo Bay with an analysis of stomach contents and a measurement of digestion times. *Hydrobiologia*, **451**, 311–20.

Ishii, M., Shouji, A., Sugimoto, S., *et al.* (2005). Objective analyses of sea-surface temperature and marine meteorological variables for the 20th century using icoads and the Kobe collection. *International Journal of Climatology*, **25(7)**, 865–79.

Ishimura, G., Sumaila, R., and Herrick, S. (2008). Desperate fishing games under climate changes: non-cooperative/cooperative games for the tim-variant share of a transboundary fish stock – the case of the Pacific sardine fishery. Poster S3-P6, presented at the symposium *Coping with global change in marine social-ecological systems*, Rome, June 2008. http://web.pml.ac.uk/globec/structure/fwg/focus4/symposium/posters.htm

Ito, S. (2007). Responses of Pacific saury and herring projected under a global warming scenario. *Kaiyo Monthly*, **39**, 303–8. [in Japanese]

Ito, S.-I., Kishi, M.J., Kurita, Y., *et al.* (2004). Initial design for a fish bioenergetics model of Pacific saury coupled to a lower trophic ecosystem model. *Fisheries Oceanography*, **13(Suppl. 1)**, 111–24.

Ito, S.-I., Megrey, B.A., Kishi, M.J., *et al.* (2007). On the interannual variability of the growth of Pacific saury (*Cololabis saira*): a simple 3-box model using NEMURO. FISH. *Ecological Modelling*, **202**, 174–83.

Ito, S.-I., Rose, K.A., Miller, A.J., *et al.* (2010). Ocean ecosystem responses to future global change scenarios: a way forward. *This volume*. Oxford University Press, Oxford.

Iverson, R.L. (1990). Control of marine fish production. *Limnology and Oceanography*, **35**, 1593–604.

Jackson, J.B.C., Kirby, M.X., Berger, W.H., *et al.* (2001). Historical overfishing and the recent collapse of coastal ecosystems. *Science*, **293**, 629–37.

Jacob, S., Farmer, F.L., Jepson, M., *et al.* (2001). Landing a definition of fishing dependent communities: potential social science contributions to meeting national standard 8. *Fisheries*, **26**, 16–22.

Jacobson, L.D., Bograd, S.J., Parrish, R.H., *et al.* (2005). An ecosystem-based hypothesis for climatic effects on surplus production in California sardine (*Sardinops sagax*) and environmentally dependent surplus production models. *Canadian Journal of Fisheries and Aquatic Sciences*, **62**, 1782–96.

Jacobson, L.D. and MacCall, A.D. (1995). Stock-recruitment models for California sardine (*Sardinops sagax*). *Canadian Journal of Fisheries and Aquatic Sciences*, **52**, 566–77.

James, A.G. (1988). Are clupeoid microphagists herbivorous or omnivorous? A review of the diets of some commercially important clupeids. *South African Journal of Marine Science*, **7**, 161–77.

Jamieson, G. and Zhang, C.-I. (2005). Report of the study group on ecosystem-based management science and its application to the North Pacific. *PICES Scientific Report*, **29**, 77pp.

Jarman, S.N. and Wilson, S.G. (2004). DNA-based species identification of krill consumed by whale sharks. *Journal of Fish Biology*, **65**, 586–91.

Jarman, S.N., Deagle, B.E., and Gales, N.J. (2004). Group-specific polymerase chain reaction for DNA-based analysis of species diversity and identity in dietary samples. *Molecular Ecology*, **13**, 1313–22.

Jarman, S.N., Redd, K.S., and Gales, N.J. (2006). Technical note: group-specific primers for amplifying DNA sequences that identify Amphipoda, Cephalopoda, Echinodermata, Gastropoda, Isopoda, Ostracoda and Thoracica. *Molecular Ecology Notes*, **6**, 268–71.

Jarre, A., Shannon, L.J., Moloney, C.L., *et al.* (1998). Comparing trophic flows in the southern Benguela to those in other upwelling ecosystems. *South African Journal of Marine Science*, **19**, 391–414.

Jarre, A., Wieland, K., MacKenzie, B., *et al.* (2000). Stock-recruitment relationships for cod (*Gadus morhua callarias*) in the central Baltic Sea incorporating environmental variability. *Archive of Fishery and Marine Research*, **48(2)**, 97–123.

Jarre, A., Moloney, C.L., Shannon, L.J., *et al.* (2006). Detecting and forecasting long-term ecosystem changes., In: Shannon, L.V., Hempel, G., Malanotte-Rizzoli, P., *et al.* eds. *Benguela: predicting a large marine ecosystem*. *Large Marine Ecosystems Series* 13, pp. 239–72. Elsevier, Amsterdam, 410 pp.

Jarre, A., Paterson, B., Moloney, C.L., *et al.* (2008). Knowledge-based systems as decision support tools in an ecosystem approach to fisheries: comparing a fuzzy-logic and a rule-based approach. *Progress in Oceanography*, **79**, 390–400.

Jennings, S. and Blanchard, J.L. (2004). Fish abundance with no fishing: predictions based on macroecological theory. *Journal of Animal Ecology*, **73**, 632–42.

Jennings, S. and Kaiser, M.J. (1998). The effects of fishing on marine ecosystems. *Advances in Marine Biology: An Annual Review*, **34**, 201–352.

Jennings, S., Greenstreet, S.P.R., and Reynolds, J.D. (1999). Structural changes in an exploited fish community: a consequence of differential fishing effects on species with contrasting life histories. *Journal of Animal Ecology*, **68**, 617–27.

Jennings, S., Kaiser, M.J., and Reynolds, J.D. (2001). *Marine fisheries ecology*. Blackwell Science, Oxford. 417pp.

Jennings, S., Pinnegar, J.K., Polunin, N.V.C., *et al.* (2002). Linking size-based and trophic analyses of benthic community structure. *Marine Ecology Progress Series*, **226**, 77–85.

Jensen, A.S. (1939). Concerning a change of climate during recent decades in the Arctic and Subarctic regions from Greenland in the west to Eurasia in the east, and contemporary biological and geophysical changes. *Det Kongelige Danske Videnskabelige Selskab, Biologiske Meddelelser*, **14**, 1–77.

Jensen, A.S. (1949). Studies on the biology of the cod in Greenland waters. *Rapports et Proces-Verbaux des Reunions du Conseil International pour l'Exploration de la Mer*, **123**, 1–77.

Jentoft, S. (1993). *Dangling lines: the fisheries crisis and the future of coastal communities, the Norwegian experience*. Institute of Social and Economic Research, St. John's, Newfoundland, Canada.

Jentoft, S. (2007). Limits of governability: institutional implications for fisheries and coastal governance. *Marine Policy*, **31**, 360–70.

Ji, R., Davis, C., Chen, C., *et al.* (2008). Influence of local and external processes on the annual nitrogen cycle and primary productivity on Georges Bank: A 3-D biological-physical modeling study. *Journal of Marine Systems*, **73(1–2)**, 31–47.

Johns, D.G., Edwards, M., Greve, W., *et al.* (2005). Increasing prevalence of the marine cladoceran *Penilia avirostris* (Dana, 1852) in the North Sea. *Helgoland Marine Research*, **59**, 214–8.

Johnson, C.L., Leising, A.W., Runge, J.A., *et al.* (2008). Characteristics of *Calanus finmarchicus* dormancy patterns in the Northwest Atlantic. *ICES Journal of Marine Science*, **65(3)**, 339–50.

Jonasdottir, S.H., Trung N.H., Hansen, F., *et al.* (2005). Egg production and hatching success in the calanoid copepods *Calanus helgolandicus* and *Calanus finmarchicus* in the North Sea from March to September 2001. *Journal of Plankton Research*, **27**, 1239–59.

Jones, G.P., McCormick, M.I., Srinivasan, M., *et al.* (2004). Coral decline threatens fish biodiversity in marine reserves. *Proceedings of the National Academy of Sciences of the United States of America*, **101**, 8251–3.

Jones, R.H. and Flynn, K.J. (2005). Nutritional status and diet composition affect the value of diatoms as copepod prey. *Science*, **307**, 1457–9.

Jonsson, P.R. and Tiselius, P. (1990). Feeding behaviour, prey detection and capture efficiency of the copepod *Acartia tonsa* feeding on planktonic ciliates. *Marine Ecology Progress Series*, **60**, 35–44.

Jørgensen, C., Enberg, K., Dunlop, E.S., *et al.* (2007). Ecology: managing evolving fish stocks. *Science*, **318**, 1247–8.

Jørgensen, C., Dunlop, E.S., Opdal, A.F., *et al.* (2008). The evolution of spawning migrations: the role of individual state, population structure, and fishing-induced changes. *Ecology*, **89(12)**, 3436–48.

Jumars, P.A. (2007). Habitat coupling by mid-latitude, subtidal, marine mysids: import-subsidised omnivores. *Oceanography and Marine Biology: An Annual Review*, **45**, 89–138.

Justić, D., Rabalais, N.N., and Turner, R.E. (2005). Coupling between climate variability and coastal eutrophication: evidence and outlook for the northern Gulf of Mexico. *Journal of Sea Research*, **54**, 25–35.

Kaartvedt, S., Melle, W., Knutsen, T., *et al.* (1996). Vertical distribution of fish and krill beneath water of varying optical properties. *Marine Ecology Progress Series*, **136**, 51–8.

Kaeriyama, M., Nakamura, M., Edpalina, R., *et al.* (2004). Change in feeding ecology and trophic dynamics of Pacific salmon (*Oncorhynchus* spp.) in the central Gulf of Alaska in relation to climate events. *Fisheries Oceanography*, **13**, 197–207.

Kaiser, M.J. and de Groot, S.J. (2000). *Effects of fishing on non-target species and habitats*. Blackwell Science. 399pp.

Kaiser, M.J. and Jennings, S. (2002). Ecosystem effects of fishing. In: Hart, P.J. and Reynolds, J.D., eds. *Handbook of fish biology and fisheries, vol. 2*. Blackwell Science, Oxford, pp.342–66.

Kaiser, M.J., Spence, F.E., and Hart, P.J.B. (1999). Fishing gear restrictions and conservation of benthic habitat complexity. *Conservation Biology*, **14**, 1512–25.

Kaiser, M.J., Collie, J.S., Hall, S.J., *et al.* (2003). Impacts of fishing gear on marine benthic habitats. In: Sinclair, M. and Valdemarsson, G., eds. *Responsible fisheries in the marine ecosystem*. FAO and CABI Publishing, Rome, Italy, and Wallingford, UK, pp. 197–218.

Kang, H.K. and Poulet, S.A. (2000). Reproductive success in *Calanus helgolandicus* as a function of diet and egg cannibalism. *Marine Ecology Progress Series*, **201**, 241–50.

Kang, Y.S. and Park, M.S. (2003). Occurrence and food ingestion of the moon jellyfish (Scyphomoa: Ulmariidae: *Aurelia aurita*) in the southern coast of Korea in summer. *Journal of the Korean Society of Oceanography*, **8**, 199–202.

Kaplan, D.M., Botsford, L.W., O'Farrell, M.R., *et al.* (2009). Model-based assessment of persistence in proposed marine protected area designs. *Ecological Applications*, **19(2)**, 433–48.

Kasai, A., Kishi, M.J., and Sugimoto, T. (1992). Modelling the transport and survival of Japanese sardine larvae in and around the Kuroshio Current. *Fisheries Oceanography*, **1**, 1–10.

Kasai, H., Saito, H., Yoshimori, A., *et al.* (1997). Variability in timing and magnitude of spring bloom in the Oyashio region, the western subarctic Pacific off Hokkaido, Japan. *Fisheries Oceanography*, **6(2)**, 118–29.

Kato, Y., Takebe, T., Masuma, S., *et al.* (2008). Turbulence effect on survival and feeding of Pacific bluefin tuna *Thunnus orientalis* larvae, on the basis of a rearing experiment. *Fisheries Science*, **74(1)**, 48–53.

Katona, S.K. (1973). Evidence for sex pheromones in planktonic copepods. *Limnology and Oceanography*, **18**, 574–83.

Katsukawa, T. and Matsuda, H. (2003). Simulated effects of target switching on yield and sustainability of fish stocks. *Fisheries Research*, **60**, 515–25.

Kawasaki, T. (1983). Why do some pelagic fishes have wide fluctuations in their numbers? Biological basis of fluctuation from the viewpoint of evolutionary ecology. In: Sharp, G.D. and Csirke, J., eds. *Proceedings of the expert consultation to examine changes in abundance and species composition of neritic fish resources. San Jose, Costa Rica, 18–29 April 1983. FAO Fisheries Report*, **291**, 1065–80.

Kawasaki, T. (1991). Long-term variability in the pelagic fish populations. In: Kawasaki, T., Tanaka, S., Toba, Y., *et al.* eds. *Long-term variability of pelagic fish populations and their environment*. Pergamon Press, Oxford, pp. 47–60.

Kawasaki, T. and Omori, M. (1986). Fluctuations in three major sardine stocks in the Pacific and the global temperature. In: Wyatt, T. and Larranenta, G., eds. *Long term changes in marine fish populations*. Bayona, Imprenta REAL, pp. 37–53.

Kawasaki, T. and Omori, M. (1995). Possible mechanisms underlying fluctuations in the far eastern sardine population inferred from time series of two biological traits. *Fisheries Oceanography*, **4(3)**, 238–42.

Keister, J.E., Johnson, J.B., Morgan, C.A., *et al.* (2005). Biological indicators of the timing and direction of warm-water advection during the 1997/1998 El Niño off the central Oregon coast, USA. *Marine Ecology Progress Series*, **295**, 43–8.

Kell, L.T. and Fromentin, J.-M. (2007). Evaluation of the robustness of maximum sustainable yield based management strategies to variations in carrying capacity or migration pattern of Atlantic bluefin tuna (*Thunnus thynnus*). *Canadian Journal of Fisheries and Aquatic Sciences*, **64(5)**, 837–47.

Kell, L.T., Fromentin, J.-M, and Scott, F. (2007). From single stock to ecosystem based advice: are reference points a moving target?. *ICES Council Meeting Papers*, **CM2007/R:19**.

Kell, L.T., Pilling G.M., and O'Brien, C.M. (2004). The implications of climate change for the management of North Sea cod (*Gadus morhua*). *ICES Journal of Marine Science*, **62**, 1483–91.

Kell, L.T., Pastoors, M.A., Scott, R.D., *et al.* (2005a). Evaluation of multiple management objectives for Northeast Atlantic flatfish stocks: sustainability vs. stability of yield. *ICES Journal of Marine Science*, **62**, 1104–17.

Kell, L.T., Pilling, G.M., Kirkwood, G.P., *et al.* (2005b). An evaluation of the implicit management procedure used for some ICES roundfish stocks. *ICES Journal of Marine Science*, **62**, 750–9.

Kell, L.T., Pilling, G.M., Kirkwood, G.P., *et al.* (2006). An evaluation of multi-annual management strategies for ICES roundfish stocks. *ICES Journal of Marine Science*, **63**, 12–24.

Kell, L.T., Mosqueira, I., Grosjean, P., *et al.* (2007). FLR: an open-source framework for the evaluation and development of management strategies. *ICES Journal of Marine Science*, **64**, 640–6.

Kendall, V.J. and Haedrich, R.L. (2006). Species richness in Atlantic deep-sea fishes assessed in terms of the mid-domain effect and Rapoport's rule. *Deep-Sea Research I*, **53**, 506–15.

Kennedy, V.S. (1990). Anticipated effects of climate change on estuarine and coastal fisheries. *Fisheries Management and Ecology*, **15**, 16–24.

Kerr, R.A. (2000). A North Atlantic climate pacemaker for the centuries. *Science*, **288**, 1984–6.

Kideys, A.E. (2002). Fall and rise of the Black Sea ecosystem. *Science*, **297**, 1482–4.

Kim, J.Y., Kim, S., Choi, Y.M., *et al.* (2006). Evidence of density-dependent effects on population variation of Japanese sardine (*Sardinops melanosticta*) off Korea. *Fisheries Oceanography*, **15**, 345–9.

Kim, H.-J. and Miller, A.J. (2007). Did the thermocline deepen in the southern California Current after the 1976–77 climate regime shift? *Journal of Physical Oceanography*, **37**, 1733–9.

Kimball, M.E., Miller, J.M., Whitfield, P.E., *et al.* (2004). Thermal tolerance and potential distribution of invasive lionfish (*Pterois volitans/miles* complex) on the east coast of the United States. *Marine Ecology Progress Series*, **283**, 269–78.

Kimmerer, W.J. and McKinnon, D. (1987). Growth, mortality, and secondary production of the copepod *Acartia tranteri* in Westernport Bay, Australia. *Limnology and Oceanography*, **32**, 14–28.

Kimura, S., Tsukamoto, K., and Sugimoto, T. (1994). A model for the larval migration of the Japanese eel: roles of the trade winds and salinity front. *Marine Biology*, **119(2)**, 185–90.

Kimura, S., Nakata, H., Margulies, D., *et al.* (2004). Effect of oceanic turbulence on the survival of yellowfin tuna larvae. *Nippon Suisan Gakkaishi*, **70**, 175–8.

King, J.R., ed. (2005). Report of the Study Group on Fisheries and Ecosystem Responses to Recent Regime Shifts. *PICES Scientific Report*, **28**, 162pp.

King, J.R. and McFarlane, G.A. (2003). Marine fish life history strategies: applications to fishery management. *Fisheries Management and Ecology*, **10**, 249–64.

King, J.R. and McFarlane, G.A. (2006). A framework for incorporating climate regime shifts into the management of marine resources. *Fisheries Management and Ecology*, **13**, 93–102.

King, J.R., McFarlane, G.A. and Beamish, R.J. (2001). Incorporating the dynamics of marine systems into the stock assessment and management of sablefish. *Progress in Oceanography*, **49**, 619–39.

King, J.R., Read, D.S., Traugott, M., *et al.* (2008). Molecular analysis of predation: a review of best practice for DNA-based approaches. *Molecular Ecology*, **17**, 947–63.

Kingsley, M.C.S. (2007). The fishery for Northern shrimp (*Pandalus borealis*) off West Greenland, 1970–2007. *NAFO Scientific Council Research Document*, **07/69**.

Kiørboe, T. (2006). Sex, sex-ratios, and the dynamics of pelagic copepod populations. *Oecologia*, **148**, 40–50.

Kiørboe, T. (2007). Mate finding, mating, and population dynamics in a planktonic copepod *Oithona davisae*: There are too few males. *Limnology and Oceanography*, **52**, 1511–22.

Kiørboe, T. (2008a). *A mechanistic approach to plankton ecology*. Princeton University Press, Princeton, NJ, 228pp.

Kiørboe, T. (2008b). Optimal swimming strategies in mate searching pelagic copepods. *Oecologia*, **155(1)**, 179–92.

Kiørboe, T. and Sabatini, M. (1994). Reproductive and life cycle strategies in egg-carrying cyclopoid and free-spawning calanoid copepods. *Journal of Plankton Research*, **16**, 1353–66.

Kiørboe, T. and Saiz, E. (1995). Planktivorous feeding in calm and turbulent environments, with emphasis on copepods. *Marine Ecology Progress Series*, **122**, 135–45.

Kiørboe, T., Saiz, E., and Visser, A. (1999). Hydrodynamic signal perception in the copepod *Acartia tonsa*. *Marine Ecology Progress Series*, **179**, 97–111.

Kiørboe, T., Bagoien, E., and Thygesen, U.H. (2005). Blind dating-mate finding in planktonic copepods. II. The pheromone cloud of *Pseudocalanus elongatus*. *Marine Ecology Progress Series*, **300**, 117–28.

Kishi, M.J., Kashiwai, M., Ware, D.M., *et al.* (2007a). NEMURO-a lower trophic level model for the North Pacific marine ecosystem. *Ecological Modelling*, **202(1–2)**, 12–25.

Kishi, M.J., Megrey, B.A., and Ito, S. (2007b). Preface to the ecological modelling special issue on the NEMURO model. *Ecological Modelling*, **202(1–2)**, 3–6.

Kishi, M.J., Megrey, B.A., Ito, S., *et al.* eds. (2007c). NEMURO (North Pacific Ecosystem Model for Understanding Regional Oceanography) and NEMURO. FISH (NEMURO for Including Saury and Herring) Modeling of North Pacific marine ecosystems. *Ecological Modelling*, **202(1–2)**, 224pp.

Kleinschmidt, H., Sauer, W.H.H., and Britz, P. (2003). Commercial fishing rights allocation in post-apartheid South Africa: reconciling equity and stability. *African Journal of Marine Science*, **25**, 25–35.

Kleppel, G. (1992). Environmental regulation of feeding and egg production by *Acartia tonsa* off southern California. *Marine Biology*, **112**, 57–65.

Klimley, A.P., Voegeli, F., Beavers, S.C., *et al.* (1998). Automated listening stations for tagged marine fishes. *Marine Technology Society Journal*, **32**, 94–101.

Knapp, G. (2000). Implications of climate change for fisheries management. In: *Microbehavior and macroresults, 10th International Conference of the Institute of Fisheries Economics and Trade*. Institute of Fisheries Economics and Trade, Oregon State University, Corvallis, OR.

Knapp, G., Livingston, P., and Tyler, A. (1998). Human effects of climate-related changes in Alaska commercial fisheries. In: Weller, G. and Anderson, P.A., eds. *Proceedings of a workshop: assessing the consequences of climate change for Alaska and the Bering Sea region*. Center for Global Change and Arctic System Research, University of Alaska, Fairbanks, AK.

Knowlton, N. (1992). Threshold and multiple stable states in coral reef community dynamics. *American Zoologist*, **32**, 674–82.

Knowlton, N. (2004). Multiple 'stable' states and the conservation of marine ecosystems. *Progress in Oceanography*, **60**, 387–96.

Kobari, T. and Ikeda, T. (1999). Vertical distribution, population structure and life cycle of *Neocalanus cirstatus* (Crustacea: Copepoda) in the Oyashio region, with notes on its regional variations. *Marine Biology*, **134**, 683–96.

Kobari, T. and Ikeda, T. (2001a). Ontogenetic vertical migration and life cycle of *Neocalanus plumchrus* (Crustacea: Copepoda) in the Oyashio region, with notes on regional variations in body sizes. *Journal of Plankton Research*, **23**, 287–302.

Kobari, T. and Ikeda, T. (2001b). Life cycle of *Neocalanus flemingeri* (Crustacea: Copepoda) in the Oyashio region, western subarctic Pacific, with notes on its regional variations. *Marine Ecology Progress Series*, **209**, 243–55.

Koehl, M.A.R. and Strickler, J.R. (1981). Copepod feeding currents: food capture at low Reynolds number. *Limnology and Oceanography*, **26**, 1062–73.

Kooiman, J and Bavinck, M. (2005). The governance perspective. In: Kooiman, J., Bavinck, M., Jentoft, S., *et al*. eds. *Fish for life: interactive governance for fisheries*, pp. 11–24. Amsterdam University Press, Amsterdam, 400pp.

Köster, F.W. and Möllmann, C. (2000). Trophodynamic control by clupeid predators on recruitment success in Baltic cod? *ICES Journal of Marine Science*, **57**, 310–23.

Köster, F.W., Hinrichsen, H.-H., Schnack, D., *et al*. (2003a). Recruitment of Baltic cod and sprat stocks: identification of critical life stages and incorporation of environmental variability into stock-recruitment relationships. *Scientia Marina*, **67(Suppl. 1)**, 129–54.

Köster, F.W., Möllmann, C., Neuenfeldt, S., *et al*. (2003b). Fish stock development in the central Baltic Sea (1974–1999) in relation to variability in the environment. *ICES Marine Science Symposia*, **219**, 294–306.

Köster, F.W., Möllmann, C., Hinrichsen, H.-H., *et al*. (2005). Baltic cod recruitment and the impact of climate variability on key processes. *ICES Journal of Marine Science*, **62**, 1408–25.

Köster, F.W., Vinther, M., MacKenzie, B.R., *et al*. (2008). Environmental effects on recruitment and implications for biological reference points of eastern Baltic cod (*Gadus morhua*). *Journal of Northwest Atlantic Fisheries Science*, **41**, 205–20.

Kristfinnsson, Ö. (2001). The herring era museum in Siglufjörður, Iceland (website). http://www.siglo.is/herring/en/

Kunze, E., Dower, J.F., Beveridge, I., *et al.*. (2006). Observations of biologically generate turbulence in a coastal inlet. *Science*, **313**, 1768–70.

Kurihara, H. (2008). Effects of CO_2-driven ocean acidification on the early development stages of invertebrates. *Marine Ecology Progress Series*, **373**, 275–84.

Kurihara, H., Asai, T., Kato, S., *et al*. (2009). Effects of elevated pCO_2 on early development in the mussel *Mytilus galloprovincialis*. *Aquatic Biology*, **4**, 225–33.

Kurlansky, M. (2007). *Cod: a biography of the fish that changed the world*. Walker Publishing, New York.

Laabir, M., Poulet, S.A., and Ianora, A. (1995). Measuring production and viability of eggs in *Calanus helgolandicus*. *Journal of Plankton Research*, **17**, 1125–42.

Lackey, R.T. (1999). Radically contested assertions in ecosystem management. *Journal of Sustainable Forestry*, **9(1–2)**, 21–34.

Lagadeuc, Y., Boulé, M., and Dodson, J.J. (1997). Effect of vertical mixing on the vertical distribution of copepods in coastal waters. *Journal of Plankton Research*, **19**, 1183–204.

Lampitt, R.S. and Antia, A.N. (1997). Particle flux in deep seas: regional characteristics and temporal variability. *Deep-Sea Research I*, **44(8)**, 1377–403.

Lampitt, R.S., Bett, B.J., Kiriakoulakis, K., *et al*. (2001). Material supply to the Abyssal seafloor in the Northeast Atlantic. *Progress in Oceanography*, **50**, 27–63.

Lan, Y.-C., Lee, M.-A., Chen, W.-Y., *et al*. (2007). Seasonal relationships between the copepod community and hydrographic conditions in the southern East China Sea. *ICES Journal of Marine Science*, **65**, 462–8.

Lance, G.N. and Williams, W.T. (1967). A general theory of classificatory sorting strategies I. Hierarchical systems. *Computer Journal*, **9**, 373–80.

Landry, M.R. (1978). Population dynamics and production of a planktonic marine copepod, *Acartia clausii*, in a small temperate lagoon on San Juan Island, Washington. *Internationale Revue der gesamten Hydrobiologie*, **63**, 77–119.

Large, W.G., McWilliams, J.C., and Doney, S.C. (1994). A review and model with a non local boundary layer parameterization. *Reviews of Geophysics*, **32**, 363–403.

Lasker, R. (1978). The relation between oceanographic conditions and larval anchovy food in the California current: identification of factors contributing to recruitment failure. *Rapports et Proces-verbaux des Reunions. Conseil International pour l'Exploration de la Mer*, **173**, 212–30.

Lasker, R. and Sherman, K. (1981). The early life history of fish: recent studies. *Rapports et Proces-Verbaux des Reunions du Conseil International pour l'Exploration de la Mer*, **178**, 1–167.

Latif, M., Roeckner, E., Botzet, M., *et al*. (2004). Reconstructing, monitoring, and predicting multidecadal-scale changes in the North Atlantic thermohaline circulation with sea surface temperature. *Journal of Climate*, **17**, 1605–14.

Latour, R.J., Gartland, J., Bonzek, C.F., *et al*. (2008). The trophic dynamics of summer flounder (*Paralichthys dentatus*) in Chesapeake Bay. *Fishery Bulletin*, **106**, 47–57.

Lau, N.-C. (1997). Interactions between global SST anomalies and the midlatitude atmospheric circulation. *Bulletin of the American Meteorological Society*, **78**, 21–33.

Lavery, A.C., Wiebe, P.H., Stanton, T.K., *et al*. (2007). Determining dominant scatterers of sound in mixed zooplankton populations. *Journal of the Acoustical Society of America*, **122**, 3304–26.

Laws, R.M. (1977). Seals and whales of the Southern Ocean. *Philosophical Transactions of the Royal Society of London, Series B, Biological Sciences*, **279**, 81–96.

Laws, R.M., (1985). The ecology of the Southern Ocean. *American Scientist*, **73**, 26–40.

Lawson, G.L., Wiebe, P.H., Ashjian, C.J., *et al.* (2004). Acoustically-inferred zooplankton distribution in relation to hydrography west of the Antarctic peninsula. *Deep-Sea Research II*, **51**, 2041–72.

Lawson, G.L., Wiebe, P.H., Stanton, T.K., *et al.* (2008a). Euphausiid distribution along the Western Antarctic Peninsula, Part A: development of robust multifrequency acoustic techniques to identify euphausiid aggregations and quantify euphausiid size, abundance, and biomass. *Deep-Sea Research II*, **55**, 412–31.

Lawson, G.L., Wiebe, P.H., Ashjian, C.J., *et al.* (2008b). Euphausiid distribution along the Western Antarctic Peninsula, Part B: distribution of euphausiid aggregations and biomass, and associations with environmental features. *Deep-Sea Research II*, **55**, 432–545.

Le Boeuf, B.J., Crocker, D.E., Costa, D.P., *et al.* (2000). Foraging ecology of northern elephant seals. *Ecological Monographs*, **70**, 353–82.

Le Fèvre, J., Legendre, L., and Rivkin, R.B. (1998). Fluxes of biogenic carbon in the Southern Ocean: roles of large microphagous zooplankton. *Journal of Marine Systems*, **17(1–4)**, 325–45.

Le Masson du Parc, F. (1727). Pêches & pêcheurs du domaine marimes aquitain au XVIII siècle. Procès Verbaux des visites faites par ordre du Roy concernant la pasche en mer. Les Editions de l'Entre-deux-Mers.

Le Quéré, C., Harrison, S.P., Prentice, I.C., *et al.* (2005). Ecosystem dynamics based on plankton functional types for global ocean biogeochemistry models. *Global Change Biology*, **11**, 2016–40.

Lebel, L., Anderies, J.M., Campbell, B., *et al.* (2006). Governance and the capacity to manage resilience in regional social-ecological systems. *Ecology and Society*, **11(1)**, art:19.

Lee, C.G., Farrell, A.P., Lotto, A., *et al.* (2003). The effect of temperature on swimming performance and oxygen consumption in adult sockeye (*Oncorhynchus nerka*) and coho (*O. kisutch*) salmon stocks. *Journal of Experimental Biology*, **206**, 3239–51.

Lees, K., Pitois, S., Scot, C., *et al.* (2006). Characterizing regime shifts in the marine environment. *Fish and Fisheries*, **7**, 104–27.

Leggett, W.C. and DeBlois, E. (1994). Recruitment in marine fishes: is it regulated by starvation and predation in the egg and larval stages?. *Netherlands Journal of Sea Research*, **32**, 119–34.

Leggett, W.C. and Whitney, R.R. (1972). Water temperature and the migrations of American shad. *Fishery Bulletin*, **70**, 659–70.

Lehodey, P. (2000). Impacts of the *El Niño* Southern Oscillation on tuna populations and fisheries in the tropical Pacific Ocean. 13th Standing Committee on Tuna and Billfish, Noumea, 5–12 July 2000, *Secretariat of the Pacific Community, Noumea, Working Paper*, **RG-1**, 32pp.

Lehodey, P. (2001). The pelagic ecosystem of the tropical Pacific Ocean: dynamic spatial modelling and biological consequences of ENSO. *Progress in Oceanography*, **49**, 439–68.

Lehodey, P. (2004). Climate and fisheries: an insight from the central Pacific Ocean. In: Stenseth, N.C., Ottersen, G., Hurrell, J.W., *et al.* eds. *Marine ecosystems and climate variation: the North Atlantic, a comparative perspective*. Oxford University Press, Oxford, pp. 137–46.

Lehodey, P., Bertignac, M., Hampton, J., *et al.* (1997). *El Niño* Southern Oscillation and tuna in the western Pacific. *Nature*, **389**, 715–8.

Lehodey, P., Chai, F., and Hampton, J. (2003). Modelling climate-related variability of tuna populations from a coupled ocean-biogeochemical-populations dynamics model. *Fisheries Oceanography*, **12**, 483–94.

Lehodey, P., Alheit, J., Barange, M., *et al.* (2006). Climate variability, fish and fisheries. *Journal of Climate*, **19**, 5009–30.

Lehodey, P., Senina, I., Sibert, J., *et al.* (2007). Preliminary forecasts of population trends for Pacific bigeye tuna under the SRES A2 IPCC scenario. *1st International CLIOTOP Symposium, 3–7 Dec 2007*, La Paz, Mexico.

Lehodey, P., Senina, I. and Murtugudde R. (2008). A spatial ecosystem and populations dynamics model (SEAPODYM): tuna and tuna-like populations. *Progress in Oceanography*, **78**, 304–18.

Leibol, M.A., Chase, J.M., Shurin, J.B., *et al.* (1997). Species turnover and the regulation of trophic structure. *Annual Review of Ecology and Systematics*, **28**, 467–94.

Leis, J.M. (2006). Are larvae of demersal fishes plankton or nekton?. *Advances in Marine Biology: An Annual Review*, **51**, 57–141.

Leising, A.W., Pierson, J.J., Halsband-Lenk, C., *et al.* (2005). Copepod grazing during spring blooms: does *Calanus pacificus* avoid harmful diatoms?. *Progress in Oceanography*, **67**, 384–405.

Lennert-Cody, C.E. and Franks, P.J.S. (1999). Plankton patchiness in high-frequency internal waves. *Marine Ecology Progress Series*, **186**, 59–66.

Lenton, R. (2002). Managing natural resources in the light of climate variability. *Natural Resources Forum*, **26**, 185–94.

Letcher, B.H., Rice, J.A., Crowder, L.B., *et al.* (1996). Variability in survival of larval fishes: disentangling components of variance with a generalized individual-based model. *Canadian Journal of Fisheries and Aquatic Sciences*, **53**, 787–801.

Leterme, S.C., Seurant, L., and Edwards, M. (2006). Differential contribution of diatoms and dinoflagellates to phytoplankton biomass in the NE Atlantic Ocean and the North Sea. *Marine Ecology Progress Series*, **312**, 57–65.

Lett, C., Roy, C., Levasseur, A., *et al.* (2006). Simulation and quantification of enrichment and retention processes in the southern Benguela upwelling ecosystem. *Fisheries Oceanography*, **15**, 363–72.

Lett, C., Rose, K.A., and Megrey, B.A. (2009). Biophysical models. In: Checkley, D., Alheit, J., Oozeki, Y., *et al.* eds. *Climate change and small pelagic fish.* Cambridge University Press, Cambridge.

Levin, L.A. (2006). Recent progress in understanding larval dispersal: new directions and digressions. *Integrative and Comparative Biology*, **46**, 282–97.

Levins, R. (1966). The strategy of model building in population biology. *American Scientist*, **54**, 421–31.

Levinton, J.S. (1983). The latitudinal compensation hypothesis: growth data and a model of latitudinal growth differentiation based upon energy budgets. I. Interspecific comparison of *Ophryotrocha* (Polychaeta: Dorvilleidae). *Biological Bulletin*, **165**, 686–98.

Levitus, S., Antonov, J.I., Wang, J.L., *et al.* (2001). Anthropogenic warming of earth's climate system. *Science*, **292**, 267–70.

Levitus, S., Antonov, J.I. and Boyer, T. (2005). Warming of the world ocean, 1955–2003. *Geophysical Research Letters*, **32**, L02604, doi:10.1029/2004GL021592.

Li, X.W., McGillicuddy, D.J., Durbin, E.G., *et al.* (2006). Biological control of the vernal population increase of *Calanus finmarchicus* on Georges Bank. *Deep-Sea Research II*, **53**, 2632–55.

Lilly, G. (2003). More on Smith Sound cod mortality. *Fish Bytes*, **9(3)**, 2.

Lindeman, R. (1942). The trophic-dynamic aspect of ecology. *Ecology*, **23**, 399–418.

Lindeque, P.K., Hay, S.J., Heath, M.R., *et al.* (2006). Integrating conventional microscopy and molecular analysis to analyse the abundance and distribution of four *Calanus* congeners in the North Atlantic. *Journal of Plankton Research*, **28**, 221–38.

Lindley, J.A. (1987). Continuous Plankton Records: the geographical distribution and seasonal cycles of decapod crustacean larvae and pelagic post-larvae in the north-eastern Atlantic Ocean and the North Sea, 1981–3. *Journal of the Marine Biological Association of the United Kingdom*, **67**, 145–67.

Lindley, J.A. and Williams, R. (1994). Relating plankton assemblages to environmental variables using instruments towed by ships-of-opportunity. *Marine Ecology Progress Series*, **107**, 245–62.

Link, J.S. (2002). Ecological considerations in fisheries management: when does it matter? *Fisheries*, **27(4)**, 10–17.

Link, J.S. (2005). Translating ecosystem indicators into decision criteria. *ICES Journal of Marine Science*, **62**, 569–76.

Link, J.S. and Ford, M.D. (2006). Widespread and persistent increase of Ctenophora in the continental shelf ecosystem off NE USA. *Marine Ecology Progress Series*, **320**, 153–9.

Link, J.S. Stockhausen, W.T., and Methratta, E.T. (2005). Food-web theory in marine ecosystems. In: Belgrano, A., Scharler, U.M., Dunne, J., *et al.* eds. *Aquatic food webs: an ecosystem approach*, pp. 98–113. Oxford University Press, Oxford, 262pp.

Litchman, E. and Klausmeier, C.A. (2008). Trait-based community ecology of phytoplankton. *Annual Review of Ecology, Evolution, and Systematics*, **39**, 615–39.

Litzow, M.A. and Ciannelli, L. (2007). Oscillating trophic control induces community reorganisation in a marine ecosystem. *Ecology Letters*, **10**, 1124–34.

Liu, H. and Hopcroft, R.R. (2006a). Growth and development of *Metridia pacifica* (Copepoda: Calanoida) in the Northern Gulf of Alaska. *Journal of Plankton Research*, **28**, 769–81.

Liu, H. and Hopcroft, R.R. (2006b). Growth and development of *Neocalanus flemingeri/plumchrus* in the northern Gulf of Alaska: validation of the artificial-cohort method in cold waters. *Journal of Plankton Research*, **28**, 87–101.

Liu, H. and Hopcroft, R.R. (2007). A comparison of seasonal growth and development of the copepods *Calanus marshallae* and *C. pacificus* in the northern Gulf of Alaska. *Journal of Plankton Research*, **29**, 569–81.

Liu, H., Dagg, M.J., and Strom, S. (2005). Grazing by the calanoid copepod *Neocalanus cristatus* on the microbial food web in the coastal Gulf of Alaska. *Journal of Plankton Research*, **27**, 647–62.

Liu, Z. and Alexander, M. (2007). Atmospheric bridge, oceanic tunnel, and global climatic teleconnections. *Reviews of Geophysics*, **45**, RG2005, doi:10.1029/2005RG000172.

Livingston, P.A. and Methot, R.D. (1998). Incorporation of predation into a population assessment model of eastern Bering Sea walleye pollock. In: *Fishery stock assessment models*, pp. 663–78. *Alaska Sea Grant College Program Publication*, **AK-SG-98–01**, 1037pp.

Lluch-Belda, D., Crawford, R.J.M., Kawasaki, T., *et al.* (1989). Worldwide fluctuations of sardine and anchovy stocks: the regime problem. *South African Journal of Marine Science*, **8**, 195–205.

Lluch-Belda, D., Lluch-Cota, D.B., Hernandez-Vazquez, S., *et al.* (1991). Sardine and anchovy spawning as related to temperature and upwelling in the California Current system. *California Cooperative Oceanic Fisheries Investigations Report*, **32**, 105–11.

Lluch-Belda, D., Shwartzlose, R., Serra, R., *et al.* (1992). Sardine and anchovy regime fluctuations of abundance in four regions of the World Ocean: a workshop report. *Fisheries Oceanography*, **2**, 339–43.

Lluch-Cota, D.B., Hernández-Vázquez, S., and Lluch-Cota, S.E. (1997). Empirical investigation on the relationship between climate and small pelagic global regimes and *El Niño*-Southern Oscillation (ENSO). *FAO Fisheries Circular*, **934**, 48pp.

Lo, W.T., Purcell, J.E., Hung, J.J., *et al.* (2008). Enhancement of jellyfish (*Aurelia aurita*) populations by extensive aquaculture rafts in a coastal lagoon in Taiwan. *ICES Journal of Marine Science*, **65**, 453–61.

Loeb, V., Siegel, V., Holm-Hansen, O., *et al.* (1997). Effects of sea-ice extent and krill or salp dominance on the Antarctic food web. *Nature*, **387**, 897–900.

Loeng, H. (1979). A review of the sea ice conditions of the Barents Sea and the area west of Spitsbergen. *Fisken og Havet*, **2**, 29–75 [in Norwegian with English abstract].

Loeng, H. and Drinkwater, K. (2007). An overview of the ecosystems of the Barents and Norwegian Seas and their response to climate variability. *Deep-Sea Research II*, **54**, 2478–500.

Loeng, H., Brander, K., Carmack, E., *et al.* (2005). Chapter 8: marine systems. In: *Arctic climate impact assessment*. Cambridge University Press, Cambridge, pp. 451–538.

Logerwell, E.A. and Smith, P. (2001). Mesoscale eddies and survival of late stage Pacific sardine (*Sardinops sagax*) larvae. *Fisheries Oceanography*, **10**, 13–25.

Logerwell, E.A., Mantua, N., Lawson, P.W., *et al.* (2003). Tracking environmental processes in the coastal zone for understanding and predicting Oregon coho (*Oncorhynchus kisutch*) marine survival. *Fisheries Oceanography*, **12**, 554–68.

Longhurst, A.R. (1998). *Ecological geography of the sea*. Academic Press, London, 398pp.

Loose, C.J. and Dawidowicz, P. (1994). Trade-offs in diel vertical migration by zooplankton – the costs of predator avoidance. *Ecology*, **75**, 2255–63.

López-Urrutia, A., Harris, R.P., and Smith, T. (2004). Predation by calanoid copepods on the appendicularian *Oikopleura dioica*. *Limnology and Oceanography*, **49**, 303–7.

Loreau, M., Naeem, S., Inchausti, P., *et al.* (2001). Biodiversity and ecosystem functioning: current knowledge and future challenges. *Science*, **294**, 804–8.

Lorenzen, K. and Enberg, K. (2002). Density-dependent growth as a key mechanism in the regulation of fish populations: evidence from among-population comparisons. *Proceedings of the Royal Society of London, Series B, Biological Sciences*, **269**, 49–54.

Lotze, H.K. (2007). Rise and fall of fishing and marine resource use in the Wadden Sea, southern North Sea. *Fisheries Research*, **87**, 208–18.

Lotze, H.K., Lenihan, H.S., Bourque, B.J., *et al.* (2006). Depletion, degradation, and recovery potential of estuaries and coastal seas. *Science*, **312**, 1806–9.

Lougee, L.A., Bollens, S.M., and Avent, S.R. (2002). The effects of haloclines on the vertical distribution and migration of zooplankton. *Journal of Experimental Marine Biology and Ecology*, **278**, 111–34.

Lough, R.G. and Boltz, G.R. (1989). The movement of cod and haddock larvae onto the shoals of Georges Bank. *Journal of Fish Biology*, **35**, 71–9.

Lough, R.G. and Broughton, E.A. (2007). Development of micro-scale frequency distributions of plankton for inclusion in foraging models of larval fish, results from a video plankton recorder. *Journal of Plankton Research*, **29**, 7–17.

Lough, R.G., Buckley, L.J., Werner, F.E., *et al.* (2005). A general biophysical model of larval cod (*Gadus morhua*) growth applied to populations on Georges Bank. *Fisheries Oceanography*, **14**, 241–62.

Lough, R.G., Broughton, E.A., Buckley, L.J., *et al.* (2006). Modeling growth of Atlantic cod larvae on the southern flank of Georges Bank in the tidal front circulation during May (1999). *Deep-Sea Research II*, **53**, 2771–88.

Lough, R.G. and Mountain, D.G. (1996). Effect of small-scale turbulence on feeding rates of larval cod and haddock in stratified water on Georges Bank. *Deep-Sea Research II*, **43**, 1745–72.

Lough, R.G., Smith, W.G., Werner, F.E., *et al.* (1994). Influence of wind-driven advection on interannual variability in cod egg and larval distributions on Georges Bank: 1982 vs 1985. *ICES Marine Science Symposia*, **198**, 356–78.

Loukos, H., Monfray, P., Bopp, L., *et al.* (2003). Potential changes in skipjack tuna (*Katsuwonus pelamis*) habitat from a global warming scenario: modelling approach and preliminary results. *Fisheries Oceanography*, **12**, 474–82.

Ludwig, D., Hilborn, R., and Walters, C. (1993). Uncertainty, resource exploitation, and conservation: lessons from history. *Science*, **260**, p.17.

Luo, J., Ortner, P.B., Forcucci, D., *et al.* (2000). Diel vertical migration of zooplankton and mesopelagic fish in the Arabian Sea. *Deep-Sea Research II*, **47**, 1451–73.

Lutcavage, M.E., Brill, R.W., Skomal, G.B., *et al.* (1999). Results of pop-up satellite tagging of spawning size class fish in the Gulf of Maine: do North Atlantic bluefin tuna spawn in the mid-Atlantic?. *Canadian Journal of Fisheries and Aquatic Sciences*, **56**, 173–7.

Lynam, C.P., Hay, S.J., and Brierley, A.S. (2004). Interannual variability in abundance of North Sea jellyfish and links to the North Atlantic Oscillation. *Limnology and Oceanography*, **49**, 637–43.

Lynam, C.P., Hay, S.J., and Brierley, A.S. (2005a). Jellyfish abundance and climatic variation: contrasting responses in oceanographically distinct regions of the North Sea, and possible implications for fisheries. *Journal of the Marine Biological Association of the United Kingdom*, **85**, 435–50.

Lynam, C.P., Heath, M.R., Hay, S.J., *et al.* (2005b). Evidence for impacts by jellyfish on North Sea herring recruitment. *Marine Ecology Progress Series*, **298**, 157–67.

Lynam, C.P., Gibbons, M.J., Axelsen, B.E., *et al.* (2006). Jellyfish overtake fish in a heavily fished ecosystem. *Current Biology*, **16**, R492–3.

Lynch, D.R., Gentlemen, W.C., McGillicuddy, D.J., Jr., *et al.* (1998). Biological/physical simulations of *Calanus finmarchicus* population dynamics in the Gulf of Maine. *Marine Ecology Progress Series*, **169**, 189–210.

Lynch, D.R., McGillicuddy, D.J., and Werner, F.E. (2008). Preface. Skill assessment for coupled biological/physical models of marine systems. *Journal of Marine Systems*, doi: 10.1016/j.jmarsys.2008.05.002.

Lyons, M.M., Smolowitz, R., Dungan, C.F., *et al.* (2006). Development of a real time quantitative PCR assay for the hard clam pathogen Quahog Parasite Unknown (QPX). *Diseases of Aquatic Organisms*, **72**, 45–52.

MacCall, A. (1990). *Dynamic geography of marine fish populations*. Washington Sea Grant Program, University of Washington Press, Seattle, WA.

MacCall, A.B. (2002). Fishery management and stock rebuilding prospects under conditions of low frequency variability and species interactions. *Bulletin of Marine Science*, **70**, 613–28.

Mace, P.M. and Sissenwine, M.P. (1993). How much spawning per recruit is enough?. *Canadian Special Publication of Fisheries and Aquatic Sciences*, **120**, 101–18.

Mackas, D.L. and Tsuda, A. (1999). Mesozooplankton in the eastern and western subarctic Pacific: community structure, seasonal life histories, and interannual variability. *Progress in Oceanography*, **43**, 335–63.

Mackas, D.L., Sefton, H., Miller, C.B., *et al.* (1993). Vertical habitat partitioning by large calanoid copepods in the oceanic subarctic Pacific during spring. *Progress in Oceanography*, **32**, 259–94.

Mackas, D.L., Goldblatt, R., and Lewis, A.G. (1998). Interdecadal variation in developmental timing of *Neocalanus plumchrus* populations at Ocean Station P in the subarctic North Pacific. *Canadian Journal of Fisheries and Aquatic Sciences*, **55**, 1878–93.

Mackas, D.L., Peterson, W.T., and Zamon, J.E. (2004). Comparisons of interannual biomass anomalies of zooplankton communities along the continental margins of British Columbia and Oregon. *Deep-Sea Research II*, **51**, 875–96.

Mackas, D.L., Peterson, W.T., Ohman, M.D., *et al.* (2006). Zooplankton anomalies in the California Current system before and during the warm ocean conditions of 2005. *Geophysical Research Letters*, **33**, L22S07, doi:10.1029/2006GL027930.

Mackas, D.L., Batten, S., and Trudel, M. (2007). Effects on zooplankton of a warmer ocean: recent evidence from the North Pacific. *Progress in Oceanography*, **75**, 223–52.

MacKenzie, B.R. and Kiørboe, T. (1995a). Encounter rates and swimming behavior of pause-travel and cruise larval fish predators in calm and turbulent laboratory environments. *Limnology and Oceanography*, **40**, 1278–89.

MacKenzie, B.R. and Kiørboe, T. (1995b). Turbulence-enhanced prey encounter rates in larval fish: effects of spatial scale, larval behaviour and size. *Journal of Plankton Research*, **17**, 2319–31.

MacKenzie, B.R. and Kiørboe, T. (2000). Larval fish feeding and turbulence: a case for the downside. *Limnology and Oceanography*, **45**, 1–10.

MacKenzie, B.R. and Köster, F.W. (2004). Fish production and climate: sprat in the Baltic Sea. *Ecology*, **85**, 784–94.

MacKenzie, B.R. and Leggett, W.C. (1991). Quantifying the contribution of small-scale turbulence to the encounter rates between larval fish and their zooplankton prey – effects of wind and tide. *Marine Ecology Progress Series*, **73**, 149–60.

MacKenzie, B.R. and Myers, R.A. (2007). The development of the northern European fishery for north Atlantic bluefin tuna *Thunnus thynnus* during 1900–1950. *Fisheries Research*, **87**, 229–39.

MacKenzie, B.R., Miller, T.J., Cyr, S., *et al.* (1994). Evidence for a dome-shaped relationship between turbulence and larval fish ingestion rates. *Limnology and Oceanography*, **39**, 1790–9.

MacKenzie, B.R., Hinrichsen, H.-H., Plikshs, M., *et al.* (2000). Quantifying environmental heterogeneity: habitat size necessary for successful development of cod *Gadus morhua* eggs in the Baltic Sea. *Marine Ecology Progress Series*, **193**, 143–56.

MacKenzie, B.R., Gislason, H., Möllmann, C., *et al.* (2007a). Impact of 21st century climate change on the Baltic Sea fish community and fisheries. *Global Change Biology*, **13**, 1–20.

MacKenzie, B.R., Bager, M., Ojaveer, H., *et al.* (2007b). Multi-decadal scale variability in the eastern Baltic cod fishery 1550–1860-evidence and causes. *Fisheries Research*, **87**, 106–19.

Macpherson, E. (2003). Species range size distribution for some marine taxa in the Atlantic Ocean. Effects of latitude and depth. *Biological Journal of the Linnean Society*, **80**, 437–55.

Macpherson, E. and Duarte, C.M. (1994). Patterns of species richness, size and latitudinal range of eastern Atlantic fishes. *Ecography*, **17**, 242–8.

Madin, L.P., Bollens, S.M., Horgan, E., *et al.* (1996). Voracious planktonic hydroids: unexpected predatory impact on a coastal marine ecosystem. *Deep-Sea Research II*, **43**, 1823–9.

Mangel, M. (2000). Trade-offs between fish habitat and fishing mortality and the role of reserves. *Bulletin of Marine Science*, **66**, 663–74.

Manizza, M., LeQuéré, C., Watson, A.J., *et al.* (2005). Bio-optical feedbacks among phytoplankton, upper ocean physics and sea-ice in a global model. *Geophysical Research Letters*, **32**, 1–4.

Mann, K.H. and Lazier, J.R.N. (1996). *Dynamics of marine ecosystems: biological-physical interactions in the oceans*. Blackwell Science, Cambridge, MA, 408pp.

Mantua, N., Haidvogel, D., Kushnir, Y. *et al.* (2002). Making the climate connectons: Bridging scales of space and time in the US GLOBEC program. *Oceanography*, **15(2)**, 75–86.

Mantua, N.J., Hare, S.R., Zhang, Y., *et al.* (1997). A Pacific interdecadal climate oscillation with impacts on salmon production. *Bulletin of the American Meteorological Society*, **78**, 1069–79.

Margalef, R. (1997). Turbulence and marine life. *Scientia Marina*, **61(Suppl. 1)**, 109–23.

Mariani, P., MacKenzie, B.R., Visser, A.W., *et al.* (2007). Individual-based simulations of larval fish feeding in turbulent environments. *Marine Ecology Progress Series*, **347**, 1551–69.

Marr, J.W.S. (1962). The natural history and geography of the Antarctic krill *Euphausia superba* Dana. *Discovery Reports*, **32**, 37–465.

Marrasé, C., Costello, J.H., Granata, T., *et al.* (1990). Grazing in a turbulent environment: energy dissipation, encounter rats, and efficacy of feeding currents in *Centropages hamatus*. *Proceedings of the National Academy of Sciences of the United States of America*, **87**, 1653–7.

Marsh, R., Petrie, B., Weidman, C.R., *et al.* (1999). The Middle Atlantic Bight tilefish fill of 1882. *Fisheries Oceanography*, **8**, 39–49.

Marshall, C.T., Needle C.L., Thorsen, A., *et al.* (2006). Systematic bias in estimates of reproductive potential of an Atlantic cod (*Gadus morhua*) stock: implications for stock-recruit theory and management. *Canadian Journal of Fisheries and Aquatic Sciences*, **63**, 980–94.

Marteinsdottir, G. and Steinarsson, A. (1998). Maternal influence on the size and viability of cod (*Gadus morhua* L.) eggs and larvae. *Journal of Fish Biology*, **52**, 1241–58.

Martin, A.P. (2003). Phytoplankton patchiness: the role of lateral stirring and mixing. *Progress in Oceanography*, **57**, 125–74.

Martindale, M.Q. (1987). Larval reproduction in the ctenophore *Mniopsis mccradyi* (Order Lobata). *Marine Biology*, **94**, 409–14.

Mathew, S. (2003). Small-scale fisheries perspectives on an ecosystem-based approach to fisheries management. In: Sinclair, M. and Valdimarsson, G., eds. *Responsible fisheries in the marine ecosystem*, FAO and CAB International, Rome, Italy, and Wallingford, UK, pp. 47–63.

Matsuda, H., Wada, T., Takeuchi, Y., *et al.* (1992). Model analysis of the effect of environmental fluctuation on the species replacement pattern of pelagic fishes under interspecific competition. *Population Ecology*, **34**, 309–19.

Mauchline, J. (1998). The biology of calanoid copepods. *Advances in Marine Biology: An Annual Review*, **33**, 710pp.

Maury, O. and Lehodey, P., eds. (2005). Climate Impacts on Oceanic TOp Predators (CLIOTOP). Science Plan and Implementation Strategy. *GLOBEC Report*, **18**, ii, 42pp.

Maury, O., Faugeras, B., Shin, Y.-J., *et al.* (2007a). Modeling environmental effects on the size-structured energy flow through marine ecosystems, Part 1: the model. *Progess in Oceanography*, **74(4)**, 479–99.

Maury, O., Shin, Y.-J., Faugeras, B., *et al.* (2007b). Modelling environmental effects on the size-structured energy flow through marine ecosystems, Part 2: simulations. *Progress in Oceanography*, **74(4)**, 500–14.

May, R.M., Levin, S.A., and Sugihara, G. (2008). Complex systems: ecology for bankers. *Nature*, **451**, 893–5.

Mayor, D.J., Anderson, T.R., Irigoien, X., *et al.* (2006). Feeding and reproduction of *Calanus finmarchicus* during non-bloom conditions in the Irminger Sea. *Journal of Plankton Research*, **28**, 1167–79.

Mazzocchi, M.G. and Paffenhofer, G.A. (1999). Swimming and feeding behaviour of the planktonic copepod *Clausocalanus furcatus*. *Journal of Plankton Research*, **21**, 1501–18.

McAllister, M.K., Starr, P.J., Restrepo, V.R., *et al.* (1999). Formulating quantitative methods to evaluate fishery-management systems: what fishery process should be modelled and what tradeoffs should be made? *ICES Journal of Marine Science*, **56**, 900–16.

McCaffrey, D. (2000). Dynamique de population de *Calanus finmarchicus* (Gunnerus, 1765) dans deux fjords norvégiens. Conséquences de la prédation sur la production secondaire et la stratégie de vie. *DEA Océanologique et environnement marin*, Université Pierre et Marie Curie, Paris VI, pp. 1–32.

McCall, A.D. (2009). A short scientific history of the fisheries. In: D. Checkley, J. Alheit, Y. Oozeki *et al.* eds. *Climate change and small pelagic fish*. Cambridge University Press, Cambridge.

McCann, K.S., Hastings, A., and Huxel, G.R. (1998). Weak trophic interactions and the balance of nature. *Nature*, **395**, 794–8.

McCann, K.S., Rasmussen, J.B., and Umbanhowar, J. (2005). The dynamics of spatially coupled food webs. *Ecology Letters*, **8**, 513–23.

McConnell, B.J., Chambers, C., and Fedak, M.A. (1992a). Foraging ecology of southern elephant seals in relation to the bathymetry and productivity of the Southern Ocean. *Antarctic Science*, **4**, 393–8.

McConnell, B.J., Chambers, C., Nicholas, K.S., *et al.* (1992b). Satellite tracking of grey seals (*Halichoerus grypus*). *Journal of Zoology*, **226**, 271–82.

McFarlane, G.A. and Beamish, R.J. (2001). The re-occurrence of sardines off British Columbia characterises the dynamic nature of regimes. *Progress in Oceanography*, **49**, 151–65.

McFarlane, G.A., Smith, P.E., Baumgartner, T.R. *et al.* (2002). Climate variability and Pacific sardine populations and fisheries. In: McGinn, N.A., ed. *Fisheries in a changing climate*, pp. 195–214. American Fisheries Society Symposium **32**, Bethesda, Maryland.

McFarlane, N.A., Boer, G.J., Blanchet, J.P., *et al.* (1992). The Canadian climate centre second-generation general circulation model and its equilibrium climate. *Journal of Climate*, **5**, 1013–44.

McGillicuddy, D.J., Lynch, D.R., Wiebe, P., *et al.* (2001). Evaluating the synopticity of the US GLOBEC Georges Bank Broad-scale sampling pattern with Observational System Simulation Experiments. *Deep-Sea Research II*, **48**, 483–99.

McGinn, A.P. (1999). *Worldwatch paper No.145: safeguarding the health of oceans*. Worldwatch Institute, Washington, DC, 87pp.

McGoodwin, J.R. (1990). Crisis in the world's fisheries: people, problems, and policies. Stanford University Press, Stanford, Pao Alto, CA.

McGoodwin, J.R. (2001). Understanding the cultures of fishing communities: a key to fisheries management and food security. *FAO Fisheries Technical Paper*, **401**, FAO, Rome, 287pp.

McGoodwin, J.R. (2007). Effects of climatic variability on three fishing economies in high-latitude regions: implications for fisheries policies. *Marine Policy*, **31**, 40–55.

McGowan, J.A. and Brown, D.M. (1966). A new opening-closing paired zooplankton net. *Scripps Institute of Oceanography Reference*, **66/23**, 1–56.

McGowan, J.A. Bograd, S.J., Lynn, R.J., *et al.* (2003). The biological response to the 1977 regime shift in the California Current. *Deep-Sea Research II*, **50**, 2567–82.

McGregor, H.V., Dima, M., Fisher, H.W., *et al.* (2007). Rapid 20th-century increase in coastal upwelling off northwest Africa. *Science*, **315**, 637–9.

McKinnell, S.M. (2007). Expectations for marine survival of Chilko Lake sockeye returning in 2007. In: *State of the Pacific Ocean 2006, PSARC Ocean Status Report* **2007/001**.

McKinnell, S.M. (2008). Fraser River sockeye salmon productivity and climate; a re-analysis that avoids an undesirable property of Ricker's curve. *Progress in Oceanography*, **77**, 146–54.

McKinnell, S.M. and Mackas, D.L. (2003). Intercalibrating SCOR, NORPAC and bongo nets and the consequences for interpreting decadal-scale variation in zooplankton biomass in the Gulf of Alaska. *Fisheries Oceanography*, **12**, 126–33.

McLaren, I.A. (1963). Effects of temperature on growth of zooplankton, and the adaptive value of vertical migration. *Journal of the Fisheries Research Board of Canada*, **20**, 685–727.

McMahon, C.R., Autret, E., Houghton, J.D.R., *et al.* (2005). Animal-borne sensors successfully capture the real-time thermal properties of ocean basins. *Limnology and Oceanography Methods*, **3**, 392–8.

McManus, M.A., Alldredge, A.L., and Barnard, A.H., *et al.* (2003). Characteristics, distribution and persistence of thin layers over a 48 hour period. *Marine Ecology Progress Series*, **261**, 1–19.

McPhaden, M.J. and Picaut, J. (1990). *El Niño*-Southern Oscillation displacements of the Western Equatorial Pacific warm pool. *Science*, **50**, 1385–8.

McPhaden, M.J., Zebiak, S.E., and Glantz, M.H. (2006). ENSO as an integrating concept in earth science. *Science*, **314**, 1740–5.

McQueen, D.J., Post, J.R., and Mills, E.L. (1986). Trophic relationships in freshwater pelagic ecosystems. *Canadian Journal of Fisheries and Aquatic Sciences*, **43**, 1571–81.

McRoy, C.P., Hood, D.W., Coachman, L.K., *et al.* (1986). Processes and resources of the Bering Sea Shelf (PROBES): the development and accomplishments of the project. *Continental Shelf Research*, **5**, 5–21.

Mee, L.D. (1992). The Black Sea in crisis: a need for concerted international action. *Ambio*, **21**, 278–86.

Megrey, B.A., Hinckley, S., and Dobbins, E.L. (2002). Using scientific visualization tools to facilitate analysis of multi-dimensional data from a spatially explicit, biophysical, individual based model of marine fish early life history. *ICES Journal of Marine Science*, **59**, 203–15.

Megrey, B.A., Ito, S., Hay, D.E., *et al.* (2007a). Basin-scale differences in lower and higher trophic level marine ecosystem response to climate impacts using a coupled biogeochemical-fisheries bioenergetics model. *Ecological Modelling*, **202**, 196–210.

Megrey, B.A., Rose, K.A., Klumb, R.A., *et al.* (2007b). A bioenergetics-based population dynamics model of Pacific herring (*Clupea harengus pallasi*) coupled to a lower trophic level nutrient-phytoplankton-zooplankton model: description, calibration, and sensitivity analysis. *Ecological Modelling*, **202**, 144–64.

Melack, J.M., Dozier, J., Goldman, C.R., *et al.* (1997). Effects of climate change on inland waters of the Pacific Coastal Mountains and Western Great Basin of North America. *Hydrology Proceedings*, **11**, 971–92.

Melian, C.J. and Bascompte, J. (2004). Food web cohesion. *Ecology*, **85**, 352–8.

Melle, W., Ellertsen, B., and Skjoldal, H.R. (2004). Zooplankton: the link to higher trophic levels, chapter 6. In: Skjoldal, H.R., Saetre, R., Faerno, A., *et al.* eds. *The Norwegian Sea ecosystem*. Tapir Academic Press, Trondheim, Norway, pp. 137–202.

Ménard, F., Labrune, C., Shin, Y.-J., *et al.* (2006). Opportunistic predation in tuna: a size-based approach. *Marine Ecology Progress Series*, **323**, 223–31.

Ménard, F., Lorrain, A., Potier, M., *et al.* (2007). Isotopic evidence of distinct feeding ecologies and movement patterns in two migratory predators (yellowfin tuna and swordfish) of the western Indian Ocean. *Marine Biology*, **153(2)**, 141–52.

Mendelssohn, R. and Schwing, F.B. (2002). Common and uncommon trends in SST and wind stress in the California and Peru-Chile current systems. *Progress in Oceanography*, **53**, 141–62.

Menge, B.A. (1992). Community regulation: under what conditions are bottom-up factors important on rocky shores?. *Ecology*, **73(2)**, 755–65.

Meredith, M.P. and King, J.C. (2005). Rapid climate change in the ocean west of the Antarctic Peninsula during the second half of the 20th century. *Geophysical Research Letters*, **32**, L19604, 10.1029/2005GL024042.

Metaxas, A. (2001). Behaviour in flow: perspectives on the distribution and dispersion of meroplanktonic larvae in the water column. *Canadian Journal of Fisheries and Aquatic Sciences*, **58**, 86–98.

Metcalfe, J.D. (2006). Fish population structuring in the North Sea: understanding processes and mechanisms from studies of the movements of adults. *Journal of Fish Biology*, **69**, 48–65.

Meyer-Harms, B., Irigoien, X., Head, R., *et al.* (1999). Selective feeding on natural phytoplankton by *Calanus finmarchicus* before, during, and after the 1997 spring bloom in the Norwegian Sea. *Limnology and Oceanography*, **44**, 154–65.

Millennium Ecosystem Assessment. (2003). Concepts of ecosystem value and valuation approaches. In: Alcamo, J., *et al.*, eds. *Ecosystems and human well-being: a framework for assessment*. Island Press, Washington, DC.

Meyer-Harms, B. (2005). *Ecosystems and human well-being: synthesis*. Island Press, Washington, DC.

Miller, A.J., Cayan, D.R., Barnett, T.P., *et al.* (1994). The 1976–77 climate shift of the Pacific Ocean. *Oceanography*, **7**, 21–6.

Miller, A.J., Alexander, M.A., Boer, G.J., *et al.* (2003). Potential feedbacks between Pacific Ocean ecosystems and interdecadal climate variations. *Bulletin of the American Meteorological Society*, **84**, 617–33.

Miller, C.B. and Clemons, M.J. (1988). Revised life history analysis for large grazing copepods in the subarctic Pacific Ocean. *Progress in Oceanography*, **13**, 201–43.

Miller, C.B., Frost, B.W., Batchelder, H.P., *et al.* (1984). Life histories of large, grazing copepods in a subarctic ocean gyre: *Neocalanus plumchrus, Neolcalanus cristatus*, and *Eucalanus bungii* in the Northeast Pacific. *Progress in Oceanography*, **13**, 201–43.

Miller, C.B., Frost, B.W., Wheeler, P.A., *et al.* (1991). Ecological dynamics in the subarctic Pacific, a possibly iron-limited ecosystem. *Limnology and Oceanography*, **36(8)**, 1600–13.

Miller, C.B., Lynch, D.R., Carlotti, F., *et al.* (1998). Coupling of an individual-based population dynamic model of *Calanus finmarchicus* to a circulation model for the Georges Bank region. *Fisheries Oceanography*, **7**, 219–34.

Miller, D.C.M., Moloney, C.L., van der Lingen, C.D., *et al.* (2006). Modelling the effects of physical-biological interactions and spatial variability in spawning and nursery areas on recruitment of sardine in the southern Benguela ecosystem. *Journal of Marine Systems*, **61(3–4)**, 212–29.

Miller, K.A. (2000). Pacific salmon fisheries: climate, information and adaptation in a conflict-ridden context. *Climatic Change*, **45**, 37–61.

Miller, K.A. (2007). Climate variability and tropical tuna: management challenges for highly migratory fish stocks. *Marine Policy*, **31**, 56–70.

Miller, K.A. and Munro, G.R. (2004). Climate and cooperation: a new perspective on the management of shared fish stocks. *Marine Resource Economics*, **19**, 367–93.

Miller, R.L., Schmidt, G.A., and Shindell, D.T. (2006). Forced annular variations in the 20th century Intergovernmental Panel on Climate Change Fourth Assessment Report models. *Journal of Geophysical Research*, **111**, D18101, doi:10.1029/2005JD006323.

Miller, T.J. (2007). Contribution of individual-based coupled physical-biological models to understanding recruitment in marine fish populations. *Marine Ecology Progress Series*, **347**, 127–38.

Miller, T.J., Herra, T., and Leggett, W.C. (1995). An individual-based analysis of the variability of eggs and their newly hatched larvae of Atlantic cod (*Gadus morhua*) on the Scotian Shelf. *Canadian Journal of Fisheries and Aquatic Sciences*, 52, 1088–93.

Miller, T.W. and Brodeur, R.D. (2007). Diets of and trophic relationships among dominant marine nekton within the northern California Current ecosystem. *Fishery Bulletin*, **105**, 548–59.

Mills, C.E. (2001). Jellyfish blooms: are populations increasing globally in response to changing ocean conditions?. *Hydrobiologia*, **451**, 55–68.

Miralto, A., Barone, G., Romano, G., *et al.* (1999). The insidious effects of diatoms on copepod reproduction. *Nature*, **402**, 173–6.

Misund, O.A., Vilhjálmsson, H., and Jákupsstovu, S.H., *et al.* (1998). Distribution, migration and abundance of Norwegian spring spawning herring in relation to the temperature and zooplankton biomass in the Norwegian Sea as recorded by coordinated surveys in spring and summer 1996. *Sarsia*, **83**, 117–27.

Mitani, Y., Bando, T., Takai, N., *et al.* (2006). Patterns of stable carbon and nitrogen isotopes in the baleen of common minke whale *Balaenoptera acutorostrata* from the western North Pacific. *Fisheries Science*, **72**, 69–76.

Mohn, R.K., Fanning, L.P., and MacEachern, W.J. (1998). Assessment of 4VsW cod in 1997 incorporating additional sources of mortality. *Canadian Stock Assessment Secretariat Research Document*, **98/78**.

Mohseni, O., Stefan, H.G., and Eaton, J.G. (2003). Global warming and potential changes in fish habitat in U.S. streams. *Climate Change*, **59**, 389–409.

Moll, A. and Stegert, C. (2007). Modelling *Pseudocalanus elongatus* stage-structured population dynamics embedded in a water column ecosystem model for the northern North Sea. *Journal of Marine Systems*, **64**, 35–46.

Möller, H. (1984). Reduction of a larval herring population by jellyfish predator. *Science*, **224**, 621–2.

Möllmann, C., Diekmann, R., Müller-Karulis, B., *et al.* (2009). The reorganization of a large marine ecosystem due to atmospheric and anthropogenic pressure-a discontinuous regime shift in the central Baltic Sea. *Global Change Biology*, **15(6)**, 1377–93.

Möllmann, C., Kornilovs, G., Fetter, M., *et al.* (2004). Feeding ecology of central Baltic Sea herring and sprat. *Journal of Fish Biology*, **65**, 1563–81.

Möllmann, C. and Köster, F.W. (2002). Population dynamics of calanoid copepods and the implications of their predation by clupeid fish in the central Baltic Sea. *Journal of Plankton Research*, **24**, 959–77.

Möllmann, C., Kornilovs, G., Fetter, M., *et al.* (2005). Climate, zooplankton, and pelagic fish growth in the central Baltic Sea. *ICES Journal of Marine Science*, **62**, 1270–80.

Moloney, C.L., Jarre, A., Arancibia, H., *et al.* (2005). Comparing the Benguela and Humboldt marine upwelling ecosystems with indicators derived from inter-calibrated models. *ICES Journal of Marine Science*, **62**, 493–502.

Moloney, C.L. and Ryan, P.G. (1995). Antarctic marine food webs. *Encyclopedia of environmental biology, vol. 1*. Academic Press, San Diego, CA, pp. 53–69.

Moloney, C.L., Jarre, A., Kimura, S., *et al.* (2010). Dynamics of marine ecosystems: ecological processes. *This volume*. Oxford University Press, Oxford.

Moore, J.K., Doney, C., Kleypas, J.A., *et al.* (2001). An intermediate complexity marine ecosystem models for the global domain. *Deep-Sea Research II*, **49**, 403–62.

Morita, K., Yamamoto, S., Takashima, Y., *et al.* (1999). Effect of maternal growth on egg and size in wild white-spotted char (*Salvelinus luecomaenis*). *Canadian Journal of Fisheries and Aquatic Sciences*, **56**, 1585–9.

Morozov, A.Y., Petrovskii, S.V., and Nezlin, N.P. (2007). Towards resolving the paradox of enrichment: the impact of zooplankton vertical migrations on plankton systems stability. *Journal of Theoretical Biology*, **248**, 501–11.

Moss, J.H. and Beauchamp, D.A. (2007). Functional response of juvenile pink and chum salmon: effects of consumer size and two types of zooplankton prey. *Journal of Fish Biology*, **70**, 610–22.

Moss, J.H., Farley, E.V., Jr., Feldmann, A.M., *et al.* (2009). Spatial distribution, energetic status and food habits of eastern Bering Sea age-0 walleye pollock. *Transactions of the American Fisheries Society*, **138**, 497–505.

Mountain, D., Green J., Sibunka, J., *et al.* (2008). Growth and mortality of Atlantic cod *Gadus morhua* and haddock *Melanogrammus aeglefinus* eggs and larvae on Georges Bank, 1995 to 1999. *Marine Ecology Progress Series* 353:225–242.

Müller, U.K., Stamhuis, E.J, and Videler, J.J. (2000). Hydrodynamics of unsteady fish swimming and the effects of body size: comparing the flow fields of fish larvae and adults. *Journal of Experimental Biology*, **203**, 193–206.

Müller, W.A., Frankignoul, C., and Chouaib, N. (2008). Observed decadal tropical Pacific-North Atlantic teleconnections. *Geophysical Research Letters*, **35**, L24810, doi:10.1029/2008GL035901.

Mueter, F.J. and Litzow, M.A. (2008). Sea ice retreat alters the biogeography of the Bering Sea continental shelf. *Ecological Applications*, **18**, 309–20.

Mueter, F.J., Peterman, R.M., and Pyper, B.J. (2002). Opposite effects of ocean temperature on survival rates of 120 stocks of Pacific salmon (*Oncorhynchus* spp.) in northern and southern areas. *Canadian Journal of Fisheries and Aquatic Sciences*, **59**, 456–63.

Mueter, F.J., Boldt, J., Megrey, B.A., et al. (2007). Recruitment and survival of Northeast Pacific Ocean fish stocks: temporal trends, covariation, and regime shifts. *Canadian Journal of Fisheries and Aquatic Sciences*, **64**, 911–27.

Mueter, F.J., Broms, C., Drinkwater, K.F., et al. (2009). Ecosystem responses to recent oceanographic variability in high-latitude Northern Hemisphere ecosystems. *Progress in Oceanography*, **81**, 93–110.

Mullon, C., Cury, P., and Penven, P. (2002). Evolutionary individual-based model for the recruitment of anchovy (*Engraulis capensis*) in the southern Benguela. *Canadian Journal of Fisheries and Aquatic Sciences*, **59**, 910–22.

Mullon, C., Freon, P., Parada, C., et al. (2003). From particles to individuals: modeling the early stages of anchovy (*Engraulis capensis/encrasicolus*) in the southern Benguela. *Fisheries Oceanography*, **12**, 396–406.

Munk, P. (1992). Foraging behaviour and prey size spectra of larval herring *Clupea harengus*. *Marine Ecology Progress Series*, **80**, 149–58.

Munk, P. (2007). Cross-frontal variation in growth rate and prey availability of larval North Sea cod *Gadus morhua*. *Marine Ecology Progress Series*, **334**, 225–35.

Munk, P. and Kiørboe, T. (1985). Feeding behaviour and swimming activity of larval herring (*Clupea harengus*) in relation to density of copepod nauplii. *Marine Ecology Progress Series*, **24**, 15–21.

Murata, M. (1989). Population assessment, management and fishery forecasting for the Japanese common squid, *Todarodes pacificus*. In: Caddy, J.R., ed. *Marine invertebrate fisheries: their assessment and management*. Wiley, New York, pp. 613–36.

Murawski, S.A. (2007). Ten myths concerning ecosystem approaches to marine resource management. *Marine Policy*, **31**, 681–90.

Murawski, S.A., Maguire, J.J., Mayo, R.K., et al. (1997). Groundfish stocks and the fishing industry. In: Boreman, J., Nakashima, B.S., Wilson, J.A., et al. eds. *Northwest Atlantic groundfish: perspectives on a fishery collapse*. American Fisheries Society, Bethesda, MD.

Murphy, E.J., Cavanagh, R.D., Johnston, N.M., et al. eds. (2008). ICED Science Plan and Implementation Strategy. *GLOBEC Report*, **26**, 68pp.

Murphy, E.J. and Reid, K. (2001). Modelling Southern Ocean krill population dynamics: biological processes generating fluctuations in the South Georgia ecosystem. *Marine Ecology Progress Series*, **217**, 175–89.

Murphy, E.J., Thorpe, S.E., Watkins, J.L., et al. (2004a). Modeling the krill transport pathways in the Scotia Sea: spatial and environmental connections generating the seasonal distribution of krill. *Deep-Sea Research II*, **51**, 1435–56.

Murphy, E.J., Watkins, J.L., Meredith, M.P., et al. (2004b). Southern Antarctic Circumpolar Current Front to the northeast of South Georgia: horizontal advection of krill and its role in the ecosystem. *Journal of Geophysical Research – Oceans*, **109(C1)**, art:C01029.

Murphy, E.J., Watkins, J.L., Trathan, P.N., et al. (2007). Spatial and temporal operation of the Scotia Sea ecosystem: a review of large-scale links in a krill centred food web. *Philosophical Transactions of the Royal Society, Series B, Biological Sciences*, **362**, 113–48.

Murtugudde, R., Beauchamp, J., McClain, C.R., et al. (2002). Effects of penetrative radiation on the upper tropical ocean circulation. *Journal of Climate*, **15**, 470–86.

Myers, R.A. (1998). When do environment-recruitment correlations work?. *Reviews in Fish Biology and Fisheries*, **8**, 285–305.

Myers, R.A. and Drinkwater, K.F. (1989). The influence of Gulf Stream warm core rings on recruitment of fish in the Northwest Atlantic. *Journal of Marine Research*, **47**, 635–56.

Myers, R.A., Bowen, K.G., and Barrowman, N.J. (1999). Maximum reproductive rate of fish at low population sizes. *Canadian Journal of Fisheries and Aquatic Sciences*, **56**, 2404–19.

Myers, R.A., MacKenzie, B.R., Bowen, K.G., et al. (2001). What is the carrying capacity for fish in the ocean? A meta-analysis of populations dynamics of North Atlantic cod. *Canadian Journal of Fisheries and Aquatic Sciences*, **58**, 1464–76.

NAFO. (2007). Report of Scientific Council Meeting 24 October–1 November 2007. *NAFO Scientific Council Summary Document* **07/24** (Revised).

Naimie, C.E., Limeburner, R., Hannah, C.G., et al. (2001). On the geographic and seasonal patterns of the near-surface circulation on Georges Bank – from real and simulated drifters. *Deep-Sea Research II*, **48**, 501–18.

Nakanowatari, T., Ohshima, K.I., and Wakatsuchi, M. (2007). Warming and oxygen decrease of intermediate water in the northwestern North Pacific, originating from the Sea of Okhotsk, 1955–2004. *Geophysical Research Letters*, **34**, L04602, doi:10.1029/2006GL028243.

Napp, J.M., Hopcroft, R.R., Baier, C.T., et al. (2005). Distribution and species-specific egg production of *Pseudocalanus* in the Gulf of Alaska. *Journal of Plankton Research*, **27**, 415–26.

NAS (National Academy of Science). (1987). *Recruitment processes and ecosystem structure of the sea: a report of a workshop*. National Academy Press, Washington, DC, 44pp.

National Research Council, US. (2005). *Valuing ecosystem services: towards better environmental decision-making*. The National Academies Press, Washington, DC, 278pp.

Neira, S. (2008). Assessing the effects of internal (trophic structure) and external (fishing and environment) forcing factors on fisheries off Central Chile: basis for an ecosystem approach to management. Ph.D. thesis, University of Cape Town, Cape Town.

Neira, S., Moloney, C., Cury, P., *et al.* (in preparation). Analyzing changes in the southern Humboldt ecosystem for the period 1970–2004 by means of food web modelling.

Neis, B. and Kean, R. (2003). Why fish stocks collapse. In: Byron, R., ed. *Retrenchment and regeneration in rural Newfoundland*. University of Toronto Press, Toronto, pp. 65–102.

Nejstgaard, J.C., Hygum, B.H., Naustvoll, L.J., *et al.* (2001a). Zooplankton growth, diet and reproductive success compared in simultaneous diatom- and flagellate-microzooplankton-dominated plankton blooms. *Marine Ecology Progress Series*, **221**, 77–91.

Nejstgaard, J.C., Naustvoll, L.-J., and Sazhin, A. (2001b). Correcting for underestimation of microzooplankton grazing in bottle incubation experiments with mesozooplankton. *Marine Ecology Progress Series*, **221**, 59–75.

Nejstgaard, J.C., Frischer, M.E., Raule, C.L., *et al.* (2003). Molecular detection of algal prey in copepod guts and fecal pellets. *Limnology and Oceanography Methods*, **1**, 29–38.

Nejstgaard, J.C., Frischer, M.E., Simonelli, P., *et al.* (2007). Quantitative PCR to estimate copepod feeding. *Marine Biology*, **153(4)**, 565–77.

Nel, D.C., Cochrane, K., Petersen, S.L., *et al.*, eds. (2007). Ecological risk assessment: a tool for implementing an ecosystem approach for southern African fisheries. *WWW South Africa Report Series*, **2007/Marine/002**, 203pp.

Nelson, M.M., Phleger, C.F., Mooney, B.D., *et al.* (2000). Lipids of gelatinous Antarctic zooplankton: Cnidaria and Ctenophora. *Lipids*, **35**, 551–9.

Nesis, K.N. (1960). Variations in the bottom fauna of the Barents Sea under the influence of fluctuations in the hydrological regime. Soviet Fisheries Investigations In: *North European Seas*. VNIRO/PINRO, Moscow, pp. 129–38. [In Russian with English abstract]

Neutel, A.M., Heesterbeek, J.A.P., and de Ruiter, P.C. (2002). Stability in real food webs: weak links in long loops. *Science*, **296**, 1120–3.

Newman, M., Compo, G.P., and Alexander, M.A. (2003). ENSO-forced variability of the Pacific decadal oscillation. *Journal of Climate*, **16**, 3853–7.

Nicholas, K.R. and Frid, C.L.J. (1999). Occurrence of hydromedusae in the plankton off Northumberland (western central North Sea) and the role of planktonic predators. *Journal of the Marine Biological Association of the United Kingdom*, **79**, 979–92.

Niebauer, H.J., Bond, N.A., Yakunin, L.P., *et al.* (1999). An update on the climatology and sea ice of the Bering Sea. In: Loughlin, T.R. and Ohtani, K., eds. *Dynamics of the Bering Sea*. University of Alaska Sea Grant, Fairbanks, AK, pp. 29–59.

Niehoff, B. (2007). Life history strategies in zooplankton communities: the significance of female gonad morphology and maturation types for the reproductive biology of marine calanoid copepods. *Progress in Oceanography*, **74**, 1–47.

Niehoff, B. and Hirche, H.J. (1996). Oogenesis and gonad maturation in the copepod *Calanus finmarchicus* and the prediction of egg production from preserved samples. *Polar Biology*, **16**, 601–12.

Niehoff, B. and Runge, J.A. (2003). A revised methodology for prediction of egg production *Calanus finmarchicus* from preserved samples. *Journal of Plankton Research*, **25**, 1581–7.

Nilsson, H.C. and Rosenberg, R. (2000). Succession in marine benthic habitats and fauna inresponse to oxygen deficiency: analysed by sediment profile-imaging and by grab samples. *Marine Ecology Progress Series*, **197**, 139–49.

Nishida, H., Yatsu, A., Ishida, M. *et al.* (2006). Stock assessment and evaluation for the Pacific stock of Japanese sardine (FY 2005). In: *Marine fisheries stock assessment and evaluation for Japanese waters FY2005/2006*. Fisheries Agency and Fisheries Research Agency of Japan, pp. 11–45. [In Japanese]

Nishioka J., Ono, T., Saito, H., *et al.* (2007). Iron supply to the western subarctic Pacific: importance of iron export from the Sea of Okhotsk. *Journal of Geophysical Research*, **112**, C10012, doi:10.1029/2006JC004055.

Nogueria, E., Gonzalez-Nuevo, G., Bode, A., *et al.* (2004). Comparison of biomass and size spectra derived from optical plankton counter data and net samples: application to the assessment of mesoplankton distribution along the Northwest and North Iberian Shelf. *ICES Journal of Marine Science*, **61**, 508–17.

Norberg, J. (2004). Biodiversity and ecosystem functioning: a complex adaptive systems approach. *Limnology and Oceanography*, **49**, 1269–77.

Norrbin, M.F., Davis, C.S., and Gallager, S.M. (1996). Differences in fine-scale structure and composition of zooplankton between mixed and stratified regions of Georges Bank. *Deep-Sea Research II*, **43**, 1905–24.

North, E., Gallego, A., and Petitgas, P., eds. (2009). Manual of recommended practices for modelling physical-biological interactions during fish early life stages. *ICES Cooperative Research Report*, **295**, 114pp.

Nottestad, L., Misund, O.A., Melle, W., *et al.* (2007). Herring at the Arctic front: influence of temperature and prey on their spatio-temporal distribution and migration. *Marine Ecology: An Evolutionary Perspective*, **28**, 123–33.

NRC. (2002). *Effects of trawling and dredging on seafloor habitat*. National Academy Press, Washington, DC.

NRDC. (2008). *Fish out of water: how water management in the bay-delta threatens the future of California's salmon fishery*. Natural Resources Defense Council Issues Paper, 35pp.

O'Boyle, R. (2009). The implications of a paradigm shift in ocean resource management for fisheries stock assessment. In: Beamish, R.J. and Rothschild, B.J., eds. *Future of fisheries science in North America*. Fish and Fisheries Series **31**, 49–76. American Institute of Fishery Research Biologists, Springer, Dordrecht, The Netherlands.

O'Boyle, R., Sinclair, M., Keizer, P., *et al.* (2005). Indicators for ecosystem-based management on the Scotian Shelf: bridging the gap between theory and practice. *ICES Journal of Marine Science*, **62(3)**, 598–605.

Occhipinti-Ambrogi, A. and Savini, D. (2003). Biological invasions as a component of global change in stressed marine ecosystems. *Marine Pollution Bulletin*, **46**, 542–51.

Odum, E.P. (1969). The strategy of ecosystem development. *Science*, **164**, 262–70.

O'Farrell, M.R. and Botsford, L.W. (2006). The fisheries management implications of maternal-age-dependent larval survival. *Canadian Journal of Fisheries and Aquatic Sciences*, **63**, 2249–58.

Oguz, T. and Gilbert, D. (2007). Abrupt transitions of the top-down controlled Black Sea pelagic ecosystem during 1960–2000: evidence for regime-shifts under strong fishery exploitation and nutrient enrichment modulated by climate-induced variations. *Deep-Sea Research I*, **54(2)**, 220–42.

Oguz, T., Fach, B., and Salihoglu, B. (2008). Invasion dynamics of the alien ctenophore *Mnemiopsis leidyi* and its impact on anchovy collapse in the Black Sea. *Journal of Plankton Research*, **30(12)**, 1385–97.

Ohman, M.D. (1990). The demographic benefits of diel vertical migration by zooplankton. *Ecological Monographs*, **60**, 257–81.

Ohman, M.D. and Hirche, H.-J. (2001). Density-dependent mortality in an oceanic copepod population. *Nature*, **412**, 638–41.

Ohman, M.D. and Hsieh, C.-H. (2008). Spatial differences in mortality of *Calanus pacificus* within the California Current System. *Journal of Plankton Research*, **30(4)**, 359–66.

Ohman, M.D. and Runge, J.A. (1994). Sustained fecundity when phytoplankton resources are in short supply-omnivory by *Calanus finmarchicus* in the Gulf of St-Lawrence. *Limnology and Oceanography*, **39**, 21–36.

Ohman, M.D., Runge, J.A., Durbin, E.G., *et al.* (2002). On birth and death in the sea. *Hydrobiologia*, **480**, 55–68.

Ohman, M.D., Eiane, K., Durbin, E.G., *et al.* (2004). A comparative study of *Calanus finmarchicus* mortality patterns at five localities in the North Atlantic. *ICES Journal of Marine Science*, **61**, 687–97.

Ohman, M.D., Durbin, E.G., Runge, J.A., *et al.* (2008). Relationship of predation potential to mortality of *Calanus finmarchicus* on Georges Bank, northwest Atlantic. *Limnology and Oceanography*, **53**, 1643–55.

Ojaveer, H. and MacKenzie, B.R. (2007). Historical development of fisheries in northern Europe: reconstructing chronology of interactions between nature and man. *Fisheries Research*, **87**, 102–5.

Olsen, E.M., Heino, M., Lilly, G.R., *et al.* (2004). Maturation trends indicative of rapid evolution preceded the collapse of northern cod. *Nature*, **428**, 932–5.

Ommer, R.E., ed. (1990). *Merchant credit and labour strategies in historical perspective*. Acadiensis Press, Fredericton, NB.

Ommer, R.E. and the Coasts Under Stress Research Project Team. (2007). *Coasts under stress: restructuring and social-ecological health*. McGill-Queen's University Press, Montreal, Canada.

Ommer, R.E., Perry, R.I., and Neis, B. (2008). Bridging the gap between social and natural fisheries science: why is this necessary and how can it be done?. *American Fisheries Society Symposium*, **49**, 177–85.

Ommer, R.E., Jarre, A.C., Perry, R.I., *et al.* (2009). The human dimensions of small pelagic fisheries under global change. In: Checkley, D., Alheit, J., Oozeki, Y., *et al.* eds. *Climate change and small pelagic fish*. Cambridge University Press, Cambridge.

Omori, M and Ikeda, T. (1984). *Methods in marine zooplankton ecology*. Wiley, New York, 332pp.

Ono, T., Tadokoro, K., Midorikawa, T., *et al.* (2002). Multiple-decadal decrease of net community production in western subarctic North Pacific. *Geophysical Research Letters*, **29**, doi:10.1029/2001GL014332.

Oosterveer, P. (2008). Governing global fish provisioning: ownership and management of marine resources. *Ocean and Coastal Managament*, **51(12)**, 797–805.

Opdal, A.F., Vikebø, F., and Fiksen, Ø. (2008). Relationships between spawning ground identity, latitude and early life thermal exposure in Northeast Arctic cod. *Journal of Northwest Atlantic Fishery Science*, **41**, 13–22.

Orcutt, J.D., Jr. and Porter, K.G. (1983). Diel vertical migration by zooplankton: constant and fluctuating effects on life history parameters of *Daphnia*. *Limnology and Oceanography*. **28**, 720–30.

Orr, J.C., Fabry, V.J., Aumont, O., *et al.* (2005). Anthropogenic ocean acidification over the twenty-first century and its impact on calcifying organisms. *Nature*, **437**, 681–6.

Osborn, T.J. and Briffa, K.R. (2006). The spatial extent of the 20th-century warmth in the context of the past 1200 years. *Science*, **311**, 841–4.

Österblom, H., Hansson, S., Larsson, U., *et al.* (2007). Human-induced trophic cascades and ecological regime shifts in the Baltic Sea. *Ecosystems*, **10**, 877–89.

Otterlei, E., Nyhammer, G., Arild Folkvord, A., *et al.* (1999). Temperature- and size-dependent growth of larval and early juvenile Atlantic cod (*Gadus morhua*): a comparative study of Norwegian coastal cod and northeast Arctic cod. *Canadian Journal of Fisheries and Aquatic Sciences*, **56**, 2099–111.

Ottersen, G. (1996). *Environmental impact on variability in recruitment, larval growth and distribution of Arcto-Norwegian cod*. Ph.D. Thesis, University of Bergen, Bergen, Norway.

Ottersen, G. and Loeng, H. (2000). Covariability in early growth and year-class strength of Barents Sea cod, haddock, and herring: the environmental link. *ICES Journal of Marine Science*, **57**, 339–48.

Ottersen, G. and Stenseth, N.C. (2001). Atlantic climate governs oceanographic and ecological variability in the Barents Sea. *Limnology and Oceanography*, **46**, 1774–80.

Ottersen, G., Michalsen, K., and Nakken, O. (1998). Ambient temperature and distribution of North-east Arctic cod. *ICES Journal of Marine Science*, **55**, 67–85.

Ottersen, G., Alheit, J., Drinkwater, K., *et al.* (2004a). The responses of fish populations to ocean climate fluctuations. In: Stenseth, N.C., Ottersen, G., Hurrell, J.W., *et al.* eds. *Marine ecosystems and climate variation: the North Atlantic, a comparative perspective*. Oxford University Press, Oxford, pp. 73–94.

Ottersen, G., Stenseth, N.C., and Hurrell, J.W. (2004b). Climatic fluctuations and marine systems: a general introduction to the ecological effects. In: Stenseth, N.C., Ottersen, G., Hurrell, J.W., *et al.* eds. *Marine ecosystems and climate variation: the North Atlantic, a comparative perspective*. Oxford University Press, Oxford, pp. 3–14.

Ottersen, G., Hjermann, D.Ø., and Stenseth, N.C. (2006). Changes in spawning stock structure strengthen the link between climate and recruitment in a heavily fished cod (*Gadus morhua*) stock. *Fisheries Oceanography*, **15(3)**, 230–43.

Ottersen, G., Kim, S., Huse, G., *et al.* (2010). Major pathways by which climate signals may force marine populations. *Journal of Marine Systems*, **79**, 343–60.

Overholtz, W.J. (2002). The Gulf of Maine-Georges Bank Atlantic herring (*Clupea harengus*): spatial pattern analysis of the collapse and recovery of a large marine fish complex. *Fisheries Research*, **57**, 237–54.

Overland, J.E. and Wang, M. (2007a). Future climate of North Pacific Ocean. *EOS*, **88**, 178,182.

Overland, J.E. and Wang, M. (2007b). Future regional Arctic sea ice declines. *Geophysical Research Letters*, **34**, L17705, doi:10.1029/2007GL030808.

Overland, J.E., Alheit, J., Bakun, A., *et al.* (2010). Climate controls on marine ecosystems and fish populations. *Journal of Marine Systems*, **79**, 305–15.

Overland, J.E., Rodionov, S., Minobe, N., *et al.* (2008). North Pacific regime shifts: definitions, issues and recent transitions. *Progress in Oceanography*, **77**, 92–102.

Owen, R.W. (1989). Microscale and finescale variations of small plankton in coastal and pelagic environments. *Journal of Marine Research*, **47**, 197–240.

Pace, M.L., Cole, J.J., Carpenter, S.R., *et al.* (1999). Trophic cascades revealed in diverse ecosystems. *Trends in Ecology and Evolution*, **14(12)**, 483–8.

Paffenhöfer, G.-A. and Lewis, K.D. (1989). Feeding behavior of nauplii of the genus *Eucalanus* (Copepoda, Calanoida). *Marine Ecology Progress Series*, **57**, 129–36.

Paffenhöfer, G.-A. and Lewis, K.D. (1990). Perceptive performance and feeding behavior of calanoid copepods. *Journal of Plankton Research*, **12**, 933–46.

Paffenhöfer, G.-A., Strickler, J.R., and Alcaraz, M. (1982). Suspension-feeding by herbivorous calanoid copepods: a cinematographic study. *Marine Biology*, **67**, 193–99.

Page, F.H., Sinclair, M., Naimie, C.E., *et al.* (1999). Cod and haddock spawning on Georges Bank in relation to water residence times. *Fisheries Oceanography*, **8**, 212–26.

Paine, R.T. (1969). A note on trophic complexity and community stability. *American Naturalist*, **103**, 91–3.

Pakhomov, E.A., Froneman P.W., and Perissinotto R. (2002). Salp/krill interactions in the Southern Ocean: spatial segregation and implications for the carbon flux. *Deep-Sea Research II*, **49**, 1881–907.

Pakhomov, E.A., Atkinson, A., Meyer, B., *et al.* (2004). Daily rations and growth of larval krill *Euphausia superba* in the Eastern Bellingshausen Sea during austral autumn. *Deep-Sea Research II*, **51**, 2185–98.

Palacios, D.P., Bograd, S.J., Mendelssohn, R., *et al.* (2004). Long-term and seasonal trends in stratification in the California Current, 1950–1993. *Journal of Geophysical Research-Oceans*, **109**, doi:10.1029/2004JC002380.

Palmer, C.T. and Sinclair, P.R. (1997). When the fish are gone: ecological disaster and fishers in northwest Newfoundland. Fernwood Publishing, Halifax, NS.

Palumbi, S.R. (2004). Marine reserves and ocean neighbourhoods: the spatial scale of marine populations and their management. *Annual Reviews of Environmental Resources*, **29**, 31–68.

Parada, C., van der Lingen, C.D., Mullon, C., *et al.* (2003). Modelling the effect of buoyancy on the transport of anchovy (*Engraulis capensis*) eggs from spawning to nursery grounds in the southern Benguela: an IBM approach. *Fisheries Oceanography*, **12(3)**, 170–84.

Paris, C.B., Cowen, R.K., Claro, R., *et al.* (2005). Larval transport pathways from Cuban snapper (Lutjanidae) spawning aggregations based on biophysical modelling. *Marine Ecology Progress Series*, **296**, 93–106.

Parrish, R.H., Schwing, F.B., and Mendelssohn, R. (2000). Midlatitude wind stress: the energy source for climatic regimes in the North Pacific Ocean. *Fisheries Oceanography*, **9**, 224–38.

Parsons, T.R. and Lalli, C.M. (1988). Comparative ocean ecology of the plankton communities of the subarctic Atlantic and Pacific Oceans. *Oceanography and Marine Biology: An Annual Review*, **26**, 317–59.

Parsons, T.R. and Lalli, C.M. (2002). Jellyfish population explosions: revisiting a hypothesis of possible causes. *La Mer*, **40**, 111–21.

Paterson, B., Jarre, A., Moloney, C.L., *et al.* (2007). A fuzzy-logic tool for multi-criteria decision-making in the Southern Benguela. *Journal of Marine and Freshwater Research*, **58(11)**, 1056–68.

Pauly, D. (1995). Anecdotes and the shifting baseline syndrome of fisheries. *Trends in Ecology and Evolution*, **10(10)**, p. 430.

Pauly, D. (2003). Ecosystem impacts of the world's marine fisheries. *Global Change Newsletter*, **55**, 21–3.

Pauly, D. and Palomares, M.L. (2005). Fishing down marine food web: it is far more pervasive than we thought. *Bulletin of Marine Science*, **76(2)**, 197–211.

Pauly, D., Christensen, V., Dalsgaard, J.P.T., *et al.* (1998). Fishing down the marine food webs. *Science*, **279**, 860–3.

Pauly, D., Christensen, V., Guénette, S., *et al.* (2002). Towards sustainability in world fisheries. *Nature*, **418**, 689–95.

Pauly, D., Libralato, S., Morisette, L., *et al.* (2009). Jellyfish in ecosystems, online databases and ecosystem models. *Hydrobiologia*, **616**, 67–85.

Pavlov, V.K. and Pavlov, P.V. (1996). Oceanographic description of the Bering Sea. In: Mathisen, O.A. and Coyle, K.O., eds. *Ecology of the bering sea: a review of the Russian literature. Alaska Sea Grant College Program Report*, **96–01**. University of Alaska, Fairbanks, AK, 95pp.

Payne, A., Cotter, J., and Potter, T., eds. (2008). Advances in fisheries science: 50 years on from Beverton and Holt. Blackwell Publishing, Oxford, xxi + 547pp.

Pearcy, W.G. (1992). *Ocean ecology of Pacific salmonids*. University of Washington Press, Seattle, WA, 179pp.

Pearcy, W.G. (2005). Marine nekton off Oregon and the 1997–98 El Niño. *Progress in Oceanography*, **54**, 399–403.

Pease, C.H. (1980). Eastern Bering Sea ice processes. *Monthly Weather Review*, **108**, 2015–23.

Peck, M.A., Buckley L.J., and Bengtson D.A. (2005). Effects of temperature, body size and feeding on rates of metabolism in young-of-the-year haddock. *Journal of Fish Biology*, **66**, 911–23.

Peck, M.A., Buckley L.J., and Bengtson D.A. (2006). Effects of temperature and body size on the swimming speed of larval and juvenile Atlantic cod (*Gadus morhua*): implications for individual-based modeling. *Environmental Biology of Fishes*, **75**, 419–29.

Pécseli, H.L. and Trulsen, J. (2007). Turbulent particle fluxes to perfectly absorbing surfaces: a numerical study. *Journal of Turbulence*, **8(42)**, 1–25.

Pederson, J.A. and Blakeslee, A.M.H. (2008). Fifth international conference on marine bioinvasions: introduction. *ICES Journal of Marine Science*, **65**, 713–5.

Perez, T., Garrabou, J., Sartoretto, S., *et al.* (2000). Mortalité massive d'invertébrés marins: un événement sans précédent en Méditerranée nord-occidentale. *Compte Rendu de l'Academie des Sciences, Sci Vie*, **323**, 853–65.

Perry, A.L., Low, P.J., Ellis, J.R., *et al.* (2005). Climate change and distribution shifts in marine species. *Science*, **308**, 1912–5.

Perry, M.J. and Rudnick, D.L. (2003). Observing the oceans with autonomous and Lagrangian platforms and sensors: the role of ALPS in sustained ocean observing systems. *Oceanography*, **16**, 31–6.

Perry, R.I. and Ommer, R. (2003). Scale issues in marine ecosystems and human interactions. *Fisheries Oceanography*, **12**, 513–22.

Perry, R.I. and Sumaila, U.R. (2007). Marine ecosystem variability and human community responses: the example of Ghana, West Africa. *Marine Policy*, **31**, 125–34.

Perry, R.I., Batchelder, H.P., Mackas, D.L., *et al.* (2004). Identifying global synchronies in marine zooplankton populations: issues and opportunities. *ICES Journal of Marine Science*, **61**, 445–56.

Perry, R.I., Cury, P., Brander, K., *et al.* (2010a). Sensitivity of marine systems to climate and fishing: concepts, issues and management responses. *Journal of Marine Systems*, **79**, 427–35.

Perry, R.I., Ommer, R.E., Allison, E., *et al.* (2010b). Interactions between changes in marine ecosystems and human communities. *This volume*. Oxford University Press, Oxford.

Perry, R.I., Ommer, R.E., Jentoft, S., *et al.* (in revision). Interactive responses of natural and human systems to marine ecosystem changes. *Fish and Fisheries*.

Pershing, A.J., Greene, C.H., Planque, B., *et al.* (2004). The influence of climate variability on North Atlantic zooplankton populations. In: Stenseth, N.C., Ottersen, G., Hurrell, J.W., *et al.* eds. *Marine ecosystems and climate variation: the North Atlantic, a comparative perspective*. Oxford University Press, Oxford, pp. 59–69.

Petereit, C., Haslob, H., Kraus, G., *et al.* (2008). The influence of temperature on the development of Baltic Sea sprat (*Sprattus sprattus*) eggs and yolk sac larvae. *Marine Biology*, **154**, 295–306.

Peterman, R.M. and Bradford, M.J. (1987). Wind speed and mortality of a marine fish, the Northern Anchovy (*Engraulis mordax*). *Science*, **235**, 354–6.

Peterman, R.M., Pyper, B.J., and Grout, J.A. (2000). Comparison of parameter estimation methods for detecting climate-induced changes in productivity of Pacific salmon (*Oncorhynchus* spp.). *Canadian Journal of Fisheries and Aquatic Sciences*, **57(1)**, 181–91.

Peters, F. and Marrasé, C. (2000). Effects of turbulence on plankton: an overview of experimental evidence and some theoretical considerations. *Marine Ecology Progress Series*, **205**, 291–306.

Peters, J., Renz, J., van Beusekom, J., *et al.* (2006). Trophodynamics and seasonal cycle of the copepod *Pseudocalanus acuspes* in the Central Baltic Sea (Bornholm Basin): evidence from lipid composition. *Marine Biology*, **149**, 1417–29.

Peters, J., Dutz, J., and Hagen, W. (2007). Role of essential fatty acids on the reproductive success of the copepod *Temora longicornis* in the North Sea. *Marine Ecology Progress Series*, **342**, 153–63.

Peters, R.H. (1983). *The ecological implications of body size.* Cambridge University Press, Cambridge.

Petersen, J.E., Sanford, L.P., and Kemp, W.M. (1998). Coastal plankton responses to turbulent mixing in experimental ecosystems. *Marine Ecology Progress Series*, **171**, 23–41.

Peterson, W.T. and Keister, J.E. (2003). Interannual variability in copepod community composition at a coastal station in the northern California current: a multivariate approach. *Deep-Sea Research II*, **50**, 2499–517.

Peterson, W.T. and Schwing, F.B. (2003). A new climate regime in northeast Pacific ecosystems. *Geophysical Research Letters*, **30**, doi: 10.1029/2003GL017528.

Peterson, W.T. website: http://www.nwfsc.noaa.gov, click on "ocean index tools".

Peterson, W.T., Miller, C.B., and Hutchinson, A. (1979). Zonation and maintenance of copepod populations in the Oregon upwelling zone. *Deep-Sea Research I*, **26(5)**, 467–94.

Peterson, W.T., Gomez-Guteirrez, J., and Morgan, C.A. (2002). Cross-shelf variation in calanoid copepod production during summer 1996 off the Oregon coast, USA. *Marine Biology*, **141**, 353–65.

Petit, J.R., Jouzel, J., Raynaud, D., *et al.* (1999). Climate and atmospheric history of the past 420,000 years from the Vostok ice core, Antarctica. *Nature*, **399**, 429–36.

Phleger, C.F., Nelson, M.M., Mooney, B., *et al.* (2000). Lipids of Antarctic salps and their commensal hyperiid amphipods. *Polar Biology*, **23**, 329–37.

PICES. (2004). Marine ecosystems of the North Pacific. *PICES Special Publication*, **1**, 280pp.

Pichegru, L., Ryan, P.G., van der Lingen, C.D., *et al.* (2007). Foraging behaviour and energetics of Cape gannets *Morus capensis* feeding on live prey and fishery discards in the Benguela upwelling system. *Marine Ecology Progress Series*, **350**, 127–36.

Pierson, J.J., Halsband-Lenk, C., and Leising, A.W. (2005). Reproductive success of *Calanus pacificus* during diatom blooms in Dabob Bay, Washington. *Progress in Oceanography*, **67**, 314–31.

Pikitch, E.K., Santora, C., Babcock, E.A., *et al.* (2004). Ecosystem-based fishery management. *Science*, **305**, 346–7.

Pimm, S.L., Russell, G.J., Gittleman, J.L., *et al.* (1995). The future of biodiversity. *Science*, **269**, 347–50.

Pinchuk, A.I. and Hopcroft, R.R. (2006). Egg production and early development of *Thysanoessa inermis* and *Euphausia pacifica* (Crustacea: Euphausiacea) in the northern Gulf of Alaska. *Journal of Experimental Marine Biology and Ecology*, **332**, 206–15.

Pinchuk, A.I. and Hopcroft, R.R. (2007). Seasonal variations in the growth rates of euphausiids (*Thysanoessa inermis, T. spinifera,* and *Euphausia pacifica*) from the northern Gulf of Alaska. *Marine Biology*, **151**, 257–69.

Pitt, K.A., Kingsford, M.J., Rissik, D., *et al.* (2007). Jellyfish modify the response of planktonic assemblages to nutrient pulses. *Marine Ecology Progress Series*, **351**, 1–13.

Pitt, K.A., Clement, A.L., Connolly, R.M., *et al.* (2008). Predation by jellyfish on large and emergent zooplankton: implications for benthic-pelagic coupling. *Estuarine, Coastal and Shelf Science*, **76**, 827–33.

Plagányi, E. (2007). Models for an ecosystem approach to fisheries. *FAO Fisheries Technical Paper*, **477**, 108pp.

Plagányi, E. and Butterworth, D.S. (2005). Modelling the impact of krill fishing on seal and penguin colonies. *CAMLR WG-EMM* **05/14**. CCAMLR, Hobart, Australia.

Plagányi, E., Rademeyer, R.A., Butterworth, D.S., *et al.* (2007). Making management procedures operational – innovations implemented in South Africa. *ICES Journal of Marine Science*, **64**, 626–32.

Planque, B. and Fox, C.J. (1998). Interannual variability in temperature and the recruitment of Irish Sea cod. *Marine Ecology Progress Series*, **172**, 101–5.

Planque, B. and Frédou, T. (1999). Temperature and the recruitment of Atlantic cod (*Gadus morhua*). *Canadian Journal of Fisheries and Aquatic Sciences*, **56**, 2069–77.

Planque, B. and Ibañez, F. (1997). Long-term time series in *Calanus finmarchicus* abundance – a question of space?. *Oceanologica Acta*, **20**, 159–64.

Planque, B. and Reid, P.C. (1998). Predicting *Calanus finmarchicus* abundance from a climatic signal. *Journal of the Marine Biological Association of the United Kingdom*, **78**, 1015–8.

Planque, B. and Reid, P.C. (2002). What have we learned about plankton variability and its physical controls from 70 years of CPR records. *ICES Marine Science Symposia*, **215**, 237–46.

Planque, B., Hays, G.C., Ibanez, F., *et al.* (1997). Large scale spatial variations in the seasonal abundance of *Calanus finmarchicus*. *Deep-Sea Research I*, **44**, 315–26.

Planque, B., Fromentin, J.-M., Cury, P., *et al.* (2010). How does fishing alter marine populations and ecosystems sensitivity to climate?. *Journal of Marine Systems*, **79**, 403–17.

Platt, T. (1985). The structure of the marine ecosystem: its allometric basis. In: Ulanowicz, R.E. and Platt, T., eds. Ecosystem theory for biological oceanography. *Canadian Bulletin of Fisheries Aquatic Sciences*, **213**, 55–64.

Platt, T. and Denman, K.L. (1975). Spectral analysis in ecology. *Annual Review of Ecology and Systematics*, **6**, 189–210.

Platt, T., Yaco, C.-F., and Frank, K.T. (2003). Spring algal bloom and larval fish survival. *Science*, **423**, 398–9.

Plikshs, M., Kalejs, M., and Grauman, G. (1993). The influence of environmental conditions and spawning stock size on the year-class strength of the Eastern Baltic cod. *ICES Council Meeting Papers*, **CM 1993/J:22**.

Plourde, S., Joly, P., Runge, J.A., *et al.* (2001). Life cycle of *Calanus finmarchicus* in the lower St. Lawrence Estuary: the imprint of circulation and late timing of the spring phytoplankton bloom. *Canadian Journal of Fisheries and Aquatic Sciences*, **58**, 647–58.

Pohnert, G., Lumineau, O., Cueff, A., *et al.* (2002). Are volatile unsaturated aldehydes from diatoms the main line of chemical defence against copepods?. *Marine Ecology Progress Series*, **245**, 22–45.

Polis, G.A., Myers, C.A., and Holt, R.D. (1989). The ecology and evolution of intraguild predation: potential competitors that eat each other. *Annual Review of Ecology and Systematics*, **20**, 297–330.

Pollnac, R.B. and Carmo, F. (1980). Attitudes toward cooperation among small-scale fishermen and farmers in the Azores. *Anthropological Quarterly*, **53(1)**, 12–19.

Pollnac, R.B. and Poggie, J.J. (2008). Happiness, well-being and psychocultural adaptation to the stresses associated with marine fishing. *Human Ecology Review*, **15**, 194–200.

Polovina, J.J. (1984). Model of a coral reef ecosystem: I. The ECOPATH model and its application to French frigate shoals. *Coral Reefs*, **3**, 1–11.

Polovina, J.J. (2005). Climate variation, regime shifts, and implications for sustainable fisheries. *Bulletin of Marine Science*, **76**, 233–44.

Polovina, J.J. and Howell, E.A. (2005). Ecosystem indicators derived from satellite remotely sensed oceanographic data for the North Pacific. *ICES Journal of Marine Science*, **62**, 319–27.

Polovina, J.J., Kobayashi, D.R., Parker, D.M., *et al.* (2000). Turtles on the edge: movement of loggerhead turtles (*Caretta caretta*) along oceanic fronts, spanning longline fishing grounds in the central North Pacific, 1997–1998. *Fisheries Oceanography*, **9**, 71–82.

Polovina, J.J., Chai, F., Kobayashi, D.R., *et al.* (2008a). Ecosystem dynamics at a productivity gradient: a study of the lower trophic dynamics around the northern atolls in the Hawaiian Archipelago. *Progress in Oceanography*, **77**, 217–24.

Polovina, J.J., Howell, E.A., and Abecassis, M. (2008b). Ocean's least productive waters are expanding. *Geophysical Research Letters*, **35**, L03618, doi: 10.1029/2007/GL031745.

Pomeroy, L.S. (1974). The ocean's food web: a changing paradigm. *BioScience*, **24**, 499–504.

Pomeroy, L.S. and Berkes, F. (1997). Two to tango: the role of government in fisheries co-management. *Marine Policy*, **21**, 465–80.

Pomeroy, L.S., Katon, B.M., and Harkes, I. (2001). Conditions affecting the success of fisheries co-management: lessons from Asia. *Marine Policy*, **25**, 197–208.

Pörtner, H.O. (2001). Climate change and temperature-dependent biogeography: oxygen limitation of thermal tolerance in animals. *Naturwissenschaften*, **88**, 137–46.

Pörtner, H.O. (2002a). Physiological basis of temperature-dependent biogeography: trade-offs in muscle design and performance in polar ectotherms. *Journal of Experimental Biology*, **205**, 2217–30.

Pörtner, H.O. (2002b). Climate variations and the physiological basis of temperature dependent biogeography: systemic to molecular hierarchy of thermal tolerance in animals. *Comparative Biochemistry and Physiology, Part A*, **132**, 739–61.

Pörtner, H.O. and Kunst, R. (2007). Climate change affects marine fishes through the oxygen limitation of thermal tolerances. *Science*, **315**, 95–7.

Pörtner, H.O., Berdal, B., Blust, R., *et al.* (2001). Climate induced temperature effects on growth performance, fecundity and recruitment in marine fish: developing a hypothesis for cause and effect relationships in Atlantic cod (*Gadus morhua*) and common eelpout (*Zoarces viviparous*). *Continental Shelf Research*, **21**, 1975–97.

Pörtner, H.O., Storch, D., and Heilmayer, O. (2005). Constraints and trade-offs in climate dependent adaptation: energy budgets and growth in a latitudinal cline. *Scientia Marina*, **69(Suppl. 2)**, 271–85.

Posgay, J.A. and Marak, R.R. (1980). The MARMAP bongo zooplankton samplers. *Journal of Northwest Atlantic Fisheries Science*, **1**, 91–9.

Potier, M., Marsac, F., Cherel, Y., *et al.* (2007). Forage fauna in the diet of three large pelagic fish (lancetfish, swordfish, and yellowfin tuna) in the western equatorial Indian Ocean. *Fisheries Research*, **83**, 60–72.

Poulet, S.A., Laabir, M., Ianora, A., *et al.* (1995). Reproductive response of *Calanus helgolandicus*. I. Abnormal embryonic and naupliar development. *Marine Ecology Progress Series*, **120**, 85–95.

Poulsen, B., Holm, P., and MacKenzie, B.R. (2007). A long-term (1667–1860) perspective on impacts of fishing and environmental variability on fisheries for herring, eel, and whitefish in the Limfjord, Denmark. *Fisheries Research*, **87**, 181–95.

Poulsen, R.T., Cooper, A.B., Holm, P., *et al.* (2007). An abundance estimate of ling (*Molva molva*) and cod (*Gadus morhua*) in the Skagerrak and the northeastern North Sea, 1872. *Fisheries Research*, **87**, 196–207.

Power, M.E. (1992). Top-down and bottom-up forces in food webs: do plants have primacy?. *Ecology*, **73**, 733–46.

Power, M.E., Tilman, D., Estes, J.A., *et al.* (1996). Challenges in the quest for keystones. *BioScience*, **46**, 609–20.

Price, H.J. and Paffenhöfer, G.-A. (1984). Effects of feeding experience in the copepod *Eucalanus pileatus*: a cinematographic study. *Marine Biology*, **84**, 35–40.

Price, H.J., Paffenhöfer, G.-A. and Strickler, J.R. (1983). Modes of cell capture in calanoid copepods. *Limnology and Oceanography*, **28**, 116–23.

Pringle, J.M. (2007). Turbulence avoidance and the wind-driven transport of plankton in the surface Ekman layer. *Continental Shelf Research*, **27**, 670–8.

Proctor, R. and Howarth, M.J. (2003). The POL Coastal Observatory. In: *Building the European capacity in operational oceanography, Proceedings of the 3rd International Conference on EuroGOOS, 3–6 December 2002*, Elsevier Oceanography Series, *Athens, Greece*. **69**, 548–53.

Proctor, R. and Howarth, M.J. (2008). Coastal observatories and operational oceanography: a European perspective. *Marine Technology Society Journal*, **42(3)**, 10–13.

PROMPEX. (2006). *Exportaciones Peruanas y Mercados de los Productos de la Acuicultura*. Convención Nacional Oportunidades de Negocios en Acuicultura, Abril 2006, Lima, Peru.

PSAT (Puget Sound Action Team). (2007a). *2007 Puget Sound update: ninth report of the Puget Sound assessment and monitoring program*. Puget Sound Action Team, Olympia, WA, 260pp.

PSAT (Puget Sound Action Team). (2007b). *State of the Sound 2007*. Puget Sound Action Team, Office of the Governor, State of Washington, WA, PSAT 07–01. 93pp.

Punt, A.E. (1992). Management procedures for Cape hake and baleen whale resources. *Benguela Ecology Programme Report*, **23**, 689pp.

Punt, A.E. (2006). The FAO precautionary approach after almost 10 years: have we progressed towards implementing simulation-tested feedback-control management systems for fisheries management? *Natural Resource Modeling*, **19**, 441–64.

Punt, A.E. and Smith, A.D.M. (1999). Harvest strategy evaluation for the eastern gemfish (*Rexea solandri*). *ICES Journal of Marine Science*, **56**, 860–75.

Punt, A.E., Campbell, R.A., and Smith, A.D.M. (2001). Evaluating empirical indicators and reference points for fisheries management: application to the broadbill swordfish fishery off eastern Australia. *Marine and Freshwater Research*, **52**, 819–32.

Purcell, J.E. (1985). Predation on fish eggs and larvae by pelagic cnidarians and ctenophores. *Bulletin of Marine Science*, **37**, 739–55.

Purcell, J.E. (2005). Climate effects on formation of jellyfish and ctenophore blooms: a review. *Journal of the Marine Biological Association of the United Kingdom*, **85**, 461–76.

Purcell, J.E. and Mills, C.E. (1988). The correlation between nematocyst types and diets in pelagic Hydrozoa. In: Hessinger, D.A. and Lenhoff, H.M., eds. *The biology of nematocysts*. Academic Press, New York, pp. 463–86.

Purcell, J.E. and Sturdevant, M.V. (2001). Prey selection and dietary overlap among zooplanktivorous jellyfish and juvenile fishes in Prince William Sound, Alaska. *Marine Ecology Progress Series*, **210**, 67–83.

Purcell, J.E., Breitburg, D.L., Decker, M.B., *et al.* (2001a). In: Rabalais, N.N. and Turner, R.E., eds. *Coastal hypoxia: consequences for living resources and ecosystems*. American Geophysical Union, Washington, DC, pp. 77–100.

Purcell, J.E., Shiganova, T.A., Decker, M.B., *et al.* (2001b). The ctenophore *Mnemiopsis* in native and exotic habitats: US estuaries versus the Black Sea basin. *Hydrobiologia*, **451**, 145–76.

Purcell, J.E., Uye, S., and Lo, W.T. (2007). Anthropogenic causes of jellyfish blooms and their direct consequences for humans: a review. *Marine Ecology Progress Series*, **350**, 153–74.

Queiroga, H., Almeida, M.J., Alpuim, T., *et al.* (2006). Tide and wind control of megalopal supply to estuarine crab populations on the Portugese west coast. *Marine Ecology Progress Series*, **307**, 21–36.

Quéro, J.-C. (1998). Changes in the Euro-Atlantic fish species composition resulting from fishing and ocean warming. *Italian Journal of Zoology*, **65**, 493–9.

Quéro, J.-C., Du Buit, M.-H., and Vayne, J.-J. (1998). Les observations de poissons tropicaux et le réchauffement des eaux dans l'Atlantique européen. *Oceanologica Acta*, **21**, 345–51.

Quetin, L.B. and Ross, R.M. (2001). Environmental variability and its impact on the reproductive cycle of Antarctic krill. *American Zoologist*, **41**, 74–89.

Quetin, L.B., Ross, R.M., Fritsen, C.H., *et al.* (2007). Ecological responses of Antarctic krill to environmental variability: can we predict the future? *Antarctic Science*, **19**, 253–66.

Quiñones, R., Serra, R., Nuñez, S., *et al.* (1997). Relación espacial entre el jurel (*Trachurus symmetricus murphyi*) y sus presas en la zona centro-sur de Chile. *Gestión de sistemas oceanograficos del Pacífico oriental*. E. Tarifeño, Comisión Oceanográfica Intergubernamental de la UNESCO, IOC/INF-1046, 187–202.

Quiñones, R., Barriga, O., Dresdner, J., *et al.* (2003). Análisis económico, social y biológico de la crisis pesquera de la VIII Región (1997–2002). *Informe Final Proyecto Análisis biológico, económico y social de las pesquerías de la VIII Región. Fondo Nacional de Desarrollo Regional de la Región del Bio Bio*, Código BIP 120183334-0. Facultad de Ciencias Naturales y Oceanográficas, Universidad de Concepción, Concepción, Chile, 566pp. + annexes.

Raakjaer Nielsen, J. and Hara, M. (2006). Transformation of South African industrial fisheries. *Marine Policy*, **30**, 43–50.

Radach, G. and Moll, A. (2006). Review of three-dimensional ecological modelling related to the North Sea shelf system, Part 2: model validation and data needs. *Oceanography and Marine Biology: An Annual Review*, **44**, 1–60.

Rademeyer, R.A., Plagányi, É.E., and Butterworth, D.S. (2007). Tips and tricks in designing management procedures. *ICES Journal of Marine Science*, **64**, 618–25.

Railsback, S.F., Lamberson, R.H., Harvey, B.C., *et al.* (1999). Movement rules for individual-based models of stream fish. *Ecological Modelling*, **123**, 73–89.

Rand, P.S., Hinch, S.G., Morrison, J., *et al.* (2006). Effects of river discharge, temperature, and future climates on energetics and mortality of adult migrating Fraser River sockeye salmon. *Transactions of the American Fisheries Society*, **135**, 655–67.

Raskoff, K.A. (2001). The impact of *El Niño* events on populations of mesopelagic hydromedusae. *Hydrobiologia*, **451**, 121–9.

Rau, G.H., Ohman, M.D., and Pierrot-Bults, A. (2003). Linking nitrogen dynamics to climate variability off central California: a 51 year record based on $^{15}N/^{14}N$ in CalCOFI zooplankton. *Deep-Sea Research II*, **50**, 2431–47.

Ravier, C. and Fromentin, J.M. (2001). Long-term fluctuations in the eastern Atlantic and Mediterranean bluefin tuna population. *ICES Journal of Marine Science*, **58**, 1299–317.

Ravier, C. and Fromentin, J.M. (2004). Are the long-term fluctuations in Atlantic bluefin tuna (*Thunnus thynnus*) population related to environmental changes?. *Fisheries Oceanography*, **13(2)**, 145–60.

Rayner, N.A., Brohan, P., Parker, D.E., *et al.* (2006). Improved analyses of changes and uncertainties in sea surface temperature measured *in situ* since the mid-nineteenth century: the HadSST2 dataset. *Journal of Climate*, **19**, 446–69.

Reid, P.C., Battle, E.J.V., Batten, S.D., *et al.* (2000). Impacts of fisheries on plankton community structure. *ICES Journal of Marine Science*, **57**, 495–502.

Reid, P.C., Borges, M.D., and Svendsen, E. (2001a). A regime shift in the North Sea circa 1988 linked to changes in the North Sea horse mackerel fishery. *Fisheries Research*, **50**, 163–71.

Reid, P.C., Holliday, N.P., and Smyth, T.J. (2001b). Pulses in eastern margin current with higher temperatures and North Sea ecosystem changes. *Marine Ecology Progress Series*, **215**, 283–7.

Reid, P.C., Colebrook, J.M., Mathews, J.B.L., *et al.* (2003a). The continuous plankton recorder: concepts and history, from plankton indicator to undulating recorders. *Progress in Oceanography*, **58**, 117–73.

Reid, P.C., Edwards, M., Beaugrand, G., *et al.* (2003b). Periodic changes in the zooplankton of the North Sea during the twentieth century linked to oceanic inflow. *Fisheries Oceanography*, **12(4–5)**, 260–9.

Reid, W.V., Mooney, H.A., Cropper, A. *et al.*, eds. (2005). *Ecosystems and human well-being. Millennium Ecosystem Assessment Synthesis Report*. Island Press, Washington, 137pp.

Reiss, C.S., Anis, A., Taggart, C.T., *et al.* (2002). Relationships among vertically structured *in situ* measures of turbulence, larval fish abundance and feeding success and copepods on Western Bank, Scotian Shelf. *Fisheries Oceanography*, **11**, 156–74.

Renaud, M.L. and Carpenter, J.A. (1994). Movements and submergence patterns of loggerhead turtles (*Caretta caretta*) in the Gulf of Mexico determined through satellite telemetry. *Bulletin of Marine Science*, **55**, 1–15.

Renz, J., Peters, J., and Hirche, H.J. (2007). Life cycle of *Pseudocalanus acuspes* giesbrecht (Copepoda, Calanoida) in the central Baltic Sea: II. Reproduction, growth and secondary production. *Marine Biology*, **151**, 515–27.

Renz, J., Mengedoht, D., and Hirche, H.J.R. (2008). Reproduction, growth and secondary production of *Pseudocalanus elongatus* Boeck (Copepoda, Calanoida) in the southern North Sea. *Journal of Plankton Research*, **30**, 511–28.

Ressler, P.H., Brodeur, R.D., Peterson, W.T., *et al.* (2005). The spatial distribution of euphausiid aggregations in the northern California Current during August 2000. *Deep-Sea Research II*, **52**, 89–108.

Rex, M.A., Stuart, C.T., and Coyne, G. (2000). Latitudinal gradients of species richness in the deep-sea benthos of the North Atlantic. *Proceedings of the National Academy of Sciences of the United States of America*, **97**, 4082–5.

Rey, C., Harris, R., Irigoien, X., *et al.* (2001). Influence of algal diet on growth and ingestion of *Calanus helgolandicus* nauplii. *Marine Ecology Progress Series*, **216**, 151–65.

Rey, F. and Loeng, H. (1985). The influence of ice and hydrographic conditions on the development of phytoplankton in the Barents Sea. In: J.S. Gray and M.E. Christiansen, eds. *Marine biology of polar regions and effects of stress on marine organisms*. Wiley, New York, pp. 49–64.

Rey, F., Skjoldal, H-R., and Slagstad, D. (1987). Primary productivity in connection with climatic changes in the Barents Sea. In: *The effect of climatic conditions on distribution and populational dynamics of the Barents Sea commercial fishes. Proceedings of III Soviet-Norwegian Symposium.* Murmansk, Russia, pp. 28–55.

Rey-Rassat, C., Irigoien, X., Harris, R., *et al.* (2002a). Energetic cost of gonad development in *Calanus finmarchicus* and *C. helgolandicus*. *Marine Ecology Progress Series*, **238**, 301–6.

Rey-Rassat, C., Irigoien, X., Harris, R., *et al.* (2002b). Growth and development of *Calanus helgolandicus* reared in the laboratory. *Marine Ecology Progress Series*, **238**, 125–38.

Riandey, V., Champalbert, G., Carlotti, F., *et al.* (2005). Zooplankton distribution related to the hydrodynamic features in the Algerian Basin (western Mediterranean Sea) in summer 1997. *Deep-Sea Research I*, **52**, 2029–48.

Ribic, C.A., Chapman, E., Fraser, W.R., *et al.* (2008). Top predators in relation to bathymetry, ice and krill during austral winter in Marguerite Bay, Antarctica. *Deep-Sea Research II*, **55(3–4)**, 485–99.

Rice, J. (1995). Food web theory, marine food webs, and what climate change may do to northern marine fish populations. In: Beamish, R.J., ed. *Climate Change and Northern Fish Populations, Canadian Special Publication of Fisheries and Aquatic Sciences*, **121**, 561–68.

Richardson, A.J. (2008). In hot water: zooplankton and climate change. *ICES Journal of Marine Science*, **65**, 279–95.

Richardson, A.J. and Gibbons, M.J. (2008). Are jellyfish increasing in response to ocean acidification?. *Limnology and Oceanography*, **53**, 2040–5.

Richardson, A.J. and Schoeman, D.S. (2004). Climate impact on plankton ecosystems in the northeast Atlantic. *Science*, **305**, 1609–12.

Richardson, A.J., Verheye, H.M., Mitchell-Innes, B.A., *et al.* (2003). Seasonal and event-scale variation in growth

of *Calanus agulhensis* (Copepoda) in the Benguela upwelling system and implications for spawning of sardine *Sardinops sagax*. *Marine Ecology Progress Series*, **254**, 239–51.

Richardson, A.J., Bakun, A., Hays, G.C., *et al.* (2009). The jellyfish joyride: causes, consequences and management responses to a more gelatinous future.*Trends in Ecology and Evolution*, **24**, 312–22.

Rignot, E. and Kanagaratnam, P. (2006). Changes in the velocity structure of the Greenland ice sheet. *Science*, **311**, 986–90.

Rikardsen, A.H., Diserud, O.H., Elliott, J.M., *et al.* (2007). The marine temperature and depth preferences of Arctic charr (*Salvelinus alpinus*) and sea trout (*Salmo trutta*), as recorded by data storage tags. *Fisheries Oceanography*, **16**, 436–47.

Roberts, C.M. (1997). Connectivity and management of Caribbean coral reefs. *Science*, **278**, 1454–7.

Roberts, C.M., McClean, C.J., Veron, J.E.N., *et al.* (2002). Marine biodiversity hotspots and conservation priorities for tropical reefs. *Science*, **295**, 1280–4.

Robinson, A.R. and Brink, K., eds. (1998). *The sea: ideas and observations on progress in the study of the seas, vol. 11: the global coastal ocean: regional studies and syntheses.* Wiley, Toronto, 1062pp.

Robinson, A.R. and Lermusiaux, P.F.J. (2002). Data assimilation for modeling and predicting coupled physical-biological interactions in the sea. Chapter 12. In: Robinson, A.R., McCarthy, J.J. and Rothschild, B.J., eds. *The sea, vol. 12.* Wiley, New York, pp. 475–1026.

Rochet, M.-J. and Rice, J.C. (2005). Do explicit criteria help in selecting indicators for ecosystem-based fisheries management? *ICES Journal of Marine Science*, **62**, 528–39.

Rochet, M.-J. and Trenkel, V.M. (2003). Which community indicators can measure the impact of fishing? A review and proposals. *Canadian Journal of Fisheries and Aquatic Sciences*, **60**, 86–99.

Roderfeld, H., Blyth, E., Dankers, R., *et al.* (2008). Potential impact of climate change on ecosystems of the Barents Sea Region. *Climate Change*, **87**, 283–303.

Rodionov, S. and Overland, J.E. (2005). Application of a sequential regime shift detection method to the Bering Sea ecosystem. *ICES Journal of Marine Science*, **62**, 328–32.

Rodriguez-Sanchez, R., Lluch-Belda, D., Villalobos, H., *et al.* (2002). Dynamic geography of small pelagic fish populations in the California Current System on the regime time scale (1931–1997). *Canadian Journal Fisheries Aquatic Sciences*, **59**, 1980–8.

Rodwell, L.D., Barbier, E.B., Roberts, C.M., *et al.* (2003). The importance of habitat quality for marine reserve-fishery linkages. *Canadian Journal of Fisheries and Aquatic Sciences*, **60**, 171–81.

Roemmich, D. and McGowan, J. (1995). Climatic warming and the decline of zooplankton in the California Current. *Science*, **267**, 1324–6.

Roessig, J.M., Woodley, C.M., Cech, J.J., *et al.* (2004). Effects of global climate change on marine and estuarine fishes and fisheries. *Reviews in Fish Biology and Fisheries*, **14**, 251–75.

Rooney, N., McCann, K., Gellner, G., *et al.* (2006). Structural asymmetry and the stability of diverse food webs. *Nature*, **442**, 265–9.

Rose, G.A. (2004). Reconciling overfishing and climate change with stock dynamics of Atlantic cod (*Gadus morhua*) over 500 years. *Canadian Journal of Fisheries and Aquatic Sciences*, **61**, 1553–7.

Rose, G.A. and O'Driscoll, R.L. (2002). Capelin are good for cod: can the northern stock rebuild without them? *ICES Journal of Marine Science*, **59**, 1018–26.

Rose, K.A. (2000). Why are quantitative relationships between environmental quality and fish populations so elusive?. *Ecological Applications*, **10**, 367–85.

Rose, K.A. and Cowan, J.H., Jr. (2003). Data, models and decision making in U.S. fisheries management: lessons for ecologists. *Annual Review of Ecology, Evolution, and Systematics*, **34**, 127–51.

Rose, K.A. and Sable, S.E. (2009). Multispecies modeling of fish populations. In: Megrey, B.A. and Moksness, E., eds. *Computers in fisheries research*, 2nd edition. Springer, New York, pp. 373–97.

Rose, K.A., Cowan, J.H., Winemiller, K.O., *et al.* (2001). Compensatory density-dependence in fish populations: importance, controversy, understanding, and prognosis. *Fish and Fisheries*, **2**, 293–327.

Rose, K.A., Megrey, B.A., Hay, D., *et al.* (2007). Climate regime effects on Pacific herring growth using coupled nutrient-phytoplankton-zooplankton and bioenergetics models. *Transactions of the American Fisheries Society*, **137**, 278–97.

Rosenberg, A., Bigford, T.E., Leathery, S., *et al.* (2000). Ecosystem approaches to fishery management through essential fish habitat. *Bulletin of Marine Science*, **66**, 535–42.

Rosenberg, A., Bolster, W.J., Alexander, K.E., *et al.* (2005). The history of ocean resources: modeling cod biomass using historical records. *Frontiers in Ecology and the Environment*, **3**, 78–84.

Rosenberg, A., Mooney-Seus, M.L., Kiessling, I., *et al.* (2008). Lessons from national-level implementation in North America and beyond. In: McLeod, K. and Leslie, H., eds. *Managing for resilience: new directions for marine ecosystem – based management*. Island Press, Washington, DC.

Ross, R.M., Quetin, L.B., Newberger, T., *et al.* (2004). Growth and behavior of larval krill (*Euphausia superba*)

under the ice in late winter 2001 west of the Antarctic Peninsula. *Deep-Sea Research II*, **51**, 2169–84.

Rothschild, B.J. and Osborn, T.R. (1988). Small-scale turbulence and plankton contact rates. *Journal of Plankton Research*, **10**, 465–74.

Rothschild, B.J., Brink, K.H., Dickey, T.D., *et al.* (1989). GLOBEC global ecosystem dynamics. *EOS*, **70(6)**, 82–4.

Rothstein, L.M., Cullen, J.J., Abott, M., *et al.* (2006a). Modeling ocean ecosystems-The PARADIGM program. *Oceanography*, **19(1)**, 16–45.

Rothstein, L.M., Allen, J., Chai, F., *et al.* (2006b). *Report of the ORION Modeling Committee*, April 17 2006, 15pp.

Roux, J.-P. and Shannon, L.J. (2004). Ecosystem approach to fisheries management in the northern Benguela: the Namibian experience. *African Journal of Marine Science*, **26**, 70–93.

Roy, C., Weeks, S.C., Rouault, M., *et al.* (2001). Extreme oceanographic events recorded in the southern Benguela during the 1999–2000 summer season. *South African Journal of Science*, **97**, 465–71.

Rudnick, D.L. and Davis, R.E. (2003). Red noise and regime shifts. *Deep-Sea Research I*, **50**, 691–9.

Runge, J.A. (1984). Egg production of the marine plankton copepod, *Calanus pacificus* Brodsky: laboratory observations. *Journal of Experimental Biology and Ecology*, **74**, 53–66.

Runge, J.A. (1985a). Egg production rates of *Calanus finmarchicus* in the sea of Nova Scotia. *Archiv fur Hydrobiologie, Beihefte, Ergebnisse der Limnologie*, **21**, 33–40.

Runge, J.A. (1985b). Relationship of egg-production of *Calanus pacificus* to seasonal-changes in phytoplankton availability in Puget-Sound, Washington. *Limnology and Oceanography*, **30**, 382–96.

Runge, J.A. and Myers, R.A. (1986). Constraints on the evolution of copepod body size. *Syllogeus*, **58**, 443–7.

Runge, J.A. and Plourde, S. (1996). Fecundity characteristics of *Calanus finmarchicus* in coastal waters of eastern Canada. *Ophelia*, **44**, 171–87.

Runge, J.A. and Roff, J.C. (2000). The measurement of growth and reproductive rates. In: Harris, R.P., Wiebe, P.H., Lenz, J., *et al.* eds. *ICES zooplankton methodology manual*. Academic Press, London, pp. 401–54.

Runge, J.A., Franks, P.J.S., Gentleman, W.C., *et al.* (2005). Diagnosis and prediction of variability in secondary production and fish recruitment processes: developments in physical-biological modelling. In: Robinson, A.R. and Brink, K., eds. *The sea, vol. 13*. Harvard University Press, Cambridge, MA, pp. 413–73.

Runge, J.A., Plourde, S., Joly, P., *et al.* (2006). Characteristics of egg production of the planktonic copepod, *Calanus finmarchicus*, on Georges Bank: 1994–1999. *Deep-Sea Research II*, **53**, 2618–31.

Ryther, J.H. (1969). Photosynthesis and fish production in the sea. *Science*, **166**, 72–6.

Sabine, C.L., Feely, R.A., Gruber, N., *et al.* (2004). The ocean sink for anthropogenic CO_2. *Science*, **305**, 367–71.

Sæmundsson, B. (1934). Probable influence of changes in temperature on the marine fauna of Iceland. *Rapports et Proces-Verbaux des Reunions du Conseil International pour l'Exploration de la Mer*, **86(1)**, 1–6.

Saila, S.B., Kocik, V.L., and McManus, J.W. (1993). Modelling the effects of destructive fishing practices on tropical coral reefs. *Marine Ecology Progress Series*, **94**, 51–60.

Sainsbury, K.J. (1988). The ecological basis of multispecies fisheries, and management of a demersal fishery in tropical Australia. In: Gulland, J.A., ed. *Fish population dynamics*, 2nd edition. Wiley, Chichester, UK, pp. 349–82.

Sainsbury, K.J. (1991). Application of an experimental approach to management in a tropical multispecies fishery with highly uncertain dynamics. *ICES Marine Science Symposia*, **193**, 301–20.

Sainsbury, K.J., Campbell, R.A., Lindholm, R., *et al.* (1997). Experimental management of an Australian multispecies fishery: examining the possibility of trawl-induced habitat modification. *American Fisheries Society Symposia*, **20**, 107–12.

Sainsbury, K.J., Punt, A.E., and Smith, A.D.M. (2000). Design of operational management strategies for achieving fishery ecosystem objectives. *ICES Journal of Marine Science*, **57**, 731–41.

Saiz, E. (1994). Observations of the free-swimming behavior of *Acartia tonsa*: effects of food concentration and turbulent water motion. *Limnology and Oceanography*, **39**, 1566–78.

Saiz, E. and Alcaraz, M. (1992). Free-swimming behavior of *Acartia clausi* (Copepoda, Calanoida) under turbulent water movement. *Marine Ecology Progress Series*, **80**, 229–36.

Saiz, E. and Kiørboe, T. (1995). Predatory and suspension feeding of the copepod *Acartia tonsa* in turbulent environments. *Marine Ecology Progress Series*, **122**, 147–58.

Saiz, E., Alcaraz, M. and Paffenhofer, G.A. (1992). Effects of small-scale turbulence on feeding rate and gross-growth efficiency of three *Acartia* species (Copepoda, Calanoida). *Journal of Plankton Research*, **14**, 1085–97.

Saiz, E., Tiselius, P., Jonsson, P.R., *et al.* (1993). Experimental records of the effects of food patchiness and predation on egg production of *Acartia tonsa*. *Limnology and Oceanography*, **38**, 280–9.

Saiz, E., Calbet, A., Trepat, I., *et al.* (1997). Food availability as a potential source of bias in the egg production method for copepods. *Journal of Plankton Research*, **19**, 1–14.

Saiz, E., Calbet, A., and Broglio, E. (2003). Effects of small-scale turbulence on copepods: the case of *Oithona davisae*. *Limnology and Oceanography*, **48**, 1304–11.

Saiz, E., Calbet, A., Atienza, D., *et al.* (2007). Feeding and production of zooplankton in the Catalan Sea (NW Mediterranean). *Progress in Oceanography*, **74**, 313–28.

Sakamoto, T.T., Hasumi, H., Ishii, M., *et al.* (2005). Responses of the Kuroshio and the Kuroshio extension to global warming in a high-resolution climate model. *Geophysical Research Letters*, **32**, L14617, doi:10.1029/2005GL023384.

Sakshaug, E. (1997). Biomass and productivity distributions and their variability in the Barents Sea. *ICES Journal of Marine Science*, **54**, 341–50.

Sakshaug, E. and Skjoldal, H.R. (1989). Life at the ice edge. *Ambio*, **18**, 60–7.

Sakurai, Y. (2007). An overview of the Oyashio system. *Deep-Sea Research II*, **54**, 2526–42.

Sakurai, Y., Bower, J.R., Nakamura, Y., *et al.* (1996). Affect of temperature on development and survival of *Todarodes pacificus* embryos and paralarvae. *American Malacological Bulletin*, **13**, 89–95.

Sakurai, Y., Bower, J.R., and Ikeda, Y. (2003). Reproductive characteristics of the ommastrephid squid *Todarodes pacificus*. *Fisken og Havet*, **12**, 105–15.

Sakurai, Y., Kiyofuji, H., Saitoh, S., *et al.* (2000). Changes in inferred spawning areas of *Todarodes pacificus* (Cephalopoda: Ommastrephidae) due to changing environmental conditions. *ICES Journal of Marine Science*, **57**, 24–30.

Sakurai, Y., Kiyofuji, H., Saitoh, S., *et al.* (2002). Stock fluctuations of the Japanese common squid, *Todarodes pacificus*, related to recent climate changes. *Fisheries Science*, **68(Suppl. I)**, 226–9.

Sala, E. (2006). Top predators provide insurance against climate change. *Trends in Ecology and Evolution*, **21(9)**, 479–80.

Sala, E. and Knowlton, N. (2006). Global marine biodiversity trends. *Annual Review of Environment and Resources*, **31**, 93–122.

Sambroto, R.N., Niebauer, H.J., Goering, J.J., *et al.* (1986). Relationships among vertical mixing, nitrate uptake and phytoplankton growth during the spring bloom in the southeast Bering Sea middle shelf. *Continental Shelf Research*, **5**, 161–98.

Sameoto, D.D., Jaroszynski, L.O., and Fraser, W.B. (1980). BIONESS, a new design in multiple net zooplankton samplers. *Canadian Journal of Fisheries and Aquatic Sciences*, **37**, 722–4.

Sánchez, F. and Olaso, I. (2004). Effects of fisheries on the Cantabrian Sea shelf ecosystem. *Ecological Modelling*, **172**, 151–74.

Sandweiss, D.H., Maasch, K.A., Chai, F., *et al.* (2004). Geoarchaeological evidence for multidecadal natural climatic variability and ancient Peruvian fisheries. *Quaternary Research*, **61**, 330–4.

Sano, M. (2004). Short-term effects of a mass coral bleaching event on a reef fish assemblage at Iriomote Island, Japan. *Fisheries Science*, **70**, 41–6.

Sara, G. and Sara, R. (2007). Feeding habits and trophic levels of bluefin tuna *Thunnus thynnus* of different size classes in the Mediterranean Sea. *Journal of Applied Ichthyology*, **23**, 122–7.

Sarmiento, J.L., Slater, R., Barber, R., *et al.* (2004). Response of ocean ecosystems to climate warming. *Global Biogeochemical Cycles*, **18**, GB3003, doi: 10.1029/2003GB002134, 23pp.

Scheffer, M., Baveco, J.M., DeAngelis, D.L., *et al.* (1995). Super-individuals a simple solution for modelling large populations on an individual basis. *Ecological Modelling*, **80**, 161–70.

Scheffer, M., and Carpenter, S., (2003). Catastrophic regime shifts in ecosystems: linking theory to observation. *Trends in Ecology and Evolution*, **18**, 648–656.

Scheffer, M., Carpenter, S., and deYoung, B. (2005). Cascading effects of overfishing marine systems. *Trends in Ecology and Evolution*, **20(11)**, 579–81.

Schell, D.M. (2000). Declining carrying capacity in the Bering Sea: isotopic evidence from whale baleen. *Limnology and Oceanography*, **45**, 459–62.

Schiermeier, Q. (2004). Climate findings let fishermen off the hook. *Nature*, **428**, p. 4.

Schindler, D.E., Rogers, D.E., Scheuerell, M.D., *et al.* (2005). Effects of changing climate on zooplankton and juvenile sockeye salmon growth in southwestern Alaska. *Ecology*, **86**, 198–209.

Schmidt, J.O., Floeter, J., Hermann, J.P., *et al.* (2003). Small scale distribution of reproducing female *Pseudocalanus* sp. in the Bornholm Basin (central Baltic Sea) during two contrasting hydrographic regimes in spring 2002 and 2003. *ICES Council Meeting Papers*, **CM 2003/P:37**.

Schmidt, K., Atkinson, A., Petzke, K.J., *et al.* (2006). Protozoans as a food source for Antarctic krill, *Euphausia superba*: complementary insights from stomach content, fatty acids, and stable isotopes. *Limnology and Oceanography*, **51**, 2409–27.

Schmidt, K., Atkinson, A., Stubing, D., *et al.* (2003). Trophic relationships among Southern Ocean copepods and krill: some uses and limitations of a stable isotope approach. *Limnology and Oceanography*, **48**, 277–89.

Schmidt, K., McClelland, J.W., Mente, E., *et al.* (2004). Trophic-level interpretation based on delta N-15 values: implications of tissue-specific fractionation and amino acid composition. *Marine Ecology Progress Series*, **266**, 43–58.

Schmittner, A. (2005). Decline of the marine ecosystem caused by a reduction in the Atlantic overturning circulation. *Nature*, **434**, 628–33.

Schmittner, A., Latif, M., and Schneider, B. (2005). Model projections of the North Atlantic thermohaline circulation for the 21st century assessed by observations. *Geophysical Research Letters*, **32**, L23710, doi:10.1029/2005GL024368.

Schmitz, W.J. (1996). On the world ocean circulation, vol. II: the Pacific and Indian Oceans/a global update. *Woods Hole Oceanographic Institution Technical Report*, **WHOI-96–08**, 237pp.

Schnute, J.T., Maunder, M.N., and Ianelli, J.N. (2007). Designing tools to evaluate fishery management strategies: can the scientific community deliver?. *ICES Journal of Marine Science*, **64**, 1077–84.

Schrank, W.E. (2005). The Newfoundland fishery: ten years after the moratorium. *Marine Policy*, **29**, 407–20.

Schwartzlose, R.A., Alheit, J., Bakun, A., *et al.* (1999). Worldwide large-scale fluctuations of sardine and anchovy populations. *South African Journal of Marine Science*, **21**, 289–347.

Schweder, T., Hagen, G.S., and Hatlebakk, E. (1998). On the effect on cod and herring fisheries of retuning the revised management procedure for minke whaling in the greater Barents Sea. *Fisheries Research*, **37**, 77–95.

Schwing, F.B. (2008). Regime shifts, physical forcing. In: Steele, J.H., Thorpe, S.A., and Turekian, K.K., ed. *Encyclopedia of ocean sciences*: Elsevier. pp. 4345–52.

Schwing, F., Murphree, T., deWitt, L., *et al.* (2002). The evolution of oceanic and atmospheric anomalies in the northeast Pacific during the *El Niño* and *La Niña* events of 1995–2001. *Progress in Oceanography*, **54**, 459–91.

Schwing, F., Bond, N.A., Bograd, S.J., *et al.* (2006). Delayed coastal upwelling along the U.S. west coast in 2005: a historical perspective. *Geophysical Research Letters*, **33**, L22S01.

Schwing, F., Mendelssohn, R., Bograd, S.J., *et al.* (2010). Climate change, teleconnection patterns, and regional processes forcing marine populations in the Pacific. *Journal of Marine Systems*, **79**, 245–57.

Selander, E., Thor, P., Toth, G., *et al.* (2006). Copepods induce paralytic shellfish toxin production in marine dinoflagellates. *Proceedings of the Royal Society of London, Series B, Biological Sciences*, **273**, 1673–80.

Senina, I., Sibert, J., and Lehodey, P. (2008). Parameter estimation for basin-scale ecosystem-linked population models of large pelagic predators: application to skipjack tuna. *Progress in Oceanography*, **78**, 319–35.

Senjyu, T. and Watanabe, T. (2004). Decadal signal in the sea surface temperature off the San'in coast in the southwestern Japan Sea. *Reports of the Research Institute for Applied Mechanics, Kyushu University*, **127**, 49–53.

Sette, O.E. (1950). Biology of the Atlantic Mackerel (*Scomber scombrus*) of North America, Part II: migrations and habits. *Fishery Bulletin*, **49**, 251–358.

Seuront, L., Schmitt, F.G., Brewer, M.C., *et al.* (2004a). From random walk to multifractal random walk in zooplankton swimming behavior. *Zoological Studies*, **43**, 498–510.

Seuront, L., Yamazaki, H., and Souissi, S. (2004b). Hydrodynamic disturbance and zooplankton swimming behavior. *Zoological Studies*, **43**, 376–87.

Shaffer, S.A. and Costa, D.P. (2006). A database for the study of marine mammal behavior: gap analysis, data standardization, and future directions. *Oceanic Engineering, IEEE Journal of Ocean Engineering*, **31**, 82–6.

Shaffer, S.A., Tremblay, Y., Weimerskirch, H., *et al.* (2006). Migratory shearwaters integrate oceanic resources across the Pacific Ocean in an endless summer. *Proceedings of the National Academy of Sciences of the United States of America*, **103**, 12799–802.

Shanks, A.L. and Brink, L. (2005). Upwelling, downwelling, and cross-shelf transport of bivalve larvae: test of a hypothesis. *Marine Ecology Progress Series*, **302**, 1–12.

Shannon, L.J, Cochrane, K.L., Moloney, C.L., *et al.* (2004a). Ecosystem approach to fisheries management in the southern Benguela: a workshop overview. *African Journal of Marine Science*, **26**, 1–8.

Shannon, L.J, Christensen, V., and Walters, C. (2004b). Modelling stock dynamics in the southern Benguela ecosystem for the period 1978–2002. In: Shannon, L.J., Cochrane, K.L., and Pillar, S.C., eds. *Ecosystem approaches to fisheries in the Southern Benguela. African Journal of Marine Science*, **26**, 179–96.

Shannon, L.J, Field, J.G., and Moloney, C.L. (2004c). Simulating anchovy-sardine regime shifts in the southern Benguela ecosystem. *Ecological Modelling*, **172**, 269–81.

Shannon, L.J, Cury, P.M., Nel, D., *et al.* (2006a). How can science contribute to an ecosystem approach to pelagic, demersal and rock lobster fisheries in South Africa? *African Journal of Marine Science*, **28**, 115–57.

Shannon, L.J, Hempel, G., Malanotte-Rizzoli, P., *et al.* eds. (2006b). *Benguela: predicting a large marine ecosystem.* Elsevier, Large Marine Ecosystems Series 14. 410pp.

Shannon, L.J, Neira, S., and Taylor, M. (2008a). Comparing internal and external drivers in the southern Benguela and the southern and northern Humboldt upwelling ecosystem. *African Journal of Marine Science*, **30(1)**, 63–84.

Shannon, L.J, Coll, M., Neira, S., *et al.* (2009). Impacts of fishing and climate change explored using trophic models. In: Checkley, D., Alheit, J., Oozeki, Y., *et al.* eds. *Climate change and small pelagic fish.* Cambridge University Press, Cambridge.

Shannon, L.J, Jarre, A., and Schwing, F.B. (2008b). Regime shifts, ecological aspects. In: Steele, J.H., Thorpe, S.A., and Turekian, K.K., ed. Encyclopedia of ocean sciences. Elsevier. pp 4329–39.

Shaul, L., Weitcamp, L., Simpson, K., *et al.* (2007). Trends in abundance and size of coho salmon in the Pacific Basin. *North Pacific Anadromous Fish Commission Bulletin*, **4**, 93–104.

Shchepetkin, A.F. and McWilliams, J.C. (2005). Regional ocean model system: a split-explicit ocean models with a free surface and topography – following vertical coordinate. *Ocean Modelling*, **9**, 347–404.

Shelton, P.A., Sinclair, A.F., Chouinard, G.A., *et al.* (2006). Fishing under low productivity conditions is further delaying recovery of Northwest Atlantic cod (*Gadus morhua*). *Canadian Journal of Fisheries and Aquatic Sciences*, **63**, 235–8.

Sheppard, S.K. and Harwood, J.D. (2005). Advances in molecular ecology: tracking trophic links through predator prey food-webs. *Functional Ecology*, **19**, 751–62.

Sherman, K., Lasker, R., Richards, W., *et al.* (1983). Ichthyoplankton and fish recruitment studies in large marine ecosystems. *Marine Fisheries Review*, **45**, 1–45.

Shiganova, T.A. (1998). Invasion of the Black Sea by the ctenophore *Mnemiopsis leidyi* and recent changes in pelagic community structure. *Fisheries Oceanography*, **7**, 305–10.

Shiganova, T.A. and Panov, V.E. (2003). *Mnemiopsis leidyi* A. Agassiz, 1865. Compiled for the Regional Biological Invasions Center. Available at: http://www.zin.ru/projects/invasions/gaas/mnelei.htm

Shiganova, T.A., Bulgakova, Y.V., Volovik, S.P., *et al.* (2001). The new invader *Beroe ovata* Mayer 1912 and its effect on the ecosystem in the northeastern Black Sea. *Hydrobiologia*, **451**, 187–97.

Shin, K., Jang, M.C., Jang, P.K., *et al.* (2003). Influence of food quality on egg production and viability of the marine planktonic copepod *Acartia omorii*. *Progress in Oceanography*, **57**, 265–77.

Shin, Y. and Cury, P. (2001). Exploring fish community dynamics through size-dependent trophic interactions using a spatialized individual-based model. *Aquatic Living Resources*, **14**, 65–80.

Shin, Y. and Cury, P. (2004). Using an individual-based model of fish assemblages to study the response of size spectra to changes in fishing. *Canadian Journal of Fisheries and Aquatic Sciences*, **61**, 414–31.

Shin, Y., Rochet, M.-J., Jennings, S., *et al.* (2005). Using size-based indicators to evaluate the ecosytems effects of fishing. *ICES Journal of Marine Science*, **62**, 384–96.

Shiomoto, A., Tadokoro, K., Nagasawa, K., *et al.* (1997). Trophic relations in the subarctic North Pacific ecosystem: possible feeding effect from pink salmon. *Marine Ecology Progress Series*, **150**, 75–85.

Shurin, J.B., Borer, E.T., Seabloom, E.W., *et al.* (2002). A cross-system comparison of the strength of trophic cascades. *Ecology Letters*, **5**, 785–91.

Sibert, J., Hampton, J., Kleiber, P., *et al.* (2006). Biomass, size, and trophic status of top predators in the Pacific Ocean. *Science*, **314**, 1773–6.

Siegel, D.A. (2001). Oceanography: the Rossby rototiller. *Nature*, **409**, 576–7.

Siegenthaler, U., Stocker, T.F., Monnin, E., *et al.* (2005). Stable carbon cycle-climate relationship during the late Pleistocene. *Science*, **310**, 1313–7.

Sieracki, C.K., Sieracki, M.E., and Yentsch, C.S. (1998). An imaging-in-flow system for automated analysis of marine microplankton. *Marine Ecology Progress Series*, **168**, 285–96.

Simonelli, P., Troedsson, C., Nejstgaard, J.C., *et al.* (in preparation). New protocol to quantitatively extract prey DNA from the plankton and inside copepod guts.

Sims, D.W., Genner, M.J., Southward, A.J., *et al.* (2001). Timing of squid migration reflects the North Atlantic climate variability. *Proceedings of the Royal Society of London, Series B, Biological Sciences*, **268**, 2607–11.

Sims, D.W., Wearmouth, V.J., Genner, M.J., *et al.* (2004). Low-temperature-driven early spawning migration of a temperate marine fish. *Journal of Animal Ecology*, **73**, 333–41.

Sinclair, M. (1988). *Marine populations: an essay on population regulation and speciation.* Washington Sea Grant/University of Washington Press, Seattle, Washington, 252pp.

Sinclair, M. and Iles, T.D. (1989). Population regulation and speciation in the oceans. *Journal du Conseil International pour l'Exploration de la Mer*, **45**, 165–75.

Sinclair, M. and Page, F. (1995). Cod fishery collapses and North Atlantic GLOBEC. *US GLOBEC News*, **8**, 1–16.

Sinclair, M., Arnason, R., Csirke, J., *et al.* (2002). Responsible fisheries in the marine ecosystem. *Fisheries Research*, **58**, 255–65.

Singarajah, K.V. (1969). Escape reactions of zooplankton: the avoidance of a pursuing siphon tube. *Journal of Experimental Marine Biology and Ecology*, **3**, 171–8.

Singarajah, K.V. (1975). Escape reactions of zooplankton: effects of light and turbulence. *Journal of the Marine Biological Association of the United Kingdom*, **55**, 627–39.

Sirabella, P., Giuliani, A., Colosimo, A., *et al.* (2001). Breaking down the climate effects on cod recruitment by principal component analysis and canonical correlation. *Marine Ecology Progress Series*, **216**, 213–22.

Sissener, E. and Bjørndal, T. (2005). Climate change and the migratory pattern for Norwegian spring-spawning herring – implications for management. *Marine Policy*, **29**, 299–309.

Sissenwine, M.P. and Shepherd, J.G. (1987). An alternative perspective on recruitment overfishing and biological reference points. *Canadian Journal of Fisheries and Aquatic Sciences*, **44**, 913–8.

Skjoldal, H.R. and Rey, F. (1989). Pelagic production and variability of the Barents Sea ecosystem. In: Sherman, K. and Alexander, L.M., eds. *Biomass yields and geography of large marine ecosystems.* AAAS Selected Symposium 111, Westview Press, Boulder, CO, pp. 241–86.

Skjoldal, H.R., Hassel, A., Rey, F., *et al.* (1987). Spring phytoplankton development and zooplankton reproduction in the central Barents Sea in the period 1977–84. In: Loeng, H., ed. *Proceedings of the 3rd Soviet-Norwegian Symposium, Murmansk, 1986.* Institute of Marine Research, Bergen, Norway, pp. 59–89.

Gjøsaeter, H. and Loeng, H. (1992). The Barents Sea ecosystem in the 1980s: ocean climate, plankton, and capelin growth. *ICES Marine Science Symposia*, **195**, 278–90.

Skogen, M.D. (2005). Clupeoid larval growth and plankton production in the Benguela upwelling system. *Fisheries Oceanography*, **14**, 64–70.

Slagstad, D. and Tande, K.S. (2007). Structure and resilience of overwintering habitats of *Calanus finmarchicus* in the Eastern Norwegian Sea. *Deep-Sea Research II*, **54**, 2702–15.

Slagstad, D. and Wassmann, P. (1997). Climate change and carbon flux in the Barents Sea: 3-D simulations of ice-distribution, primary production and vertical export of particulate organic matter. *Memoirs of the National Institute of Polar Research*, Special Issue **51**, 119–41.

Slagstad, D., Tande, K., and Wassmann, P. (1999). Modelled carbon fluxes as validated by field data on the north Norwegian shelf during the productive period in 1994. *Sarsia*, **84**, 303–17.

Smetacek, V. and Nicol, S. (2005). Polar ocean ecosystems in a changing world. *Nature*, **437**, 362–8.

Smith, A.D.M., Punt, A.E., Wayte, S.E., *et al.* (1996). Evaluation of harvesting strategies for eastern gemfish (*Rexea solandri*) using Monte Carlo simulation. In: Smith, A.D.M., ed. *Evaluation of harvesting strategies for Australian fisheries at different levels of risk from economic collapse.* CSIRO, Hobart, Australia, pp. 120–60.

Smith, A.D.M., Sainsbury, K.J., and Stevens, R.A. (1999). Implementing effective fisheries-management systems-management strategy evaluation and the Australian partnership approach. *ICES Journal of Marine Science*, **56**, 967–79.

Smith, A.D.M., Fulton, E.J., Hobday, A.J., *et al.* (2007a). Scientific tools to support the practical implementation of ecosystem-based fisheries management. *ICES Journal of Marine Science*, **64**, 633–9.

Smith, T.M., Peterson, T.C., Lawrimore, J.H., *et al.* (2005). New surface temperature analyses for climate monitoring. *Geophysical Research Letters*, **32(14)**, L13712.

Smith, W.O., Ainley, D.G., and Cattaneo-Vietti, R. (2007b). Trophic interactions within the Ross Sea continental shelf ecosystem. *Philosophical Transactions of the Royal Society of London, Series B, Biological Sciences*, **362**, 95–111.

Sørnes, T. and Aksnes, D.L. (2006). Concurrent temporal patterns in light absorbance and fish abundance. *Marine Ecology Progress Series*, **325**, 181–6.

Sosik, H.M. and Olson, R.J. (2007). Automated taxonomic classification of phytoplankton sampled with imaging in-flow cytometry. *Limnology and Oceanography Methods*, **5**, 204–16.

Soutar, A. and Isaacs, J.D. (1974). Abundance of pelagic fish during the 19th and 20th centuries as recorded in anaerobic sediment off the Californias. *Fishery Bulletin*, **72**, 257–94.

Sparholt, H. (1994). Fish species interactions in the Baltic Sea. *Dana*, **10**, 131–62.

Sparre, P. (1991). Introduction to multispecies virtual population analysis. *ICES Marine Science Symposia*, **193**, 12–21.

Speirs, D.C., Gurney, W.S.C., Heath, M.R., *et al.* (2005). Modelling the basin-scale demography of *Calanus finmarchicus* in the north-east Atlantic. *Fisheries Oceanography*, **14**, 333–58.

Speirs, D.C., Gurney, W.S.C., Heath, M.R., *et al.* (2006). Ocean-scale modelling of the distribution, abundance, and seasonal dynamics of the copepod *Calanus finmarchicus*. *Marine Ecology Progress Series*, **313**, 173–92.

Spencer, P.D. (1997). Optimal harvesting of fish populations with nonlinear rates of predation and autocorrelated environmental variability. *Canadian Journal of Fisheries and Aquatic Sciences*, **54**, 59–74.

Stabeno, P.J., Schumacher, J.D., Davis, R.F., *et al.* (1998). Under-ice observations of water column temperature, salinity, and spring phytoplankton dynamics: eastern Bering Sea shelf. *Journal of Marine Research*, **56**, 239–55.

Stabeno, P.J., Bond, N.A., Kachel, N.B., *et al.* (2001). On the temporal variability of the physical environment of the south-eastern Being Sea. *Fisheries Oceanography*, **10**, 81–98.

Stachowicz, J.J., Terwin, J.R., Whitlatch, R.B., *et al.* (2002). Linking climate change and biological invasions: ocean warming facilitates nonindigenous species invasions. *Proceedings of the National Academy of Sciences of the United States of America*, **99**, 15497–500.

Stammler, D., Wunsch, C., Giering, R., *et al.* (2002). Global ocean circulation during 1992–1997 estimate from ocean observations and a general circulation model. *Journal of Geophysical Research-Oceans*, **107**, 3118, doi:10.1029/2001JC000888.

Stanton, T.K. (1985a). Volume scattering: echo peak PDF. *Journal of the Acoustical Society of America*, **77**, 1358–66.

Stanton, T.K. (1985b). Density estimates of biological sound scatterers using sonar echo peak PDFs. *Journal of the Acoustical Society of America*, **78**, 1868–73.

Starkey, D.J., Holm, P., and Barnard, M., eds. (2008). *Oceans past: management insights from the history of marine animal populations*. Earthscan Research Editions, London.

Steele, J.H. (1974). *The structure of marine ecosystems*. Harvard University Press, Cambridge, MA, x,128pp.

Steele, J.H. (1985). A comparison of terrestrial and marine ecological systems. *Nature*, **313**, 355–8.

Steele, J.H. (2004). Regime shifts in the ocean: reconciling observations with theory. *Progress in Oceanography*, **60**, 135–41.

Steele, J.H. and Henderson, E.W. (1992a). A simple model for plankton patchiness. *Journal of Plankton Research*, **10**, 1397–403.

Steele, J.H. and Henderson, E.W. (1992b). The role of predation in plankton models. *Journal of Plankton Research*, **14**, 157–72.

Steele, J.H., Carpenter, S.R., Cohen, J.E., *et al.* (1993). Comparing terrestrial and marine ecological systems. In: Leven, S.A., Powell, T.M., and Steele, J.H., eds. *Lecture Notes in Biomathematics 96*, pp. 1–12. Springer-Verlag, 307pp.

Steele, J.H., Collie, J.S., Bisagni, J.J., *et al.* (2007). Balancing end-to-end budgets of the Georges Bank ecosystem. *Progress in Oceanography*, **74**, 423–48.

Steele, J.H., Collie, J.S., Bisagni, J.J., *et al.* (2008). Reconciling concepts in biological oceanography. *GLOBEC International Newsletter*, **14(1)**, 3–7.

Steffen, W., Crutzen, P.J., and McNeill, J.R. (2007). The Anthropocene: are humans now overwhelming the great forces of Nature? *Ambio*, **36**, 614–21.

Stegert, C., Kreus, M., Carlotti, F., *et al.* (2007). Parameterisation of a zooplankton population model for *Pseudocalanus elongatus* using stage durations from laboratory experiments. *Ecological Modelling*, **206**, 213–30.

Stegmann, P.M., Quinlan, J.A., Werner, F.E., *et al.* (1999). Atlantic menhaden recruitment to a southern estuary: defining potential spawning regions. *Fisheries Oceanography*, **8**, 111–23.

Steig, T.W. and Greene, C.H. (2006). Real time three-dimensional behavioral results of acoustically tagged shrimp. *EOS, Transactions, American Geophysical Union*, **87**, 36.

Steingrund, P. and Gaard, E. (2005). Relationship between phytoplankton production and cod production on the Faroe shelf. *ICES Journal of Marine Science*, **62**, 163–76.

Stenevik, E.K. and Sundby, S. (2007). Impacts of climate change on commercial fish stocks in Norwegian waters. *Marine Policy*, **31**, 19–31.

Stenevik, E.K., Melle, W., Gaard, E., *et al.* (2007). Egg production of *Calanus finmarchicus*-A basin-scale study. *Deep-Sea Research II*, **54**, 2672–85.

Stenseth, N.C., Ottersen, G., Hurrell, J.W., *et al.* (2003). Studying climate effects on ecology through the use of climate indices: the North Atlantic oscillation, El Niño-southern oscillation and beyond. *Proceedings of the Royal Society of London, Series B, Biological Sciences*, **270**, 2087–96.

Stenseth, N.C., Ottersen, G., Hurrell, J.W., *et al.*, eds. (2004). *Marine ecosystems and climate variation: the North Atlantic, a comparative perspective*. Oxford University Press, Oxford.

Stephenson, D.B., Pavan, V., Collins, M., *et al.* Participating CMIP2 Modelling Groups. (2006). North Atlantic oscillation response to transient greenhouse gas forcing and the impact on European winter climate: a CMIP2 multi-model assessment. *Climate Dynamics*, **27**, 401–20.

Stibor, H., Vadstein, O., Lippert, B., *et al.* (2004). Calanoid copepods and nutrient enrichment determine population dynamics of the appendicularian *Oikopleura dioica*: a mesocosm experiment. *Marine Ecology Progress Series*, **270**, 209–15.

Stocker, T.F. and Johnsen, S.J. (2003). A minimum thermodynamic model of the bipolar seesaw. *Paleoceanography*, **18**, art: 1087.

Stow, C.A., Jolliff, J., McGillicuddy, D.J., *et al.* (2009). Skill assessment for coupled biological/physical models of marine systems. *Journal of Marine Systems*, **76**, 4–16.

Strand, E., Huse, G., and Giske, J. (2002). Artificial evolution of life history and behavior. *American Naturalist*, **159**, 624–44.

Strathmann, R.R., Hughes, T.P., Kuris, A.M., *et al.* (2002). Evolution of local recruitment and its consequences for marine populations. *Bulletin of Marine Science*, **70**, S377–96.

Strickler, J.R. (1982). Calanoid copepods, feeding currents, and the role of gravity. *Science*, **218**, 158–60.

Strickler, J.R. (1984). Sticky water: a selective force in copepod evolution. In: Meyers, D.G. and Strickler, J.R., eds. *Trophic interactions within aquatic ecosystems*. Westview Press, Boulder, CO, pp. 187–239.

Strickler, J.R. (1985). Feeding currents in calanoid copepods: two new hypotheses. In: Laverack, M.S., ed. *Physiological adaptations of marine animals. Symposia of the Society for Experimental Biology*, **39**, 459–85.

Strickler, J.R. (1998). Observing free-swimming copepods mating. *Philosophical Transactions of the Royal Society of London, Series B, Biological Sciences*, **353**, 671–80.

Stroeve, J., Holland, M.M., Meier, W., *et al.* (2007). Arctic sea ice decline: faster than forecast, *Geophysical Research Letters*, **34**, L09501, doi: 10.1029/2007GL029703.

Strong, D.R. (1992). Are trophic cascades all wet? Differentiation and donor-control in speciose ecosystems. *Ecology*, **73**, 747–54.

Strong, D.R. (2002). Populations or entomopathogenic nematodes in food webs. In: Gaugler, R., ed. *Entomopathogenic nematology*. CABI Publishing, New York.

Strub, P.T., Batchelder, H.P., and Weingartner, T.J. (2002). US GLOBEC Northeast Pacific Program: overview. *Oceanography*, **15**, 30–5.

Struck, U., Altenbach, A.V., Emeis, K.C., *et al.* (2002). Changes of the upwelling rates of nitrate preserved in the delta N-15-signature of sediments and fish scales from the diatomaceous mud belt of Namibia. *Geobios*, **35**, 3–11.

SUBPESCA. (2007). Cuota global anual de captura de jurel, año 2008. Subsecretaría de Pesca, Ministerio de economía, Gobierno de Chile. *Informe Técnico (R. Pesq.)*, **86/2007**, 48pp.

Suchman, C.l., Daly, E.A., Keister, J.E., *et al.* (2008). Feeding patterns and predation potential of scyphomedusae in a highly productive upwelling area. *Marine Ecology Progress Series*, **358**, 161–72.

Sünksen, K. (2007). Discarded by-catch in shrimp fisheries in Greenlandic offshore waters 2006–2007. *NAFO Scientific Council Research Document*, **07/88**.

Sumaila, U.R. (2004). Intergenerational cost benefit analysis and marine ecosystem restoration. *Fish and Fisheries*, **5**, 329–43.

Sumaila, U.R. (2008). Getting values and valuation right: a must for reconciling fisheries with conservation. *American Fisheries Society Symposium*, **49**, 707–12.

Sumaila, U.R. and Walters, C. (2005). Intergenerational discounting: a new intuitive approach. *Ecological Economics*, **52**, 135–42.

Sundby, S. (1997). Turbulence and ichthyplankton: influence on vertical distributions and encounter rates. *Scientia Marina*, **61**, 159–76.

Sundby, S. (2000). Recruitment of Atlantic cod stocks in relation to temperature and advection of copepod populations. *Sarsia*, **85**, 277–98.

Sundby, S. (2006). Klimavariasjoner, klimaendringer og virkninger på marine økosystemer (Climate variations, climate change and impacts on marine ecosystems). *CICERONE*, **4/2006**, 37–9.

Sundby, S. and Drinkwater, K. (2007). On the mechanisms behind salinity anomaly signals of the northern North Atlantic. *Progress in Oceanography*, **73**, 190–202.

Sundby, S. and Fossum, P. (1990). Feeding conditions of Arcto-norwegian cod larvae compared with the

Rothschild-Osborn theory on small-scale turbulence and plankton contact rates. *Journal of Plankton Research*, **12**, 1153–62.

Sundby, S. and Nakken, O. (2008). Spatial shifts in spawning habitats of Arcto-Norwegian cod related to multi-decadal climate oscillations and climate change. *ICES Journal of Marine Science*, **65**, 953–62.

Sundby, S., Ellertsen, B., and Fossum, P. (1994). Encounter rates between first-feeding cod larvae and their prey during moderate to strong turbulent mixing. *ICES Marine Science Symposium*, **198**, 393–405.

Suthers, I.M. and Sundby, S. (1996). Role of the midnight sun: comparative growth of pelagic juvenile cod (*Gadus morhua*) from the Arcto-Norwegian and a Nova Scotian stock. *ICES Journal of Marine Science*, **53**, 827–36.

Sutor, M., Cowles, T.J., Peterson, W.T., *et al.* (2005). Acoustic observations of finescale zooplankton distributions in the Oregon upwelling region. *Deep-Sea Research II*, **52**, 109–21.

Sutton, R.T. and Hodson, D.L.R. (2005). Atlantic Ocean forcing of North American and European summer climate. *Science*, **309**, 115–8.

Svensen, C. and Kiørboe, T. (2000). Remote prey detection in *Oithona similis*: hydromechanical versus chemical cues. *Journal of Plankton Research*, **22**, 1155–66.

Sverdrup, H.U. (1953). On conditions for the vernal blooming of phytoplankton. *Journal du Conseil International pour l'Exploration de la Mer*, **11**, 287–95.

Swartzman, G., Hickey, B., Kosroc, P.M., *et al.* (2005). Poleward and equatorward currents in the Pacific Eastern 2005. Boundary Current in summer 1995 and 1998 and their relationship to the distribution of euphausiids. *Deep-Sea Research II*, **52**, 73–88.

Sydeman, W.J., Bradley, R.W., Warzybok, P., *et al.* (2006). Planktivorous auklet, *Ptychoranphus aleuticus*, responses to ocean climate, 2005: unusual atmospheric blocking?. *Geophysical Research Letters*, **33**, L22S09.

Symes, D., ed. (1998). *Property rights and regulatory systems in fisheries*. Fishing News Books, Oxford.

Symes, D., ed. (2000). *Fisheries dependent regions*. Fishing News Books, Oxford.

Tadokoro, K., Chiba, S., Ono, T., *et al.* (2005). Interannual variation in *Neocalanus* biomass in the Oyashio waters of the western North Pacific. *Fisheries Oceanography*, **14**, 210–22.

Takasuka, A., Oozeki, Y., and Aoki, I. (2007). Optimal growth temperature hypothesis: why do anchovy flourish and sardine collapse or vice versa under the same ocean regime? *Canadian Journal of Fisheries and Aquatic Sciences*, **64**, 768–76.

Tande, K. and Miller, C.B. (1996). Preface to: Trans-Atlantic Study of *Calanus finmarchicus*. Proceedings of a workshop. *Ophelia*, **44**, 1–3.

Tang, X., Stewart, W.K., Vincent, L., *et al.* (1998). Automatic plankton image recognition. *Artificial Intelligence Review*, **12**, 177–99.

Taylor, J.E., III. (1999). *Making salmon: an environmental history of the northwest fisheries crisis*. University of Washington Press, Seattle, WA and London, 421pp.

Taylor, M.H., Tam, J., Blaskovic, V., *et al.* (2009). Trophic modelling of the northern Humboldt current ecosystem, Part II: elucidating ecosystem dynamics from 1995 to 2004 with a focus on the impact of ENSO. *Progress in Oceanography*, **79(2–4)**, 366–78.

Teo, S.L.H., Boustany, A., Blackwell, S.B., *et al.* (2004). Validation of geolocation estimates based on light level and sea surface temperature from electronic tags. *Marine Ecology Progress Series*, **283**, 81–98.

Teo, S.L.H., Botsford, L.W., and Hastings, A. (2009). Spatio-temporal covariability in coho salmon (*Oncorhynchus kisutch*) survival, from California to Southeast Alaska. *Deep-Sea Research II*. **56**, 2570–8.

Thiele, D., Chester, E.T., Moore, S.E., *et al.* (2004). Seasonal variability in whale encounters in the Western Antarctic Peninsula. *Deep-Sea Research II*, **51**, 2311–25.

Thomas, K. (1983). *Man and the natural world*. Penguin Books, Harmondworth, UK.

Thorne, R.E. and Thomas, G.L. (2008). Herring and the "Exxon Valdez" oil spill: an investigation into historical data conflicts. *ICES Journal of Marine Science*, **65**, 44–50.

Thorpe, S.E., Heywood, K.J., Stevens, D.P., *et al.* (2004). Tracking passive drifters in a high resolution ocean model: implications for interannual variability of larva; krill transport to South Georgia. *Deep-Sea Research I*, **51**, 909–20.

Thorpe, S.E., Murphy, E.J., and Watkins, J.L. (2007). Circumpolar connections between Antarctic krill (*Euphausia superba* Dana) populations: investigating the roles of ocean and sea ice transport. *Deep-Sea Research II*, **54**, 792–810.

Tian, Y., Kidokoro, H., Watanabe, T., *et al.* (2008). The late 1980s regime shift in the ecosystem of Tsushima warm current in the Japan/East Sea: evidence from historical data and possible mechanisms. *Progress in Oceanography*, **77**, 127–45.

Tilman, D., Kilham, S.S., and Kilham, P. (1982). Phytoplankton community ecology: the role of limiting nutrients. *Annual Review of Ecology and Systematics*, **13**, 349–72.

Timmermann, A. and Jin, F.F. (2002). Phytoplankton influences on tropical climate. *Geophysical Research Letters*, **29**, 2104, doi:10.1029/2002GL015434.

Tiselius, P. (1992). Behavior of *Acartia tonsa* in patchy food environments. *Limnology and Oceanography*, **37**, 1640–51.

Tiselius, P. and Jonsson, P.R. (1990). Foraging behaviour of six calanoid copepods: observations and hydrodynamic analysis. *Marine Ecology Progress Series*, **66**, 23–33.

Tiselius, P. and Jonsson, P.R. and Verity, P.G. (1993). A model evaluation of the impact of food patchiness on foraging strategy and predation risk in zooplankton. *Bulletin of Marine Science*, **53**, 247–64.

Titelman, J. (2001). Swimming and escape behavior of copepod nauplii: implications for predator-prey interactions among copepods. *Marine Ecology Progress Series*, **213**, 203–13.

Titelman, J. and Fiksen, O. (2004). Ontogenetic vertical distribution patterns in small copepods: field observations and model predictions. *Marine Ecology Progress Series*, **284**, 49–63.

Titelman, J. and Kiørboe, T. (2003). Motility of copepod nauplii and implications for food encounter. *Marine Ecology Progress Series*, **247**, 123–35.

Tittensor, D.P., deYoung, B., and Tang, C.L. (2003). Modelling the distribution, sustainability and diapause emergence timing of the copepod *Calanus finamarchicus* in the Labrador Sea. *Fisheries Oceanography*, **12(4–5)**, 299–316.

Tjelmeland, S. and Bogstad, B. (1998). MULTSPEC – a review of multispecies modelling project for the Barents Sea. *Fisheries Research*, **37**, 127–42.

Tjelmeland, S. and Lindstrøm, U. (2005). An ecosystem element added to the assessment of Norwegian spring spawning herring: implementing predation by minke whales. *ICES Journal of Marine Science*, **62**, 285–94.

Tokar, J.M. and Dickey, T.D. (2000). Chemical sensor technology: current and future applications. In: Varney, M.S., ed. *Chemical sensors in oceanography*. Gordon and Breach Scientific Publishers, Amsterdam, pp. 303–29.

Toresen, R. and Østvedt, L.J. (2000). Variation in abundance of Norwegian spring-spawning herring (*Clupea harengus*, Clupeidae) throughout the 20th century and the influence of climatic fluctuations. *Fish and Fisheries*, **1**, 231–56.

Torgersen, T., Kaartvedt, S., Melle, W., *et al.* (1997). Large scale distribution of acoustic scattering layer at the Norwegian continental shelf and eastern Norwegian Sea. *Sarsia*, **82**, 87–96.

Townsend, C.R., Begon, M., and Harper, J.L. (2000). *Essentials of ecology*, 2nd edition. Blackwell Publishing, Oxford, 544pp.

Townsend, D.W. and Thomas, A.C. (2002). Winter-spring transition of phytoplankton chlorophyll and inorganic nutrients on Georges Bank. *Marine Ecology Progress Series*, **228**, 57–74.

Travers, M., Shin, Y.-J., Jennings, S., *et al.* (2007). Towards end-to-end models for investigating the effects of climate and fishing in marine ecosystems. *Progress in Oceanography*, **75**, 751–70.

Tremblay, M.J., Loder, J.W., Werner, F.E., *et al.* (1994). Drift of sea scallop larvae *Placopecten magellanicus* on Georges Bank: a model study of the roles of mean advection, larval behavior and larval origin. *Deep-Sea Research II*, **41**, 7–49.

Trenberth, K.E. (1997). The use and abuse of climate models. *Nature*, **386**, 131–3.

Trenberth, K.E., Stepaniak, D.P., and Caron, J.M. (2002). Interannual variations in the atmospheric heat budget. *Journal of Geophysical Research -Oceans*, **107C**, 4066, doi:10.1029/2000JD000297.

Trenberth, K.E., Jones, P.D., Ambenje, P., *et al.* (2007). Observations: surface and atmospheric climate change. In: Solomon, S., Qin, D., Manning, M., *et al.*, eds. *Climate change 2007: The physical science basis. Contribution of Working Group I to the Fourth Assessment Report of the Intergovernmental Panel on Climate Change*. Cambridge University Press, Cambridge and New York.

Trippel, E.A., Kraus, G., and Koster, F.W. (2005). Maternal and paternal influences on early life history traits and processes of Baltic cod *Gadus morhua*. *Marine Ecology Progress Series*, **303**, 259–67.

Trites, A.W., Christensen, V., and Pauly, D. (2006). Effects of fisheries on ecosystems: just another top predator?. In: Boyd, I., Wanless, S., and Camphuysen, C.J., eds. *Top predators in marine ecosystems: their role in monitoring and management*. Cambridge University Press, Cambridge, pp. 11–27.

Troadec, J.-P. (2000). Adaptation opportunities to climate variability and change in the exploitation and utilization of marine living resources. *Environmental Monitoring and Assessment*, **61**, 101–12.

Troedsson, C., Lee, R.F., Stokes, V., *et al.* (2008a). Development of a Denaturing High-Performance Liquid Chromatography (DHPLC) method for detection of protist parasites of metazoans. *Applied and Environmental Microbiology*, **74**, 4336–45.

Troedsson, C., Lee, R.F., Stokes, V., *et al.* (2008b). Detection and discovery of crustacean parasites in blue crabs (*Callinectes sapidus*) by 18S rDNA targeted Denaturing High-Performance Liquid Chromatography (DHPLC). *Applied and Environmental Microbiology*, **74**, 4346–53.

Trudel, M. (2007). Average growth conditions for coho salmon of the west coast of Vancouver Island. In: *State of the Pacific Ocean 2006, Pacific Stock Assessment Review Committee Ocean Status Report* **2007/001**.

Trudel, M. Theis, M.E., Bucher, C., *et al.* (2007). Regional variation in the marine growth and energy accumulation of juvenile Chinook salmon and coho salmon along the west coast of North America. In: Grimes, C.B., Brodeur, R.D., Halderson, L.J., *et al.* eds. *The ecology of juvenile salmon in the northeast Pacific Ocean. American Fisheries Society, Symposium* **87**, Bethseda, MD.

Tsuda, A. and Miller, C.B. (1998). Mate-finding behaviour in *Calanus marshallae* Frost. *Philosophical Transactions of* the Royal Society of London, Series B, Biological Sciences, **353**, 713–20.

Tsuda, A., Saito, H., and Kasai, H. (1999). Life histories of *Neocalanus flemingeri* and *Neocalanus plumchrus* (Calanoida: Copepoda) in the western subarctic Pacific. *Marine Biology*, **135**, 533–44.

Tsuda, A., Saito, H., and Kasai, H. (2004). Life histories of *Eucalanus bungii* and *Neocalanus cristatus* (Copepoda: Calanoida) in the western subarctic Pacific Ocean. *Fisheries Oceanography*, **13**, 10–20.

Turner, J.T. (2004). The importance of small planktonic copepods and their roles in pelagic marine food webs. *Zoological Studies*, **43**, 255–66.

Turner, J.T., Overland, J.E., and Walsh, J.E. (2007). An Arctic and Antarctic perspective on recent climate change. *International Journal of Climatology*, **27**, 277–93.

Turner, N.J, Boelscher Ignace, M., and Ignace, R. (2000). Traditional ecological knowledge and wisdom of aboriginal peoples in British Columbia. *Ecological Applications*, **10**, 1275–87.

Tyler, J.A. and Rose, K.A. (1994). Individual variability and spatial heterogeneity in fish population models. *Reviews in Fish Biology and Fisheries*, **4**, 91–123.

Tynan, C.T. (1998). Ecological importance of the southern boundary of the Antarctic Circumpolar current. *Nature*, **392**, 708–10.

Tynan, C.T., Ainley, D.G., Barth, J.A., *et al.* (2005). Cetacean distributions relative to ocean processes in the northern California Current System. *Deep-Sea Research II*, **52**, 145–67.

Tyrrell, M.C., Link, J.S., Moustahfid, H., *et al.* (2008). Evaluating the effect of predation mortality on forage species population dynamics in the Northeast US continental shelf ecosystem using multispecies virtual population analysis. *ICES Journal of Marine Science*, **65**, 1689–700.

Uchima, M. and Hirano, R. (1988). Swimming behavior of the marine copepod *Oithona davisae*: internal control and search for environment. *Marine Biology*, **99**, 47–56.

Ulanowicz, R.E. (1986). Growth and development: ecosystem phenology. Springer-Verlag, New York, 203pp.

Ulanowicz, R.E. and Platt, T., eds. (1985). Ecosystem theory for biological oceanography. *Canadian Bulletin of Fisheries and Aquatic Sciences*, **213**, 260pp.

Ulltang, O. (1996). Stock assessment and biological knowledge: can prediction uncertainty be reduced. *ICES Journal of Marine Science*, **53(4)**, 659–75.

Ulltang, O. (2003). Fish stock assessments and predictions: integrating relevant knowledge. An overview. *Scientia Marina*, **67**, 5–12.

Umlauf, L. and Burchard, H. (2005). Second-order turbulence closure models for geophysical boundary layers. A review of recent work. *Continental Shelf Research*, **25**, 795–827.

Uriarte, A., Prouzet, P., and Villamor, B. (1996). Bay of Biscay and Ibero Atlantic anchovy populations and their fisheries. *Scientia Marina*, **60(Suppl. 2)**, 237–55.

US GLOBEC. (1988). *Report of a workshop on Global Ocean Ecosystems Dynamics, Wintergreen, Virginia, May 1988.* Joint Oceanographic Institutions, Inc., 131pp.

US GLOBEC. (1991). GLOBEC workshop on acoustical technology and the integration of acoustical and optical sampling methods. *US GLOBEC Report*, **4**, 69pp.

US GLOBEC. (1995). Secondary production modelling workshop report. *US GLOBEC Report*, **13**, 17pp.

US GLOBEC. (2007). Strategies for pan-regional synthesis in US GLOBEC. *US GLOBEC Report*, **21**, 50pp.

USCOP. (2004). *US Commission on ocean policy. An ocean blueprint for the 21st Century.* 676pp. (http://oceancommission.gov/documents/full_color_rpt/welcome.html#full)

Uye, S. and Shimauchi, H. (2005). Population biomass, feeding, respiration and growth rates, and carbon budget of the scyphomedusa *Aurelia aurita* in the Inland Sea of Japan. *Journal of Plankton Research*, **27**, 237–48.

Uye, S. and Ueta, Y. (2004). Recent increase of jellyfish in Hiroshima Bay based on the poll survey for fishermen [in Japanese with English abstract]. *Bulletin of Japanese Society of Fisheries and Oceanography*, **68**, 9–19.

Uz, B.M., Yoder, J.A., and Osychny, V. (2001). Pumping of nutrients to ocean surface waters by the action of propagating planetary waves. *Nature*, **409**, 597–600.

Vadas, F. (2007). *Modelling climate-forced dynamics of the Peruvian scallop fishery. M.Sc. Thesis.* Faculty of Biology and Chemistry, University of Bremen, Bremen, 86pp.

Vaillancourt, R.D., Marra, J., Seki, M.P., *et al.* (2003). Impact of a cyclonic eddy on phytoplankton community structure and photosynthetic competency in the subtropical North Pacific Ocean. *Deep-Sea Research I*, **50**, 829–47.

Valdés, J., Ortlieb, L., Gutiérrez, D., *et al.* (2008). 250 years of sardine and anchovy scale deposition record in Mejillones Bay, northern Chile. *Progress in Oceanography*, **79(2–4)**, 198–207.

Valdés, L., O'Brien, T., and López-Urrutia, A. (2006). Zooplankton monitoring results in the ICES area, summary status report 2004/2005. *ICES Cooperative Research Report*, **281**, 43pp.

Vallina, S.M. and Le Quere, C. (2008). Preferential uptake of NH_4^+ over NO_3^- in marine ecosystem models: a simple and more consistent parameterization. *Ecological Modelling*, **218**, 393–7.

van der Lingen, C.D., Hutchings, L., and Field, J.G. (2006a). Comparative trophodynamics of anchovy *Engraulis encrasicolus* and sardine *Sardinops sagax* in the southern

Benguela: are species alternations between small pelagic fish trophodynamically mediated? *African Journal of Marine Science*, **28(3/4)**, 465–77.

van der Lingen, C.D., Shannon, L.J., and Cury, P., *et al.* (2006b). Chapter 8 Resource and ecosystem variability, including regime shifts, in the Benguela Current system. In: Shannon, L.V., Hempel, G., Malanotte-Rizzoli, P., *et al.* eds. *Benguela: predicting a large marine ecosystem*, pp. 147–84. Elsevier, Large Marine Ecosystems Series 14. 410pp.

van der Lingen, C.D., Freon, P., Fairweather, T.P., *et al.* (2006c). Density-dependent changes in reproductive parameters and condition of southern Benguela sardine *Sardinops sagax*. *African Journal of Marine Science*, **28(3/4)**, 625–36.

van der Lingen, C.D., Bertrand, A., Bode, A., *et al.* (2009). Trophic dynamics. In: D. Checkley, J. Alheit, Y. Oozeki *et al.* eds. *Climate change and small pelagic fish*. Cambridge University Press, Cambridge.

Van Dolah, F.M. (2000). Marine algal toxins: origins, health effects, and their increased occurrence. *Environmental Health Perspectives*, **108**, 133–41.

van Duren, L.A. and Videler, J.J. (1995). Swimming behaviour of developmental stages of the calanoid copepod *Temora longicornis* at different food concentrations. *Marine Ecology Progress Series*, **126**, 153–61.

van Duren, L.A. (1996). The trade-off between feeding, mate seeking and predator avoidance in copepods: behavioural responses to chemical cues. *Journal of Plankton Research*, **18**, 805–18.

van Franecker, J.A. (1992). Top predators as indicators for ecosystem events in the confluence zone and the marginal ice zone of the Weddell and Scotia seas, Antarctica, November 1988-January 1989. *Polar Biology*, **12**, 93–102.

Vargas, C.A., Escribano, R., and Poulet, S. (2006). Phytoplankton food quality determines time windows for successful zooplankton reproductive pulses. *Ecology*, **87**, 2992–9.

Vasseur, D.A. and Yodzis, P. (2004). The color of environmental noise. *Ecology*, **85**, 1146–52.

Vaughan, D.G., Marshall, G.J., Connolley, W.M., *et al.* (2003). Recent rapid regional climate warming on the Antarctic Peninsula. *Climatic Change*, **60**, 243–74.

Vecchi, G.A., Soden, B.J., Wittenberg, A.T., *et al.* (2006). Weakening of tropical Pacific atmospheric circulation due to anthropogenic forcing. *Nature*, **441**, 73–6.

Veit, R.R., Silverman, E.D., and Everson, I. (1993). Aggregation patterns of pelagic predators and their principal prey, Antarctic krill, near South Georgia. *Journal of Animal Ecology*, **62**, 551–64.

Verheye, H.M., Richardson, A.J., Hutchings, L., *et al.* (1998). Long-term trends in the abundance and community structure of ciastal zooplankton in the southern Benguela system, 1951–1996. In: Pillar, S.C., Moloney, C.L., Payne, A.I.L., *et al.* eds. *Benguela dynamics: impacts of variability on shelf-sea environments and their living resources. South African Journal of Marine Science*, **19**, 317–32.

Vestheim, H., Edvardsen, B., and Kaardvedt, S. (2005). Assessing feeding of a carnivorous copepod using species-specific PCR. *Marine Biology*, **147**, 381–5.

Vezina, A.F. and Hoegh-Guldburg, O. (2008). Effects of ocean acidification on marine ecosystems Introduction. *Marine Ecology Progress Series*, **373**, 199–201.

Viitasalo, M., Kiørboe, T., Flinkman, J., *et al.* (1998). Predation vulnerability of planktonic copepods: consequences of predator foraging strategies and prey sensory abilities. *Marine Ecology Progress Series*, **175**, 129–42.

Vikebø, F., Sundby, S., Ådlandsvik, B., *et al.* (2005). The combined effect of transport and temperature on distribution and growth of larvae and pelagic juveniles of Arcto-Norwegian cod. *ICES Journal of Marine Science*, **62**, 1375–86.

Vikebø, F., Jørgensen, C., Kristiansen, T., *et al.* (2007a). Drift, growth, and survival of larval Northeast Arctic cod with simple rules of behaviour. *Marine Ecology Progress Series*, **347**, 207–19.

Vikebø, F., Sundby, S., Adlandsvik, B., *et al.* (2007b). Impacts of a reduced thermohaline circulation on transport and growth of larvae and pelagic juveniles of Arcto-Norwegian cod (*Gadus morhua*). *Fisheries Oceanography*, **16**, 216–28.

Vilhjálmsson, H. (1997a). Climatic variations and some examples of their effects on the marine ecology of Icelandic and Greenlandic waters, in particular during the present century. *Journal of the Marine Research Institute*, **15(1)**, 9–29.

Vilhjálmsson, H. (1997b). Climatic variations and some examples of their effects on the marine ecology of Icelandic and Greenland waters, in particular during the present century. *Rit Fiskideildar*, **40**, 7–29.

Vilhjálmsson, H., Hoel, A.H., Agnarsson, S., *et al.* (2005). Fisheries and aquaculture. In: Symon, C., Arris, L., and Heal, B. eds. *Arctic climate impact assessment*. Cambridge University Press, Cambridge, pp. 692–780.

Vincent, A. and Meneguzzi, M. (1991). The spatial structure and statistical properties of homogeneous turbulence. *Journal of Fluid Mechanics*, **225**, 1–20.

Vinje, T. and Kvambekk, A. (1991). Barents Sea drift ice characteristics. *Polar Research*, **10**, 59–68.

Visser, A.W. (2007a). Biomixing of the oceans. *Science*, **316**, 838–9.

Visser, A.W. (2007b). Motility of zooplankton: fitness, foraging and predation. *Journal of Plankton Research*, **29**, 447–61.

Visser, A.W. and Kiørboe, T. (2006). Plankton motility patterns and encounter rates. *Oecologia*, **148**, 538–46.

Visser, A.W. and Thygesen, U.H. (2003). Random motility of plankton: diffusive and aggregative contributions. *Journal of Plankton Research*, **25**, 1157–68.

Visser, A.W. Saito, H., Saiz, E., et al. (2001). Observations of copepod feeding and vertical distribution under natural turbulent conditions in the North Sea. *Marine Biology*, **138**, 1011–9.

Visser, A.W., Mariani, P., and Pigolotti, S. (2009). Swimming in turbulence: zooplankton fitness in terms of foraging efficiency and predation risk. *Journal of Plankton Research*, **31**, 121–33.

Vlietstra, L.S., Coyle, K.O., Kachel, N.B., et al. (2005). Tidal front affects the size of prey used by a top marine predator, the short-tailed shearwater (*Puffinus tenuirostris*). *Fisheries Oceanography*, **14(Suppl. 1)**, 196–211.

Voss, R., Hinrichsen, H.H., and St. John, M. (1999). Variations in the drift of larval cod (*Gadus morhua* L.) in the Baltic Sea: combining field observations and modelling. *Fisheries Oceanography*, **8**, 199–211.

Waggett, R.J. and Buskey, E.J. (2006). Copepod sensitivity to flow fields: detection by copepods of predatory ctenophores. *Marine Ecology Progress Series*, **323**, 205–11.

Wagner, M., Durbin, E., and Buckley, L. (1998). RNA:DNA ratios as indicators of nutritional condition in the copepod *Calanus finmarchicus*. *Marine Ecology Progress Series*, **162**, 173–81.

Wagner, M., Campbell, R.G., Boudreau, C.A., et al. (2001). Nucleic acids and growth of *Calanus finmarchicus* in the laboratory under different food and temperature conditions. *Marine Ecology Progress Series*, **221**, 185–97.

Wainwright, T.C., Feinberg, L.R., Hooff, R.C., et al. (2007). A comparison of two lower trophic models for the California Current System. *Ecological Modelling*, **202**, 120–31.

Walker, B., Holling, C.S., Carpenter, S.R., et al. (2004). Resilience, adaptability and transformability in social-ecological systems. *Ecology and Society*, **9**, art:5.

Walsh, J.J. and McRoy, C.P. (1986). Ecosystem analysis in the southeastern Bering Sea. *Continental Shelf Research*, **5**, 259–88.

Walstad, L.J. and McGillicuddy, D.J. (2000). Data assimilation for coastal observing systems. *Oceanography*, **13(1)**, 47–53.

Walter, E.E., Scandol, J.P., and Healey, M.C. (1997). A reappraisal of the ocean migration patterns of Fraser River sockeye salmon (*Oncorhynchus nerka*) by individual-based modelling. *Canadian Journal of Fisheries and Aquatic Sciences*, **54**, 847–58.

Walters, C. and Kitchell, J.F. (2001). Cultivation/depensation effects on juvenile survival and recruitment: implications for the theory of fishing. *Canadian Journal of Fisheries and Aquatic Sciences*, **58(1)**, 39–50.

Walters, C., Pauly, D., Christensen, V., et al. (2000). Representing density dependent consequences of life history strategies in aquatic ecosystems: Ecosim II. *Ecosystems*, **3**, 7–83.

Walters, C.J. and Collie, J.S. (1988). Is research on environmental factors useful to fisheries management. *Canadian Journal of Fisheries and Aquatic Sciences*, **45**, 1848–54.

Walters, C.J. and Juanes, F. (1993). Recruitment limitation as a consequence of natural selection for use of restricted habitats and predation risk taking by juvenile fishes. *Canadian Journal of Fisheries and Aquatic Sciences* **50**, 2058–70.

Walters, C.J. and Ludwig, D. (1981). Effects of measurement errors on the assessment of stock-recruitment relationships. *Canadian Journal of Fisheries and Aquatic Sciences*, **38**, 704–10.

Walters, C.J. and Parma, A.M. (1996). Fixed exploitation rate strategies for coping with effects of climate change. *Canadian Journal of Fisheries and Aquatic Sciences*, **53**, 148–58.

Wang, R., Zuo, T., and Wang, K. (2003). The Yellow Sea Cold Bottom Water-an oversummering site for *Calanus sinicus* (Copepoda, Crustacea). *Journal of Plankton Research*, **25**, 169–83.

Ward, T.M., Hoedt, F., McLeay, L., et al (2001). Effects of the 1995 and 1998 mass mortality events on the spawning biomass of sardine, *Sardinops sagax*, in South Australian waters. *ICES Journal of Marine Science*, **58**, 865–75.

Ware, D.M. (1977). Spawning time and egg size of Atlantic mackerel, *Scomber scombus*, in relation to the plankton. *Journal of the Fisheries Research Board of Canada*, **34**, 2308–15.

Ware, D.M. and McFarlane, G.A. (1989). Fisheries production domains in the Northeast Pacific Ocean. In: Beamish, R.J. and McFarlane, G.A., eds. *Effects of ocean variability on recruitment and an evaluation of parameters used in stock assessment models. Canadian Special Publication in Fisheries and Aquatic Sciences*, **108**, 359–79.

Ware, D.M. and Thomson, R.E. (1991). Link between long-term variability in upwelling and fish production in the Northeast Pacific Ocean. *Canadian Journal of Fisheries and Aquatic Sciences*, **48**, 2296–306.

Ware, D.M. and Thomson, R.E. (2005). Bottom-up ecosystem trophic dynamics determine fish production in the Northeast Pacific. *Science*, **308**, 1280–4.

Waring, G.T., Pace, R.M., Quintal, J.M., *et al.* (2004). US Atlantic and Gulf of Mexico marine mammal stock assessments-2003. *NOAA Technical Memorandum*, **NMFS-NE-182**.

Warren, J.D., Stanton, T.K., Wiebe, P.H., *et al.* (2003). Inference of biological and physical parameters in an internal wave using multiple-frequency, acoustic-scattering data. *ICES Journal of Marine Science*, **60**, 1033–46.

Warren, W.G. (1997). Changes in the within-survey spatio-temporal structure of the northern cod (*Gadus morhua*) population, 1985–1992. *Canadian Journal of Fisheries and Aquatic Sciences*, **54(Suppl. 1)**, 139–48.

Wassmann, P., Ratkova, T., Andreassen, I., *et al.* (1999). Spring bloom development in the marginal ice zone and the central Barents Sea. *Marine Ecology (Pubblicazioni della Stazione Zoologicz di Napoli)*, **20**, 321–46.

Wassmann, P., Slagstad, D., Riser, C.W., *et al.* (2006). Modelling the ecosystem dynamics of the Barents Sea including the marginal ice zone II. Carbon flux and interannual variability. *Journal of Marine Systems*, **59**, 1–24.

Watanabe, H. and Kawaguchi, K. (2003a). Decadal change in abundance of surface migratory myctophid fishes in the Kuroshio region from 1957 to 1994. *Fisheries Oceanography*, **12**, 100–11.

Watanabe, H., Kawaguchi, K., and Kawaguchi, K. (2003b). Decadal change in the diets of the surface migratory myctophid fish *Myctophum nitidulum* in the Kuroshio region of the western North Pacific: predation on sardine larvae by myctophids. *Fisheries Science*, **69**, 716–21.

Watanabe, Y., Zenitani, H., and Kimura, R. (1996). Offshore expansion of spawning of the Japanese sardine *Sardinops melanostictus* and its implications for egg and larval survival. *Canadian Journal of Fisheries and Aquatic Sciences*, **53**, 55–61.

Watanabe, Y., Ono, T., Shimamoto, A., *et al.* (2001). Possibility of a reduction in the formation rate of the subsurface water in the North Pacific, *Geophysical Research Letters*, **28**, 3285–8.

Watanabe, Y., Ishida, H., Nakano, T., *et al.* (2005). Spatiotemporal decreases of nutrients and chlorophyll-a in the surface mixed layer of the western North Pacific from 1971 to 2000. *Journal of Oceanography*, **61**, 1011–16.

Watermeyer, K., Shannon, L.J., and Griffiths, C.L. (2008a). Changes in the trophic structure of the southern Benguela before and after the onset of industrial fishing. *African Journal of Marine Science*, **30(2)**, 351–82.

Watermeyer, K., Shannon, L.J., Roux, J.-P., *et al.* (2008b). Changes in the trophic structure of the northern Benguela before and after the onset of industrial fishing. *African Journal of Marine Science*, **30(2)**, 383–403.

Watters, G.M., Hinke, J.T., Hill, S.H., *et al.* (2005). A krill-predator-fishery model for evaluating candidate management procedures. *CCAMLR WG-EMM* **05/13**, CCAMLR, Hobart, Australia.

Weatherby, T.M. and Lenz, P.H. (2000). Mechanoreceptors in calanoid copepods: designed for high sensitivity. *Arthropod Structure and Development*, **29**, 275–88.

Weikert, H. and John, H.-C. (1981). Experiences with a modified Be´ multiple opening-closing plankton net. *Journal of Plankton Research*, **3**, 167–76.

Weimerskirch, H., Salamolard, M., Sarrazin, F., *et al.* (1993). Foraging strategy of Wandering albatrosses through the breeding season: a study using satellite telemetry. *Auk*, **110**, 325–42.

Weimerskirch, H., Wilson, R.P., Guinet, C., *et al.* (1995). Use of seabirds to monitor sea-surface temperatures and to validate satellite remote-sensing measurements in the Southern Ocean. *Marine Ecology Progress Series*, **126**, 299–303.

Weimerskirch, H., Guionnet, T., Martin, J., *et al.* (2000). Fast and fuel efficient? Optimal use of wind by flying albatrosses. *Proceedings of the Royal Society of London, Series B, Biological Sciences*, **267**, 1869–74.

Weimerskirch, H., Le Corre, M., Jaquemet, S., *et al.* (2004). Foraging strategy of a top predator in tropical waters: great frigatebirds in the Mozambique Channel. *Marine Ecology Progress Series*, **275**, 297–308.

Weissburg, M.J., Doall, M.H., and Yen, J. (1998). Following the invisible trail: kinematic analysis of mate-tracking in the copepod *Temora longicornis*. *Philosophical Transactions of the Royal Society of London, Series B, Biological Sciences*, **353**, 701–12.

Welch, D.W., Ishida, Y., and Nagasawa, K. (1998). Thermal limits and ocean migrations of sockeye salmon (*Oncorhynchus nerka*): long-term consequences of global warming. *Canadian Journal of Fisheries and Aquatic Sciences*, **55**, 937–48.

Weng, K.C., Castilho, P.C., Morrissette, J.M., *et al.* (2005). Satellite tagging and cardiac physiology reveal niche expansion in salmon sharks. *Science*, **310**, 104–6.

Werner, F.E., Page, F.H., Lynch, D.R., *et al.* (1993). Influence of mean 3-D advection and simple behavior on the distribution of cod and haddock early life history stages on Georges Bank. *Fisheries Oceanography*, **2**, 43–64.

Werner, F.E., Perry, R.I., Lough, R.G., *et al.* (1996). Trophodynamic and advective influences on Georges Bank larval cod and haddock. *Deep-Sea Research II*, **43**, 1793–822.

Werner, F.E., Quinlan, J.A., Lough, R.G., *et al.* (2001a). Spatially-explicit individual based modeling of marine populations: a review of the advances in the 1990s. *Sarsia*, **86**, 411–21.

Werner, F.E., MacKenzie, B.R., Perry, R.I., *et al.* (2001b). Larval trophodynamics, turbulence, and drift on Georges Bank: a sensitivity analysis of cod and haddock. *Scientia Marina*, **65(Suppl. 1)**, 99–115.

Werner, F.E., Ito, S., Megrey, B.A., *et al.* (2007a). Synthesis of the NEMURO model studies and future directions of marine ecosystem modeling. *Ecological Modelling*, **202**, 211–23.

Werner, F.E., Cowen, R.K., and Paris, C.B. (2007b). Coupled biological and physical models: present capabilities and necessary developments for future studies of population connectivity. *Oceanography*, **20(3)**, 54–69.

Westin, L. and Nissling, A. (1991). Effects of salinity on spermatozoa motility, percentage of fertilized eggs and egg development of Baltic cod *Gadus morhua*, and implications for cod stock fluctuations in the Baltic. *Marine Biology*, **108**, 5–9.

Whitehouse, M.J., Korb, R.E., Atkinson, A., *et al.* (2008). Formation, transport and decay of an intense phytoplankton bloom within the High-Nutrient Low-Chlorophyll belt of the Southern Ocean. *Journal of Marine Systems*, **70**, 150–67.

Whitfield, P.E., Gardner, T., Vives, S.P., *et al.* (2002). Biological invasion of the Indo-Pacific lionfish *Pterois volitans* along the Atlantic coast of North America. *Marine Ecology Progress Series*, **235**, 289–97.

Whitfield, P.E., Hare, J.A., David, A.W., *et al.* (2006). Abundance estimates of the Indo-Pacific lionfish *Pterois volitans*/miles complex in the Western North Atlantic. *Biological Invasions*, **9(1)**, 53–64.

Whitney, F.A., Crawford, W.R., and Harrison, P.J. (2005). Physical processes that enhance nutrient transport and primary productivity in the coastal and open ocean of the subarctic NE Pacific. *Deep Sea Research II*, **52(5–6)**, 681–706.

Whittington, R.J., Crockford, M., Jordan, D., *et al.* (2008). Herpesvirus that caused epizootic mortality in 1995 and 1998 in pilchard, *Sardinops sagax neopilchardus* (Steindachner), in Australia is now endemic. *Journal of Fish Diseases*, **31**, 97–105.

Wichard, T., Poulet, S.A., Boulesterix, A.-L., *et al.* (2008). Influence of diatoms on copepod reproduction: II. Uncorrelated effects of diatom-derived alpha, beta, gamma, delta-unsaturated aldehydes and polyunsaturated fatty acids on *Calanus helgolandicus* in the field. *Progress in Oceanography*, **77**, 30–44.

Wiebe, P.H. and Benfield, M.C. (2003). From the Hensen Net toward four-dimensional biological oceanography. *Progress in Oceanography*, **56**, 7–136.

Wiebe, P.H. and Greene, C.H. (1994). The use of high frequency acoustics in the study of zooplankton spatial and temporal patterns. *Proceedings of the NIPR Symposium on Polar Biology*, **7**, 133–57.

Wiebe, P.H., Morton, A.W., Bradley, A.M., *et al.* (1985). New developments in the MOCNESS, an apparatus for sampling zooplankton and micronekton. *Marine Biology*, **87**, 313–23.

Wiebe, P.H., Mountain, D., Stanton, T.K., *et al.* (1996). Acoustical study of the spatial distribution of plankton on Georges Bank and the relation of volume backscattering strength to the taxonomic composition of the plankton. *Deep-Sea Research II*, **43**, 1971–2001.

Wiebe, P.H., Stanton, T.K., Benfield, M.C., *et al.* (1997). High-frequency acoustic volume backscattering in the Georges Bank coastal region and its interpretation using scattering models. In: Shallow water acoustics, geophysics, and oceanography. *IEEE Journal of Oceanic Engineering*, **22**, 445–64.

Wiebe, P.H., Beardsley, R., Mountain, D., *et al.* (2002a). U.S. GLOBEC Northwest Atlantic/Georges Bank Program. *Oceanography*, **15**, 13–29.

Wiebe, P.H., Stanton, T.K., Greene, C.H., *et al.* (2002b). BIOMAPER II: an integrated instrument platform for coupled biological and physical measurements in coastal and oceanic regimes. *IEEE Journal of Oceanic Engineering*, **27**, 700–16.

Wiebe, P.H., Ashjian, C., Gallager, S., *et al.* (2004). Using a high powered strobe light to increase the catch of Antarctic krill. *Marine Biology*, **144**, 493–502.

Wiebe, P.H., Harris, R.P., St. John, M.A., *et al.* (eds.). (2007). BASIN. Basin-scale Analysis, Synthesis, and INtegration. *GLOBEC Report*, **23**, 56pp.

Wiebe, P.H., Harris, R.P., St. John, M.A., *et al.* eds. (2009). BASIN: Basin-scale Analysis, Synthesis and INtegration. Science plan and implentation strategy. (2008). *GLOBEC Report*, **27**, 53pp.

Wieland, K., Waller, U., and Schnack, D. (1994). Development of Baltic cod eggs at different levels of temperature and oxygen content. *Dana*, **10**, 163–77.

Wieland, K., Jarre, A., and Horbowa, K. (2000). Changes in the timing of spawning of Baltic cod: possible causes and implications for recruitment. *ICES Journal of Marine Science*, **57**, 452–64.

Wieland, K., Storr-Paulsen, M., and Sünksen, K. (2007). Response in stock size and recruitment of Northern shrimp (*Pandalus borealis*) to changes in predator biomass and distribution in West Greenland waters. *Journal of Northwest Atlantic Fishery Science*, **39**, 21–33.

Wiggert, J.D., Hofmann, E.E., and Paffenhöfer, G.-A. (2008). A modelling study of developmental stage and environmental variability effects on copepod foraging. *ICES Journal of Marine Science*, **65**, 1–20.

Wilderbuer, T.K., Hollowed, A.B., Ingraham, W.J., Jr., *et al.* (2002). Flatfish recruitment response to decadal climate variability and ocean conditions in the eastern Bering Sea. *Progress in Oceanography*, **55**, 235–47.

Wilson, R.W., Millero, F.J., Taylor, J.R., *et al.* (2009). Contribution of fish to the marine inorganic carbon cycle. *Science*, **323**, 359–62.

Wilson, S. (2000). Launching the Argo armada. *Oceanus*, **42**, 17–19.

Wiltshire, J.C. (2001). Mineral extraction, authigenic minerals. In: Steele, J., Turekian, K., Thorp, S., eds. *Encyclopedia of ocean sciences, vol. 3.* Academic Press, San Diego, CA, pp. 1821–30.

Wiltshire, K.H. and Manly, B.F.J. (2004). The warming trend at Helgoland Roads, North Sea: phytoplankton response. *Helgoland Marine Research*, **58**, 269–73.

Wirsing, A.J., Heithaus, M.R., Frid, A., *et al.* (2008). Seascapes of fear: evaluating sub-lethal predator effects experienced and generated by marine mammals. *Marine Mammal Science*, **24(1)**, 1–15.

Wolff, M. and Mendo, J. (2000). Management of the Peruvian bay scallop metapopulation with regard to environmental change. *Aquatic Conservation Marine Freshwater Ecosystems*, **10**, 117–26.

Wolff, M., Taylor, M.H., Mendo, J., *et al.* (2007). A catch forecast model for the Peruvian scallop (*Agropecten purpuratus*) based on estimators of spawning stock and settlement rate. *Ecological Modelling*, **209**, 333–41.

Wolter, K. and Timlin, M.S. (1998). Measuring the strength of ENSO-How does 1997/98 rank?, *Weather*, **53**, 315–24.

Wood, S.N. (1994). Obtaining birth and mortality patterns from structured population trajectories. *Ecological Monographs*, **64**, 23–44.

Woodson, C.B., Webster, D.R., Weissburg, M.J., *et al.* (2005). Response of copepods to physical gradients associated with structure in the ocean. *Limnology and Oceanography*, **50**, 1552–64.

Woodward, E.M.S. and Rees, A.P. (2001). Nutrient distributions in an anticylonic eddy in the northeast Atlantic Ocean, with reference to nanonmolar concentrations. *Deep-Sea Research II*, **48**, 775–93.

World Bank. (2000). *Cities, seas, and storms: managing change in Pacific Island economies, volume IV: adapting to climate change.* Papua New Guinea and Pacific Country Unit, World Bank, Washington, DC.

World Resource Institute, ed. (2005). Millennium ecosystem assessment: ecosystems and human well-being: synthesis. Island Press, St. Louis. 137pp.

Worm, B. and Myers, R.A. (2003). Meta-analysis of cod-shrimp interactions reveals top-down control in oceanic food webs. *Ecology*, **84**, 162–73.

Worm, B., Barbier, E.B., Beuamont, N., *et al.* (2006). Impacts of biodiversity loss on ocean ecosystem services. *Science*, **314**, 787–90.

Wulff, F., Field, J.G., and Mann, K.H., eds. (1989). *Network analysis in marine ecology. Methods and applications.* Springer-Verlag, Berlin, 284pp.

Wunsch, C. (1996). *The ocean circulation inverse problem.* Cambridge University Press, Cambridge. 442pp.

Yamada, K. and Ishizaka, J. (2006). Estimation of interdecadal change of spring bloom timing, in the case of the Japan Sea. *Geophysical Research Letters*, **33**, LO2608, doi:10.1029/2005GL024792.

Yamamoto, J., Masuda, S, Miyashita, K., *et al.* (2002). Investigation on the early stages of the ommastrephid Squid *Todarodes pacificus* near the Oki Islands (Sea of Japan). *Bulletin of Marine Science*, **71**, 987–92.

Yamamoto, J., Shimura, T., Uji, R., *et al.* (2007). Vertical distribution of *Todarodes pacificus* (Cephalopoda: Ommastrephidae) paralarvae near the Oki Islands, southwestern Sea of Japan. *Marine Biology*, **153**, 7–13.

Yamamura, O. (2004). Trophodynamic modeling of walleye pollock (*Theragra chalcogramma*) in the Doto area, northern Japan: model description and baseline simulations. *Fisheries Oceanography*, **13**, 138–54.

Yamazaki, H. (1993). Lagrangian study of planktonic organisms: perspective. *Bulletin of Marine Science*, **53**, 265–78.

Yamazaki, H., Mackas, D., and Denman, K. (2002). Coupling small scale physical processes to biology: towards a Lagrangian approach. In: A.R. Robinson, J.J. McCarthy, and B.J. Rothschild, eds, *The sea, vol. 12: biological-physical interactions in the sea.* Wiley, New York, pp. 51–112.

Yamazaki, H., Osborn, T.R., and Squires, K.D. (1991). Direct numerical simulation of planktonic contact in turbulent flow. *Journal of Plankton Research*, **13**, 629–43.

Yatsu, A., Nagasawa, K., and Wada, T. (2003). Decadal changes in abundance of dominant pelagic fishes and squids in the Northwestern Pacific Ocean since the 1970s and implications on fisheries management. *American Fisheries Society Symposia*, **38**, 675–84.

Yebra, L., Harris, R.P., and Smith, T. (2005). Comparison of five methods for estimating growth of *Calanus helgolandicus* later development stages (CV-CVI). *Marine Biology*, **147**, 1367–75.

Yen, J. (1988). Directionality and swimming speeds in predator-prey and male-female interactions of *Euchaeta rimana*, a subtropical marine copepod. *Bulletin of Marine Science*, **43**, 395–403.

Yen, J. and Fields, D.M. (1992). Escape responses of *Acartia hudsonica* (Copepoda) nauplii from the flow field of *Temora longicornis* (Copepoda). *Archiv fur Hydrobiologie, Beihefte, Ergebnisse der Limnologie*, **36**, 123–34.

Yen, J. and Nicoll, N.T. (1990). Setal array on the first antennae of a carnivorous marine copepod, *Euchaeta norvegica*. *Journal of Crustacean Biology*, **10**, 218–24.

Yen, J., Lenz, P.H., Gassie, D.V., *et al.* (1992). Mechanoreception in marine copepods: electrophysiological studies on the first antennae. *Journal of Plankton Research*, **14**, 495–512.

Yen, J., Sanderson, B., Strickler, J.R., *et al.* (1991). Feeding currents and energy dissipation by *Euchaeta rimana*, a subtropical pelagic copepod. *Limnology and Oceanography*, **36**, 362–9.

Yen J., Weissburg M.J., Doall, M.H. (1998). The fluid physics of signal perception by mate-tracking copepods. *Philosophical Transactions of the Royal Society of London Series B. Biological Sciences*, **353**, 787–804.

Yoshida, T., Jones, L.E., Ellner, S.P., *et al.* (2003). Rapid evolution drives ecological dynamics in a predator-prey system. *Nature*, **424**, 303–6.

Yoshie, N., Fujii, M., and Yamanaka, Y. (2005). Ecosystem changes after the SEED iron fertilization in the western North Pacific simulated by a one-dimensional ecosystem model. *Progress in Oceanography*, **64**, 283–306.

Young, J.W., Bradford, R., Lamb, T.D., *et al.* (2001). Yellowfin tuna (*Thunnus albacares*) aggregations along the shelf break off south-eastern Australia: links between inshore and offshore processes. *Marine and Freshwater Research*, **52**, 463–74.

Yunev, O.A., Vedernikov, V.I., Basturk, O., *et al.* (2002). Long-term variations of surface chlorophyll *a* and primary production in the open Black Sea. *Marine Ecology Progress Series*, **230**, 11–28.

Zenitani, H. and Yamada, S. (2000). The relation between spawning area and biomass of Japanese pilchard, *Sardinops melanostictus*, along the Pacific coast of Japan. *Fishery Bulletin*, **98(4)**, 842–8.

Zhou, M. (2006). What determines the slope of a plankton biomass spectrum? *Journal of Plankton Research*, **28**, 437–48.

Zhou, M. and Dorland, R.D. (2004). Aggregation and vertical migration behavior of *Euphausia superba*. *Deep-Sea Research II*, **51**, 2119–37.

Zhou, M. and Huntley, M.E. (1997). Population dynamics theory of plankton based on biomass spectra, *Marine Ecology Progress Series*, **159**, 61–73.

Zhou, M. and Tande, K., eds. (2002). Report of Optical Plankton Counter workshop 17–20 June 2001, Tromsø, Norway. *GLOBEC Report*, **17**, 67pp.

Zhou, M. and Zhu, Y. (2002). Mesoscale zooplankton distribution and its correlation with physical and fluorescence fields in the California Current in 2000. *EOS, Transactions, American Geophysical Union*. **83(4)**, Ocean Sciences Meet. Suppl. Abstract OS21N-04.

Zhou, M., Zhu, Y., and Peterson, J.O. (2004). *In situ* growth and mortality of mesozooplankton during the austral fall and winter in Marguerite Bay and its vicinity. *Deep-Sea Research II*, **51**, 2099–118.

Index